Pro SQL Server 2012 BI Solutions

Randal Root
Caryn Mason

Apress·

President and Publisher: Paul Manning
Lead Editor: James Markham
Technical Reviewers: Maradelyn Taylor-Root and Robert Glancy
Editorial Board: Steve Anglin, Ewan Buckingham, Gary Cornell, Louise Corrigan, Morgan Ertel, Jonathan Gennick, Jonathan Hassell, Robert Hutchinson, Michelle Lowman, James Markham, Matthew Moodie, Jeff Olson, Jeffrey Pepper, Douglas Pundick, Ben Renow-Clarke, Dominic Shakeshaft, Gwenan Spearing, Matt Wade, Tom Welsh
Coordinating Editors: Corbin Collins and Mark Powers
Copy Editor: Kim Wimpsett
Compositor: SPi Global
Indexer: SPi Global
Artist: SPi Global
Cover Designer: Anna Ishchenko

This book is dedicated to my friends and family. Writing does not come easily to me, and without their support, I doubt that I would have survived the process! Specifically, I would like to thank my coauthor and friend, Caryn, who made this project manageable and more fun than it would have been had I tried doing it solo. Finally, I am especially grateful to Margot Alice for her help as a professional editor. She not only edited this book but also provided encouragement and guidance when my path was uncertain. A journey, however hard, is best shared with friends and family!
—Randal Root

For my two amazing children, Kaylie and Collin, who are my world. May your lives be beautiful, wonderous, and delightfully fulfilling, just as you make my life—every day!

And in honor of Anthony David, who would have been so proud.
—Caryn Mason

Contents at a Glance

Contents

Foreword

I first met Randal as one of his programming students. His teaching style really opened up the subject for me in ways that I had not previously understood. He taught the class from the perspective of someone who had once started from square one himself. And because of this, he was able to convey challenging subjects in a simplified manner. I was the type to ask a million questions, and I began to realize that he was exceedingly knowledgeable about the subjects he taught. It ended up becoming the beginning of a great friendship.

Later he asked me to work with him on projects. He knew that my forté was writing, and we learned that we also worked well together as a team. When he asked me to join him in writing this book, it seemed rather overwhelming to me to even consider it. But he believed in my ability to grasp complicated topics and knew that I would do well in helping translate high-level concepts into readable text. And my need to ask many questions seemed to help the project along as well.

Because the idea for this book began with Randal, I asked him to convey his vision in his own words:

> "I've read many books that talk about creating BI projects. They would start with a database project and move onto various topics from there, but all of these books left out fundamental things such as the planning and documentation. I thought there was a need to break it down into more bite-sized pieces and explain the creation of a BI solution in my own words. I felt that having taught for over 10 years and seeing the reaction of students to the various ways of discussing/teaching the subject gave me a unique perspective, because I was able to see my students' reactions. I'm not sure that many authors have had that opportunity.
>
> I think that there is a misconception in the industry that BI solutions are difficult and expensive. I wanted to make sure that people understood that this wasn't the case; BI can be inexpensive and provide immediate benefits, even for very small companies.
>
> I've seen a lot of students struggle with topics such as data warehouse design, ETL processing, or OLAP cubes. I was hoping I could do a better job at making complex concepts seem simple. I've always liked to believe that I had good skills in that regard."

I agree that Randal is excellent at this, because I have seen it for myself. And because of this, I chose to join him in writing this book. It has been quite the undertaking but well worth it. I hope we have been able help a BI developer or two along the way!

Caryn Mason

About the Authors

 **Randal Root** is a senior consultant specializing in .NET programming, SQL Server BI solutions, and technical education. Although he has worked in the industry as a network administrator, DBA, and programmer since the 1980s, for the last 10 years he has focused on providing technical training for businesses and schools such as Microsoft, the University of Washington, and Bellevue College. Randal has now authored two books, *Pro SQL Server 2012 BI Solutions* (Apress) and *A Tester's Guide to .NET Programming* (Apress), and obtained several Microsoft professional certifications including MCSE, MCP+I, MCTS, MCDBA, MCAD, and MCT.
RandalRoot@NorthwestTech.org

 **Caryn Mason** is a content developer and computer programmer with more than 15 years of technical writing experience in a variety of industries, including IT and software. She studied at Bellevue College where she obtained certifications in programming writing and computer programming, specializing in web development. She is especially adept at taking technical concepts and making them understandable by both novices and professionals alike. Caryn has a passion for writing, whether it be technical or fiction, and often finds her inspiration through her two children. In addition to coauthoring *Pro SQL Server BI Solutions 2012* (Apress), Caryn works as an independent technical writing consultant and is working on a young-adult novel.
CarynMason@KeystrokePublications.com

About the Technical Reviewers

Maradelyn Taylor-Root has more than a decade of varied experience in several industries including the insurance, airline, automobile, retail, and financial industries. She has worked as the senior business analyst in the Enrollment Planning Department at Seattle public schools for many years but has recently accepted a position as a business analyst consultant at Microsoft with the Business Excellence Group. Maradelyn's exposure to this group has given her new abilities to see the big picture and ascertain the best way to improve current processes and create new ones using business intelligence tools. In her spare time, she teaches evenings for the University of Washington, Bellevue College, and Cascadia Community College. She has a bachelor's degree in computer science and an MCDBA certification.

Robert Glancy is a product data quality specialist working with Microsoft's SQL Server and Access databases. He is skilled in creating professional BI reports using C#, SSRS, SQL, and VBA. Robert has a bachelor's degree in mathematics from the University of Washington.

Acknowledgments

Wow, I cannot believe that it is finally done! I started writing this book about a year and a half ago, and at that time I figured I could complete it in about six to eight months. Oh, how much I have learned!

Chronologically, I would like to thank Jonathan Gennick, our editorial director at Apress. He helped me get started on this project and provided much guidance along the way.

As time went on, I realized that doing the project on my own was too much and that the project needed an actual technical writer involved. So, I turned to my friend and professional writer, Caryn Mason, who thankfully agreed to coauthor the book with me. Thanks to her, we were able to take what I believed was good content and turn it into meaningful information.

Beginning with the early drafts of the book, we abused the good natures of our technical reviewers. Both Maradelyn Taylor-Root and Robert Glancy were invaluable at proofreading and testing our exercises. We think you will find that their hard work makes the exercises smoother and more enjoyable, which in turn makes the learning process more effective.

As we completed our first drafts, more people at Apress became involved, and I would like to thank them for their time and hard work: Corbin Collins, Tracy Brown Collins, Kim Wimpsett, James Markham, and Mark Powers. All worked hard on the project, but Mark especially made a difference in coordinating the final stages of the book. Thanks, all!

Also, I would like to thank my many students who helped the project by reading our draft chapters and Margot Alice, for her work as our final proof-reader and editor.

—Randal Root

First and foremost I would like to thank Randal Root for bringing me in on this project with him. Your lighthearted nature and supportive personality make you such a joy to work with. I always said you have performed the work of at least five BI professionals in writing this book. That may have been an understatement! You put your heart into what you do, and you do it right. It has been an honor.

Thanks to our technical editors; Maradelyn Taylor-Root jumped in—no holds barred—and worked particularly hard on ensuring this book was done right, and Robert Glancy went beyond the call of duty to work all of the material from start to finish.

I would also like to thank everyone at Apress. Specifically, our editorial director Jonathan Gennick has been a pillar from the start. Your unexpected sense of humor is refreshing. I also want to thank our development editor James Markham, our managing coordinating editor Corbin Collins who oversaw the core of this project, Tracy Brown Collins who handled the times of transition, and Mark Powers who brought the project to a smooth finale.

Thank you to our copy editor, Kim Wimpsett, who made me a little insane trying to decide what to capitalize and which word to use at times. Microsoft can be rather ambiguous, and the challenge was quite the learning experience!

Thank you to Margot Alice for her professional editing. For you I know it was a labor of love, and for that I am so grateful. Thanks to James Mason for the endurance and patience required to see me through this project. And I also wanted to thank Collette Steinwert for being such an inspiration to me and for providing me with a benchmark to attain to.

Thanks to all my wonderful friends and family who continue to believe in me and have shown support on this book. You have kept me going not just on this project but on everything while this book sometimes took center stage. Thanks for all your love and support!

—Caryn Mason

CHAPTER 1

Business Intelligence Solutions

Business intelligence (BI) solutions are all the buzz as of late, and BI developers are highly sought after. Considering the amount of data that needs to be tracked to run a business successfully, it is no wonder. When an employee has been with a company for 20 years, how will management be notified? Perhaps staffing is suffering because of vacation trends or sales need to be tracked after targeted advertising. Maybe product preordering for a sales event needs to be estimated, or who sold what and when needs to be documented for an upcoming contest.

There is no end to how much data needs to be managed, and countless hours, money, and resources are wasted in attempts to research the information, often with minimal results, multiple errors, and missed opportunities in decision making. And when more than one employee needs access to the same information, the errors are often multiplied.

With a well-designed BI solution, important data can be called up instantly in a user-friendly manner. Calculations are made with a click of a button, and reports are easily generated. No longer will that 20-year employee be unrecognized for such a long duration of loyalty and service. Staffing can be more properly managed, advertising can be better targeted to the proper demographic, and so on.

This book shows how to build a successful BI solution step-by-step. We cover the entire process from initial preparations and planning to complex layers of designing and configuring your project, and from creating reports to drafting user instructions, and releasing your project. This book is simple in its approach. If you are new to BI solutions, you will find the instructions thorough and easy to follow with clear images to demonstrate the process. Yet, it is fast-paced and rich enough in information for even the most advanced database professional to learn from.

Who Should Read This Book?

This book is for each professional who works with the many aspects of BI solutions. These include database administrators, project managers, testers, support techs, report developers, and many others.

This book is not a sales pitch for the latest features of SQL server. Nor is it focused on technologies designed only for very large companies. Instead, this book is about how small, medium, and large companies, as well as departments within those companies, can take advantage of Microsoft SQL Server's effective and inexpensive BI software. This book defines the glue that is used to bind all four of Microsoft's BI servers (MSSS, SSIS, SSAS, and SSRS) together into a BI solution.

After reading this book and working through the recommended exercises, you will have the tools to build your own BI solutions, as well as interact with other BI team members with a greater understanding of their roles within the BI solution process.

What Is a Business Intelligence Solution?

A BI solution is a collection of objects that allows data to be turned into useful information. These objects must be designed, created, tested, and ultimately approved to create a working BI solution.

When creating a BI solution, it is important first to understand what that solution consists of, how each component is combined to create the whole, and finally, how to recognize when you have achieved your goal.

Knowing where to begin is vital to the success of your project. In Figure 1-1 we have outlined eight steps to use as a guideline. We progress through each of these steps and explain them in detail throughout this book. We also develop working BI solutions in the exercises within each chapter to gain the skills necessary to complete increasingly complex solutions in your future. Chapter 2 provides an overview of the entire process.

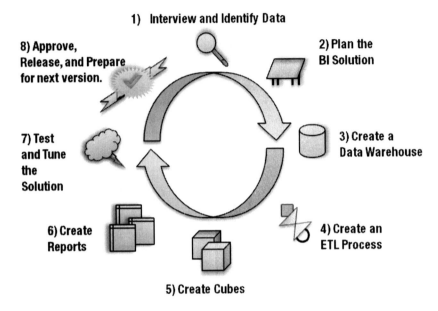

Figure 1-1. *The BI solution life cycle*

We chose to represent the tasks in Figure 1-1 as a circle, because the nature of a BI solution is one of continual change. As time goes by, a company's requirements change, the data that is available changes, and the technology to bring these two aspects together changes. Because of this, the process of creating a BI solution can often begin with the continuation of a prior solution, with each successive iteration refining and extending the current solution.

Perhaps the first step is to define the questions that your BI solution will answer. An example might be, how are our products selling? Another question might be, how often do people use our website?

One common misconception about BI solutions is that they are useful only to large corporations. This is simply not true. Clients as seemingly dissimilar as a dentist and a horse breeder will find they need to keep detailed records of important information, from patient visits to horse lineage. This information is used to determine their future plans or review past activities. Every business, group, and individual who needs to keep track of data will have questions they would like to have answered that a BI solution can provide. Formulating these questions and determining what to do with them lead us to the first step in developing a BI solution.

Step 1: Interview and Identify Data

The process of designing your solution begins with interviewing your client to determine what type of information is needed. Chapter 3 discusses the types of questions to ask and what the interview process entails.

The answers to these questions allow you to better locate the data necessary for your solution. Data can be found in many forms, and you may use one or more types to fill your requirements.

Some common data sources include the following:

- Spreadsheets

- Existing databases

- Simple text files

- Log files

- XML text files

- Paper documents

Once the data is located, the next step is to decide how much of it is relevant to your needs. You also need to decide whether your data's current location is sufficient for your BI solution's needs or whether you must copy some or all of the data to a more appropriate location. This leads us to step 2.

Step 2: Plan the BI Solution

Few developers relish creating extensive documentation before building a project. And yet, just as it is necessary for blueprints to be drawn up and approved before a home is built, projects must be planned and documented before creating a working BI solution.

In Chapter 4 we discuss creating a description of what your solution will accomplish, documenting the source and the destination objects, and beginning the formal documentation. A solution's formal document can be laid out with common tools such as Microsoft Excel or even Microsoft Word. These Excel or Word documents can then be taken back to the client for approval. Once approved, these documents will become an outline that can be worked with much like a blueprint. You then create Visual Studio projects that become the building blocks of your BI solution from these blueprints.

Step 3: Create a Data Warehouse

Your BI solution data will typically end up stored in a data warehouse database. Microsoft's SQL Server 2012 makes this very easy and cost efficient. Microsoft's SQL Server takes time and effort to master, yet the vast majority of tasks required to build your solution are performed using tools that are as simple to use as Microsoft's user-friendly Access database application.

In Chapters 4 and 5, we show how to design and implement a data warehouse database yourself, regardless of your level of experience with Microsoft's SQL Server. Various design options are demonstrated in these chapters, such as star versus snowflake dimensions and how to create fact and dimension tables. Once complete, you will understand the design differences between online transaction processing (OLTP) and data warehouse tables similar to those shown in Figure 1-2.

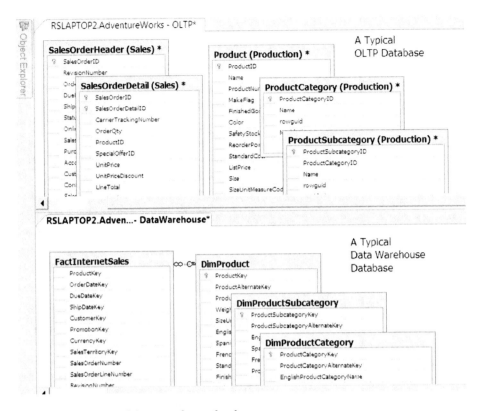

Figure 1-2. *OLTP and data warehouse databases*

Step 4: Create an ETL Process

Getting data from the original source to your data warehouse entails extracting the data from its original location, transforming the data to be consistent with your new data warehouse design, and loading the data into the new data warehouse location. This ETL process is discussed in great detail in Chapters 6, 7, and 8.

Although this process can be one of the most in-depth and complicated tasks in developing your BI solution, Microsoft SQL Server 2012 provides invaluable tools to help you accomplish it, saving time and simplifying the process for you. Using a combination of SQL programming and SQL Server's Integration Server (SSIS), you will create an ETL process much like the one shown in Figure 1-3.

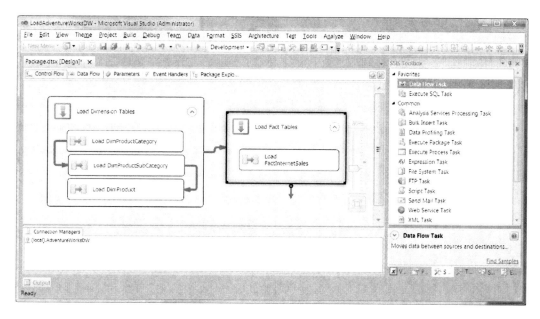

Figure 1-3. *Working with SSIS*

Step 5: Create Cubes

Microsoft SQL Server 2012 includes an additional high-performance server for hosting OLAP cube databases called SQL Server Analysis Services (SSAS).

Both the standard, relational data warehouse, and the SSAS cube databases have their place in BI solutions. The relational data warehouse contains a set of one or more tables and is by far the most commonly used database type. We work with this relational type of database extensively in Chapters 4 and 5. The second type of database contains one or more cubes instead of tables. You can think of these cubes as a set of report tables combined into a single object. Figure 1-4 illustrates how a cube is configured using an SSAS project in Visual Studio 2010. We discuss constructing and configuring cubes in Chapters 9 through 12.

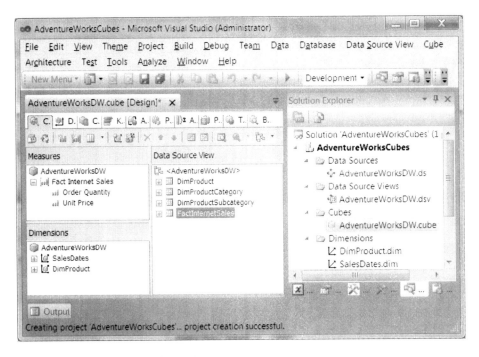

Figure 1-4. *Configuring a cube in SSAS*

Step 6: Create Reports

Once you have your data loaded into a data warehouse and/or cube, you need to create preliminary reports to continue your work. These may be your first reports for your BI solution, but they certainly will not be the last. The end goal of a BI solution is to convert data into usable information, and that information is routinely represented within reports.

The term *BI solution* is not very self-explanatory. It might be better if the industry as a whole changed the term *business intelligence solutions* to *business reporting solutions*. Even *make life easier on managers solutions* might be more descriptive than *business intelligence solutions*.

■ **Note**   About a year ago, Randal performed a casual experiment to see how many of his co-workers within the IT industry understood what the term *BI solution* meant. As he expected, 90% did not know. Some guesses were pretty comical. A favorite was "intelligent robots for businesses." But many guesses were nothing more than a long string of verbs in search of a definition. As you might imagine, only about 10% of his co-workers had a problem figuring out what a reporting solution was.

No matter what you call your BI solution, the most common output is a set of reports that present meaningful information to your users. You have many reporting tool options from which to choose. In this book, we focus on using the most readily available Microsoft technologies to create your BI reports, including Excel and SQL Server Reporting Services (SSRS).

Deciding what type of data source the reports will use is an important aspect of reporting. A typical pattern in the industry begins with simple solutions and moves progressively toward more complex ones over time (outlined in Figure 1-5).

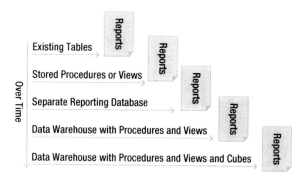

Figure 1-5. *An example of how reporting data sources change over time*

Many companies begin by selecting report data directly from OLTP relational tables. Quite often, they come to regret this choice when performance issues occur and maintenance costs rise. It has long been considered a poor choice to do so, yet this is still happening in businesses today.

An improvement on this design, and what is considered to be "best practice," is to create views or stored procedures that select data from one or more OLTP tables and use these as the source for all of your reports. Many reports can then be created against a single view or stored procedure, which makes maintaining your reports much easier over time. For example, consider a scenario where a decision has been made that all tables must be renamed to start with the letters *tbl_*. All that you need to do to keep your reports working properly is change the table names in the select statements within the view or procedure to reflect the new table names, while maintaining the same output from the view or procedure. With this simple step, your reports will continue to work as they always have. Chapter 13 of this book shows how easy it is to create both views and stored procedures.

Stored procedures and views can access data in the same database, across databases, and even across different database servers. You will gain better performance, however, when you query data from a dedicated reporting database, otherwise known as a *data warehouse*. These report databases are designed to provide simple and efficient reporting. Once the data warehouse has been created, you need an ETL process to copy the data from its original locations to the new reporting data warehouse database.

Note The term *data warehouse* can have a number of meanings. In this book, a database designed for reporting with one or more centralized fact tables containing measured data such as sales quantities, with zero or more supporting dimension tables containing additional measured data descriptions, is considered a data warehouse. You may hear this type of database referred to as a *data mart*, *data silo*, *data factory*, and a host of other names. However, Microsoft documents refer to it as a data warehouse, so we do too.

Additional report performance is provided by using SSAS cubes. This performance increase, however, is at the cost of your solution becoming more complex. The most common complexity is that cube databases use different programming languages than relational databases. We discuss the most common of these programming languages, known as MDX, in Chapter 14.

To round out your report-building skills, we present report-building applications in Chapter 15. We work with Microsoft's desktop-based reporting application, Excel 2010. Then, in Chapters 16 and 17, we create reports using Microsoft's server-based reporting application, Reporting Services 2012.

Step 7: Test and Tune the Solution

Once you have built your first reports, you need to test those reports for accuracy, visual consistency, and performance. The most important of the three is accuracy. If the reports are slow or do not look professional, it is indeed cause for concern, but if your reports are inaccurate, your entire BI solution will fail! We cover a number of ways to plan and implement testing procedures in Chapter 18. We also include important performance-tuning techniques in Chapter 18 to insure your reports run quickly for your end users.

Step 8: Approve, Release and Prepare

At the end of the solution development cycle, you need to package and deploy your documents, scripts, databases, and reports. You also need to create user documentation, as well as train your users to use your newly developed BI solution. These topics are discussed in the last chapter of this book, Chapter19.

Practice Exercises and More

Rather than just talking about all of these subjects, the chapters in this book offer detailed instructions on how to perform your BI solution tasks with step-by-step practice exercises that build upon each other from one chapter to the next. We created simple, easy-to-follow examples that outline key principles applicable to both large and small BI solutions.

We also offer "Learn by Doing" activities at the end of each chapter. These activities provide an outline and hints indicating which course of action to take, but they allow you a chance to practice your skills without such detailed instructions. Table 1-1 describes the exercises within this book.

Table 1-1. *Exercises in This Book*

Exercise Type	Description	Instructions
Exercises	Detailed, progressive, step-by-step instructions that correspond with the subject matter within each chapter. A complete and functioning BI solution is created by the end of this book.	Detailed instructions are included within each chapter.
Learn by Doing	A simple outline of the steps required to implement a BI task that corresponds to the subject of each chapter.	Outlined instructions are within folders included in the downloadable book content. See this book's catalog page at `www.apress.com/9781430234883`.

You are given the opportunity to accomplish multiple BI tasks by the end of each chapter. The goal is to help you master the steps involved in building your own real-world BI solutions.

Downloadable Content

All example projects, exercises, and scripts have been organized into folders by chapter and compressed into zip files. This downloadable content includes all of the BI solution files and information pertaining to the locations of the original databases to make these files work.

You may at times need a hint on how to complete a task. Not to worry, help is available in the form of completed and commented solutions to each standard exercise and "Learn by Doing" exercise.

All of this and more can be found on the Apress website: `www.apress.com`. See the catalog page for this book at `www.apress.com/9781430234883`.

In addition, there is even more content available on each of the author's websites: `http://NorthwestTech.org/ProBISolutions` and `www.keystrokepublications.com`. Here you will find things that just could not fit within this one book such as articles, demos, templates, and videos!

Our Example Scenarios

We work on two BI solution scenarios in this book. Each scenario is based on a sample database created by Microsoft for demonstration purposes. The databases are as follows:

> *The Publications BI solution*: The Pubs database was created in the 1980s for both Sybase's and Microsoft's SQL Server demonstrations. Pubs has a number of flaws in its design, naming conventions, and datatyping. This provides an opportunity to remedy the flaws during the creation of the data warehouse and the ETL process, just as you would find in a real-world scenario. This database also has a number of archetypal data structures useful for highlighting advanced dimensional structures. Another advantage to the Pubs database is that it is the simplest Microsoft demonstration database available. Because of all of these features, we use it as the focal point for the in-chapter practice exercises.

> *The Northwind Foods BI solution*: Made in the early 1990s, the Northwind database is larger and slightly more complex than the Pubs database. It was also created for demonstrations by Microsoft and has numerous design flaws that are discussed and addressed in our data warehouse and ETL processes. This database is used to frame the "Learn by Doing" exercises for each chapter.

All of these databases are readily available and have been used as examples in hundreds of books. Because of this, you may already be familiar with these databases, and you can easily find additional information and code samples to enhance your understanding.

Setup Instructions

Although we have tried to keep the setup requirements as light as possible, there are still a number of complex tasks that need to be performed before you can get the full benefit of this book. You need the following items:

- A full install of SQL Server 2012 developer edition, with all of its supporting servers (SSIS, SSAS, and SSRS)

- The Pubs and Northwind databases

- Administrator-level access to SQL Server and its supporting servers (SSIS, SSAS, and SSRS)

We included setup instructions, files, and videos in a single folder called `_SetupFiles` that is included as part of the downloadable content from the Apress website, `www.apress.com`. Therefore, you have only one downloadable file to worry about. This folder is inside the same zip file as the exercises.

Of course, you have to unzip the file before you can use it. We include detailed instructions on how to copy it to the root of your C:\ drive in Chapter 2, but you can unzip the downloadable content anywhere you want until then. On a Windows 7 PC, the typical location would be the `Downloads` folder.

In Figure 1-6, we have unzipped the file and copied the resulting folders to the location described in Exercise 2-1.

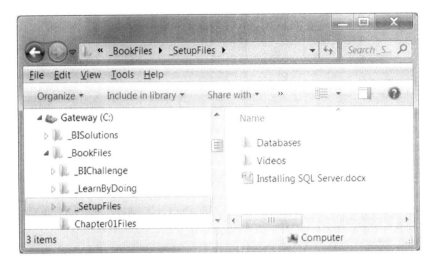

Figure 1-6. *Setup files and folders*

Please review the files in this folder before you start to go through this book. Full instructions are found inside the _SetupFiles folder.

■ **Tip** We have included additional videos and links that can help you tackle the installation if you still feel overwhelmed. These are found on one of the author's websites at www.NorthwestTech.org/InstallingSoftware.

Think Small, Win Big

Creating BI solutions has never been easier. The tools that many vendors offer have become more refined and user-friendly than was dreamed of a decade ago. Still, even with good and inexpensive tools, a BI solution can go horribly wrong if it is not planned and implemented properly.

In the past, a number of approaches have been attempted to ensure that BI solutions have a big impact on a business. One early approach was to include everything that was needed by the business into one master solution. These solutions often took years to complete and were not always consistent with a company's current needs by the time they were finished. This led to a number of issues that have now become widely believed misconceptions about BI solutions. These misconceptions include the following:

- They take years to implement before anything useful is available to the end users.

- They take months of planning before they even get started.

- They cost a lot of time and money.

- They are a luxury, applicable only to large companies with large budgets and large development teams.

Large and long-term solutions have their place, but they are not always necessary. Many companies can benefit immediately from small, quickly designed, and quickly developed solutions. We even go as far as to say that most BI solutions will easily fit this pattern.

A number of changes in IT over the past decade have allowed small BI solutions to become viable. The computers and the software that we run on them are more powerful and less expensive. Something as simple as a Microsoft Excel spreadsheet, for example, can now work with millions of rows at once, allowing you to create very simple BI solutions starting with that tool alone. Microsoft's SQL Server, which has always been reasonably priced, can now work with many terabytes of data, run distributed queries among a collection of servers, and comes with powerful BI tools such as Integration Services, Analysis Services and Reporting Services, at no extra cost. To see what we mean, compare earlier versions of Microsoft Excel and SQL Server. You will see that the cost to purchase these tools, without all of these new features, was roughly the same in the 1990s as it is today, not even taking into account the difference due to inflation.

The combination of more powerful computers and inexpensive software add up to a big win for small to midsize businesses. These businesses can now afford to perform BI tasks that traditionally only their larger competitors were capable of.

The following examples give an idea of how small BI reporting solutions can provide a big win to any type of business:

- Monthly sales reports for a gift shop

- Reports on a development team's projects over time

- Reports that track medication dispensed within a medical clinic

- An auto part store's inventory reports

- Reports that track support calls to a call center

Considering how reporting solutions can be beneficial to companies with 10 employees or 10,000 employees, it is no wonder that BI is such an expanding aspect of our IT industry.

Rapid Application Development for BI Solutions

Once you have established the need for BI solutions, how do you successfully plan, start, and complete them? Although there is no single answer, experience has shown that completing simple, fast, and extensible solutions are the most likely to provide the best cost-to-benefit ratio.

One of the more popular ways to initialize the development process is by using the techniques associated with rapid application development (RAD). In RAD, you start with a short planning phase, followed by a short development phase working on a simple prototype. You then test your prototype for accuracy, consistency, and performance. Once the testing phase has passed, the next step is to release the prototype for comments and prepare to start the next iteration of your solution. This next version of your solution takes comments about the existing features into account and extends the previous solution with new ones. The cycle continues, providing increasing benefit to your users over time.

RAD will not work for all projects, but it will work for a majority of them. This is one of the more successful techniques in the industry today; therefore, we focus on building solutions based on this methodology.

Moving On

In this chapter, we have outlined the steps needed to create a BI solution and discussed the subject matter covered in this book. In Chapter 2, we take a more in-depth look at the entire process by building a very simple BI solution. We start with gathering solution requirements and end with a simple, functioning prototype BI solution. It is time now to get your hands dirty and start work!

What's Next?

In each chapter, we have made our best attempt to focus on what is essential knowledge for every BI professional. We realize that this topic is much too complex for any one book and our essentials may not cover all you need to know. To help further your understanding of the topic within each chapter, we have included reading suggestions for further study.

For more information on RAD, we recommend the book *Rapid Development: Taming Wild Software Schedules* by Steve McConnell (Microsoft Press).

■ ■ ■

CHAPTER 2

A Big-Picture Overview

Eschew the monumental. Shun the Epic. All the guys who can paint great big pictures can paint great small ones.

—Ernest Hemingway

Having a clear overview of how a BI solution is constructed can be one of the most important tasks to ensure a BI solution's success. To understand how a BI solution works, it is important not only to understand its individual projects but to comprehend how these projects integrate into a solution. Jumping into the intricate details without having a full understanding of where each piece of the puzzle fits is setting yourself up for failure. In other words, before starting work on any part of the solution, you need to see and comprehend the big picture.

The process of learning to create a BI solution is not much different. Therefore, to avoid the mistake of jumping in to create the individual projects that make up the solution, in this chapter we walk you through an entire BI solution from start to finish. You will see how each component is integrated and how they function together as a complete solution. Later, as you progress through the other chapters of this book, you will delve deeply into each of the component projects. This overview will help you understand the big picture.

The 10,000-Foot View

To start, let us list the steps that you will be performing in this solution. You begin building the solution by looking at the solution requirements and isolating the data you will be working with. You then move onto documenting the requirements and building your data warehouse. When the data warehouse is complete, you fill it up with data using a SQL Server Integration Service (SSIS) package. After filling the data warehouse, you create a cube and finally a report against the cube you have created.

Figure 2-1 shows a representation of these components. There are icons in the upper left of the figure representing the original source of the data. These original sources may be database tables or files, but in either case, you must review these objects in order to isolate the data you need for your particular BI solution. Afterward, move the data from its original source location into a data warehouse database.

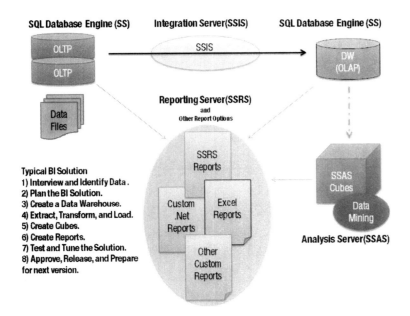

Figure 2-1. *A BI solution overview*

The data warehouse you create is designed on the principles of an online analytical processing (OLAP) style of database utilizing dimensional and fact tables. This is a different style of design than the online transactional processing (OLTP) databases that most developers are familiar with. The design differences are based upon their purpose. Databases that focus on gathering new data are designed around the OLTP format. The OLAP format focuses on providing information from existing data. As you will see later, you work with both OLTP and OLAP databases in a BI solution. OLAP databases come in two common forms: relational databases and cube databases. Relational databases use tables to contain the reporting data, while cube databases use cubes instead. This makes sense when you remember that the terms *tables* and *relations* are synonymous in database terminology.

In a BI solution, data warehouses are created using relational databases in an OLAP format. Nevertheless, you may also create an OLAP cube in addition to the data warehouse. Note that in Figure 2-1, we have displayed this connection between these two objects with a dotted line, indicating that the cube database represents an optional component.

Not all BI solutions need a cube database. In fact, many companies choose to create reports using the data warehouse alone. In Figure 2-1, the thin lines from the data warehouse to the reporting options represent this standard scenario. In addition, it is still possible to pull report data from the original online transaction-processing (OLTP) databases when needed, indicated in Figure 2-1.

The data warehouses and cubes provide additional options that make these structures desirable. For instance, because SSAS cubes host data mining capabilities, you can pull data mining results to your reports through your cubes. Another advantage of having a cube is that a variety of reporting applications are available designed to work with cubes alone.

Interviewing and Isolating Data

In any BI solution, the first course of action is interviewing the client or company owner that needs the solution. Because we do not have real life clients to interview, we describe the scenario here. Let's consider the following as our letter of engagement:

Dear Consultant,

I need reports that will give me information about weather patterns. Currently, I have been collecting data in the format shown in Table 2-1. I track the dates, the maximum and minimum temperatures, and the events of that day. Could you please create an example of what you do for customers like me?

Sincerely,
A Typical Client

And as the letter promises, Table 2-1 shows an example of the data.

Table 2-1. *The Data in the WeatherHistory.txt File*

Date	Max TemperatureF	Min TemperatureF	Events
1/23/2011	48	43	Rain
1/24/2011	51	46	Rain
1/25/2011	52	38	Rain/Sun
1/26/2011	53	41	Rain/Sun
1/27/2011	50	37	Fog

The client has not provided much detail, which is consistent with what you are likely to see in a real-life scenario. Yet once you review the data, you will find you have enough to get started. Besides, creating the prototype solution is often better than asking for more details when you are first trying to understand a client's needs. You are more likely to understand what questions to ask and be able to extract more information from the client in a second interview after you have created a simple prototype.

If you want to understand what is needed in a BI solution, start by understanding its data. For example, look at the range of values and data types noted in Table 2-1. You can see under the date column, for example, that the customer is using days, months, and years, but not hours or seconds. You can see whole values without decimal points under the maximum temperature column. You can also see that the client is using text descriptions in the Events column.

These facts give you vital clues about what your solution can accomplish. For instance, you will be able to create reports that tell you it was raining on a particular day, but not whether it was raining at noon on that day.

Once you have evaluated the data and identified what is available, you can begin the planning phase for the solution.

Plan the Solution

In each BI solution, you should create a document describing what you are trying to accomplish. Creating this document is the first part of the planning phase.

You also need to decide on a place to store your documentation. This location should be readily accessible to any team member working on the project. In this book, we use a subfolder in a Visual Studio solution folder as our document repository. This is convenient, because we are going to create several Visual Studio projects, and each of these projects will be added to the same Visual Studio solution as our documentation folder. Once complete, all of the projects and the documentation that defines those projects will be included under a single Visual Studio solution folder on the hard drive.

Creating Planning Documents

We created two tables (Tables 2-2 and 2-3) that document information about the client's data and what we know about it so far.

Table 2-2 lists the data source combined with descriptive names in one column and the data types in the other. Because all the data is coming from a text file rather than an existing database table, the data types are all strings.

Table 2-2. *Documenting the Source*

Data Source	Source Data Type
FlatFile.Date	String
FlatFile.Max TemperatureF	String
FlatFile.Min TemperatureF	String
FlatFile.Events	String

In Table 2-3, you see a listing of the destination columns, destination data types, any transformations we can expect to use, and an example of the outcome of those transformations. The purpose of this is to document the design of the destination tables, so we have listed the appropriate data types.

Table 2-3. *Documenting the Destination*

Data Destination	Destination Data Type	Transformations	Example
DimDates.DateName	datetime	add zero as needed and cast to datetime	01/23/2011
FactWeather.MaxTempF	int	cast to int	48
FactWeather.MinTempF	int	cast to int	43
DimEvents.EventName	varchar(50)	n/a	Rain

We often informally record source and destination information using a Microsoft Excel spreadsheet. From this informal evaluation, we then proceed to create more formalized documents toward the end of the solution life cycle. The formal documents will become a part of the BI solution we deliver to a client, while the informal spreadsheet is for development.

One advantage of using Excel is that it may be used to outline many parts of the solution using the different worksheets within one workbook.

As an example, one of the worksheets can include the informal information we have laid out in Tables 2-2 and 2-3, which defines the Extract Transform and Load (ETL) process in a solution. Figure 2-2 shows that we have recorded the need to extract dates from the flat file and convert the string data into a datetime data type, on a worksheet called ETL Planning.

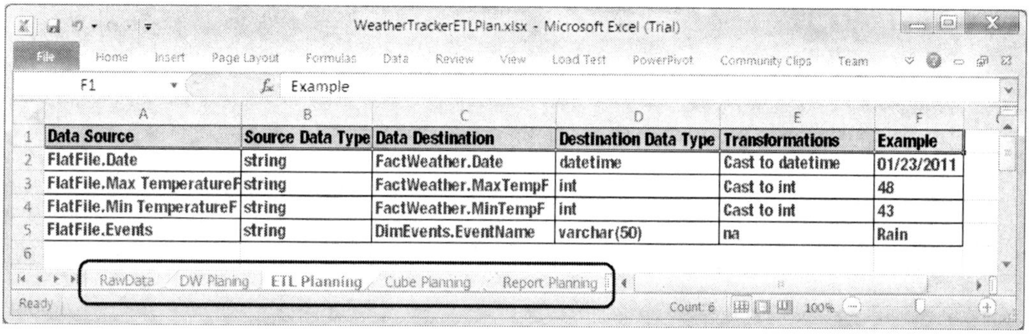

Figure 2-2. *Documenting the plan*

During the planning phase, researching how to accomplish the types of transformations you need during the ETL process helps us estimate what needs to be done during the ETL process. It also lets us contact the client earlier if we discover a problem. Although you do not actually create the ETL process yet, you do want to feel confident that you can accomplish the task when the time comes.

Listing 2-1 shows SQL code that takes a date as a string of 11 characters like those found in the text file and converts them into datetime data. One of the transformations listed in the Excel file in Figure 2-2 requires this change; thus, we can test how this is accomplished and whether this data will be clean enough to use for the ETL process we perform later.

Listing 2-1. Sample ETL Code

```
-- Convert the string to datetime
Declare @Date Char(11)
Set @Date = '1/23/2011'
Select @Date; -- Outcome = 1/23/2011
Select Convert(datetime, @Date) -- Outcome = 2011-01-23 00:00:00.0
```

Adding Documents to Visual Studio

At this point, we have two documents that outline the BI solution: the original file and our Excel workbook. We should now think about organizing our work by grouping the documents in some manner. As we mentioned earlier, we are placing the documents into a folder that will be added to a Visual Studio solution.

If you are not familiar with Visual Studio already, you should know that it organizes projects and code files under a structure Microsoft calls a *solution*. These Visual Studio solutions consist of a folder with a set of XML files that identify which projects and files are part of the solution.

Creating Visual Studio Solutions and Projects

You can create a Visual Studio solution in a couple of ways. For example, if you create a Visual Studio project, a Visual Studio solution will automatically be created for you. If you are not ready to make a project yet, you can also create a blank solution and add projects to it later. In both cases, you can add documentation and script files to the solution folder at any time.

Each project you make in Visual Studio uses a predefined template. These templates are part of various plug-ins to Visual Studio. Once a project plug-in installs, it becomes part of Visual Studio, similar to how the Adobe's Flash plug-in becomes part of your web browser.

The Visual Studio plug-in that comes with SQL Server is either SQL Server Data Tools (SSDT) or Business Intelligence Development Studio (BIDS) depending on which version of SQL Server you install. As of SQL 2012, BIDS is a subset of SSDT, but in earlier versions it was a stand-alone plug-in. You may find the terms BIDS and SSDT used interchangeably on the Internet, but do not let it worry you too much. Think of SSDT as the newer version of BIDS instead of its replacement, and you will be fine. As you read though this book, you will notice we usually refer to both generically as Visual Studio.

With the BIDS/SSDT plug-in to Visual Studio, you can design SQL Server Integration Services (SSIS), SQL Server Analysis Services (SSAS), and SQL Server Reporting Services (SSRS) projects using templates. These install automatically into Visual Studio 2010 when you install SQL Server 2012. In fact, if you do not have Visual Studio 2010 already, the SQL Server installation will install it as well.

If it still seems confusing, consider the following:

- Visual Studio is a host for development tools.

- If we were to install Microsoft's C# development tools, for example, it would install Visual Studio and the C# development plug-in for Visual Studio.

- If we decided later to add Microsoft's Visual Basic .NET, it only needs to install the plug-in to the already installed Visual Studio.

If we decided later to add Microsoft's SQL Server Data Tools, the installation checks to see whether a compatible version of Visual Studio is already installed: if not, it will install it for you. If it already is installed, it just adds the SSDT plug-in as an additional development tool. Either way, the BIDS/SSDT plug-in becomes part of Visual Studio 2010.

■ **Note**   We provide a lot of detail about how to use these project templates throughout the book, so don't be intimidated by the sudden inundation of acronyms. In this chapter, we created all the projects for you as part of the downloadable content. All you need to do is review these projects as we continue through this chapter.

Using Visual Studio

Visual Studio 2010 can be accessed either through SQL Server's menu item (Windows Start Button ➤ All Programs Microsoft SQL Server 2012 ➤ SQL Server Data Tools) or under the Visual Studio menu item (Windows Start Button ➤ All Programs ➤ Microsoft Visual Studio 2010). Both options open Visual Studio 2010 and present a selection of project templates in the New Project dialog window (Figure 2-3).

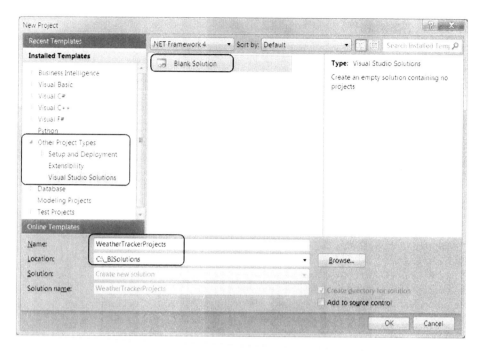

Figure 2-3. *Creating a blank Visual Studio solution*

In Figure 2-3, under Installed Templates, you can see Business Intelligence, Visual Basic, C#, and other categories listed. Beneath these template categories are the templates themselves. To select a template, click a category listed in the treeview and then choose a template in the center of the dialog window.

Each category can have many templates, so Microsoft includes subcategories to help organize the templates. For instance in Figure 2-3, you can see that there is only one template called Blank Solution, under Other Project Types ➤ Visual Studio Solutions.

Be warned, you may not see the same categories and templates on every computer! Those shown in Figure 2-3 appear because the screenshot was taken on a computer that had all of these plug-ins installed. If, however, you only have SQL Server installed, then you will not see the Visual Basic or C# plug-ins on your computer. Instead, you will see only BI projects. That is not a problem, of course, because that is exactly the type of project we want to create.

Creating a Blank Solution

In Visual Studio, new solutions that do not use a project template are referred to as *blank solutions*. Creating a blank solution is quite easy. This is done by selecting File ➤ New ➤ Project from the file menu option.

When the new project dialog window opens, a list of project types is displayed on the left side of the screen. Expanding the Other Project Types by clicking the small arrow (or triangle) allows you to select the Visual Studio Solutions option (Figure 2-3).

Because a Visual Studio solution is a collection of one or more projects, we have named the solution WeatherTrackerProjects in the Name textbox. Place the solution folder somewhere that is easy to find. In Figure 2-3, we typed C:_BISolutions into the Location textbox. (This naming convention corresponds to the downloadable content and the step-by-step guide within the exercises.)

Working with the Blank Solution

After you have chosen a template and configured both the name and location, click OK to close the dialog window and begin working with the new solution. Behind the scenes, Visual Studio creates a number of files and folders, but all you see is a single folder displayed in a treeview-based window called Solution Explorer. This is the main window used to work with Visual Studio solutions.

⬛ **Note**   Visual Studio automatically generates new subfolders for each project within the solution folder. In addition, because we specified a nonexistent folder (C:_BISolutions), Visual Studio creates both the _BISolutions folder and the WeatherTrackerProjects solution folder for us. In this book, we use the _BISolutions folder to organize all of our solutions folders under one principal folder.

When you create a blank solution, Visual Studio shows the solution name in the Solution Explorer window but not much else. We are going to add a new solution folder specifically to hold our solution documents by clicking the Add New Solution Folder button circled in Figure 2-4. Once you click this button, a new folder is created instantly, and the text is highlighted to enable you to rename it easily.

You should rename your folder to something appropriate. This solution folder will hold a collection of documents for our solution, so a name such as SolutionDocuments is appropriate (Figure 2-4).

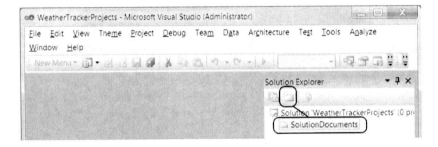

Figure 2-4. *Creating a solution folder*

Once the folder is created and renamed, you can then add documents you have created or collected to it. Simply click the new SolutionDocuments folder you created, which highlights the folder. Then right-click the folder and select Add ➤ Existing Item from the context menu, as shown in Figure 2-5.

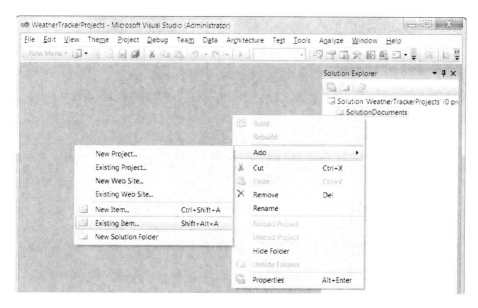

Figure 2-5. *Adding existing files to a solution folder*

Another method is to highlight the new Solution Documents you have created, and from Visual Studio's main menu, select Project ➤ Add Existing Item. Both allow you to navigate to where the files are located and add them to your solution.

After you have selected your files and added them to your blank solution, Visual Studio will either copy the file to your solution folder or reference the file from its existing location.

■ **Important** About 90% of the time Visual Studio will copy the file instead of making a reference to it. It is always important to verify whether a reference or copy was made. Using references can cause major problems because any changes to the files in your Visual Studio solution will change what you believed to be a copy. You can tell where a file is located by right-clicking the file and selecting Properties from the context menu (similar to Figure 2-5), and the file's path will be displayed in a Property window. In cases where Visual Studio creates a reference, when what you really wanted was a copy, you need to use Windows Explorer to copy the file to the solution folder on your hard drive yourself and then make a reference to the newly copied file.

One of the primary goals of this book is to give you a chance to practice the art of creating BI solutions. To keep things simple, we have created all the WeatherTracker BI solution documents and BI projects for you in this example. This provides you with a quick introduction to the anatomy of a BI solution and introduces you to organizing your projects using Visual Studio, without having to explain how to create these projects and files. Don't worry! We explain how those items are created in the other chapters of this book.

EXERCISE 2-1. PREPARING THE SOLUTION FILES

In this exercise, you add the downloadable book files to your C: drive. You then create a blank Visual Studio solution to hold BI solution documents, and connect to the files you downloaded within the new solution. This step is the foundation of all your future exercises and must be completed for future exercises to work properly.

Later in this chapter, you will add SSIS, SSAS, and SSRS projects to this Visual Studio solution. Figure numbers provide hints for most steps.

Install the Book Files

The files for this exercise, as well as all of the exercises throughout this book, are available in the downloadable book content and need to be installed on your computer before you can continue.

1. If you have not done so, download the book files from the Apress website. These files are in a zipped format.

2. Create a folder on your C:\ drive called _BookFiles and unzip the downloadable files into it.

Each operating system unzips files in a different manner; therefore, we are not including step-by-step instructions on how to unzip these files. Just make sure that the book files are on the root of your C:\ drive using the name _BookFiles. We strongly recommend you use this name, because we continually reference it in the book.

Once unzipped, this folder includes all the files and projects you need to complete each exercise within this book.

Create a Folder for all BI Solutions

We want to have one place where all of the BI solutions you create in this book are stored, so let's create one now.

1. Inside the _BookFiles folder, locate the Chapter02Files folder and open it.

2. Find a subfolder called _BISolutions and use Windows Explorer to copy this folder to the root of C:\ drive, as shown in Figure 2-6. (To open Windows Explorer, access your computer's Start menu, and click Computer. In the left column select Local Disk (C:). On the right, where all of your C:\ files are listed, paste the entire _BISolutions folder here.)

You now have a second folder on your hard drive called C:_BISolutions.

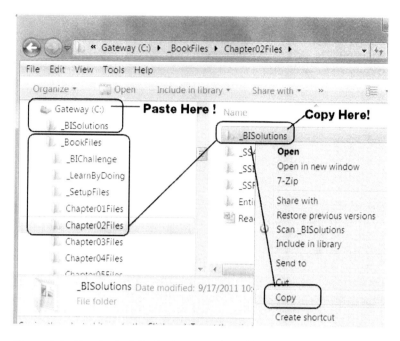

Figure 2-6. *Now both _BISolutions and _BookFiles are on the C:\ drive*

Review the WeatherTrackerProjects Files

After you have copied the _BISolutions folder to its new location, you may want to look inside it to see exactly what you just copied. Inside this folder is a subfolder called WeatherTrackerProjects. Within that folder are four documents we use in this chapter's BI solution:

- InstWeatherTrackerDW.sql (SQL code to create a data warehouse)

- SQLTransformations.sql (SQL code for the ETL process)

- WeatherHistory.txt (a text file with the client's data)

- WeatherTrackerETLPlan.xls (an Excel file that outlines the solution plan)

1. Verify that you have both the _BookFiles (the original folder that holds all of our chapter files and demos) as well as the _BISolutions folder (the folder you will place your work in) directly on your hard drive, as shown in Figure 2-6.

Placing these files directly on your C:\ drive makes it much easier to navigate to the files you need for these exercises, and leaves less room for confusion later, because we access this folder quite often throughout this book.

Open Visual Studio

You now need to open Visual Studio and create a new solution. The following steps walk you through the process. Visual Studio opens from either the Microsoft Visual Studio 2010 or the SQL Server Data Tools menus. We have chosen to use the Visual Studio option for simplicity, but either menu item works.

1. Open Visual Studio 2010. You can do so by clicking on the Start button and navigating to All Programs ➤ Microsoft Visual Studio 2010 ➤ r. Right-click Microsoft Visual Studio 2010 to see an additional context menu (Figure 2-7). Then, click on the Run as Administrator menu item.

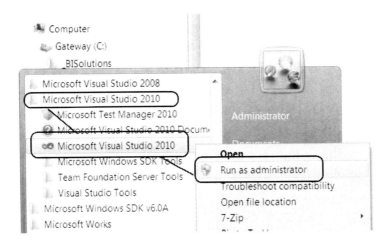

Figure 2-7. *Opening Visual Studio and running as admininstrator*

2. If the User Account Control (UAC) message box appears asking "Do you want the following program to make changes to this computer?" click Yes (or Continue depending upon your operating system) to accept this request.

3. When Visual Studio opens, select File ➤ New ➤ Project from the menu. (Do not use the Create: Project option from the Start Page as you may have done in the past with other types of solutions.)

4. When the New Project dialog window opens, on the left side of your screen click the arrow to expand Other Project Types and select Visual Studio Solutions, as shown in Figure 2-3.

5. In the templates section, select Blank Solution, as shown in Figure 2-3. The Name and Location textboxes are filled in with a default name, but we change these in the next step.

6. In the Name textbox, at the bottom of the screen, type the name **WeatherTrackerProjects**. In the Location textbox, type **C:_BISolutions** (as shown previously in this chapter in Figure 2-3), and finally, click OK.

Once the solution is created, it appears in the Solution Explorer window of Visual Studio, on the right side of your screen.

Review the Files Created by Visual Studio

1. Right-click the solution WeatherTrackerProjects icon, and select the Open Folder in Windows Explorer menu item (Figure 2-8). Windows Explorer will open to the location of your solution folder.

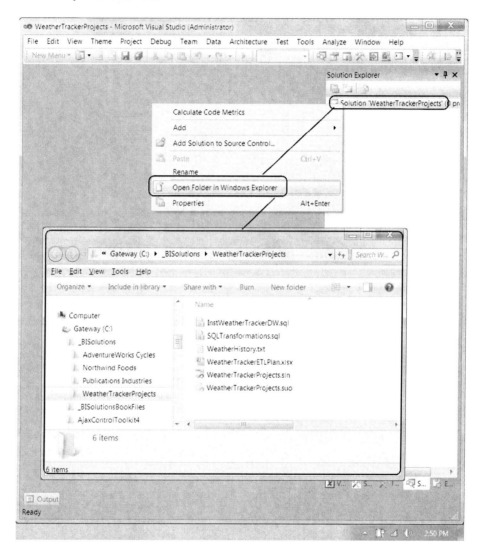

Figure 2-8. *Viewing the files within the solution folder*

2. Review the files in this folder and notice that they are not currently showing in the Solution Explorer window. (Solution Explorer shows only a minimum of files from a solution or project folder, but we can change this by adding the file to the solution as existing items, as we do in just a moment.)

3. Now that you have seen what is in the folder, close Windows Explorer.

Add a New Solution Folder

You can organize your files into logical groups using a solution folder. To do so, you must first create a folder and then add files to it. Let's do that now.

1. Add a new solution folder to your Visual Studio Solution by clicking the Add New Solution Folder button at the top of the Solution Explorer window. (This step is illustrated in Figure 2-4 of this chapter.)

2. Rename the new folder as SolutionDocuments (shown previously in Figure 2-4).

3. Right-click the new SolutionDocuments folder and select Add ➤ Existing Items from the context menu (Figure 2-5). A dialog window will open.

4. In the left column, select your Local Disk (C:). On the right, first double-click _ BISolutions, and then double-click the WeatherTrackerProjects subfolder to access the files within it.

5. While holding down the control button, click and select the following files: InstWeatherTrackerDW.sql, SQLTransformations.sql, WeatherHistory.txt, and WeatherTrackerETLPlan.xls (Figure 2-9).

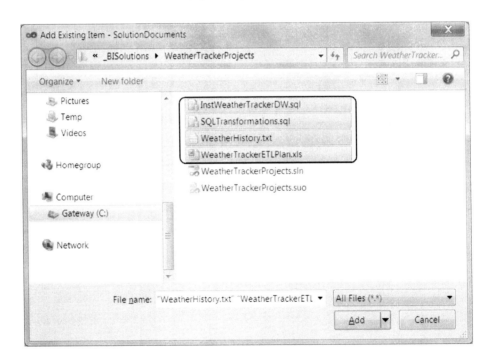

Figure 2-9. *Selecting your BI solution files*

Note: Windows by default hides the file extensions. So, you may see only the first part of the name of each of these files. We recommend turning off this feature when you get the chance. It will often be very helpful to know the extensions of all your files. To find more of this feature, search the Web for "Windows Show Extensions."

6. Click the Add button to add the highlighted items to your SolutionDocuments folder. (Visual Studio uncharacteristically creates a reference instead of a copy when you add an existing item to a solution folder.)

7. Visual Studio will open each of the files in Visual Studio as well as Excel so that you can see their content. We are not making any changes to these files. Look at them if you like, but then close them by clicking Excel's closing X and the X on each Visual Studio tab, as shown in Figure 2-10. Note that Microsoft hides the closing X on a tab until it has been selected.

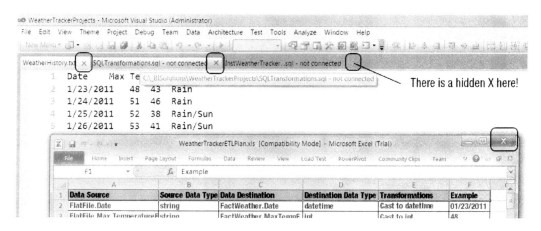

Figure 2-10. *Closing your BI solution files*

8. Use Visual Studio's File menu to save your work by selecting the Save All option.

9. Leave Visual Studio open for now because we continue to work with it in the next exercise.

In this exercise, you created a blank solution and added documents that will be used for creating your SSIS, SSAS, and SSRS projects. We refer to these documents in future exercises.

Creating the Data Warehouse

Once you have assembled the documents that outline your solution plan and after you have added those documents to a Visual Studio solution, it is time to create the BI solution projects starting with the data warehouse. Let's begin this process with an overview of what a data warehouse is and how it is created. Then we provide you with code that creates the data warehouse, and finally, we add that code to a new Visual Studio solution folder called DWWeatherTracker.

An Example Data Warehouse

In this book, we describe a data warehouse as a collection of one or more data marts. These data marts consist of one or more fact tables and their supporting dimension tables. In Figure 2-11 you see a design with a single fact table called FactWeather and a one-dimensional table called DimEvents. Notice the correlation between the notation in the Excel spreadsheet of Figure 2-2 and the design of these tables in Figure 2-11.

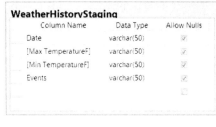

Figure 2-11. *The data warehouse tables*

These tables represent a very minimal design. As shown in Chapter 4, there are typically several dimension tables in a data warehouse, not just one. For now, though, let's keep focusing on the big picture and come back to the details later.

Using SQL Code to Create a Data Warehouse

One of the solution documents, `InstWeatherTrackerDW.sql`, has SQL code that creates the DWWeatherTracker data warehouse for you when it is executed in SQL Server Management Studio. Before we have you execute this code, let's review what it does.

■ **Note** The code file `InstWeatherTrackerDW.sql` can be found as one of the documents you added to your Visual Studio solution in Exercise 2-1. It opens within Visual Studio if you double-click the file. In the next exercise, we open and run the code in SQL Server Management Studio, so you will become used to working with both tools.

Create the Database

The first set of tasks that the SQL code tackles is checking to see whether the database already exists and, if so, drop it. We labeled the first tasks Step 1 in our code (Listing 2-2). After that, in Step 2, the code creates the database and tells SQL Server to use the new database for all the commands that come next.

Listing 2-2. Drop and Create the Database

```
--Step 1) Drop the database as needed
Use Master
Go
If ( exists( Select Name from SysDatabases Where name = 'DWWeatherTracker' ) )
  Begin
    Alter Database [DWWeatherTracker] Set single_user With rollback immediate

    Drop Database [DWWeatherTracker]

  End
```

```
Go
-- Step 2) Create Data Warehouse Database
```

Create Database DWWeatherTracker

```
Go
```

Use DWWeatherTracker

```
Go
```

Create the Tables

The next three steps outlined in the InstWeatherTrackerDW.sql code file creates three tables (Listing 2-3).The first table is to hold raw data imported from the text file WeatherHistory.txt. The second table, DimEvents, is our one and only dimension table in this example. The third table, FactWeather, is our fact table.

Listing 2-3. Creating Three Tables

```
-- Step 3) Create a Staging table to hold imported ETL data
```

CREATE TABLE [WeatherHistoryStaging]

```
( [Date] varchar(50)
, [Max TemperatureF] varchar(50)
, [Min TemperatureF] varchar(50)
, [Events] varchar(50)
)
-- Step 4) Create Dimension Tables
```

Create Table [DimEvents]

```
( [EventKey] int not null Identity
, [EventName] varchar(50) not null
)
Go

-- Step 5) Create Fact Tables
```

Create Table [FactWeather]

```
( [Date] datetime not null
, [EventKey] int not null
, [MaxTempF] int not null
, [MinTempF] int not null
)
```

In step 4, the DimEvents dimension table is created (Figure 2-11). In this table, we have both a key column and a name column. This is characteristically the minimum design seen in real-life examples. In most cases, however, there are also additional descriptive columns in the table.

Using the Identity Option

In Listing 2-3, we included an identity attribute on the EventKey column. In SQL Server, a column marked with an identity attribute automatically adds incremental integer values to the column each time a row of data is inserted into the table. In other words, because we have configured the EventKey column to be an identity column, adding a new event name to the DimEvents table will automatically insert an integer of "1" into the EventKey column. When we add another event name, an integer of "2" is inserted for the second row, and so on.

Adding Primary Key Constraints

You should include primary key constraints in all of your dimension and fact tables because they keep your data ordered and free of duplicate values. In most dimension tables, you add a primary key constraint to its single key column. But in fact tables, you add a primary key constraint to multiple key columns, because it is the combination of key values that distinguishes one row from another. When a primary key constraint is associated with multiple columns, these columns form a *composite primary key*.

As an example, there are two key columns in the FactWeather table, the Date and EventKey, both of which refer to dimensional tables. The other two columns in the table are MaxTempF and MinTempF, both of which are measure columns. The multiple dimensional key columns form a composite primary key for a fact table.

The code in Listing 2-4 creates a primary key constraint on the DimEvents and FactWeather tables. Adding the constraint to the table identifies which column or columns are part of the primary key and enforces uniqueness of values across these columns.

Listing 2-4. Adding the Primary Keys

```
-- Step 6) Create Primary Keys on all tables
Alter Table DimEvents Add Constraint
  PK_DimEvents Primary Key ( [EventKey] )
Go
Alter Table FactWeather Add Constraint
  PK_FactWeathers Primary Key ( [Date], [EventKey] )
Go
```

Looking back at Figure 2-11, you can see the primary key icons are on both the Date and EventKey columns, which indicates that both columns are part of a composite primary key. Look for these icons, or something similar, in any database diagram you review.

Adding Foreign Key Constraints

Notice in Figure 2-11 that both the fact table and the dimension table have a column called EventKey. In the fact table, the EventKey column forms a foreign key relationship back to the DimEvents dimensional table. The code in Listing 2-5 adds a foreign key constraint to enforce this relationship and will not allow you to enter key values in the fact table if they do not first exist in the dimension table. For instance, if you try to insert an EventKey value of 42 to the fact table, the constraint would check to see whether an EventKey value of 42 exists in the dimension table. If not, the database engine generates an error message and the insert fails!

Listing 2-5. Adding the Foreign Keys

```
-- Step 7) Create Foreign Keys on all tables
Alter Table FactWeather Add Constraint
  FK_FactWeather_DimEvents Foreign Key( [EventKey] )
  References dbo.DimEvents ( [EventKey] )
Go
```

▥ **Note** Many exercises in this book are written in a way that assumes you have some familiarity with SQL programming. We have tried to make our code simple enough for all levels of developers, but some of this subject matter may be difficult if you have never used SQL before. To help you become more familiar with this language, we recommend checking out the excellent, and free, SQL tutorial on the website www.w3schools.com.

Running SQL Code from Visual Studio

You can manage and execute your database scripts using Visual Studio even if it is not obvious how to do so. In the next exercise, you have an opportunity to do just that. We provided step-by-step instructions on how to do so.

EXERCISE 2-2. CREATING THE DATA WAREHOUSE

In this exercise, you create the data warehouse and the tables within it. You can do this by using the code found in the `InstWeatherTrackerDW.sql` file. Once that is accomplished, you create a new solution folder in Visual Studio and move the `InstWeatherTrackerDW.sql` file to the new folder. Figure numbers provide hints for most of these steps.

Completion of this exercise is required to be able to complete future exercises throughout this chapter.

Open Visual Studio (Optional)

1. Visual Studio should still be open from the previous exercise. If it is not, please open it. Please remember to run Visual Studio as an administrator by right-clicking the menu item and selecting the Run as Administrator option.

2. With Visual Studio open, access the WeatherTrackingProject solution from the File ➤ Recent Projects and Solutions menu.

Connect to SQL Server and Execute the Code

1. Double-click the file `InstWeatherTrackerDW.sql` file in Solution Explorer. The SQL code you see in Figure 2-12 opens in your main window.

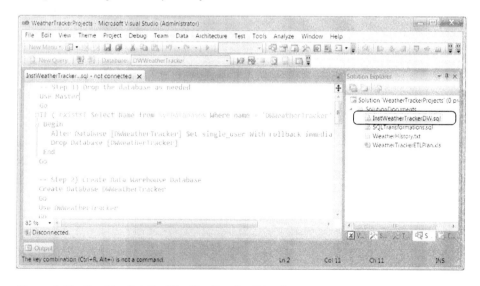

Figure 2-12. *Double-click* `InstWeatherTrackerDW.sql` *to open it.*

2. Without selecting any of the SQL code, right-click a blank area of the query window (the window where the SQL code is) and then click the Connection ➤ Connect menu option (Figure 2-13). The Connect to Database Engine dialog window appears.

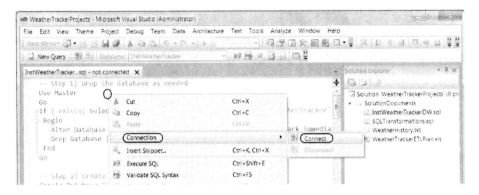

Figure 2-13. *Connecting to SQL Server from Visual Studio*

3. In the Connect to Database Engine dialog window, type in the name of your computer or use the alias of (local) in the Server Name textbox (Figure 2-14); then click the Connect button to make the connection. If you have trouble with this step, see the upcoming "Important" note.

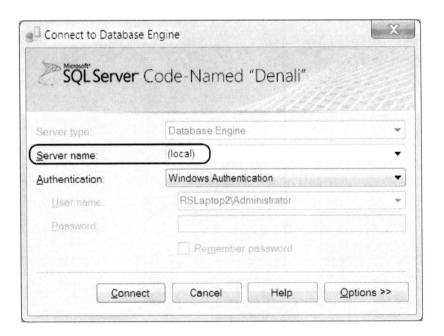

Figure 2-14. *Entering the server name*

Important: If you have installed SQL Server on the same computer multiple times or if you named the instance of your single install, your SQL 2012 installation may be called a name other than (local). For example, Randal has SQL installed as (local)\SQL2012, and Caryn references her server as (local)\Denali. (When you install SQL, you get to make up the name!) If a named instance is used, you must connect to SQL Server using the instance name in the "Server name" textbox. For more information, search the Web for "SQL Server Named Instances."

4. After connecting to the database, execute the SQL code by right-clicking an area in
    the SQL code window to bring up the context menu, and select Execute SQL
    (Figure 2-15).

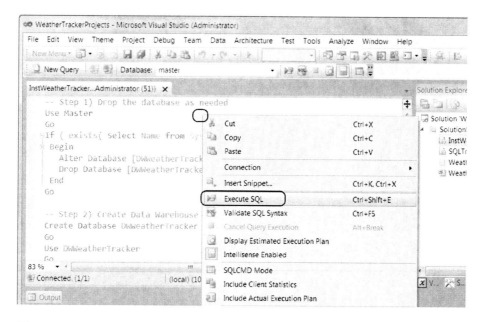

Figure 2-15. *Executing your code*

5. The code in the SQL file should complete successfully in only a few seconds. You
    will know that it has worked when the message displayed in Figure 2-16 appears.
    This is a good sign, but you should also verify that the database was created by
    connecting to it. We will do that next.

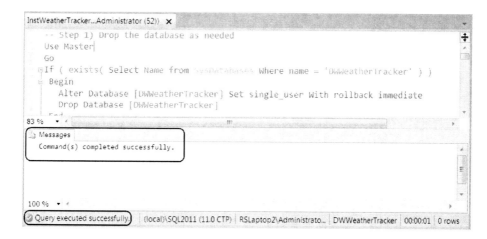

Figure 2-16. *The SQL code executed successfully*

Verify That The Database Was Made

1. Open the Server Explorer window of Visual Studio. You can do so by using the View ➤ Server Explorer menu item (Figure 2-17). Be careful, because it is easy to click the Solution Explorer item by mistake. Server Explorer should display on the left side of Visual Studio.

Figure 2-17. *Displaying Server Explorer*

2. In Server Explorer, right-click the Data Connections icon, and select Add Connection from the context menu (Figure 2-18). The Add Connection dialog window appears (Figure 2-19).

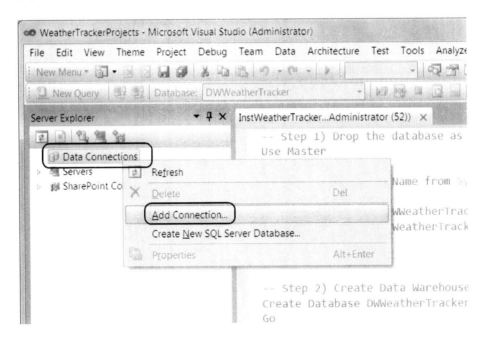

Figure 2-18. *Connecting to the SQL database engine from Server Explorer*

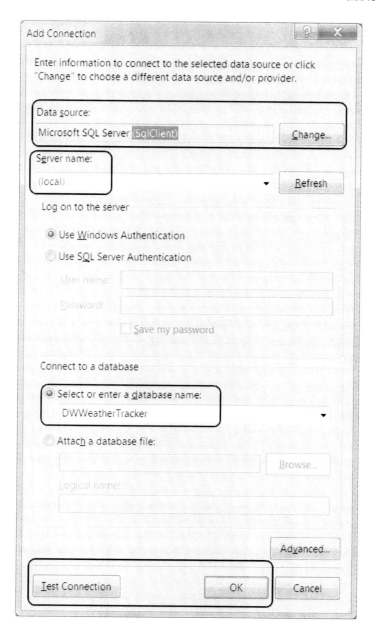

Figure 2-19. *Configuring the connection*

3. In the Choose Data Source dialog window (Figure 2-20), set the data source to Microsoft SQL Server (SQLClient). If you need to change this setting, click the Change button.

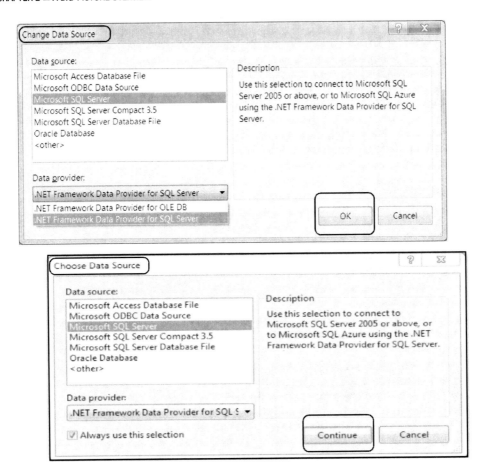

Figure 2-20. *Setting the connection to use SQL Server as the data source*

Important: Depending on a combination of things, Visual Studio will sometimes display the Choose Data Source window, shown at the bottom of Figure 2-20, instead of the Add Connection window. Both windows look almost identical, and we cannot be sure which will open on your computer. You can use either one to select your data provider.

If the Choose Data Source window appears before the Add Connection window on your computer, just select the Microsoft SQL Server data source and the .NET Framework Data Provider for the SQL Server data provider; then click the Continue button, and the Add Connection window appears. Add the Microsoft SQL Server (SQLClient) setting to the Data Source dropdown box, as shown in Figure 2-19.

If the Add Connection window appears but the Microsoft SQL Server (SQL Client) setting is not in the Data Source dropdown box, then click the Change button, and it will open the Change Data Source window, also shown in Figure 2-20.

4. In the Add Connection dialog window (Figure 2-19), set the server name to either the name of your computer or the alias of (local) in the Server name dropdown box. If you are using a SQL named instance, then you have to include that name as well (for more information, search the Web for "SQL Server named instances").

5. In the "Select or enter database name" dropdown box, select the DWWeatherTracker database (Figure 2-19).

6. Now, test your connection using the Test button, and when it succeeds, close it by clicking OK. Then click OK to close the Add Connection dialog window.

7. Click the small arrow next to the DWWeatherTracker database icon to expand Server Explorer's list of database objects (Figure 2-21).

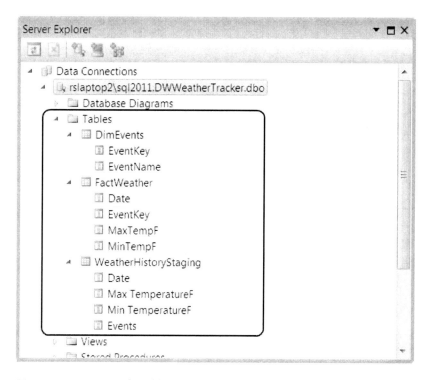

Figure 2-21. *Viewing the tables and columns in DWWeatherTracker*

8. Click the small arrow next to the Tables folder to expand Server Explorer's list of tables.

9. Click the small arrow next to the DimEvents table to expand the column listing.

10. Double-click the file WeatherTrackerETLPlan.xls in Solution Explorer. This opens an Excel window.

11. Verify that the tables are properly made by comparing them to the Excel spreadsheet, as shown in Figure 2-22.

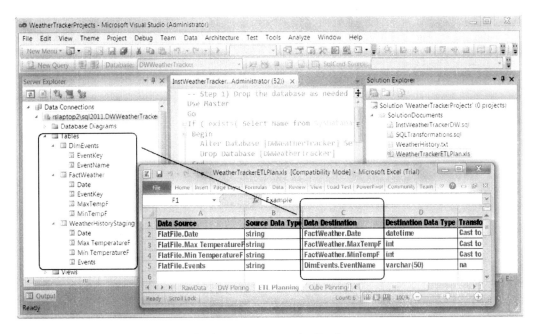

Figure 2-22. *Verifying the tables and columns in DWWeatherTracker*

12. All should be correct, so you can close the Excel spreadsheet
    (`WeatherTrackerETLPlan.xls`), the SQL file (`InstWeatherTrackerDW.sql`), and
    Server Explorer.

Place the Code File in a New Solution Folder

Now that the data warehouse is built and you have verified that it is correct, let's move the SQL code file to a
new Visual Studio solution folder.

1. In Solution Explorer, select the solution WeatherTrackerProjects icon by clicking
   it and then click the Add New Solution Folder button to create a new solution
   folder. (Be sure you do not have the Solution Documents folder highlighted,
   or the new folder will be created as a subfolder of it instead of a subfolder
   of the entire solution.) Select the title and then click it again to rename it
   WeatherTrackerDataWarehouse.

2. Click the `InstWeatherTrackerDW.sql` file in the `Solution Documents` folder and
   drag it into your new `WeatherTrackerDataWarehouse` folder. Note that this does not
   move the file to a new location on the hard drive; it just adds a new visual reference
   to the file in Solution Explorer.

3. Now you have two references to this file, one in each solution folder. We now
   remove the original. Right-click the `WeatherTrackerDW.sql` file that is within
   `SolutionDocuments` and select Remove from the context menu. This does not
   delete the file but only removes that reference. Verify that what you see in Solution
   Explorer looks similar to Figure 2-23.

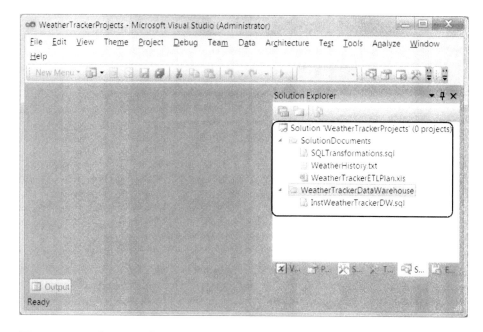

Figure 2-23. *Solution Explorer at the end of Exercise 2-2*

4. Leave Visual Studio open for now because we continue working with it in the next exercise.

In this exercise, you executed a SQL script that created a data warehouse. You then placed this script into a new solution folder using Visual Studio. Soon, you will be adding additional projects to the solution. Your ultimate goal is to place all the code and projects you need for the WeatherTracker BI solution into this one Visual Studio solution. We continue this process in the next exercise by adding a SQL Server Integration Services project.

Create the ETL Process

With the data warehouse created, it is time to start the extract, transform, and load (ETL) process. During this phase of a BI solution, just as the title of the process states, you must first extract the source data, then transform it as necessary, and finally load it into the data warehouse tables. In the WeatherTracker project, the text file called WeatherHistory.txt is the source of the data. The destinations are the tables you created in Exercise 2-2.

The transformations needed for the WeatherTracker ETL process are listed in the Excel spreadsheet we created earlier (Figure 2-2). To create the ETL process using SSIS, we examine this spreadsheet making special note of column names, data types, and transformations listed. Let's review an SSIS project based on our recorded plan.

ETL with an SSIS Project

SSIS represents Microsoft's premier ETL tool. It is one of Microsoft's business intelligence servers, and it is one of the project types available in Visual Studio. To create an SSIS project, start Visual Studio and select File ➤ Add ➤ New Project from its main menu, as shown in Figure 2-24. Doing so forces the Add New Project dialog window to appear (Figure 2-25).

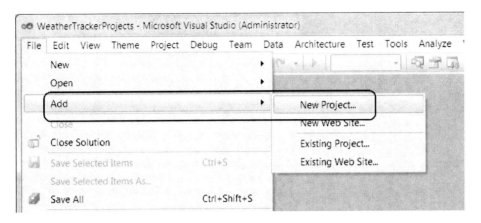

Figure 2-24. *Adding another project to the current Visual Studio solution*

In the Add New Project dialog window, you have selected the type of project you want it to make, so expand the Business Intelligence category and select the Integration Services subcategory, as shown in Figure 2-25. Then select the Integration Services Project template, available in the center of the dialog window.

Before you click the OK button to close the dialog window, you should set the project's name to something appropriate and verify the project's location. In Figure 2-25, we have set the Name textbox to WeatherTrackerETL and left the Location textbox set at (C:_BISolutions\WeatherTrackerProjects). After configuring these settings and clicking the OK button, a new subfolder is created under the WeatherTrackersProjects solution folder containing the SSIS project.

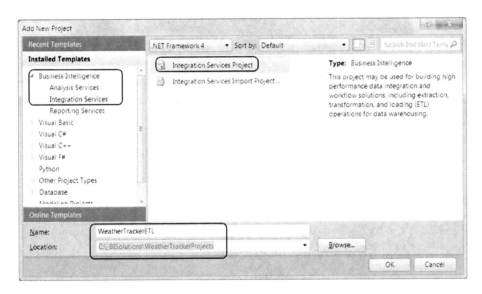

Figure 2-25. *Selecting the Integration Services Project template and naming the new project*

■ **Note** Although it is possible to place the project in a location other than a subfolder of the solution, doing so would make it more difficult to locate all of the projects for a given solution. Therefore, we recommend you always keep your projects for a particular BI solution together under one Visual Studio solution folder.

Creating an SSIS Package

An SSIS project consists of one or more package files. When you first create an SSIS project, its template includes an empty SSIS package called `Package.dtsx`. The file displays in the Solution Explorer window (Figure 2-26). Unless you are creating a throwaway demo package, you should rename the file to something indicating its purpose. You can do so by right-clicking the file and selecting the Rename menu item from the context menu.

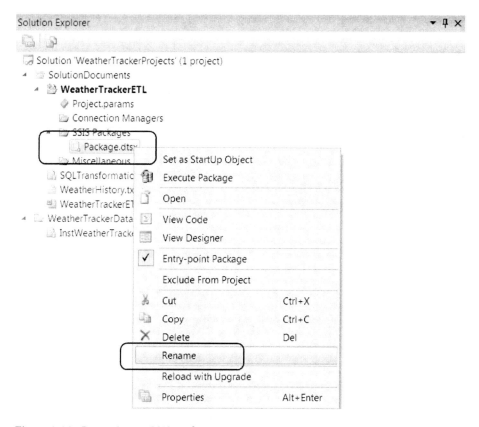

Figure 2-26. *Renaming an SSIS package*

41

THE GETTING STARTED (SSIS) WINDOW

In SQL 2012, SSIS has a new start-up window that provides helpful links to videos and articles (Figure 2-27). At some point, you may want to view these videos, but most of the time you can simply close this window.

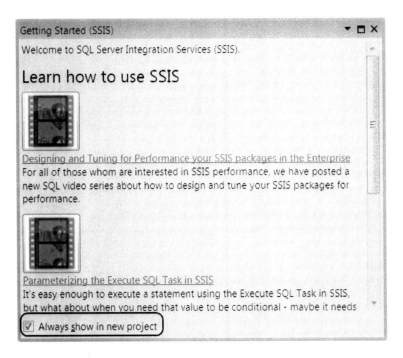

Figure 2-27. *The new Getting Started (SSIS) window*

You can close this window each time you make a new SSIS project, or you can click the "Always show in new project" checkbox to change this behavior. If you want to see this window at another time, there is a Getting Started menu item under the SSIS main menu that displays it again.

An SSIS package file is essentially a text file formatted as in the XML language. While this file can be manually programmed, you likely will let Visual Studio do the coding for you. To use this feature, you drag and drop items from the SSIS Toolbox onto a package's design surface. This act invisibly writes your XML code for you. (We will elaborate on this in a moment.)

As of SQL 2012, SSIS now includes a dedicated Visual Studio Toolbox in addition to the standard Visual Studio Toolbox (Figure 2-28).

Each item in the Toolbox represents a set of SSIS commands. For example, Figure 2-28 shows a Data Flow task icon and an Execute SQL Task icon within the Toolbox. These tasks represent a collection of individual SSIS programming commands used during ETL processing.

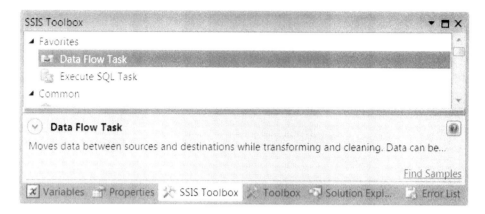

Figure 2-28. *The new SSIS Toolbox for SQL 2012*

Outlining the Control Flow Tasks

To configure a new SSIS package, we recommend outlining what you intend to accomplish by adding tasks from the Toolbox onto the package's designer interface. We show an example of what this looks like in Figure 2-29.

The designer interface is separated by tabs. SSIS tasks are created and configured on the Control Flow tab (Figure 2-29).

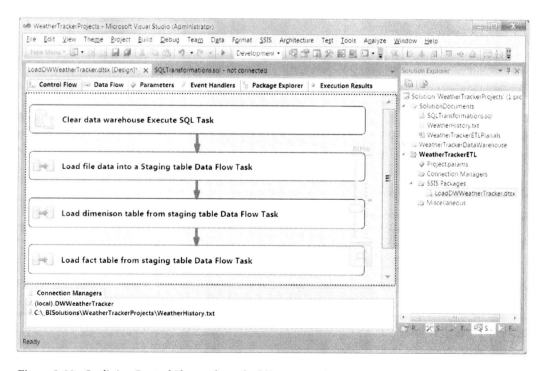

Figure 2-29. *Outlining Control Flow tasks and adding connections*

Each task must be added from the Toolbox onto the Control Flow surface and then configured. One of the first things to configure are the names of the tasks. Notice that we have configured our tasks to have a uniquely descriptive name. This is an important step because each package can have a large number of tasks within it. Without proper naming of both the package file and the tasks within the package, it will be confusing to you as well as to anyone who will be maintaining the package over time.

The tasks shown in Figure 2-29 include an Execute SQL task and three Data Flow tasks. There are also three Precedence Constraints shown, indicated by the arrows in Figure 2-29. Precedence Constraints arrows represent the flow of the tasks, that is, which task will run first, next, and last. You can create a precedence constraint by clicking one task and dragging the resulting—magically appearing—arrow to another task.

SSIS Connections

Each SSIS package needs one or more connection objects to perform the ETL processing. When a package is first made, it does not include any connections, but they can be added to the package from the Connection Manager tab (Figure 2-30). After you outline your SSIS package, you will have a good idea of what connections you will need and can begin to create the connection objects for your tasks. Connection objects can be created by clicking in the Connection Managers area at the bottom of the screen (Figure 2-30) and choosing a connection type from the context menu that appears.

Figure 2-30. *Adding Connection objects*

■ **Note** We go into more detail about how to make connections in Chapter 7.

Configuring a Flat File Connection

SSIS can connect to text files, databases, and even web services. In our example, we connect to both the WeatherHistory.txt file, that contains the client's data, and the DWWeatherTracker database, which we created in the previous exercise. Note that each connection is also named accordingly.

To configure a flat file connection, use the Flat File Connection Manager Editor dialog window. All of these connection dialog windows have one or more pages. The pages are listed on the left side of the dialog window and the configurations for each page are displayed on the right (Figure 2-31). This is a common pattern throughout SSIS.

For example, in Figure 2-31, you can see that we configured the File name property, on the General page, of a Flat File Connection Manager Editor window. This is how SSIS knows which file to import data from.

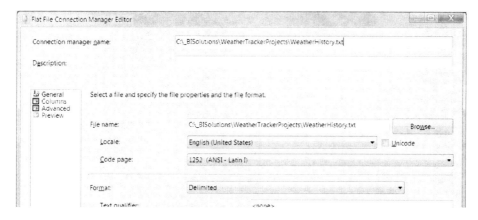

Figure 2- 31. *Configuring a flat file connection*

Configuring a SQL Server Connection

When configuring a SQL Server connection, the editing window allows you to identify the server name as well as the database name, as shown in Figure 2-32. This dialog window is almost identical to the one you used while connecting to the data warehouse in Exercise 2-2. Microsoft reuses this same dialog window in all of the BI projects, so expect to see it a number of times as you proceed through the book.

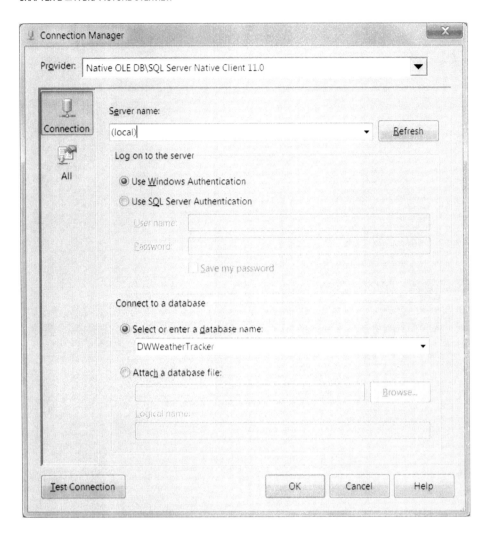

Figure 2-32. *Configuring a SQL Server connection*

After you have created and configured the connection, you can configure the SSIS tasks. We typically do so in the order defined by the precedence constraints. The first task in our package is an Execute SQL task, so we start there.

Configuring an Execute SQL Task

As the name implies, an Execute SQL task allows you to run SQL statements from SSIS packages. They are often used to clear out the data warehouse tables so they can be refilled with new data. This "flush and fill" technique works only with smaller data warehouses tables, but because it is the simplest technique, we will use it in our WeatherTracker ETL project.

Tables can be cleared by using a set of Delete From < table name > SQL commands like the ones shown in Figure 2-33. When SSIS runs an Execute SQL task, these SQL statements are executed on the connected SQL server.

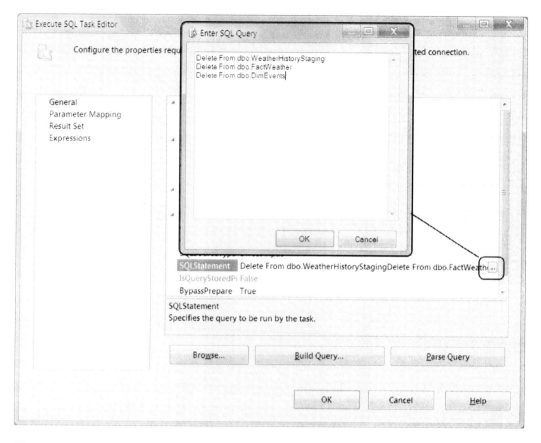

Figure 2-33. *Editing an Execute SQL task*

Configuring Data Flow Tasks

In addition to the Execute SQL tasks, Data Flow tasks are something you will use on a regular basis. Their purpose is to transfer data from one location to another.

Once you have placed a Data Flow task onto the Control Flow surface, you need to configure it. To configure a Data Flow task, you either double-click the Data Flow or simply highlight it and click the Data Flow tab, as shown in Figure 2-34. Both options take you to the same location and allow you to edit the Data Flow task.

Data Flows are somewhat unique in that they have their own editing tab and their own set of Toolbox items. If you watch the Toolbox as you switch between the Control Flow tab and the Data Flow tab, you can see Toolbox items change.

The Data Flow's Toolbox items are specifically designed to move data from one location to another and to apply transformations as the data moves from point to point. There is a large set of Toolbox items to choose from, and they can be grouped into categories. The first category is Data Flow Sources. These items allow you to pull data from text files or database tables.

The next category is Data Flow Transformations. These optional tasks provide ways to manipulate the data as it moves from the source to the destination.

The final category is Data Flow Destinations. These are used to connect to files or database tables that you want to fill with data. In summary, each Data Flow task consists of at least one source and one destination and optionally one or more transformations.

Because each data flow always has a source and destination, you can start outlining one source task and one destination task. In our example, we need a flat file source and a SQL Server destination (Figure 2-34).

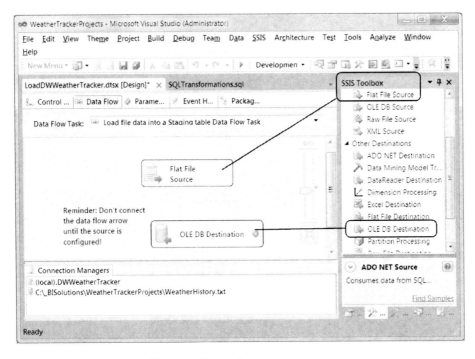

Figure 2-34. *Outlining the first Data Flow task*

When configuring a Data Flow task, it is important to configure the data source first before you configure the data destination. One common mistake is to outline the process of your data flow by putting a source and destination task onto the data flow surface and then connecting the tasks immediately. You do not want to connect the destination until after you have configured the data source. Doing it out of order will automatically, but improperly, configure the data destination.

⬛ **Note**   If you mistakenly edit the destination task before you configure and connect the data source to it, the data destination becomes corrupt. The simplest way to resolve this is to delete the data destination and replace it with a new one from the Toolbox. Afterward, configure and connect the data source before you attempt to edit the destination task.

Configuring Additional Data Flows

Since a single SSIS package can consist of many Data Flow tasks, Microsoft made it easy for you to switch between them. At the top of the Data Flow tab there is a dropdown box labeled Data Flow Task. You can select between the individual Data Flow tasks that are part of your SSIS package using this dropdown box (Figure 2-35).

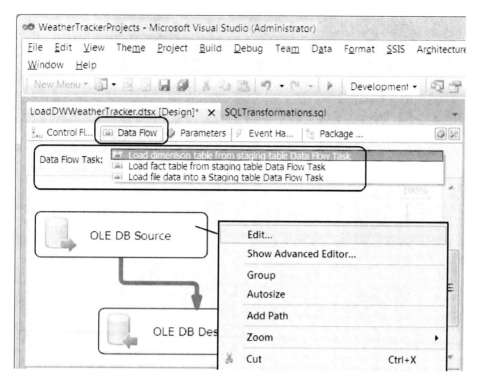

Figure 2-35. *Navigating between Data Flow tasks*

Once you focus on a selected Data Flow task, you can edit its SSIS items by either double-clicking them or using the Edit option from the context menu that appears when you right-click an item (Figure 2-35).

Many items that you configure have both a standard and advanced editor. Each has its own dialog window. The standard dialog windows have all the settings that you would commonly use, and in our example, they contain all the settings we need.

Configuring a Data Source

To use a data source, you must first configure it. For example, in an OLE DB source, there are three basic configurations you need to adjust: the OLE DB connection manager, the data access mode, and the SQL command text.

In Figure 2-36, you can see that we configured the OLE DB source to use the (local).DWWeatherTracker connection manager, one of the two connection objects that we created earlier (Figure 2-32).

We also configure to use a SQL command as the data access mode, which allows you to use a SQL statement instead of just the name of a table in a database. Using a SQL statement is preferred since you can filter out columns or rows you do not want. You can also apply transformations, such as data conversions, when the SQL code executes. This is a simple and effective way to transform the data as it is retrieved (Figure 2-36).

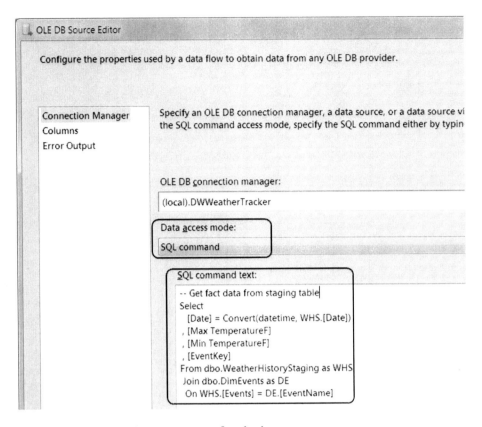

Figure 2-36. Using a SQL statement to refine the data source

Executing an SSIS Task

Once you have created and configured an SSIS package, test your work by executing the package. To do this, right-click the package in Solution Explorer and select Execute from the context menu, as shown in Figure 2-37.

Executing the SSIS package may take a while as your installed SSIS service reads the underlying XML code instructions, attempts to make the connections to the text file and the database, and then performs the extraction, transformation, and loading tasks.

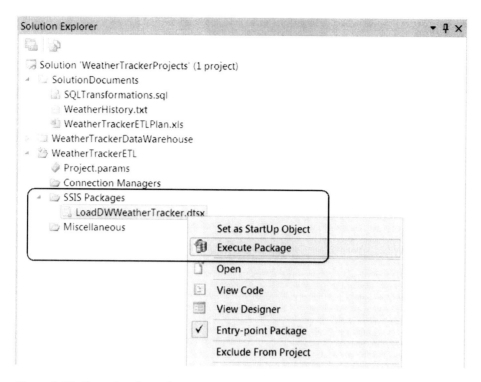

Figure 2-37. *Executing the package*

Completing the Package Execution

While a task is executing, it shows an indicator icon on the right side of the task (Figure 2-38). As each individual task processes, the icon changes from a yellow wheel icon to a green check mark icon once it completes successfully.

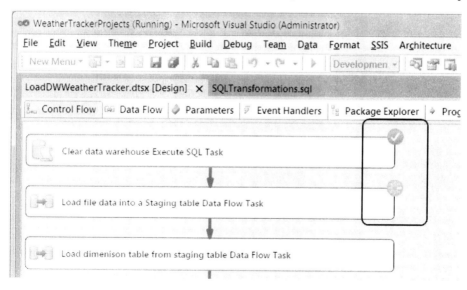

Figure 2-38. *The package while it is executing your code*

If SSIS encounters an error, the task that is causing the problem displays a red X icon, and the execution of the package comes to a halt.

When all of the tasks complete, successfully or not, manually stop its debugging process using the Debug menu at the top of the Visual Studio window or by selecting the stop debugging hyperlink at the bottom of the Visual Studio window (Figure 2-39).

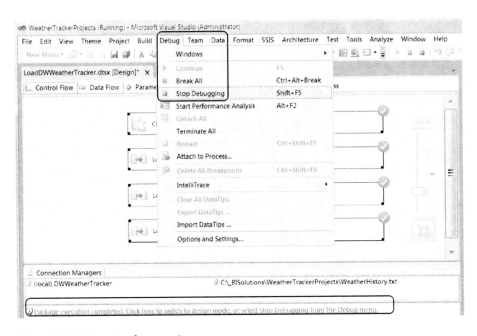

Figure 2-39. *Stopping the execution*

You now have an overview of how to create, configure, and execute an SSIS package. In Chapter 7 we look at this in depth, but even at this level you should have a pretty good feel for the process. It is now time to get some practice by doing another exercise in which you add our existing package to a new SSIS project you create. You then verify its configurations and finally execute it to fill up your data warehouse.

EXERCISE 2-3. ADDING AN SSIS PROJECT TO YOUR SOLUTION

In this exercise, you make a new SSIS project and add it to your current Visual Studio Solution. You then add an existing SSIS package to the project that is included in the downloadable content. After the package has been added to your solution, verify that the connections are configured correctly and finally execute the package. Completion of this exercise is required to complete the other exercises throughout this chapter.

1. Visual Studio should still be open from the previous exercise, but if not, please open it and access the WeatherTrackingProject solution from the File ➤ Recent Projects and Solutions menu. Remember to always use Visual Studio as an administrator by right-clicking the menu item, selecting Run as Administrator, and then answering Yes to close the UAC pop-up window.

Add a New Project to a Current Solution

First, we add a new project to the WeatherTrackerProjects solution. The following steps guide you through this process.

1. With the WeatherTrackerProjects Visual Studio solution open, click the solution 'WeatherTrackerProjects' icon in Solution Explorer. (The Visual Studio menus are context sensitive, and we want to be working with the solution itself and not any of the solution folders.)

2. Use the File menu to add a new project to your current Visual Studio solution by clicking File ➤ *Add* ➤ New Project. (Make sure you do not choose File ➤ New ➤ Project. The two options are easily confused.) See Figure 2-24.

3. When the Add New Project dialog window opens, select the Business Intelligence ➤ Integration Services option on left side of this window, and on the right side, select Integration Services Project from the options. See Figure 2-25.

4. At the bottom of this dialog window, in the Name textbox, type **WeatherTrackerETL**. Verify that the Location textbox reads C:_BISolutions\WeatherTrackerProjects. See Figure 2-25.

5. Click OK to close this dialog box.

Add an Existing SSIS Package to the Project

Once the WeatherTrackerETL project has been created, it appears in your Solution Explorer window. We are now going to add an existing package to the WeatherTrackerETL project.

1. Select the SSIS Packages folder and right-click the `Package.dtsx` file to access the context menu. This is similar to Figure 2-26, but we choose a different option from the context menu.

2. We do not need the `Package.dtsx` file, so click the Delete context menu option. A pop-up message box appears asking whether you are sure you want to permanently delete this. Click OK to continue.

3. Now we add a premade `.dtsx` file to the project from the downloadable book files. Right-click the SSIS Packages folder and select the Add Existing Package option from the context menu. See Figure 2-40. The Add a Copy of Existing Package dialog window appears (Figure 2-41).

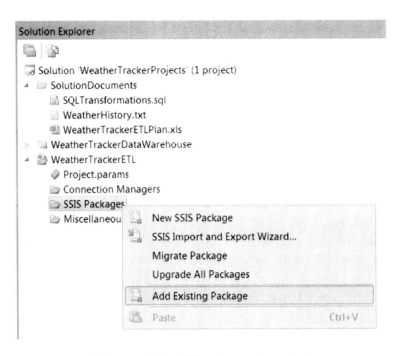

Figure 2-40. *Adding an existing SSIS package to the project*

4. Within the Add Copy of Existing Package dialog window, select File System from the Package Location drop-down box (Figure 2-41).

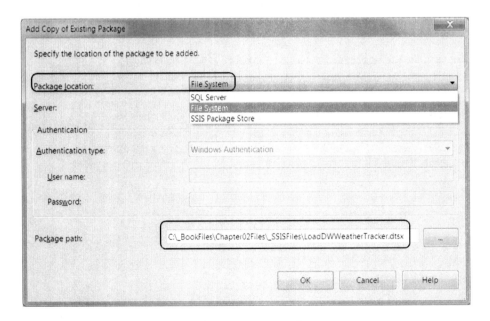

Figure 2-41. *Identifying the path to the existing SSIS package*

5. For the Package Path textbox, click the ellipsis button, browse to the file
`C:\_BookFiles\Chapter02Files\_SSISFiles\LoadDWWeatherTracker.dtsx`, and
click Open. Then click the OK button to close the dialog window and have Visual Studio
copy the `.dtsx` file to your SSIS project. The file should now appear in Solution Explorer.

6. Once the file is added, verify in the Visual Studio Properties window that the package
file was copied from the proper file. (If you do not see the Properties window,
you can open it by right-clicking `LoadDWWeatherTracker.dtsx` and selecting
Properties.) The path should show the `_BISolutions\...` path (not `_BookFiles`).

Verify the Connections

Whenever you are working with someone else's SSIS package, there is a chance that the SQL server
instance name on their computer is not the same as the one on your computer. At this point, we need to
make sure that the connections are correct for your computer by opening and editing the package.

1. To open and edit the package, double-click the `LoadDWWeatherTracker.dtsx`
package file in Solution Explorer or right-click the file and choose Open from the
context menu. The package opens, and the Control Flow tab displays the four tasks
shown in Figure 2-29.

2. Locate the two connections under the Connection Manager tab at the bottom of
Visual Studio, as shown in Figure 2-30.

3. Right-click the connection called (local).DWWeatherTracker (Figure 2-42), and
uncheck the Work Offline option from the context menu, as needed.

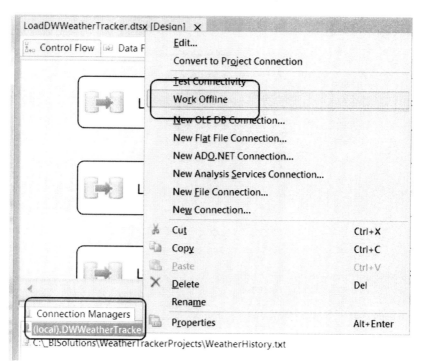

Figure 2-42. *Unchecking the Work Offline option*

4.   Double-click the first connection called (local).DWWeatherTracker (Figure 2-42).

5.   When the Connection Manager dialog window appears, click the test connection button
     at the bottom of the window to verify that you can connect to the data warehouse
     database. A window will open with the text Test Connection Succeeded (Figure 2-32).
     If this does not occur, change the name of the server in the Server Name dropdown
     box to match the name of your SQL server installation and try the test button again.

Important: You do not need or want to change the SSIS connect name even if you change the SQL Server
name.

6.   Double-click the second connection called WeatherHistoryTextFile. When the Flat File
     Connection Manager editing window appears, verify that the file name is pointing
     to the C:_BISolutions\WeatherTrackerProjects\WeatherHistory.txt
     text file (Figure 2-31).

7.   Click the Preview page on the left side of the dialog window. After verifying that the
     data looks like Figure 2-43, click OK to close it.

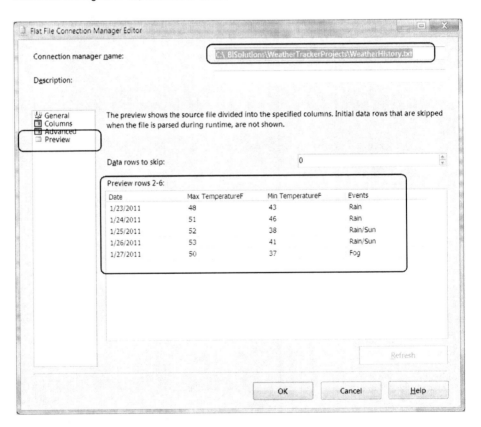

Figure 2-43. *Verifying the file connection*

Important: If the file connection does not work, please check that you correctly placed the _BISolutions folder
on the C:\ drive. (You can review Exercise 2-1 for more information.) Click the Data Flow tab at the top of your

screen and review the three data flows that are listed in the Data Flow Task drop-down box within this tab (Figure 2-35). In future chapters, we show you how to create these types of tasks yourself. At this point, you are just reviewing these to familiarize yourself with the process.

Execute the SSIS Package

After you have reviewed the Data Flow tasks and connections, it is time to run the package and fill the data warehouse.

1. While in the Control Flow tab, right-click the LoadTablesFromFiles.dtsx package in the Solution Explorer and select Execute Package from the context menu (Figure 3-37).

2. The package will run, and all the tasks on the Control Flow designer window should finish with a green checkmark icon (Figures 2-38 and 2-39).

3. Once the package has completed running, use the Debug ➤ Stop Debugging menu option to stop the package execution (Figure 2-39).

In this exercise, you created an SSIS project within your Visual Studio solution. You then added a preexisting package from the downloadable content included with this book. Then you verified that the package connections were valid, reviewed the tasks that were configured, and executed the package.

In the next exercise, you will add a premade SSAS cube project to this same Visual Studio solution, so please leave Visual Studio open for now.

Creating a Cube

Great! So now, you have a data warehouse filled with data. It is time to create a cube that uses it! The most common way to make a cube is by using the SQL Server Analysis Server (SSAS) template in Visual Studio.

Like SSIS, you can add an Analysis Server project to an existing Visual Studio solution. Do this using the File ➤ Add ➤ New Project menu item within Visual Studio. When the Add New Project dialog window appears, select the Analysis Server Project template from among the Business Intelligence projects, as shown in Figure 2-44.

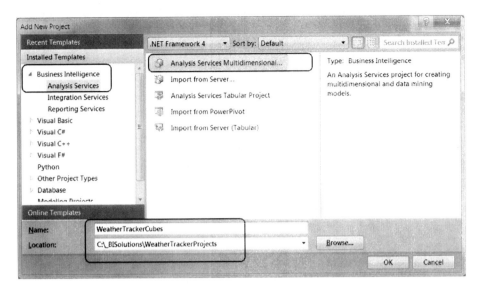

Figure 2-44. *Adding a new SSAS project to the WeatherTrackerProjects solution*

In this dialog window, you have a choice of several options. The choice we focus on in this chapter is the Analysis Services Multidimensional Project (Figure 2-44).

As always, you should give the project a descriptive name that indicates what the project is used for. In Figure 2-44, we have named the project WeatherTrackerCubes.

After clicking the OK button, a new SSAS project appears in Solution Explorer, as shown in Figure 2-45. Note that you now have two BI projects and two Solution folders in this one Visual Studio solution.

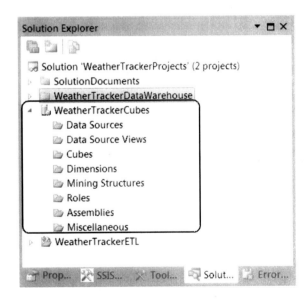

Figure 2-45. *The new SSAS project in Solution Explorer*

Making a Connection to the Data Warehouse

You must have a source of data to make a cube. Most often, this source is a data warehouse such as the one that we built at the beginning of this chapter. In an SSAS project, you are able to connect to the data warehouse by making a data source object. A data source can be created by right-clicking the folder named Data Sources in the Solution Explorer window and selecting New Data Source from the context menu, as shown in Figure 2-46.

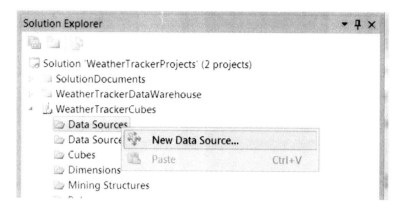

Figure 2-46. *Adding a data source*

While working in SSAS, a wizard will help you create a data connection. As we will see in Chapters 9 through 12, most of the objects in SSAS are created using one wizard or another.

The Data Source Wizard starts with the welcome screen, but it does not provide much information. Clicking Next, however, moves you to the next page, which then allows you to either create a new data connection or use an existing one. Because SSAS is a project inside a Visual Studio solution, if you have created a previous connection to a database, the existing connections listed are ones that Visual Studio remembered from previous projects. As you can see in Figure 2-47, the DWWeatherTracker database is an example of a previously created connection. Therefore, you will be able to select it as an existing data connection on the Data Source Wizard's second page.

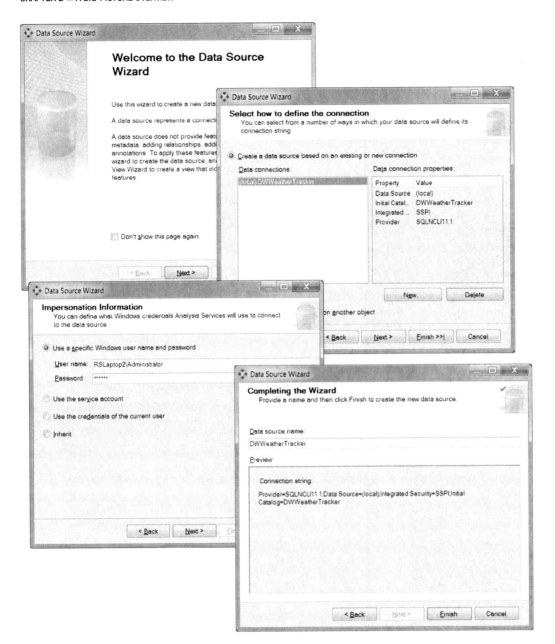

Figure 2-47. *Creating an SSAS data source*

As you proceed to the next page, you are asked for impersonation information. This is an important screen, because it defines the account that will be used to connect to the Analysis Server from Visual Studio and also from the Analysis Server to the data warehouse. This is another topic that is discussed further in Chapter 9.

The best choice is to enter your computer name followed by a slash and a personal account name. Make sure the account you use has access to both the Analysis Server and the SQL Server where your data warehouse is hosted. In Figure 2-48, Randal has an account on his laptop computer called Administrator,

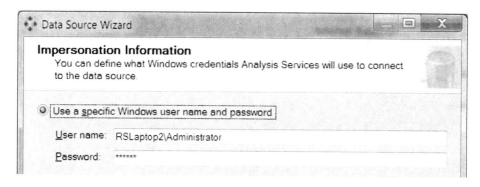

Figure 2-48. *Configuring the impersonation information*

so he typed in **RSLaptop2\Administrator**. Note the direction of the slash. On web addresses you use a forward slash, but here it must be a back slash.

■ **Note** Using a full administrator account is not recommended in a production environment, but for demonstration purposes, it is a good choice because it resolves a lot of issues that can be difficult to troubleshoot. In this book, we use an administrator account for all of our exercises. If you are particularly familiar with Windows, SQL Server, and Analysis Server security, you can use an account that has restrictive privileges. Otherwise, we highly recommend using an administrator account for the book exercises. Search the Web for "True Windows Administrator" for more information.

The next screen of the wizard allows you to name the connection and finish the wizard. The wizard puts spaces between any words it believes to be a concatenation of multiple words. It does this simply by looking for uppercase and lowercase letters. Although this may make it more readable, our experience has shown that spaces in names cause problems for some computer programs. Therefore, we can either take out the spaces that the wizard puts in or just change the name altogether.

Creating a Data Source View

After the connection is made, you need to create a data source view. A data source view provides the foundation of any cubes or dimensions you create. Since Analysis Server 2005, you no longer build cubes as dimensions directly against the data source. Instead, Microsoft provides an abstraction layer represented by the data source view (Figure 2-49).

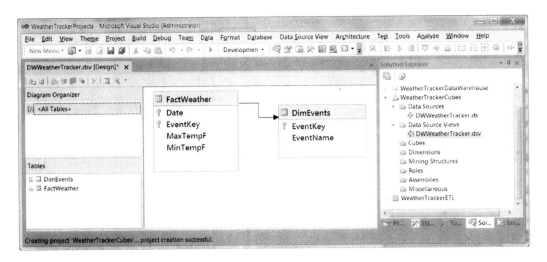

Figure 2-49. *An SSAS data source view*

In the data source view, you can add any tables that you want to use for the creation of both cubes and dimensions. In this example, we need only our single fact table and the single dimension table, but normally you would have many dimension tables associated with one or more fact tables.

One important aspect of the data source view is the relationship line between the fact tables and the dimension tables. This is similar to a foreign key relationship. If your data warehouse has a foreign key relationship between the tables, the wizard automatically adds relationship lines for you. Otherwise, you have to add them yourself.

It may be perplexing that the relationship lines do not visually attach to the columns between the tables, but don't worry. The columns are connected even though the data source does not display it that way. If you were to double-click the relationship line, you would see a dialog box that opens indicating which columns are connected.

As with most things in SSAS, you use a wizard to create the data source view. To start the wizard, right-click the Data Source Views folder in Solution Explorer and select New Data Source View from the context menu.

When the wizard launches, it shows a welcome screen (Figure 2-50). Click the Next button at the bottom of this screen to proceed to the screen you are able to configure. The configuration on the next page involves selecting the data source you will be using. If there is only one data source, the choice is pretty obvious. Therefore, simply click the next button to move to the next page.

On this page, select the tables that are to be included in the data source view. As we mentioned previously, you need to select only the tables required for your cubes. In the case of the WeatherTracker project, this would be the DimEvents and FactWeather tables.

The fourth page of the wizard allows you to name your data source view and finish the wizard. Once again, the wizard provides extra spaces in the name, and you may want to remove them.

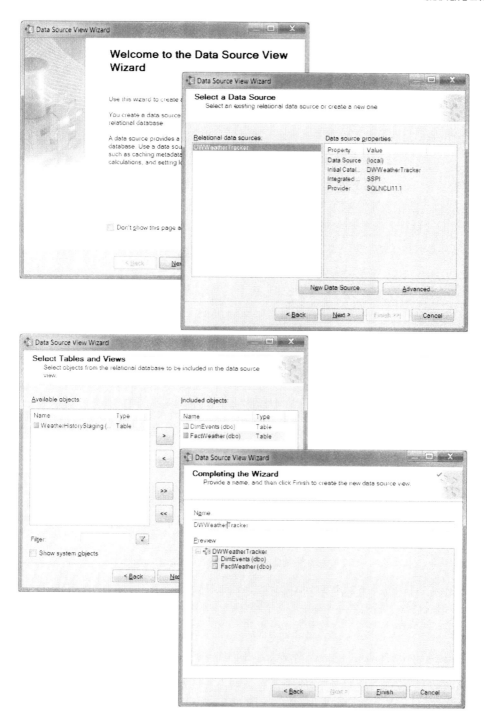

Figure 2-50. *Creating an SSAS data source view*

Creating Dimensions

It is probably not surprising that you use a wizard to create dimensions as well. Start the wizard by right-clicking the Dimensions folder in Solution Explorer. Right-clicking this folder brings up a context menu, and from there you can select the New Dimension option (Figure 2-51).

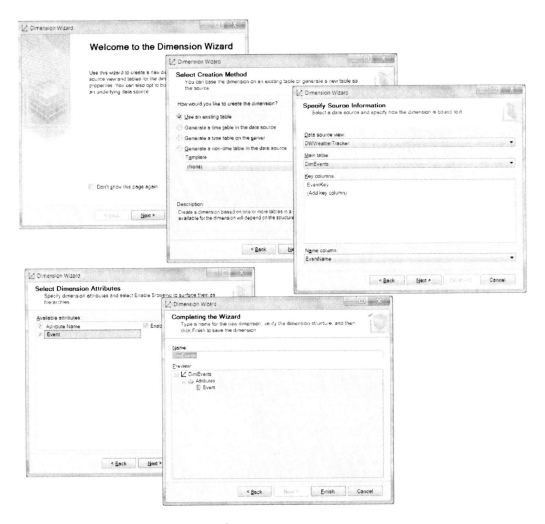

Figure 2-51. *Using the Dimension Wizard*

When the wizard launches, you are given a choice to use existing tables or create new tables within the data warehouse. The most common option is to use the tables you created in an existing data warehouse.

The third page of the wizard allows you to select which data warehouse dimension table you would like to base the new SSAS dimension upon. In our example, we use the DimEvents table. While choosing a table, you also configure the Key and Name columns by using the drop-down boxes on the wizard page. We can use the EventID, for example, as the key column and the EventName for the name column. SSAS uses the key column internally, but it displays the name column in the reporting applications, such as Microsoft's Excel or Reporting Services.

After clicking the Next button, you have a chance to rename your Event ID column to a friendlier name. This is how the column will appear in reporting applications. Because the reporting application will see the value of the name column, keeping the name EventId is not logical, so we change it to Event.

Clicking the Next button once again brings you to the final page of the wizard, which allows you to name the dimension and finish the wizard.

We need to create a dimension for the dates as well. To do this, we once again start the wizard and proceed through its pages, but this time we use the fact table as a dimension table. We do this because all the dates we need are in the fact table. As you can see from this example, a fact table can occasionally be used as a dimension table. We will tell you more about this in Chapters 4 and 9.

Creating Cubes

Creating a data source, a data source view, and the dimensions are all preparatory to creating a cube. The act of creating a cube is done using another wizard (Figure 2-52).

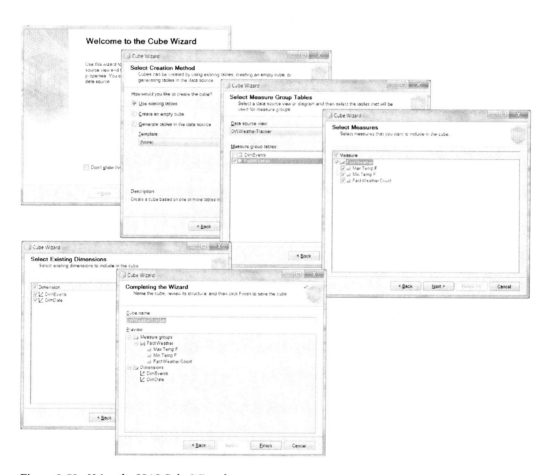

Figure 2-52. *Using the SSAS Cube Wizard*

We cover all of the different pages of this wizard in Chapter 10, but here is a quick overview:

- The first page of the wizard is a welcome page.

- The second page allows you to use existing tables or create new tables similar to the Dimension Wizard.

- On the third page, the wizard allows you to select which tables represent your fact tables.

- The fourth page allows you to select your measures.

- On the fifth page of the wizard, you are allowed to select your existing dimensions.

- The final page of the wizard allows you to name the cube and finish the wizard.

Deploying and Processing

With the completion of this wizard, all of the code needed to create the cube, dimensions, data source views, and data connection will be in a set of Visual Studio XML files. Analysis Server projects use an XML-based programming language that is similar to Integration Services. Each SSAS object created using the wizard creates XML files that represent those objects.

For the XML code to be translated into something that the Analysis Server can understand, you must build the Visual Studio solution. This building process creates a master XML file that can then be uploaded to the Analysis Server.

To upload the file to the Analysis Server, you need to deploy it. This act of deploying is effortlessly accomplished using Visual Studio's menu options. When the XML code deploys to the Analysis Server, the data source, data source view, dimensions, and cubes are all created on that SSAS server.

The cubes and dimensions will not have any data yet. To fill the cubes and dimensions with data, you must process both of them. The act of processing copies data from the data warehouse to the SSAS objects. This is accomplished from the Visual Studio's menu options.

We realize that was a very fast overview and that you likely have many questions, but we discuss all of this again in Chapter 10 in more detail. For now, let's have you add the SSAS project we created for this chapter to your current BI solution in this next exercise.

EXERCISE 2-4. ADDING AN SSAS PROJECT TO YOUR SOLUTION

In this exercise, you add the authors' completed Analysis Server project to your own Visual Studio solution. You verify that the connection in the data source is valid for your machine and then review the data source view, dimensions, and cube. You then build the XML code in project files into a master XML code file, deploy the master XML code file to your SSAS server, and finally process the data from the data warehouse into the SSAS objects.

Completion of this exercise is required to complete the other exercises throughout this chapter.

1. Visual Studio should still be open from the previous exercise, but if not, open it running as administrator (right-click the Visual Studio menu item, select Run as Administrator, and answer Yes to close the UAC pop-up window), and access the WeatherTrackingProject solution from the File ➤ Recent Projects and Solutions menu.

Add an Existing Project to a Current Solution

The first thing we do is let Visual Studio copy all of the SSAS project files from their current location in the downloadable book files folder, `C:\_BookFiles`, to your solution folder. We do that by adding an existing project to the current solution.

1. Using Visual Studio's menu, select the File ➤ Add ➤ Existing Project option (Figure 2-53), and the Add Existing Project dialog window appears.

Figure 2-53. *Adding an existing project to your Visual Studio solution*

2. Locate the WeatherTrackerCubes project folder that you unzipped from the authors' downloadable files (Figure 2-54). They are at this location on your hard drive: `C:\_BookFiles\Chapter02Files\_SSASFiles\WeatherTrackerCubes`.

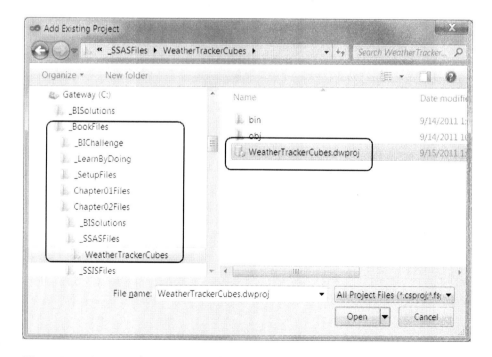

Figure 2-54. *Locating the SSAS project file*

3. Once you have located this folder, look inside to find the Analysis Server Project File, WeatherTrackerCubes.dwproj. Note that the file extension may not show in the title, depending on your operating system's settings, but the icon and type description will look like Figure 2-54.

4. Click this file; then click the Open button to close the dialog window. (A warning may appear to remind you to only open projects from a trustworthy source. If it does, click OK to continue.)

5. The downloadable SSAS project files will now be added to your Visual Studio solution.

Configure the Data Connection

Whenever you use a Visual Studio Project that was prepared on someone else's computer, you will likely need to change, or at least check, that the connections are appropriate to your own computer. Let's do that now.

1. After Visual Studio finishes adding the project to your WeatherTrackerProjects solution, it appears in the Solution Explorer window. If it is not expanded, click the arrow to the left of the SSAS project name to expand the Solution Explorer tree, as shown in Figure 2-49.

2. Locate the WeatherTrackerDW.ds file under the Data Sources folder. Right-click this file and choose the View Designer option from the context menu. The Data Source Designer dialog window appears, as shown in Figure 2-55.

3. Once the dialog window appears, click the Edit button to display the Connection Manager dialog window (Figure 2-55).

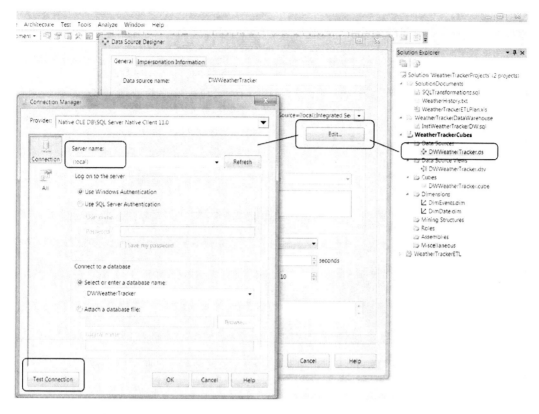

Figure 2-55. *Verifying the connection*

4. Once this dialog window appears, verify that the server name is appropriate for your computer by reviewing the Server Name textbox. If the server name is incorrect, change it to the correct one now.

Important: Remember, if you have installed SQL server on the same computer multiple times, your SQL 2012 installation may be called a name other than (local). For more information, search the Web for "SQL Server Named Instances."

5. Confirm that the DWWeatherTracker database is selected in the "Select or enter database name" dropdown box (Figure 2-55).

6. Click the Test Connection button to verify that your connection works (Figure 2-55).

7. Click OK to close the Connection Manager window and return to the Dialogue Source Designer window (Figure 2-55).

8. Please leave the Dialogue Source Designer dialog window open for now.

Change the Impersonation Settings

Each SSAS project uses an impersonation account to deploy and process its cubes and dimensions. Now configure the connection to use the impersonation information suitable for your computer.

Important: The account you use must have access to both Analysis Server and SQL Server. On a nonproduction machine, we recommend using an administrator account. In addition, a blank password will not work here. If you try to use one, you will find later that the cube deployment will fail.

1. From the Data Source Designer dialog window, select the Impersonation Information tab, as shown in Figure 2-56.

2. Change the username and password (do not leave the password blank) to an account that will work on your computer (Figure 2-56). For example, on Randal's laptop, he uses RSlaptop2\Administrator for the user account, because this account has full admin rights on both SQL Server and Analysis Server. You need to change it to your < computer name > \Administrator.

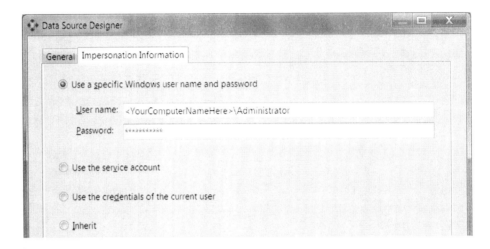

Figure 2-56. *Setting the impersonation name and password*

3. Click OK to close the Data Source Designer dialog window.

Review the Project Files

In the next series of steps, simply review the downloaded project files to help familiarize yourself with how these files are used. In Chapter 9, we discuss what each file does and how to configure the files.

1. Double-click the WeatherTrackerDW.dsv file to open a data source view designer tab in the center of your screen. This file can be located under the Data Source Views folder in Solution Explorer.

2. Review the tables displayed in this designer window and note the relationship arrow that connects the two tables. Double-click the arrow (your cursor will turn into a double-ended arrow when you hover over it) to cause the Edit Relationship dialog window to appear.

3. Verify that the EventKey columns are connected via this relationship, and click the OK button to close the dialog window.

4. Expand the cubes folder in Solution Explorer and locate the DWWeatherTracker.cube file within the cube file. Double-click this file to bring up a cube designer window.

5. This window is divided into the Measures and Dimensions pane on the left side of the screen. Click the small expansion + symbol on each of the objects on the left of your screen to expand each treeview.

6. With the treeview expanded, note that the measures and dimensional attributes are displayed. This indicates which measures and dimensional attributes are parts of the cube.

7. In Solution Explorer, expand the Dimensions folder and note that the DimTime dimension and the DimEvents dimension have their own program files.

8. Leave open or close the windows you just reviewed if you want, but do not close Visual Studio yet.

Deploy the Cube and Dimensions

From here, we proceed to building, deploying, and processing the cube and dimensions. Although there should not be any problems at this time, if there are errors, an Error List window will appear. Try your best to read any error message that may occur, resolve the error and build it again until the build succeeds. The most common errors involve an incorrect name or password used for impersonation (Figure 2-56).

1. Locate Visual Studio's Build menu and select the Build WeatherTrackerCubes option (Figure 2-57). Visual Studio will build all the files in this project into a deployment file and present a "Build succeeded" message at the bottom of Visual Studio's window.

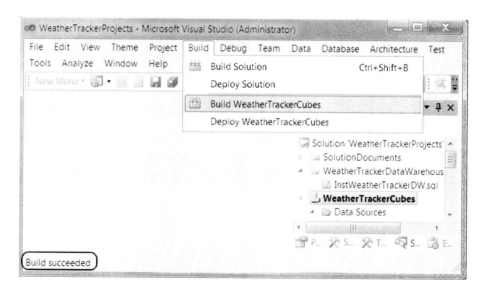

Figure 2-57. *Building the SSAS project*

2. Locate the WeatherTrackerCubes project icon in Solution Explorer. Right-click this icon and select Properties from the context menu. The WeatherTrackerCubes Property Pages dialog window will appear (Figure 2-58).

3. Select the Deployment page under Configuration Properties on the left side of this dialog window, and verify that the Target Server name is correct for your computer. In most cases, the word *localhost*, or (*local*), will work for the server name. If this is not the case on your computer, please change it accordingly (Figure 2-58).

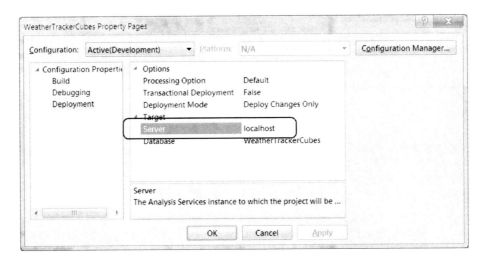

Figure 2-58. *Configuring the server property for deployment*

4. Click OK to close this dialog window.

5. Using Visual Studio's build menu, select the Deploy WeatherTrackerCubes option. This is similar to what you see in Figure 2-57, except you should select the Deploy option rather than the Build option.

6. A dialog window will appear to ask if you would like to build and deploy. Click Yes to continue (Figure 2-59).

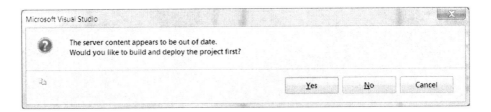

Figure 2-59. *Confirm that you want to build and deploy*

Important: Visual Studio may act differently on your computer and not display the message in Figure 2-59. Not to worry; it still performed the action in the background. In addition, the message may or may not appear whenever you are processing or deploying the cube or dimensions.

7. Visual Studio will connect your analysis server and upload the master XML file that was created during the build process, and a Deployment Progress window will appear. When it has completed deployment, Visual Studio will indicate that it was successful, as shown in Figure 2-60.

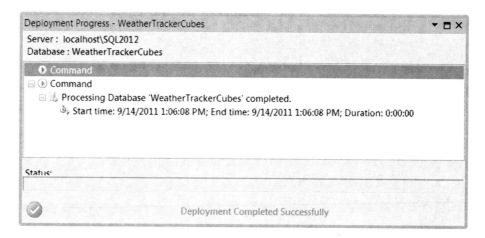

Figure 2-60. The Deployment Progress window

8. When the deployment has completed successfully, close the Deployment Progress window.

Process the Database

At this point, SSAS has created the cube and dimensions on the SSAS server, but the data has yet to be copied from the data warehouse into either of these objects. To do that, you must process them. Let's do that next.

1. Right-click the WeatherTrackerCubes project icon in Solution Explorer, and select Process from the context menu. The Process Database dialog window appears (Figure 2-61). (This menu is context sensitive, so make sure you are on the project icon before you access the menu, or it will not say *Process Database*.)

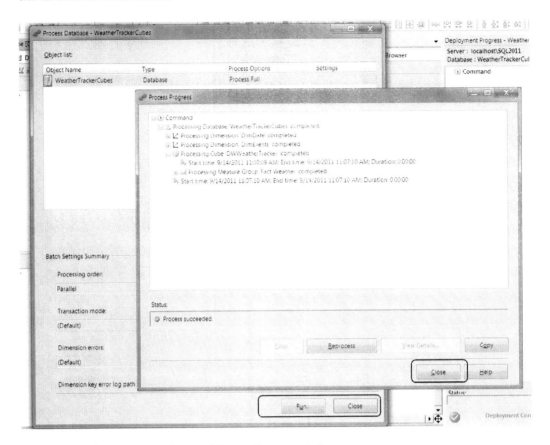

Figure 2-61. *The Process Database and Process Progress windows*

2. Click Run. Visual Studio will tell SSAS to start processing data from the data warehouse into your SSAS objects. Visual Studio will display a Process Progress dialog window.

3. When the processing completes, a status message of Process Succeeded will appear. Click the Close button to exit the Process Progress dialog window, and click Close again to exit the Process Cube dialog window.

In this exercise, you added an Analysis Server project to your existing Visual Studio solution and then built, deployed, and processed the SSAS database. Since the SSAS database now has data, you can create reports based on it. The next step of a BI solution is to create a report and verify that the data is clean, consistent and useful.

Creating Reports

Both the cube and the data warehouse now have data, so let's create some reports. Microsoft has a number of reporting tools that can be used, but we take a look at its server-based reporting tool, SQL Server Reporting Services (SSRS).

SSRS is part of Microsoft's BI application stack and has its own template in Business Intelligence Development Studio. You can create an SSRS project by opening Visual Studio and selecting a project template, just as you did with SSIS and SSAS.

Two templates are associated with the Reporting Server: the first one is the Report Server Project Wizard, and the second is the Report Server Project (Figure 2-62). Both of these templates create report server projects that look identical to each other, but one launches the report creation wizard. We will let you guess which one.

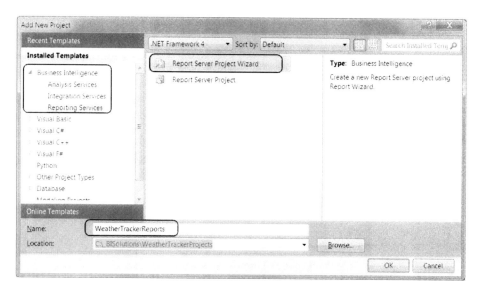

Figure 2-62. *Adding an SSRS project to your solution*

Using the SSRS Wizard

The simplest choice of how to start a new Report Server project is to use the wizard. Once you choose this project template, the Report Wizard launches. The first few pages of the wizard ask you to define the connection to either your data warehouse or your cube. These pages look like Figure 2-63.

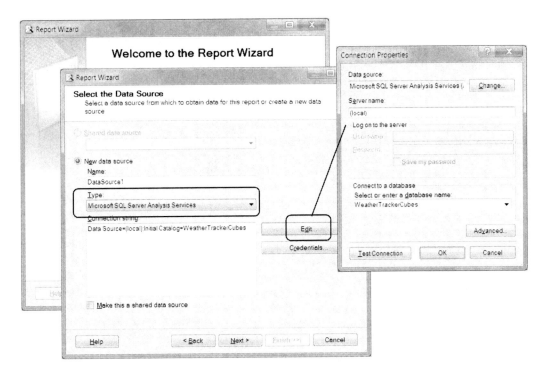

Figure 2-63. *The first two pages of the Report Wizard*

After you have created a connection, the next page of the wizard allows you to create a report query. Microsoft added a graphical way to build programming statements to make it easier for developers who are not SQL or MDX programmers. If you choose a relational database for the connection, you see a SQL query builder. However, we chose an SSAS connection, so the wizard will display an MDX query builder. You need to click the Query Builder button before the Query Builder editor displays (Figure 2-64).

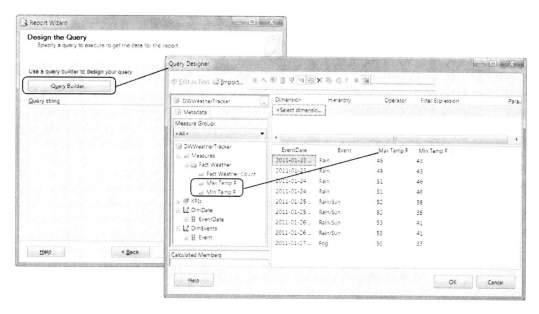

Figure 2-64. *The third page of the Report Wizard*

You can drag and drop elements in the MDX query designer from the treeview on the left side of this dialog window onto the report display section in the center of the dialog window. MDX code is written behind the scenes every time you add or subtract an item from the report display section. When you click the OK button and return to the previous dialog window, the MDX code is displayed.

After you create a report query, the wizard moves you through three pages that define the style of the report. In Figure 2-65, we configure the wizard to create a tabular-stepped report with data grouped by events.

You are able to choose between tabular or matrix formats. The tabular format looks very similar to an Excel spreadsheet, and the matrix format looks similar to an Excel pivot table. We cover more on this subject in Chapters 16 and 17.

Depending on your choice, you are asked to further clarify how you want the data presented in your report. Because we chose a tabular format, the next screen of the wizard (as shown in Figure 2-65) allows us to display data in groups, and the next screen allows a choice between a data-stepped or blocked format.

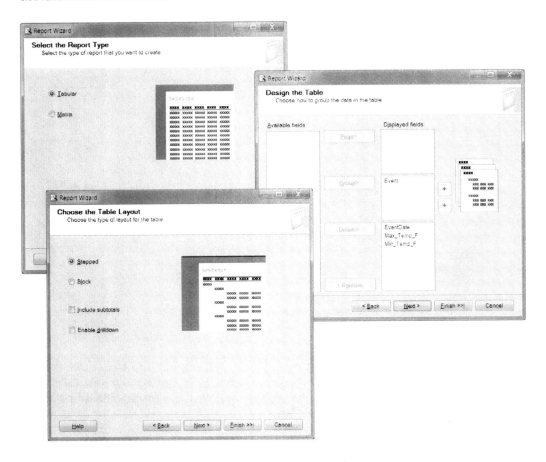

Figure 2-65. *The fourth, fifth, and sixth pages of the Report Wizard*

The last two pages of the wizard allow you to choose some basic colors for the report and summarize the choices you have made throughout the wizard (Figure 2-66).

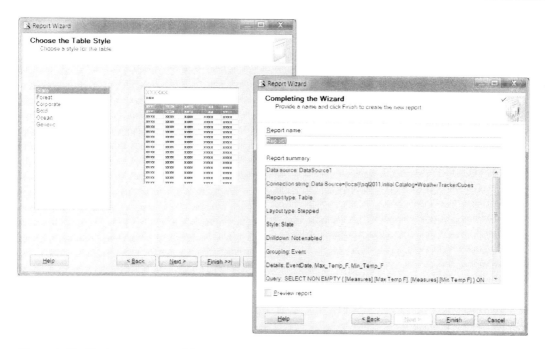

Figure 2-66. *The last two pages of the Report Wizard*

Once the wizard is completed, you end up with a single report page that includes all of the fundamental report data. From there, your task is to continue refining the look and feel of the report. If you would like more reports, this same wizard can easily be launched from Visual Studio by just adding a new report to the project.

The report editor displays in the center of Visual Studio (Figure 2-67). This editor has two options to choose from: Design and Preview.

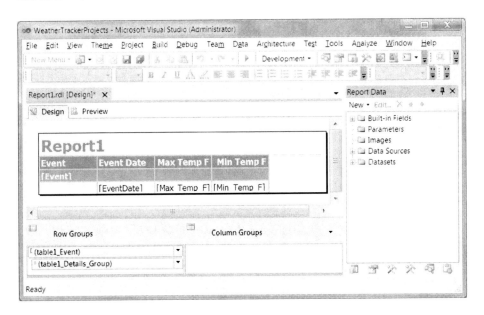

Figure 2-67. *The SSRS Report Wizard is complete.*

When displayed in Design mode, the report can be modified. When displayed in Preview mode, it displays the final colors and appearance that the client can expect to see. You can switch between these two display options to complete the report configuration until you are satisfied that the information you want to share is clearly and professionally presented, as shown in Figure 2-68.

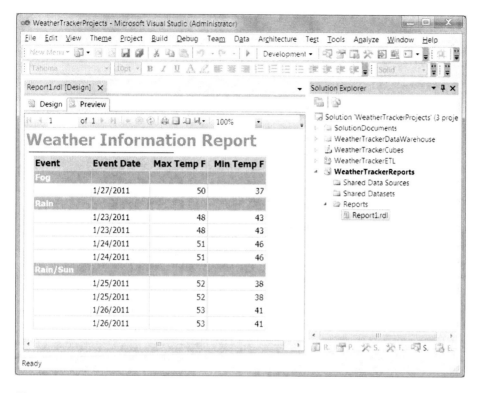

Figure 2-68. *Previewing a report*

Manually Creating SSRS Reports

When creating an SSRS object, you are not as reliant upon a wizard as you are with an SSAS object. You typically create your SSRS reports manually once you become familiar with this method.

To manually create a report, you begin by adding a data source object to the report. This is very similar to how it is done in SSIS and SSAS. Once the data source object has been added, you must add one or more SSRS datasets to the report to define which data you will use. An SSRS dataset consists of an associated data source connection and a programming query, typically in SQL or MDX code. This code is either entered manually or by the same query-building tool that is utilized by the SSRS wizard.

There are a few other ways to create SSRS reports beyond these two options. In addition to developing reports, there is a management aspect that you need to know about. All of this is discussed in Chapter 16. For now, let's do an exercise to create an SSRS report.

EXERCISE 2-5. CREATING A REPORT

In this exercise, you create an SSRS report using the Report Wizard.

1. Visual Studio should still be open from the previous exercise, but if not, please open it and access the WeatherTrackingProject solution from the File ➤ Recent Projects and Solutions menu. (Remember to always run Visual Studio as an administrator by right-clicking the Visual Studio menu item, selecting Run as Administrator, and then answering Yes to close the UAC popup window.

Create a Report with the Report Wizard

1. Using the Visual Studio menu, select the File ➤ *Add* ➤ New Project option. The Add New Project dialog window appears.

2. Select the Business Intelligence option on the left side of this window and the Report Server Project Wizard on the right side (Figure 2-62).

3. Locate the Name textbox and change the name to *WeatherTrackerReports*. Verify that this project will be created in the `C:\_BISolutions\WeatherTrackerProjects` folder (Figure 2-62).

4. Click OK to close the dialog window. You should see the new SSRS project added to your solution in Solution Explorer, and the Report Wizard welcome screen will appear.

5. You can read what is on this page or just click Next to move to the next page (Figure 2-63).

6. The next page of the wizard is the Select the Data Source page. Change the value in the Name textbox to *WeatherTrackerCubes*.

7. Locate the Type dropdown box and use it to select Microsoft's SQL Server Analysis Services connection type (Figure 2-63).

8. Locate the Edit button and click it to bring up the Connection Properties dialog window.

9. When the Connection Properties dialog window appears, type in the name of your SSAS server in the "Server name" textbox. This is usually your computer name, localhost or the word (local) with parentheses; then click OK (Figure 2-63).

10. Click Next to go to the next page of the Report Wizard dialog window (Figure 2-64). This page, called Design the Query, allows you to create SQL, MDX or DMX queries using a query-building tool. Click the Query Builder button. MDX is the language used to query Analysis Server cubes, so clicking this button will bring up an editing tool that creates MDX code, and the Query Designer dialog window will appear (Figure 2-64).

11. On the left side of this dialog window is a representation of the DWWeatherTracker cube. Click the + symbol to expand the Measures, and then click the Fact Weather treeview icons.

12. Locate the Max Temp F icon. Drag this icon to the center of the dialog window. This measure is displayed as a representation of the report data.

13. Locate the Min Temp F icon. Drag this icon to the center of the dialog window. This measure is also displayed as a representation of report data.

14. Locate the Event icon directly under the DimEvents icon. Drag this icon to the center of the dialog window. This dimensional attribute is now displayed along with the measures.

15. Locate the EventDate icon directly under the DimDate icon. Drag this icon to the center of the dialog window. This dimensional attribute is now displayed as well.

16. Click OK to close the Query Designer dialog window.

17. You will see the MDX code that was created when the Query Designer window closes. Click Next to continue to the next page of the Report Wizard.

18. The next page of the wizard, shown in Figure 2-65, allows you to select between two report formats. In this case, you use the Tabular format. So, verify that tabular is selected, and click Next.

19. The next page allows you to group report data together. Click the Event field in the Available Fields window pane on the left side of this dialog window. Then click the Group button to add the event field to the Group window pane (Figure 2-65).

20. Click the date field and then click the Details button to add this field to the Details window pane. Continue adding MinTempF and the MaxTempF fields by selecting each and clicking the Details button so that all three fields are seen in the Details window pane (Figure 2-65). Click Next to continue.

21. On this page, verify that the Stepped radio button is selected, and check the Enable drilldown checkbox. This allows the report user to expand and contract the report items much like the treeview in Solution Explorer. Click Next to continue (Figure 2-65).

22. On this page, you can choose between different table styles. These styles only affect the visual look of the report and not its functionality. Because the look is not that important at this time, choose one that you like and click Next to continue (Figure 2-66).

Important: If this is the first time a report has been created in a Visual Studio solution, the wizard will move to the Choose Deployment Location page. You can accept any default values and click Next to move to the last screen of the wizard. (If you have not set up your Report Server, Chapter 16 walks you through the process. Until then, you will not be able actually to deploy the report, but that is not necessary in this chapter.)

23. On this last page of the wizard, supply a name for your report and review the choices you have made. Change the name to *WeatherEventsReport*, and click finish to create the SSRS report file (Figure 2-66).

24. When the Report Wizard completes, you should see the `WeatherEventReport.rdl` file in Solution Explorer. You will also see the report in a report designer window (Figure 2-67).

25. Click the Preview tab to preview the report that you have made with the SSRS Report Wizard. Your report should look similar to Figure 2-68.

At this point, the report has been created, but it may not look perfect. You can adjust the report by navigating between the design and preview windows, changing the color scheme, column widths, and value formats until you are satisfied with the look and feel of your report. This is covered more extensively in Chapters 16 and 17, so for now we leave the report looking as it is.

In this exercise, you created a report against the SSAS cube you created in Exercise 2-4. In future chapters, we show you how to create more complex reports using Microsoft's Reporting Server and Excel.

Testing the Solution

Testing the solution is much like editing a book. The tester must understand the basic look and feel of the content, verify that the information presented is accurate in a format prescribed by the business requirements and confirm that the end product has not deviated too far from its original specification. Having someone edit your work is always preferable to doing your own editing, because it is quite easy for you to overlook mistakes in your own work.

As a developer, you can help this process with documentation and consistency. These tools play a major part in making any solution easy to review and validate. The Excel spreadsheet in Figure 2-2, for example, can be used to document the data source columns and data destination columns for testing purposes. Using this, the tester can verify that the column names, data types, and transformations outlined in the Excel spreadsheet are the same at the end of the project as they were at the beginning.

Any deviations from this plan can be questioned, and responses to why the deviation occurred can be answered. These answers are recorded in hopes that the changes created during the first iteration can be anticipated in the next step of the BI solution.

We delve deeper into the details of how to test a BI solution in Chapter 18.

Approve, Release, and Prepare

At the end of the BI solution is a formal approval process. During this phase it is important to review whatever changes are found during the testing process and to document an evaluation of the findings. Typically this document is in the form of a Microsoft Word document that describes, in paragraph form, the various aspects of the individual projects that were included in the BI solution.

It does not have to be very long nor does it have to read like a literary novel. It should simply state the facts so that you can plan the next version of the solution using knowledge gleaned from the previous version. Once this is presentable, you can release the solution to the client. You may need to draft a user manual to be released at the same time. At this point, it is a good idea to gather feedback from the users.

Although there will always be another "final" version, eventually you will come to a place where a new final version will not happen nearly as frequently. Therefore, you should always plan for the version of the BI solution you are releasing to be transitory so that it can be improved upon through experience gained and through user's feedback. We discuss this process further in Chapter 18.

Moving On

In this chapter, we discussed how to create a BI solution from start to finish. We started by examining the requirements and identifying the data, and then we moved on to planning the BI solution using simple documentation. Next, we built a data warehouse in SQL Server, filled it with data using Integration Services, created a cube with Analysis Services, and made a report with Reporting Services.

Now we restart at the beginning with an in-depth look at planning your BI solution in the next chapter.

What's Next?

There are many ways to create a BI solution. Learning about other methods will supply you with additional tools to customize the steps we have laid out here to match those in your organization. We recommend the following book as a good place to start: *Delivering Business Intelligence with Microsoft SQL Server 2012* by Brian Larson (McGraw-Hill Osborne).

CHAPTER 3

Planning Solutions

He who fails to plan is planning to fail.

—Winston Churchill

The most important first step in designing a data warehouse (DW)/business intelligence (BI) system, paradoxically, is to stop.

—Ralph Kimball

Planning is fundamentally important in any undertaking. This is no less true in the case of creating a BI solution. Although no plan is perfect and a protracted planning process is the bane of many projects, even simple BI solutions can benefit from some planning. The trick is to find a balance.

In this chapter, we show you techniques to plan a BI solution. We include tips on conducting interviews, data identification, and documenting your plan as well as examples of documentation that is easy to create within a minimal amount of time. We also give you tips on building a BI solution team, defining the different roles the team will play, determining the infrastructure needs of your BI solution, and estimating the cost.

By the end of this chapter, you will have preliminary plans to start implementation on the demo BI solution used throughout this book.

Note   It is not our intent to turn this chapter into a project management book, especially because documentation alone isn't going to create the BI solution. We would much rather get to the part where we are creating our project. One problem that repeatedly presents itself, however, is stumbling across a project with no documentation at all. This is likely because many developers do not even know where to begin, from estimating how long the project is going to take to what simple documentation should look like. Most of the books we have found do not discuss this process. Therefore, we decided that we would change that with our book and give you an example of how to create some basic and relatively painless documentation.

Outline the Steps in the Process

When you start a solution, it is important to create a simple outline of what you are trying to accomplish. For example, you may even want to make yourself a flowchart, similar to one shown in Figure 3-1. We have created Figure 3-1 using Microsoft's PowerPoint, but you can use Microsoft's Visio or simply draw one with pen and paper and it will work just as well. What you use to create the outline is not as important as making sure to take the time to do it before you begin developing.

The planning process typically begins by interviewing the client (or whoever the BI solution is being developed for) and documenting what they want. In Figure 3-1, you can see that the interview process has been included at the top of the flowchart.

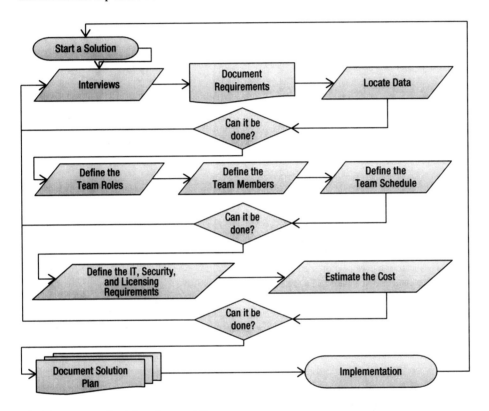

Figure 3-1. *A typical planning workflow*

Next, you need to see whether or not what they are asking for can even be accomplished. A good place to begin is to look for the data required to build the type of solution they are requesting. It is common for clients to want your BI solution to produce information for them that is not supported by their data. This is one of the first things to check for because it is such a common obstacle. If you cannot provide them with the information they want, letting them know early on will save everyone a lot of wasted effort, and it will allow you to work with the clients to revise their expectations while helping to avoid disappointment.

Also, keep in mind other important deal breakers that might kill the solution before it begins. These change for every solution but may include the following:

- Can you complete the solution with the number of staff you have?

- Can you complete the solution in the timeframe allotted?

- Can you complete the solution with the current infrastructure, software, and security?

- Can you complete the solution within the allocated budget?

If the answer is no to any of these questions, then you need to go back to the beginning and redefine the requirements of the BI solution. The process repeats until you have created a plan that balances the customer's needs with the resources at hand. As we have mentioned before, it is best to find out at the beginning whether time spent on the solution will be worthwhile and affordable for the client.

Once you have a working plan, you need to document it. The complexity of the documentation is determined by how complex the BI solution is projected to be. Common items will appear in every solution. The amount of documentation may also depend on how much documentation is required by law or by a company's business practice.

In every case, getting information into those documents is determined by how much you can extract from the objects or events on which you are modeling your solution. Therefore, the best place to start gathering data is through an interview process.

Interviewing

The term *interview* is typically thought of as a meeting where questions are asked. The purpose of the interview itself can vary; some examples are to enable a hiring decision to be made, to provide facts for a story to be written or even to provide leads on an investigation. In this case, considering this type of interview to be like an investigation might be the most accurate means of approaching this for our purposes.

We do not want to limit this process to a single conversation, nor do we want the interview itself to be our only source of information. Clients may not always know how to voice their needs, particularly when they do not know everything you are capable of doing for them.

Do a little research by taking a look at their preexisting documentation or solutions to familiarize yourself with their situation and to help you see potential solutions ahead of time. Past letters and emails with the client may contain facts that the client is assuming you are already taking into consideration or that they simply forgot to bring up at this stage of the process.

Reviewing past correspondence and asking about specifics from concerns they have already voiced can be vital to keeping your client happy. Get as much information as you can while keeping in mind that this process is not limited to verbal communication. The goal is to pinpoint the client's needs and to determine what will best help them with their business. This is also the time to determine the assumptions the client is making about what to expect. Just be sure to avoid treating the client as a hostile witness!

However you plan to get your information, you will want a list of questions answered before you proceed. Here are some that will work for most occasions:

- Why do we need it?

  - What is the goal of the BI solution?

  - What is the hoped-for outcome of the BI solution?

  - Is the project worth the estimated cost?

  - Who will use the BI solution?

- What are we building?

  - What must be in the BI solution?

  - What will be nice to have in the BI solution?

  - What will *not* be in the BI solution?

- How will we build it?
    - Can we release it in increments?
    - Will it need to have all features before it is released?
    - How will the BI solution plan be distributed to the developers?
    - How will progress be monitored?
- Who will we get to build it?
    - Who will be involved in the BI solution?
    - What roles will be needed on the project?
    - Do we have team members who can fulfill those roles?
    - Does the development team have the necessary skills?
- When will we need it?
    - What is the timeframe for the BI solution?
    - When will the BI solution be completed?
    - Who will monitor the progress of the BI solution?
    - Who will sign off on the BI solution completion?
- How will we finish it?
    - Who will document the outcome of the solution?
    - Who will test and approve the BI solution?
    - Who will train the users?
    - How can users submit questions, comments, or requests?
    - What system will be used for bug tracking?

In a perfect world, you will get all of your questions answered. In the real world, you will have to settle for a bit less. Any questions you can get answers to are extremely valuable to the outcome of your BI solution. Each answered question will help you decide how to continue or if you should continue.

Why Do We Need It?

This may be the most important question you can ask. If you do not have a clear understanding of why the BI solution is needed, then perhaps it is not needed at all. It is important that you define this need so that you can validate it against the BI solution you create. A successful BI solution is one where you manage expectations and prove that these were met. If that does not happen, you need to explain why it did not happen.

We recommend looking at what the client is currently using to stay organized and what they are using to help make proper business decisions, as well as how they manage their data. Ask the client what is working well for them within their current system. This can help you determine what is not working for them and what might be added to improve the current system. Be aware that you may have to interview a number of people before you get a clear understanding of what is going on. You can speed up the process by calling a meeting, but in a group setting, only the most vocal members of the group will give you feedback. This can be a productive setting, but sometimes it turns into an opportunity to air past grievances. Take the time to talk with individuals regardless of whether you opt for a group meeting, and you will likely get a much better understanding of what is needed. This is one of those times where technical skills are less important than people skills.

What Are We Building?

Once you have established that the BI solution is beneficial to the client, begin figuring out what is going to be a part of the solution and what is not. Create a list of all the features that your client requested to be included in the solution, and then examine existing documentation and reports for inspiration about what else should be added to the solution. It is common that the interviews with the people involved will not identify all the requirements of the solution.

When your list is created, prioritize what has to be in this version of the solution. One method of doing this is to use a technique known as *four-quadrant prioritizing*. An example of this is shown in Figure 3-2.

Identify what is the value to your client versus the difficulty and cost of an item. When an item provides substantial benefit to users at a low to medium cost during the development cycle, you can classify this item as a Must Have.

If, on the other hand, an item does not provide much benefit to the users and is difficult or costly to implement, then it should be excluded from the project at this time. It does not mean that in the future it will not be included; it just means that right now it is a Not Now item.

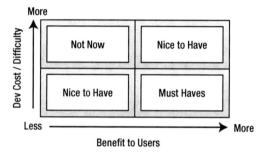

Figure 3-2. *A four-quadrant prioritizing matrix*

If an item provides a lot of benefit to the users but may be costly or difficult, then it is something that will be considered Nice to Have but may not be absolutely necessary to the solution. As such, evaluate these Nice to Haves carefully before including them in the current solution. Once again, this does not necessarily mean it will never be part of the BI solution; it just may not be part of the current version.

Be careful of the final quadrant, in the lower left of Figure 3-2. The items in this quadrant are very easy to implement but provide little to no value for the users. This is a very attractive quadrant because these items will keep you quite busy implementing them and will make you feel as if you are really accomplishing something. In the end, however, it will not benefit your BI solution much, and most clients find it frivolous. Few will feel it pertains to their needs; and consider the fact that if they are paying for your work, they do not want to pay for something for which they did not ask. This particular quadrant is the one that is responsible for the dreaded demon of developers known as *feature creep*. As we say in the field, "When in doubt, leave them out!"

Additional Considerations for Determining What You Will Build

Both the interview process and the examination of current reports and documentation should provide you with a good understanding of what will benefit the users. Identifying the cost involved, however, can be quite a bit more difficult to determine. For example, you may have to consider the ease of use of the current networking infrastructure and hardware that a client is using. Another consideration is the security and compliance restrictions or licensing issues that may be required by the company or by law.

Patient records in the medical industry, for example, must be made available but also held securely. Your ability to access these records may be severely restricted, greatly increasing the cost and difficulty of working with this type of sensitive data. Another consideration is the accessibility of the data you need. You may encounter situations where the data is accessible for only small windows of time. If that is the case, then you must plan accordingly. If accessibility is sporadic, you will find it very difficult to achieve success. You need to find out about this early on and then budget time and costs to account for this.

Another consideration is the conformity of the data. In larger companies, you may find that the same data is recorded in a number of places within the same organization or that the same information is represented in a number of ways. For example, if a company uses descriptors such as Good, Better, or Best for a line of products but then attempts to record the same information in another database using the numbers 1, 2 or 3, you must get a consensus on how this information is expected to be presented in your BI solution before you can continue. And that takes time.

Latency issues, such as the time it takes for the data to change in one part of the organization versus the time it shows up as information in your BI solution is another cost consideration. In projects where a large amount of time or high degree of latency is acceptable, it is much easier to develop a cost-efficient solution. When working with companies where only a small degree of latency can be tolerated, the development costs and urgency required for a completed BI solution may be too exorbitant to continue with your current solution design.

Planning around these obstacles can be quite challenging and may in turn reveal other obstacles such as the following:

- Do the users have the skillsets to extract information from the solution you build for them?

- Will the solution fit the corporate culture of the company with which you are working?

- Do you (or your team) have the skills to manage these types of complications?

As you can see, the interviewing process and taking the time to review the data in depth can determine the success of your BI solution. Taking the time to document a plan will often bring these types of considerations to light.

Determining Your Ability to Complete the Solution

In the end, each BI solution has its own challenges and benefits. You must do your best to evaluate what these are and realize that you will make mistakes just like every other human before you. Do not get bogged down and frozen by indecision. Just do the best you can with the tools you have available. As with all things, your first attempts will contain more errors than your later ones. But, if you never start because you are too afraid of missing something, you will never become experienced enough to know how to avoid most errors. Document your mistakes, learn from them, and move on.

One of the biggest ways to mitigate the number of mistakes you will make is to restrict the complexity of the solution. A solution should always be as complex as it needs to be to get the information the customer needs but as simple as you can possibly make it. The "keep it simple" rule of design will make your life easier and your solutions more profitable.

When you find a solution that cannot be simplified enough for your team to accomplish it, you may want to consider passing up the offer to create the solution. It may be that the solution needs a larger and more experienced team. That may be your team in the future, but perhaps not today.

If you do decide to tackle the solution, be resigned to the fact that changes to the plan are likely. It is imperative that you communicate these changes as they occur during the solution cycle. Communicating with the client and managing the users' expectations are vital to the success of any BI solution. It is better to disappoint clients expecting a particular feature at the beginning of the project than to have them wait indefinitely for the feature to become available.

Explain to the users that this version of the solution will not include a particular set of features. Then explain to them what will be included. Also let them know when the solution will be delivered. Explain that this delivery date would be unattainable if the features that are being left out were added back in. If the users push back, demanding a particular feature, then it is time to reevaluate your ability to do the BI solution at this current time in the given timeframe. Once again, this evaluation is very important to make at the beginning of the solution. It benefits no one if you start, rack up a lot of hours, and then fail to deliver what you said you could. Be up front and truthful about what you can and cannot do, and your client will appreciate your honesty.

How Long Will It Take to Build?

Close on the heels of the question, "Should we build it?" is the question, "How long will it take to build?" Ideally, the project should begin and end in the shortest time possible. That way, users can have access to the information needed, and you can move on to creating the next version of the solution. However, the definition of the words "shortest time possible" is open to interpretation. Are we talking months or simply weeks? Will a portion of the solution be available sooner, as an incremental release? Or, will users have to wait until all the components are completed before they can start utilizing it?

One means of determining how long it takes to complete a solution is to break it into its constituent parts and assign an estimated number of hours/days it will take to complete those parts. Using rapid application development (RAD) as a model, you can estimate that an employee can accomplish approximately six hours of production in each eight-hour day. (If your team is accustomed to working more or fewer hours per day, then change this number accordingly.) Therefore, for each set of six hours estimated, you can record it as one day of work on the solution. After you have totaled up the days for the different tasks, divide it by the team members you have working on the solution to find an estimated time of completion. On a very small solution, you might estimate something like this:

- Create the data warehouse (6 hours)

- Create the ETL process (24 hours)

- Create the cubes (12 hours)

- Create the reports (12 hours)

- Test the solution (12 hours)

Then add the hours for each project together to give an overall idea of days the solution will take.

- Days 11 = (6 hr + 24 hr + 12 hr + 12 hr + 12 hr)/6-hour days

Of course, we have not taken into account the time it takes to perform the interview and identify the data, plan the solution, obtain the final approval, or release tasks. Therefore, those items will have to be added in as well, in addition to special considerations with your project that are unique and are not listed here.

With regard to the development team, it is unlikely that everyone will work on the same part of the solution at the same time. Ideally, it should be a coordinated effort with the data warehouse, ETL process, cubes, and reports all developed as close to simultaneously as possible. This can be impractical, but when achievable, the turnaround between the planning phase and the sign-off phase is substantially shorter. For example, when the data warehouse is created, a few test entries can be inserted into each table. From these, a cube could be built, and reports can then be created on both the data warehouse and the cube. Keep in mind that these would be preliminary creations. Until you have data within the data warehouse, it is unwise to consider testing and sign-off preparations. However, even with these test values, you may uncover unforeseen deterrents to completing your solution as you build the preliminary prototypes. That is a good thing, because you can let the customer know about these roadblocks early on.

Transparency is crucial to team members' ability to do their jobs efficiently and in coordination with all other stakeholders. Try to set up some mechanism for distributing information about solution progress to all involved. Make sure to monitor the solution's progress and publish these results. We have used basic web pages, bulletin boards, and wiki pages in the past. All of these are simple, inexpensive, and effective.

How Will We Build It?

Building a BI solution can be approached in two ways: a top-down approach or bottom-up approach.

In the top-down approach, the needs of the company as a whole are determined, including all the data that a company uses, with the ultimate goal of creating a solution that will cover all the needs of all users as soon as the entire BI solution is published. In some cases, this is necessary. In these cases, decision makers need all available information through a single viewpoint to make decisions. Interim reports and partial updates are insufficient. Looking at a half-dozen viewpoints is not feasible either. The information from the BI solution is not just an additional tool used in the decision-making process; it is the primary tool. Here, all of the components must be present before the solution is of use.

In the bottom-up approach, the focus is on defining a particular business process and building a solution around this process. The idea is that you can add onto the solution by focusing on a different process in the next version. You continually refine the solution by analyzing the process and incorporating it in the current solution. Your BI solution is only one tool used in the decision-making process, and the solution can be functional even when it is only partially complete.

■ **Tip** The top-down versus bottom-up approaches have been defined as the two paths to choose from. Each is commonly associated with two different developers, Bill Inmon and Ralph Kimball, who are both early (and often competitive) authors on the subject of data warehousing. In the end, both are correct approaches, just not for all solutions.

Over the years, the top-down approach has lost many of its adherents. This is mainly because the approach is often extremely costly and impractical. Therefore, the recommended practice is to use the bottom-up approach whenever possible. Conversely, when the bottom-up approach is impractical, switch to the top-down approach. The trick is asking the question of what is necessary for this particular BI solution. As a consultant, you may not initially know the business you are building the solution for well enough to predetermine what is necessary, so make sure to conduct interviews with this question in mind.

As you may have guessed, in this book we focus on the bottom-up approach, using techniques associated with RAD. (We first mentioned RAD when discussing how to determine the hours it will take to build your solution.) With RAD development, the focus is on supplying the top 80% of the requested requirements. Quite often, this 80% can be accomplished in about 20% of the time it would take to do a full implementation of all requirements. Although this is not always the case, you might be surprised by how often it is accurate. When it is, you are able to supply your customers with a satisfactory BI solution at a fraction of the cost.

Who Will We Get to Build It?

Finding good people to work on a solution is easier said than done. It is a wonderful fantasy to imagine you have all the experience and all the knowledge and all the free time to perform every task the solution requires within a timely manner; however, it is unrealistic to believe that you are capable of accomplishing all of this on your own. What can be done to mitigate the reality of human frailty? In a word: teamwork!

Putting together a team of developers with compatible skills and personalities is a rewarding and lucrative endeavor. In addition to the developers, a team should consist of competent managers, testers, and interested stakeholders. We highly recommend looking for compatible personalities first and sufficient skills second. Although others may argue that we have this reversed, experience has shown that skilled team members who can work together will achieve much more than a group of even better skilled individuals who refuse to cooperate with one another. Failure to build a compatible team costs money and time, two commodities that are often in short supply.

Beyond these soft skills, development teams can fail for two reasons. The first is inexperience with the technology. This failing can be addressed through study and research. For example, learning how to use the SQL Server BI tools by reading and performing the exercises in this book greatly increases your solution success rate. (And telling you this gives us a chance to promote our book!)

The other common failure is harder to prepare for, that is, failure to understand the business process that is being reported. For example, if your solution consists of reporting against the repair rate of computers, you will need to know about the repair centers, the warranty terms, the components that are in the computer, the manufacturers that made the different components, and any internal designation of a group (generation) of components. You will also need to know whether the computers start aging once they are shipped from the factory or whether your business considers their lifetime to begin the first time they are turned on. The users expect the reports to be correct, meaning they answer the question people believe is being asked. If you misunderstand the question even though you provide a legitimate answer, you will not get credit for your answer because it is not the question that was asked.

From this single example, you can see how easy it can be for a solution to fail if your client's business processes are not understood. Once again, transparency is fundamental to the success of your solution. Keep stakeholders informed of your progress, ask for their validation, and—with the client's input—properly determine whether the generated reports do indeed meet the needs of the company. If you can catch these errors early in the development cycle, you will save both time and money.

Stakeholders can be a good choice as project sponsors. A project sponsor is responsible for communicating the progress of your solution to the end users and in turn is able to provide specific feedback to the development team about any misunderstandings. A good stream of communication with a sponsor will affect the final user acceptance at the release of your solution.

Unfortunately, this communication highway can also deliver change requests. Resist the desire to make everyone happy by changing the definition of the solution while it is in progress. If indeed you do find that one of the questions your solution is trying to answer was incorrectly interpreted, then this must be fixed, but you are not looking for new or different questions to answer. New questions can be asked in the next version of your BI solution.

In RAD development, you typically focus on small teams consisting of between six and eight people. Often you try to find team members who have a diverse set of skills so that they are assigned several roles. Some key roles to define include the following:

- Solution sponsor

- Solution manager

- User acceptance coordinator

- Solution planner

- Technical writer

- Programmer writer

- Data warehouse developer

- ETL developer

- Cube developer

- Report developer

- Tester

- Technical trainer

If you research these titles, you will find that there are many different names for these roles as well as additional roles beyond what we have listed. You will need to define your own list depending on the type of solution you are trying to create. Simple solutions do not demand as many roles as more complex ones.

Keep in mind that these represent roles rather than a specific number of people. One person may play many roles. The solution manager may also be in charge of coordinating user acceptance and working with the solution sponsor.

When Will We Need It?

Nobody likes to wait for things that they want, but it is an unpleasant fact that we usually have to. In our industry, every year seems to bring more opportunities for instant gratification. If somebody doesn't answer an email by the end of the day, we can feel slighted. It is hard to remember that only ten years ago people expected to wait for letters to come by mail over a few days' time. Because you cannot change our culture, you will need to plan for the inevitable desire for the project to be done sooner rather than later. Estimate how long the project will take, track its progress, and disclose your adherence or delinquency to the given timeframe.

Key things that you need to define at the beginning of the project include how you are going to monitor the progress and who will sign off on the solution's completion. In a fully staffed team, project managers will be involved in tracking development progress, and you may even have a liaison to interact with customers to get a sign-off document. On smaller projects, team members tend to shoulder many responsibilities that cross these boundaries. When the boundaries of job responsibilities blur, it can be easy to forget which roles are assigned to which team member. Establish these roles early, and make sure that the actions associated with these roles are carried out within the timeframe allotted for the task.

How Will We Finish It?

The completion of a BI solution is just as important as its beginning. In the beginning, you should document what you are trying to accomplish. At the end, you will need to document what has been accomplished. With this document in hand, you will be able to validate that the end users also feel you have accomplished your goal. We have found a 100% buy-in on projects to be an unrealistic target, but the 80/20 rule works pretty well. If you can get everyone to agree that you have accomplished 80% of the most important aspects of the solution and that the other 20% will be worked on in the next version of the solution, chances are that you will have happy clients. Remember that if you set expectations at the beginning of the project correctly, the end of the project is much more likely to be deemed successful.

Like most developers, including one of the authors of this book, you may hate documenting what you are doing. (Of course, this means that the other author enjoys this sort of thing.) Nevertheless, for the sake of everyone involved, even if this task is disliked, it is quite necessary. Creating good documentation can help you track what you did right and what you did wrong on a project. This information is vital for your long-term success as a BI consultant. Think of it as an investment that will pay off over the course of many years, because that is exactly what it is. The good news is that you will see dividends immediately when you begin your next solution and are able to avoid many of the pitfalls that were uncovered in the previous solution.

Another task many developers are not wildly enthusiastic about is testing. Yet testing is fundamentally important to your success. It can be very tempting to skip this portion of the BI solution. Resist this temptation! Undoubtedly, you have found mistakes already in this book. Can you imagine how many there would be if we never had an editor review it?

Even with the best intentions, an author will make mistakes. Without a good editor, these mistakes will be missed and passed on to you, the reader! This analogy mirrors the relationship between a developer and tester. The developer creates the content and the tester reviews and approves it. The feedback given from the tester to the developer makes a project better and more efficient, an investment that will pay off long-term.

The next aspect involves making sure that the end users are trained to use the BI solution you created. Making users happy is very important to finishing your solution. Moreover, their happiness is likely to be determined more by perception than fact. Even if your solution is well-made and consistent, if the client perceives that the program is unwieldy, then all the preparation and hard work you did may be for nothing. Do not skip this step as you complete the BI solution. Take some time to instruct the users how to "use" your solution efficiently, and everyone will be happier.

Striving for perfection can keep you on a never-ending chase for the unattainable. One way to avoid this is to allow users an opportunity to submit questions, comments, or requests. As always, be sure to let users know that the question and comment process is not a means of revising the previously agreed-upon solution. The purpose is to prepare for future revisions of the current solution. If users believe that their comments will change the current version, they will be dissatisfied when they find out it did not. If you let them affect the current solution, you will begin a tug-of-war where you are pulled from the left and the right to the point where you can get nothing done. Although some may be disappointed that their needs are not being met immediately, most will understand that if they are patient, you will address them as soon as possible.

Additionally, you may need someone to give the final approval on a project and advertise that approval. This person is your sponsor. Sponsors are usually managers working for the client. They filter information from your team to the users and from the users to your team. This is similar to how a wedding coordinator works with caterers, the florist, and the rental company. In this analogy, the BI solution team (including you) are the caterers, the florist, and the rental company; the sponsor is the wedding coordinator; and the report users are the happy couple.

You will need to protect the credibility of this sponsor (or sponsors) by making sure that your projects provide clear and accurate information. Not having enough information, not getting it to perform fast enough, and not making it look professional will kill your solution and the sponsor's credibility. Make sure you coordinate with the people approving your project and get their honest feedback before you publish your solution.

▨ **Note**   Remember, planning a BI solution is very complex. As mentioned earlier, this book focuses on simple solutions that can be used as building blocks for larger solutions. The planning process discussed here is not meant to be an entire course on data warehouse project management. It will, however, get you going in the right direction. Nevertheless, for very large or complex projects, you may need more research to become fully proficient in BI solution planning. You can find many excellent and in-depth articles specific to data warehousing by the Kimball Group at www.ralphkimball.com.

"Hey, Wait! I'm a Developer, Not a Manager"

We know, we know . . . This is supposed to be a developer book. So, why are we rattling on so much about project management, documentation, salesmanship, and testing? Well, the fact of the matter is, to be successful in creating BI solutions, you need team members who have these skills. In some cases, you may have sufficient skills in these areas to perform all the tasks yourself. If not, just knowing what these task are will allow you to look for and find team members who can help you in these areas or at least to understand your role on a team.

It is time to move away from all this theory and focus on creating a BI solution as an example. We keep it simple so that you are not spending the next week doing nothing but planning when what you really want is to get started creating a working example. However, we get into enough detail that you are able to get a good idea of what it takes to plan a BI solution. Don't worry, it won't hurt.

EXERCISE 3-1. THE PLANNING PHASE

In this exercise, you take on the role of a consultant performing interviews. You then outline the goals of the BI solution you are being asked to create. There is not much to do in this activity, because for the sake of providing a simple scenario, we have to give you both the questions and the answers. Your job is to follow along to get an understanding of what needs to be accomplished.

Scenario

Your new client (let's call them Publication Industries, Inc.) has a small business. They are booksellers that manage books produced by various publishers and sales to various stores. Although they do not publish books themselves, they do act as an intermediary in selling the books wholesale.

Their current method of managing their data is to track the sales of boxes of books to various stores in a SQL database. They also track additional information about the sales supplied to them from the publishers they represent.

All reports are made in an ad hoc fashion (meaning there was no prior existing system, and there is not a specific standard as to how they collect and manage the information), and clients are finding that they often create reports with conflicting information. The company wants a new business intelligence reporting solution that will provide them with accurate information about their company sales.

Some Questions Asked of the Client

Q. Why do we need it?

A. Our current reporting solution is inaccurate, and we need something better.

Q. What is the BI solution's goal?

A. To provide accurate information about our sales.

Q. What is the hoped outcome of the solution?

A. Our company will end up with one location where all the information is consistent so that everyone who needs it can access it for accurate information.

Q. Who will use the solution?

A. We have managers who read most of the reports. It is rare that all employees will read these reports.

Q. What are we building?

A. We would like several reports involving sales information. Ideally this solution will include a much cleaner database that we can develop our reports from and more accurately track our sales information.

Q. What must be in the solution?

A. Our database is a mess. We need to get it cleaned up and make it consistent.

Q. What would be nice to have in the solution?

A. We know that other companies are using cubes for their reports. We would like to have a cube also.

Q. What will *not* be in the solution?

A. We don't want anything that costs a lot of money to develop or maintain.

Q. How will we build it?

A. We already have a SQL Server that hosts our current database. We would like to use the existing server for all items.

Q. Can we release it in increments?

A. That will be acceptable as long as we start seeing a practical outcome as soon as possible.

In this exercise, you imagined having the role of an interviewer asking questions about the nature of the BI solution you are proposing to create. In the next exercise, we look at defining what will be included in and excluded from the BI solution.

Documenting the Requirements

With the interview process complete, it is time to make some notes about what you found. A simple Word document is sufficient to record these findings. You could simply start by making an itemized list, or you could use a document with tables, diagrams, and section headings. Let's look at an example of this more complex type of document, which you can see in Figure 3-3.

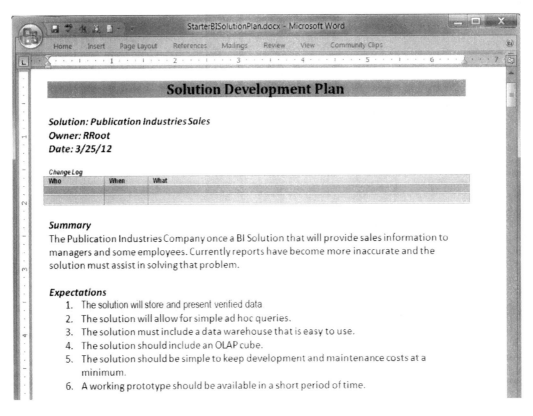

Figure 3-3. *An example of a "Solution Development Plan" document*

When you look over the documents shown in Figure 3-3, you see it has several sections. (You can also find this document in the Chapter 3 folder of the downloadable content files for a clearer view.)

At the top, it identifies the name of the solution, the owner, and the date the document started. Of course, there are other items that could be added, but in an effort to keep things simple, we have reduced this content to just these three items.

Beneath that is a section for recording changes to the document. This area is used after your initial writing of the document, which at this point will not take place until other steps are completed first.

The "Change Log" section is followed by the summary of the solutions you are going to build. This is a parenthetical description of the solution.

Following that is an outline of expectations. It is important to list what will satisfy the client once the solution is completed. You will compare the success or failure of the solution against these expectations once it is completed. Reviewing your successes and failures will allow you to plan for future solutions, allowing you to avoid the same mistakes. This is also a section used by testers to verify that your solution has accomplished its goal.

In Figure 3-4, you can see that the document includes open and closed issues sections. Developers, testers, and managers can use these sections to track questions about the solution and the answers that go with them.

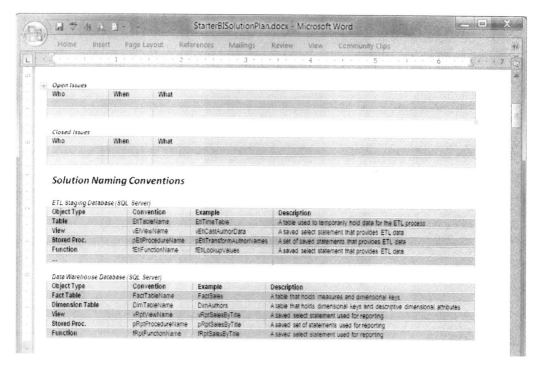

Figure 3-4. *The second page of the "Solution Development Plan" document*

As questions come up, they are recorded in the "Open Issues" section. Answers are recorded in the "Closed Issues" section. When the solution is completed, examining both these sections will help you plan for future solutions.

One section that is often omitted is the one that records the naming conventions used in your solution. Although this section is not strictly necessary, it is more professional looking when all the objects in your BI solution are named consistently. It also makes a difference in the cost of maintenance, as items are easier to find and interpret based solely on their name. For example, should the dimension tables in the data warehouse start with the prefix *Dim*, or should there be no prefix? Most developers have found that including the prefix on the dimension tables makes them easier to find and interpret their use. Therefore, the table that held a list of authors might be named DimAuthors. You would record this preference in the "Solution Naming Conventions" section. Testers of your solution will be asked to verify adherence to the naming conventions.

There are still several more sections to review, but we get to those as we start working on the items that pertain to those sections. For now, let's get organized by completing this next exercise.

EXERCISE 3-2. THE DOCUMENTS

In this exercise, you review a document that describes the BI solution you are creating and copy it from its current location to the solution folder using Windows Explorer.

Before beginning this exercise, you should have downloaded and unzipped the book files from the Apress website. If you did not do so in Chapter 2, please do so now. When the file is unzipped, you will have a folder called _BookFiles. Copy this folder to the root of your C:\ drive.

1. Navigate to the _BookFiles folder on your C:\ drive, locate the Chapter03Files subfolder, and open it (Figure 3-5).

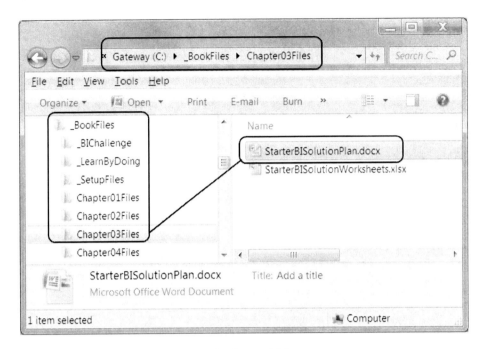

Figure 3-5. *Locating the documents in the chapter folder*

2. Locate the file titled StarterBISolutionPlan.docx within the Chapter03Files subfolder (Figure 3-5) and double-click it to open it in Microsoft Word.

3. Review the contents of this document noting the location of each section. This document is used in later exercises.

4. Close the document when you have finished reviewing it.

5. Next, while you are still in the Chapter03Files folder (C:_BookFiles\Chapter03Files), right-click the StarterBISolutionPlan.docx file, and select Copy.

6. Navigate to the C:_BISolutions folder (that you created in Chapter 2) and paste it into the PublicationIndustries folder (Figure 3-6).

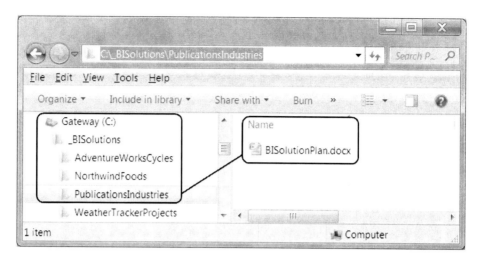

Figure 3-6. *Copying the "Solution Development Plan" document to its new location*

7. Right-click the copied file and select Rename from the context menu. Rename the file BISolutionPlan.docx, as shown in Figure 3-6.

In this exercise, you began to build the basic documents needed at the beginning of your BI solution. The next step is to locate the data and verify whether it will be feasible to create the solution. In Exercise 3-4 within this chapter, you add the file to a new Visual Studio solution using Solution Explorer.

Locating Data

Once you have created a basic outline of the BI solution, locate the data necessary to create it. This data may be in many different places, including simple text files, emails, or more typically, a database.

No matter where the data is located, you need to review what is available and decide what to include and what to discard from the current solution. As always, try to keep things as simple as possible while satisfying the requirements of the solution.

In the case of text files, review each one and decide whether the text file as a whole will be part of your BI solution. After you have categorized which files will be included and which ones will not, closely scrutinize each field within the file to determine whether it is valuable to the current solution or whether it should be ignored during this iteration.

In the case of a database, the process is quite similar. First examine the tables available and then decide whether they are important to the current BI solution. Once you have determined which ones are important and which ones are not, closely review each field within the tables in order to decide which of these should be included.

This process is much easier to learn by performing it than reading about it. So, let's do that now in this next exercise.

▧ **Note** If you are like us, you want to make your own decisions about something you are creating. Nevertheless, because these exercises do not have a real team or client and because all the chapters in this book have to work together from start to finish, we make the decisions for you. We include information from the fictitious client as we go along, as though you were in communication with them.

We are sorry for the inconvenience and know that you may not agree with some of our decisions or thought processes, but we think you will agree that it is still helpful to see how someone else approaches these tasks. In addition, it will help you understand what we are doing in future exercises of this book.

In a real situation, it is best to meet with pertinent members of your team, as well as the client for the review process. Each will often see things that the others do not, and the client will have more perspectives on what is necessary to include than you will without them.

EXERCISE 3-3. REFINING THE PLAN

In this exercise, you take on the role of a solution planner and data warehouse developer. You examine the data currently available and decide whether you will be able to create an effective data warehouse from the data. You further refine the current definition of the BI solution by removing requirements that cannot be accomplished during this iteration of the development cycle.

Reviewing the Current Plan

1. Open SQL Server Management Studio and run as administrator. This is done by clicking the Start button at the bottom left of your screen and navigating to All Programs ➤ Microsoft SQL Server 2011. Then right-click SQL Server Management Studio and select Run as Administrator from the context menu. (You will always run these programs as administrator, because of the permissions required to access databases, and so on.)

2. When SQL Server Management Studio opens, connect to your computer's database engine, and type in the name of your SQL Server. Then click the Connect button to connect to the database engine (Figure 3-7).

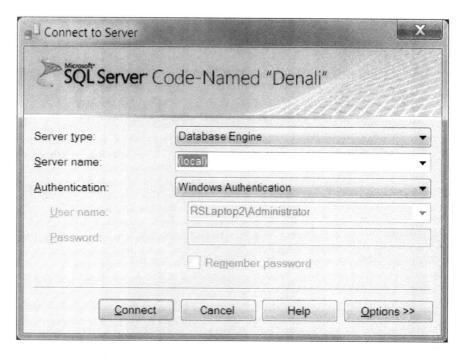

Figure 3.7. *Connecting to the database engine in SQL Server Management Studio*

Usually this will be (local) or localhost, but remember that if you are using a named instance of SQL Server, you will have to use the full name in the format Computer-Name\NamedInstance. For example, Randal uses (local)\SQL2012 and Caryn uses (local)\Denali because we each have installed SQL Server under those names, we are using a beta version, and Caryn never got around to giving her computer a real name. For further information on connecting to your computer's database, see Chapter 5 and search the Web for *SQL Server Named Instances*.

3. After the connection is made, you should see the databases icon in the Object Explorer window. Locate this icon and expand the treeview by clicking the plus (+) sign next to the word *Databases* (Figure 3-8).

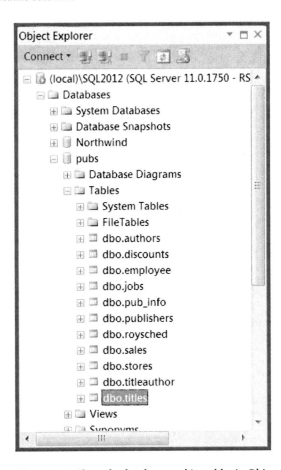

Figure 3-8. *The pubs database and its tables in Object Explorer*

4. In the list of databases, locate the pubs database and expand the treeview by clicking the plus (+) sign next to the pubs database name to see the Tables folder (Figure 3-8).

5. The pubs database was one of the databases you added as part of the setup process. If it does not exist, please review the instructions found in the folder C:_BookFiles_SetupFiles.

6. Expand the treeview by clicking the plug (+) sign next to the Tables folder to see the list of tables (Figure 3-8).

7. Review the tables noting the subject matter of each one. Figure 3-9 shows a diagram of these tables and their relation to each other.

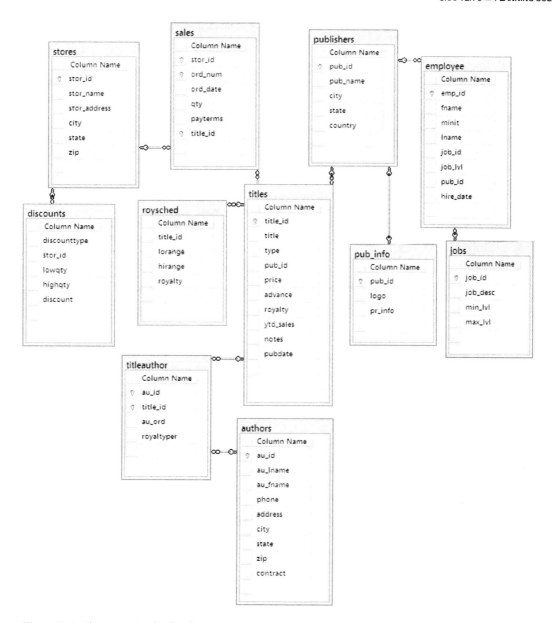

Figure 3-9. *The current pubs database*

Deciding Which Tables to Include

Although the titles table is the center of the diagram, the sales table provides us with information about the company sales. Therefore, this table is the focus of our reporting efforts. As you examine this diagram, notice that the other tables seem to support the sales information. Each table at some point was considered important, but this is not the same as being important to our BI solution. It is a best practice not to include everything from the OLTP environment into the data warehouse. Instead, each table should be reviewed to determine its relevance to the current BI solution.

At an absolute minimum, you need the information found in the sales table. Now, let's look at what else to include.

In addition to the sales information, you need descriptors of each sale. These descriptors are also referred to as *dimensional attributes*. Start by reviewing the other tables associated with the sales table; then determine whether the data inside these tables are a Must Have, a Nice to Have, or a Not Needed item. You can use the four-quadrant prioritizing technique we discussed earlier in this chapter (Figure 3-2).

The following steps take you through each of the tables and outline facts that help determine what to include:

1. Examine the dbo.Stores and dbo.Titles tables. These tables include information about which books are being sold to which stores. Therefore, both stores and titles need to be included. These tables provide a great deal of value and are easy to implement. These are easily determined to be a Must Have.

2. Examine the dbo.Publishers table. Because this fictitious company happens to wholesale books from many publishers, we also need the publisher information. The publisher's information is easy to obtain from this table and is of great use. Therefore, dbo.Publishers is categorized as a Must Have.

3. Examine the dbo.Authors table. Information about which authors write which books may not be as important to the sales reports as the name of the titles may be, but it would be nice to know. Therefore, you might consider including information about the authors as well. This information is easy to obtain and provides somewhat useful information. Therefore, dbo.Authors should be classified as Nice to Have.

4. Examine the dbo.Employee table. For some reason, the database records information about the employees known to work at various publishers. This seems to have little to do with the event of making a sale. The information is easy to obtain but of little value. As such, dbo.Employee qualifies as Not Needed.

5. Examine the dbo.Jobs table. Oddly enough, the database has information about which jobs are held by publisher employees. Exactly how this information was obtained is unimportant. What is important is that it is not necessary to our data warehouse design. So, let's classify dbo.Jobs as a Not Needed item.

6. Examine the dbo.TitleAuthors table. It tracks the royalty percentages given to various authors in the dbo.TitleAuthors table and the order in which each author is listed on each book in this same table. Royalties may have little to do with making a sale, and we doubt that this company—which is acting as a middleman wholesaler—even finds it useful. The order in which the authors' names appear can have an impact on the sales of the book, because only the first author's name may appear in certain listings. The important factor to consider is that this table represents a many-to-many relationship between dbo.Titles and dbo.Authors. Therefore, dbo.TitleAuthors is required to enable our solution to map the relationship between these tables effectively. We must classify this table as a Must Have item.

7. Examine the dbo.RoySched table. The royalty schedule table, dbo.RoySched, tracks ranges of royalties based on the amount of books sold. As more books are sold, the royalties to the authors increase. Although this influences profit, our BI solution

focuses on sales volumes. In the future we may want this, but currently it is easy to obtain but of little value to us. As such, dbo.RoySched is classified as Not Needed.

8. Examine the dbo.Discounts table. We see that some stores seem to have discounts tracked in the discounts table. This would seem to be something we would like to include as well. And yet, let's say that we asked the company owner about how the discounts were calculated. The company owner then informed us that not only are discounts no longer tracked but the data in these fields are considered iffy at best. In real life, it is common to discover something in the database that was once tracked in the past but somehow got put away in the background and ignored. As time goes by, the reason for this information is lost. This means that it is difficult to verify that the data is truly valid, and even if the table would provide good information for reports, it still should be classified as Not Needed.

9. Examine the dbo.Pub_Info table. The publisher information table, dbo.Pub_Info, would seem to be useful at first sight. On closer inspection, we see that it is probably not useful to us. In real life, it is often the case that a table looks like it will be of use at first glimpse, but upon further examination, it is not as pertinent as it first seemed. In this case, we classify this table as Nice to Have.

In the next section of this exercise, we take a look at the data in these tables to determine which columns will be included.

Deciding Which Columns to Include from Sales

At this point, you have examined the tables and categorized them by priority. It is time to do a closer inspection of the data in each of these tables. In SQL Server Management Studio, you can quickly look at the contents of a table by right-clicking the table and selecting Select Top 1000 Rows from the context menu (Figure 3-10). We start with the sales table because it is central to our BI solution.

1. Right-click the dbo.Sales table in the Object Explorer treeview and choose the Select Top 1000 Rows menu item from the context menu. SQL Server Management Studio will create and execute a SQL select statement for you and display the results in a query window, as shown in Figure 3-10.

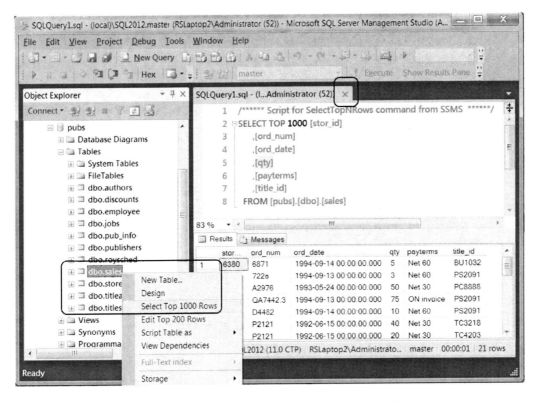

***Figure 3-10.** Displaying dbo.Sales table data in SQL Server Management Studio*

2. Examine the data in each column to determine whether the data represents a measured value or a descriptive value. In the data warehouse, measured values are translated into measures, and descriptive values are translated into dimensional attributes.

3. Close the query window by clicking the *x* on this window's tab.

The obvious measure in the sales table is the sales quantities. All other columns represent descriptive values. These current descriptive values, such as the title_id, in and of themselves hold little meaning to most clients using the reports. Therefore, you need to add dimensional attributes to the data warehouse in order to clarify what items such as title_id really indicate. It is important to include these columns as dimensional keys within your fact table.

■ **Note**   The term *dimensional key* defines a column used to identify an individual row of dimensional data. This is usually used as a primary key in the dimension tables and a foreign key in the fact table. An example here is the title_id.

Reviewing the Data in the Titles Table

We need to look at the data in the various supporting tables. We take a look at the dbo.Titles table next (Figure 3-11).

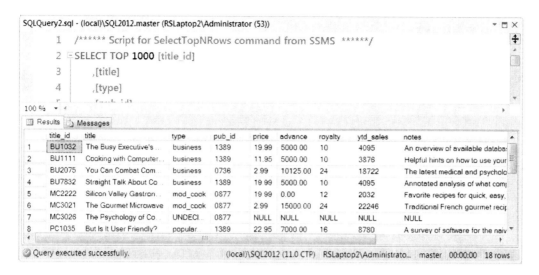

Figure 3-11. *The titles table*

4. Locate the dbo. Titles table in the treeview list, and right-click this table to access the context menu.

5. Choose Select Top 1000 Rows from the context menu. SQL Management Studio will create a query window, execute the query, and show you the results (Figure 3-11).

In the titles table, notice that along with the title there is a type for each title. This information would be quite useful in a report because it provides a way to group the titles collectively. Let's classify title type as a Must Have item.

A publisher ID is included in this table that could also be used to group titles. Let's classify this as a Must Have item.

In addition, the price of each title is listed here and could prove useful for making measured calculations. Although it does seem odd that this information was not included in the sales table when we created the data warehouse, we can rectify this. Let's classified this as a Must Have item.

Other columns may or may not be as useful. For example, it is unlikely that sales reports would need to categorize sales based on the type of advance given to each author for a given title. Let's classify this as Not Needed.

Also, the year-to-date column looks temptingly like a measure but provides aggregate values, and as we show later, aggregate values do not go into a fact table holding measured data. Let's classify this as Not Needed.

Additional auxiliary information includes a set of notes about each title and the date the titles were published. It seems somewhat obvious that the notes can be dismissed as being superficial to the sales reports. However, inclusion of the published dates is a little less clear. It may be useful to know how many sales have occurred since its published date, and this is easy to obtain. So, we include the published date in our design by classifying it as Nice to Have and exclude the notes by classifying them as Not Needed.

6. Close the query window by clicking the *x* on this window's tab.

Reviewing the Data in the Publishers Table

With the dbo.Titles table examined, let's turn our attention to the dbo.Publishers table (Figure 3-12).

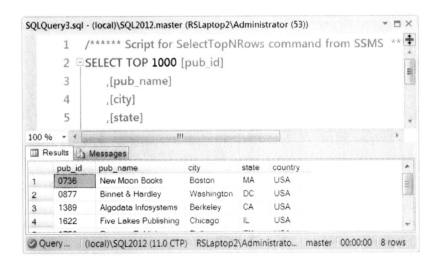

Figure 3-12. *The publishers table*

The following are the steps to open the dbo.Publishers table:

1. Locate the dbo.Publishers table in the treeview list, and right-click this table to access the context menu.

2. Choose Select Top 1000 Rows from the context menu. SQL Management Studio will create a query window, execute the query, and show you the results (Figure 3-12).

The dbo.Publishers table contains the name of the publisher, which will be useful in our sales reports.

3. This table also includes the city, state, and country where those publishers are located. At first you would likely decide that these additional columns provide information that would be useful to include in our data warehouse as dimensional attributes. Yet, looking at the values, you may notice that it is never the case that the cities or states are repeated. You would not utilize either of these columns to group publishers in your sales reports. However, there are a number of repeating values in the country column, so grouping publishers by country might prove useful. Therefore, from a practical sense, the publisher name is a Must Have item whereas country, state, and city are only Nice to Have but not required. Close the query window by clicking the x on this window's tab.

Reviewing the Data in the Authors Table

Let's see if something similar is happening in the dbo.Authors table (Figure 3-13).

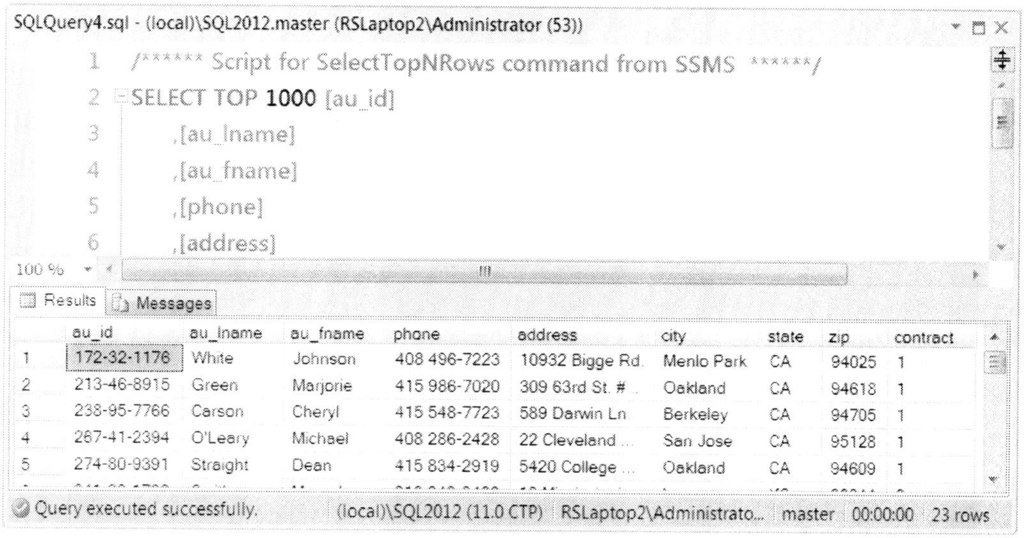

Figure 3-13. *The authors table*

Here are the steps to open the dbo.Authors table:

1. Locate the dbo.Authors table in the treeview list and right-click this table to access the context menu.

2. Choose Select Top 1000 Rows from the context menu. SQL Management Studio will create a query window, execute the query, and show you the results (Figure 3-13).

In the authors table it is pretty clear that the author's first and last name are Must Have attributes. However, the author's phone number and address data are less so. Keeping things simple, you might want to leave out these unnecessary columns. They do not take up much space; however, when you are trying to keep things as clean and uncluttered as possible, removing things that are unnecessary is the easiest way to accomplish this goal. Therefore, let's keep the author's name columns as Must Have and the other columns as only Nice to Have. (It may be that in another version of the data warehouse this priority list will change.)

3. Close the query window by clicking the *x* on this window's tab.

Reviewing the Data in the TitleAuthor Table

The table that connects the titles and authors together in a many-to-many relationship is the dbo.TitleAuthor table. Let's look at this table next (Figure 3-14).

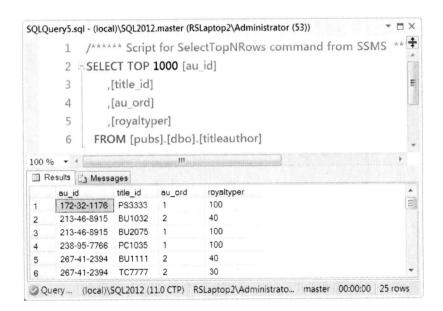

Figure 3-14. The titleauthor table

Here are the steps to open the titleauthor table:

1. Locate the dbo.TitleAuthor table in the treeview list and right-click this table to access the context menu.

2. Choose Select Top 1000 Rows from the context menu. SQL Management Studio will create a query window, execute the query, and show you the results (Figure 3-14).

This table is a Must Have for its ability to connect a many-to-many relationship between titles and authors. In order for that to be accomplished, you must include the title_id and au_id columns. However, the royalty percentages may not be useful to us and should be excluded. The author order column, au_ord, which may have an impact on sales, can be classified as a Nice to Have item.

3. Close the query window by clicking the *x* on this window's tab.

Reviewing the Data in the Stores Table

Next up is the dbo.Stores table shown in Figure 3-15.

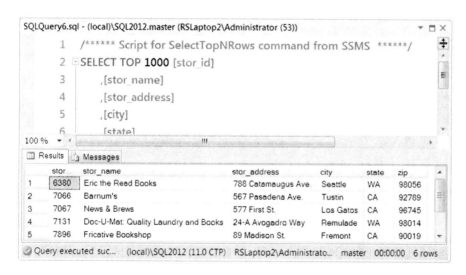

Figure 3-15. *The stores table*

Here are the steps to open the stores table:

1. Locate the dbo.Stores table in the treeview list, and right-click this table to access the context menu.

2. Choose Select Top 1000 Rows from the context menu. SQL Management Studio will create a query window, open an SQL query window, execute the query, and show you the results (Figure 3-15).

Similar to the authors table, the stores table contains the names and addresses of the stores. Once again, the names qualify as Must Have data, whereas the address information would be Nice to Have but not required. It may be interesting to examine sales on a state-by-state basis, but it is much less likely that sales would be examined based on street addresses. Cities fall somewhere in between the two, but if you look closely, you see that cities are never repeated. Therefore, it is unlikely that useful information can be extracted regarding sales based on the city in which the stores are located.

One question we forgot to ask during the interview process was how long the client has been collecting this particular set of data. If you went back and asked this question, you would find that the number of stores, publishers, and authors seldom changed over the last few years. If a lot of change had been occurring, we might have decided that the city column would become more useful as time went on. However, because change is slow or nonexistent, all the city column will do is provide additional data in the data warehouse without much additional information. Using the "keep it simple" rule, you might choose to exclude the street address, city, and zip code from this first version of the data warehouse. As time goes by, if you decide that this data becomes useful, you can add it during the creation of future versions.

Note: We are leaving each store's state data out of the first version of the design. As we see later in the book, we will come to regret this decision. But, it allows us a chance to examine what to do if something is missing from an initial version.

    3.   Close the query window by clicking the *x* on this window's tab.

Reviewing the Data in the Pub_Info Table

A number of columns in various supporting tables are not necessarily useful for this version but could be useful in the future. For example, the publishers' information table, or pub_info as it is called, has little in it that is of use to us at this time (Figure 3-16).

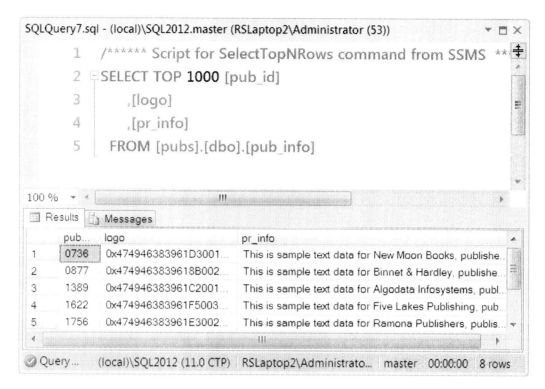

Figure 3-16. *The pub_info table*

Here are the steps to open the pub_info table:

1. Locate the dbo.Pub_Info table in the treeview list, and right-click this table to access the context menu.

2. Choose Select Top 1000 Rows from the context menu. SQL Management Studio will create a query window, execute the query, and show you the results (Figure 3-16).

Note: This table includes binary data that represents an image of the publisher companies' logo. It also holds some nonsensical text as a placeholder that has never been filled in retroactively.

3. Close the query window by clicking the *x* on this window's tab.

Reviewing the Data in the RoySched Table

If you remember, we excluded a few tables at the beginning of this exercise. While we are here, let's look at them again to make sure we made the right decision.

One of these excluded tables was the royalty schedule table, otherwise known as *roysched* (Figure 3-17).

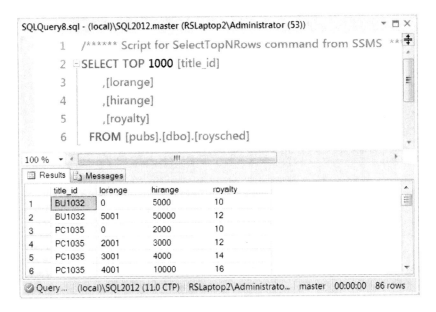

SQLQuery8.sql - (local)\SQL2012.master (RSLaptop2\Administrator (53))

```
1    /****** Script for SelectTopNRows command from SSMS  **
2  ⊟ SELECT TOP 1000 [title_id]
3        ,[lorange]
4        ,[hirange]
5        ,[royalty]
6    FROM [pubs].[dbo].[roysched]
```

100 %

Results | Messages

	title_id	lorange	hirange	royalty
1	BU1032	0	5000	10
2	BU1032	5001	50000	12
3	PC1035	0	2000	10
4	PC1035	2001	3000	12
5	PC1035	3001	4000	14
6	PC1035	4001	10000	16

Query... (local)\SQL2012 (11.0 CTP) RSLaptop2\Administrato... master 00:00:00 86 rows

Figure 3-17. *The royalty schedule table*

Here are the steps to open the roysched table:

1. Locate the dbo.RoySched table in the treeview list and right-click this table to access the context menu.

2. Choose Select Top 1000 Rows from the context menu. SQL Management Studio will create a query window, execute the query, and show you the results (Figure 3-17).

In this table you see data that represents the amount of royalties paid on a particular book based on the range of sales. When a book sells from 0 to 5,000 copies the royalty paid is 10%. However, when book sales reach between 5,001 and 50,000, the royalty percentage goes up to 12%. Because royalties are paid by the publisher to the authors and not by the wholesaling company who we are building the BI solution for, this information is irrelevant.

Reviewing the Data in the Employee Table

Another excluded table was the dbo.Employee table. At first glance, information about employees could prove useful to providing sales information. In most cases, this would be a true assumption; however, in the pubs database this is not the case. On closer inspection, we see that the employee table (Figure 3-18) provides a list of employees who only work at a particular publishing house, and the table is classified as Not Needed.

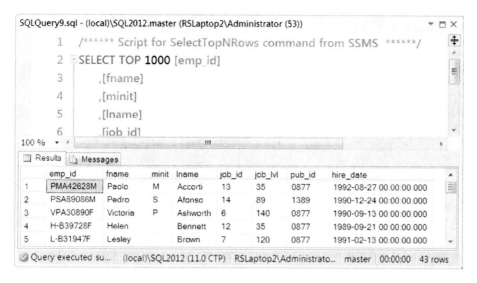

Figure 3-18. *The employee table*

Here are the steps to open the dbo.Employee table:

1. Locate the dbo.Employee table in the treeview list and right-click this table to access the context menu.

2. Choose Select Top 1000 Rows from the context menu. SQL Management Studio will create a query window, execute the query, and show you the results (Figure 3-18).

Although it may be important to have a list of employee contacts who can be reached by Publication Industries, it is not relevant to the sales of individual titles to individual stores. One wonders how this information was even collected.

Deciding to Continue with the Solution

At this point, it seems obvious that we have sufficient data to create a BI solution that will provide reporting information on sales. We may not have every single detail, but we have enough to make an initial prototype. Once the prototype is created, we can get additional feedback and start working on the next version of the BI solution that will provide even greater functionality and benefit to the customer.

In this exercise, you examined the data available to you and decided what was going to be included in the BI solution and what was going to be excluded. For that matter, you also decided whether the BI solution could even be accomplished. Because it was decided that it can be accomplished, your next step is to define the roles required to create the solution and acquire the team members that will implement these roles.

Defining the Roles

Now comes the point of defining the roles needed for the team. At a minimum, you need one team member to build the data warehouse, one team member to fill up the data warehouse, and one team member to make the initial reports. The solution must be documented as well, so you need somebody to build the documentation. In addition, the solution you come up with needs to be tested. Therefore, add a tester role to the list. They asked for a cube as well, so you need somebody to build one.

You could make a formal document describing the role and which task they are to perform; however, let's just add an additional worksheet to our Excel spreadsheet and then make a simple list.

▪ **Note** You can find this Excel file in the `C:\BookFiles\Chapter03Files` folder. We called it `StarterBISolutionWorksheets.xslx`.

You may not necessarily know who your team members are in the beginning, but we can make a blank column where their names can be entered as the roles are filled (Figure 3 -19).

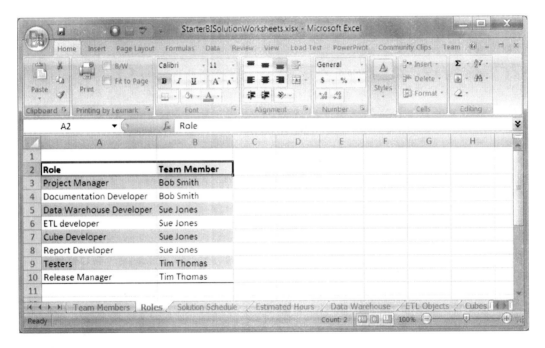

Figure 3-19. *Listing the roles*

Defining the Team

For learning purposes, you are fulfilling each role in this book, and it is your task to create each of the various elements of your BI solution. In real life, you would likely want an entire team to work on your BI solution. The old adage about two heads being better than one really does apply when you are trying to work on something as complex as a BI solution. One individual may miss seeing a possible problem that is quite obvious to another.

As stated before, documentation does not have to be exhaustive to be of use. You can simply list the team members' names, phone numbers, and email addresses, and we also recommend listing the hours they work. Three years after the project is completed, it may be useful for the BI solution users to contact members of the development team to ask questions. Offering something as simple as a list of contacts in the solution documents can make life simpler (Figure 3 -20).

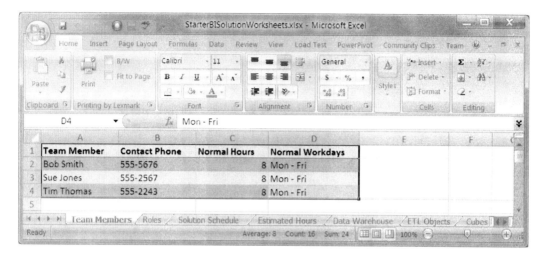

Figure 3-20. *Listing the team members*

Determining the Schedule

In order to come up with a schedule, you need to have an idea of approximately how long each task will take. If you are not very experienced with these types of tasks, such as developing cubes or ETL processes, coming up with a schedule can be difficult. As you gain experience, it will be easier to make these estimates.

To get started, identify the list of tasks that you think you will need to accomplish for your particular BI solution. All BI solutions share certain similarities; however, it is a given that there will be something distinct about each one.

Try not to micromanage each task, but sum it up into sets of one to three hours of work. In an eight-hour day, consider that breaks will be taken, emails will require answering, teammates will ask questions, phones will ring and lunches will be scheduled. In an eight-hour day, the average person will complete about six hours of work.

As you can see in Figure 3-21, the number of hours is strongly weighted toward the planning phase. This may seem out of proportion, but in reality once you have defined the names of all your tables, columns, dimensions, cubes, basic reports, and so on, their creation will be a much faster process. Therefore, much of the development time is the planning phase. This includes selecting the datatype, setting certain properties, and defining the relationships between various objects. Lastly, be sure to plan time for the communication process, which can take several hours of meetings, depending upon the size of the project.

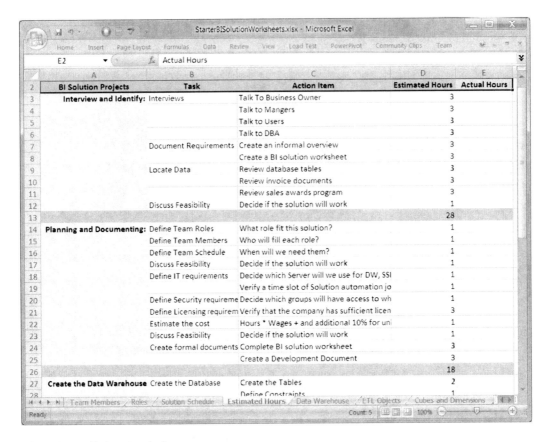

Figure 3-21. Estimating the hours

Taking the time to preplan how long a project will take and tracking hours spent will provide you with invaluable information for future projects. You will be able to estimate the cost and forecast release dates of future projects more accurately. Adding a column to your spreadsheet that tracks the hours spent is easy to do, and you will find that you refer to it often (Figure 3 -21).

Once you have estimated the hours required to complete the project, creating a timeline is relatively easy. Total up the hours for each section. If you have determined that each day represents about five hours of activity, you can make a calendar to map out how long the project will take. This calendar can be as simple as adding yet another worksheet (Figure 3-22). Keep in mind that most months have holidays, and the like.

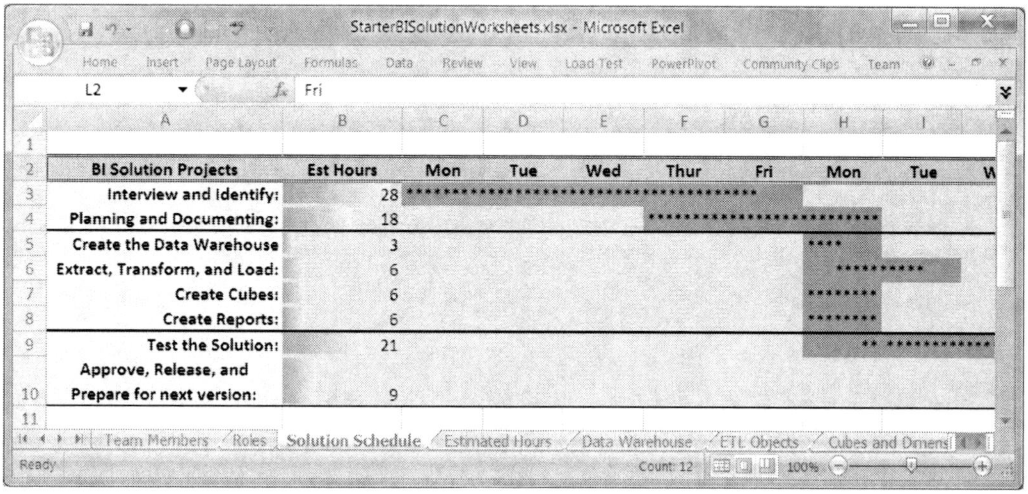

Figure 3-22. *Planning the schedule*

As with anything new, the more you practice it, the more proficient at it you will become. Keep in mind that you can track how accurate you were at planning each solution by comparing what occurred to what you had planned. An additional benefit to scheduling your project is that your teammates, who may be involved with other projects as well, can integrate their schedules with yours. In the end you will find that a simple timeline is easy to create and provides advantages that far outweigh the time it takes to implement.

The IT, Security and Licensing Requirements

As consultants, we have worked with many different companies. Each company has different needs with regard to IT, security, and licensing. It is important that you consider these requirements when you are creating a BI solution.

Defining these IT requirements includes (and is not limited to) the following:

- Network bandwidth

- Which servers are available

- The age of the servers

- The space on the servers

- The people who maintain these servers

Defining security requirements involves evaluating many factors. Everyone wants their data to be secure, but what "secure" means to one company is not at all secure to another. If an industry, such as pharmaceuticals, has governmental regulations, then additional security requirements need to be addressed.

For licensing requirements, there is also a wide range of possible scenarios. One company may have a blanket license for a lot of different software. Another company may need to purchase their software individually. It may be advantageous to use software that is flexible or comes bundled with as many features as you require to complete the project. An example of this is Microsoft SQL Server. With the purchase of SQL Server alone, you gain a whole suite of BI applications. These applications allow you to build the data warehouse, perform the ETL process, create cubes, and even create reports. In fact, it could be argued that one of the best reasons to choose Microsoft SQL Server is that, for many companies, this will be the most effective tool to implement a BI solution.

Estimating the Cost

When estimating the cost of the solution, a good place to start is by evaluating how many hours you thought each task would take and multiplying it by the hourly wages you expect to pay. Keep in mind that issues can come up that could block you from performing your tasks in a particular order or on a particular date. In a perfect world, everything runs smoothly and works as planned. Nevertheless, as we all know, that is not likely to happen for the vast majority of solutions. A common way to mitigate this is to add a percentage to the estimated cost. This additional percentage usually varies between 10% to 20% depending on how risky you feel any given project is in regard to overruns and changes.

Documenting the Solution Plan

You should have a good idea of what the solution consists of after you have examined the tasks your solution will entail. By this time, you should know the following:

- Which tables you need

- Which type of ETL tasks you need to perform

- What the name of your cube might be

- The titles of your basic prototype reports

You can choose to put this information directly in your formal Word document or record it in a simple spreadsheet. Either way, it should be recorded. This formal or informal document is then used for the creation of your BI solution. Additionally, this documentation should correspond with what is incorporated into the official contract, if one is necessary.

Assume that you will make mistakes. Assume that you will miss some items. Assume that some of your time estimates may be incorrect. In the end, it does not matter that you get everything perfect the first time out. What does matter is that you learn by your mistakes and become better at creating BI solutions as time goes on.

Let's assume you are going to use an informal Excel spreadsheet to start your solution documentation. You can add another worksheet to the spreadsheet that you have already been using to document your solution plan. Make a list of the tables that you think you need to document the data warehouse.

You can start with the name of database itself. Try to stay true to whatever naming conventions you decided on in the planning throughout the project. For example, in the exercises for this book, we determined that our naming convention indicates the object's type followed by the object's name. In this case, we recorded the data warehouse name as Data Warehouse Publication Sales. Because that title is rather long, we abbreviated it to DWPubsSales.

To continue this example, we needed a fact table to hold our sales information. Therefore, we listed the fact table under the name FactSales (Figure 3-23). Its formal name is the name of the database, followed by the name of the namespace or schema, followed by the name of the object. Therefore, we listed this in the spreadsheet as DWPubsSales.dbo.FactSales.

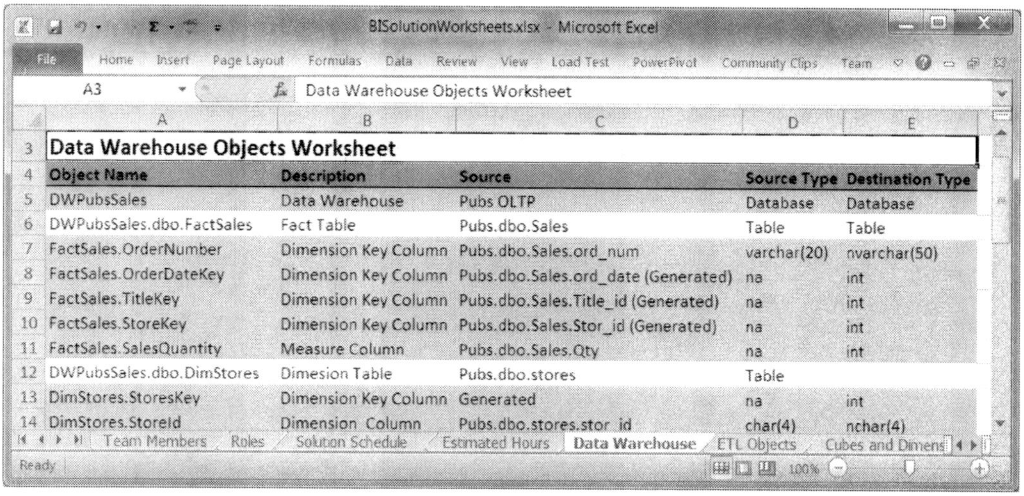

	A	B	C	D	E
3	Data Warehouse Objects Worksheet				
4	Object Name	Description	Source	Source Type	Destination Type
5	DWPubsSales	Data Warehouse	Pubs OLTP	Database	Database
6	DWPubsSales.dbo.FactSales	Fact Table	Pubs.dbo.Sales	Table	Table
7	FactSales.OrderNumber	Dimension Key Column	Pubs.dbo.Sales.ord_num	varchar(20)	nvarchar(50)
8	FactSales.OrderDateKey	Dimension Key Column	Pubs.dbo.Sales.ord_date (Generated)	na	int
9	FactSales.TitleKey	Dimension Key Column	Pubs.dbo.Sales.Title_id (Generated)	na	int
10	FactSales.StoreKey	Dimension Key Column	Pubs.dbo.Sales.Stor_id (Generated)	na	int
11	FactSales.SalesQuantity	Measure Column	Pubs.dbo.Sales.Qty	na	int
12	DWPubsSales.dbo.DimStores	Dimesion Table	Pubs.dbo.stores	Table	
13	DimStores.StoresKey	Dimension Key Column	Generated	na	int
14	DimStores.StoreId	Dimension Column	Pubs.dbo.stores.stor_id	char(4)	nchar(4)

Figure 3-23. *Planning the data warehouse*

Our naming conventions may be different from what you are accustomed to, but every company will have their own preferences. Whatever the preferences are, your documentation should reflect this.

In Exercise 3-3, you examined each table in the pubs database and determined which columns to include and which to omit. These can now be listed in the worksheet (Figure 3-23).

After the objects are listed, it can be helpful to list their description, source, source datatype, and destination datatype as well. We have found these columns to be quite useful when building the data warehouse. In fact, we use this list to build the data warehouse in Chapter 4.

We can then continue to list the objects in the SSIS ETL project, the SSAS cubes project, and the SSRS reports project. The list does not have to be perfect. It is simply a way of getting started. It has been our experience that we always miss something in the initial documentation anyway. We recommend that you use the spreadsheet as the initial documentation and update it as you find omissions and mistakes. Later, at the end of the solution development cycle, you can update the formal document that was created using Microsoft Word. Another, better option is to get a technical writer on your team who can update this to the formal document as you go. It is this formal version that is submitted to your client along with other documents throughout the course of your project such as your initial contract, changes to your contract if there are any, technical notes, observations, billing documents, any specific items requested by the client to be submitted, and any thank-you messages. For more information about formal documentation, see Chapter 19.

Implementation

Once you have formulated a plan of action, you must perform that action. In the case of a SQL Server BI solution, that means creating a data warehouse and several Visual Studio projects. As we saw in Chapter 2, one Visual Studio solution can hold many projects and even solution documents. So, perhaps the first thing to do is to create a solution in Visual Studio and add our documents to it. Let's do that now.

EXERCISE 3-4. ADDING THE DOCUMENTS TO YOUR SOLUTION

In this exercise, copy the Excel spreadsheet file to your _BISolutions folder, create a new Visual Studio solution to hold both documents you examined in this chapter and add these document to your Visual Studio solution.

Copying the Excel File to the Solution Folder

1. Inside the C:_BookFiles\Chapter03Files folder, locate the StarterBISolutionWorksheets.xlsx file.

2. Copy this file to the C:_BISolutions\PublicationsIndustries folder (Figure 3-24).

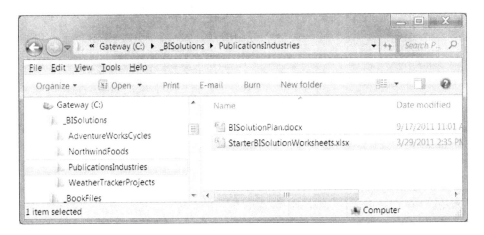

Figure 3-24. *Now both both BI planning documents are in the PublicationsIndustries folder.*

3. Rename the StarterBISolutionWorksheets.xlsx file to BISolutionWorksheets.xlsx.

Placing these files directly on your C:\ drive makes it much easier to navigate to the files you need for these exercises and leaves less room for confusion later, because we access this particular folder quite often throughout this book.

Creating a Empty Visual Studio Solution

We now need a new Visual Studio solution to hold our BI documents and projects. We do that by first creating a blank solution and then adding documents and projects to it.

1. Open Visual Studio 2010. You can do so by clicking the Start button and navigating to All Programs ➤ Microsoft Visual Studio 2010 ➤ Microsoft Visual Studio 2010; then right-click this menu item to see an additional context menu (Figure 3-25). Click the "Run as administrator" menu item. If the UAC message box appears, click Continue to accept this request.

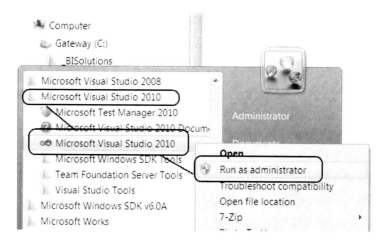

Figure 3-25. Opening Visual Studio and running as admininstrator

2. When Visual Studio opens, select File ➤ New ➤ Project from the menu. (Do not use the Create: Project option from the Start Page as you may have done in the past with other types of solutions.)

3. When the New Project dialog window opens on the left side of your screen, click the arrow to expand Other Project Types and select Visual Studio Solutions, as you shown in Figure 3-26.

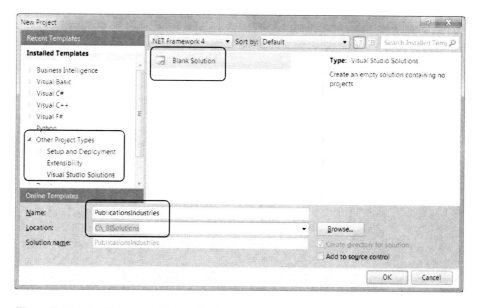

Figure 3-26. Creating a new blank solution

4. In the templates section, select Blank Solution, as shown in Figure 3-26. The Name and Location textboxes will be filled in with a default name, but we change these in the next step.

5. In the Name textbox, type the name **PublicationsIndustries**.

This is the same folder where the solution files are located. When Visual Studio makes the solution, it will use this already created folder.

6. In the Location textbox, type **C:_BISolutions**, as shown in Figure 3-26, and finally click OK.

Once the solution is created, it will appear in Solution Explorer. This will be on the right side of your screen. (This is its default location for the Solution Explorer window, but it can be moved.)

7. Right-click the solution PublicationsIndustries icon and select the Open Folder in the Windows Explorer context menu item. Windows Explorer will open to the location of your solution folder (Figure 3-27).

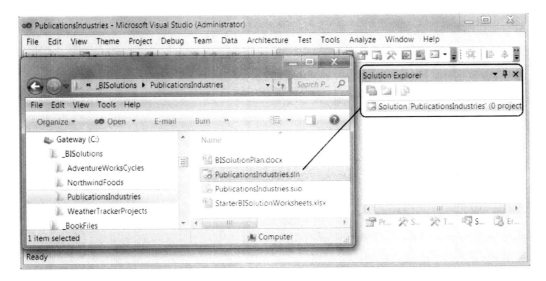

Figure 3-27. *Viewing the files in the solution folder*

8. Review the files in this folder and notice that they are not currently showing in the Solution Explorer window. Solution Explorer shows only a minimum of files from a solution or project folder, but we can change this by adding the file to the solution as existing items. (Depending on your computer settings you may not see the .suo hidden file.)

9. Close Windows Explorer.

Adding the Solution Documents

1. Add a new solution folder to your Visual Studio solution by clicking the new folder button at the top of the Solution Explorer window.

2. Rename the new folder as **SolutionDocuments**. It will appear in the Solution Explorer.

3. Right-click the new SolutionDocuments folder, and select Add ➤ Existing Items from the context menu, as shown in Figure 3-28. A dialog window will open.

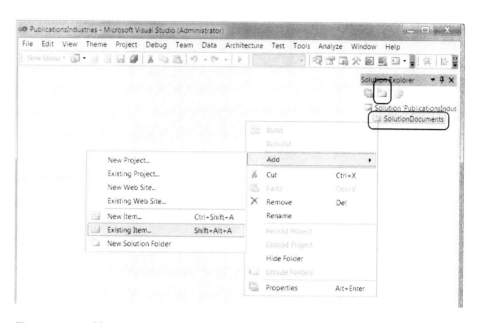

Figure 3-28. *Adding existing items to the solution folder*

4. When the Add Existing Item dialog window appears, select Local Disk (C:), and then navigate to the C:_BISolutions\PublicationIndustries folder.

5. While holding down the Control button, click to select the following files: BISolutionWorksheets.xlsx and BISolutionPlan.docx (Figure 3-29).

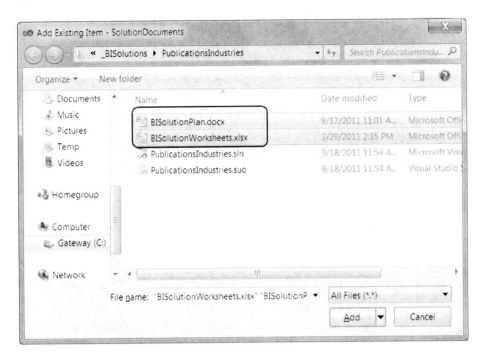

Figure 3-29. *Viewing the files in the solution folder*

Note: Windows by default hides file extensions. So, you may only see the first part of the name of each of these files. We recommend turning off this feature when you get the chance. We often find it very helpful to know the extension of all the files. To find out more about this feature, search the Web for *Windows Show Extensions*.

6. Click the Add button to add them to the SolutionDocuments folder. (Visual Studio uncharacteristically creates a reference instead of a copy when you add an existing item to a solution folder.)

7. Visual Studio will try to open each of the files so that you can see their content. We are not making any changes to these files. Review them if you like, but then close Excel and Word afterward.

8. Use Visual Studio's File menu to save your work by selecting the Save All option.

9. Close Visual Studio.

In this exercise, you created a blank solution and added documents used for creating your SSIS, SSAS, and SSRS projects. We refer to these documents in future exercises throughout the book.

Moving On

In this chapter, we have seen how to come up with a basic plan to implement a BI solution. We reviewed the basic workflow that outlines the various tasks involved in deciding whether the BI solution would be a viable project. The workflow incorporated items such as interviewing the customer, defining what would be included in the solution, reviewing the available data, determining which roles each team member would play, and creating simple documentation for tracking and planning the project.

In the next chapter, we create the data warehouse for the BI solution based on our plan. We tend to think that creating the data warehouse is a lot more fun than all this planning. But good planning can make the difference between a profitable, effective BI solution and a chaotic expensive mess. Hence, Winston Churchill's wise words on planning, once again, come to mind.

LEARN BY DOING

In this "Learn by Doing" exercise, you perform the process defined in this chapter using the Northwind database. We have included an outline of the steps you performed in this chapter and an example of how the authors handled them in two Word documents. These documents are found in the folder `C:\_BISolutionsBookFiles\_LearnByDoing\Chapter03Files`. Please see the `ReadMe.doc` file for more instructions.

What's Next?

Entire books are written about solution planning, and one chapter in this book is insufficient to make anyone an expert on the subject. By the same token, reading all the books ever written about the subject will not necessarily make you proficient at performing it, either. Becoming proficient involves a combination of practical experience and researched knowledge.

For further instruction on solution planning, we recommend these books: *The Kimball Group Reader: Relentlessly Practical Tools for Data Warehousing and Business Intelligence* by Ralph Kimball and Margy Ross (Wiley), and *Lessons in Project Management* by Jeff Mochal (Apress).

CHAPTER 4

■ ■ ■

Designing a Data Warehouse

Organizing is what you do before you do something, so that when you do it, it is not all mixed up.

—A. A. Milne

Designing a data warehouse is one of the most important aspects of a business intelligence solution. If the data warehouse is designed correctly, all other aspects of the solution will benefit. Conversely, if it is created incorrectly, it will cause no end of problems.

In this chapter, we show techniques for designing a data warehouse, including different designs and terms used in the creation process and their proper use. Our focus is on simple practical designs that will get you building your first data warehouses quickly and easily.

When you have completed this chapter, you should be able to design data warehouses using industry standards and know the common rules that are considered "best practice" for the design process.

What Is a Data Warehouse?

At its core, a data warehouse is a collection of data designed for the easy extraction of information. It can be in any form, including a series of text files, but most often it is a relational database. Because of this, most developers think of a data warehouse simply as a reporting database. And although that is not the most highbrow definition, it is fairly accurate.

Many developers will have differing opinions on what is the best way to design a data warehouse. But there are common characteristics you can expect to see in all of them. The first common characteristic is a set of values used for reports. These are called *measures*. For example, InventoryUnits and SalesDollars can be considered measures. Another common characteristic found in data warehouses is a set of dimensions. Dimensions describe the measured data. Examples of dimensions include the dates that the InventoryUnits were documented or the zip code of the customers who bought a particular product.

We discuss more on both of these subjects in the next few pages.

What Is a Data Mart?

A data mart is also a collection of data. It, too, is designed to allow for the easy extraction of information. The information in a data mart, however, is more specific than that of a data warehouse. Typically, a data mart is created for a particular process, such as a sales event or taking inventory.

Data marts can also be designed around departments within the company. But you are typically better off defining the data mart based on a process, not a department. This is because when you define it with a process, many different departments can use the same data mart. If you define it on a department, it may be isolated from being reused, and you can end up with redundant data marts that cause more confusion rather than provide accurate information.

Data marts can be thought of as a subset of a data warehouse. Because data warehouses represent multiple processes and a data mart focuses on a single process, a data warehouse can be thought of as containing one or more data marts. Therefore, if a business has both a sales or inventory process, a sales and inventory data mart would be part of that business's data warehouse. When you build your data warehouse, you create two fact tables in one database. These two tables provide the core data for the data warehouse (Listing 4-1).

Listing 4-1. *An Expression Describing a Data Mart*

```
/* Data Warehouse=a set of data marts {Sales Data Mart, Inventory Data Mart} */
Select * from FactSales
Select * from FactInventory
```

This allows for easy access to the information for a particular process and a way of managing data for all of the company's processes. In summary, a data warehouse is a collection of one or more data marts, and a data mart is a collection of data around a particular process.

Competing Definitions

The definitions we use in this book are found in many other books as well, but not all. A number of competing viewpoints exist that describe what something is called in the business intelligence world. This diversity of terms began with two different industry leaders, Bill Inmon and Ralph Kimball.

In the early 1990s, Inmon published several articles on data warehousing. Later, Kimball also published articles, as well as a famous book known as *The Data Warehouse Toolkit* in the mid-to-late 1990s.

Although both agree on the principle of data warehousing providing information to the users, they differ on how to accomplish this task. Inmon believes that a data warehouse should be very comprehensive and reflect all aspects of the business before it can truly be useful. This tactic has proven to be most effective for many large companies. But it has proven to be less useful for smaller to midsize companies that do not require such complete integration from all aspects of the company in order for their businesses to operate.

Kimball's theory is that a BI solution should focus on smaller specific topics such as business processes. As soon as you implement one process in a BI solution, it is available for reporting, and work can begin on the next process to be added to the solution. This is considered a bottom-up approach.

To paraphrase, Kimball outlines a bottom-up approach by starting small and using building blocks to create something of larger scope, which can be added to over time. Inmon's theory, on the other hand, focuses on a large holistic approach building from the top down in a comprehensive manner that requires all aspects to be incorporated before the data warehouse can function as it is designed.

Many developers have adopted one philosophy or another. Sometimes they even get a bit overzealous about it, not unlike choosing an athletic team to root for in sports. Both approaches, however, have their uses and should be considered tools that, when used appropriately, provide maximum results.

Stereotypically, larger companies will benefit from the top-down approach because they often need holistic information about the many departments that make up the company. Small to midsize companies are more likely to benefit from the bottom-up approach. These companies have few departments, and communication between these departments is easier to manage.

Keep in mind that both approaches will work for companies of all sizes, so each solution must be evaluated independently, but a bottom-up approach usually is appropriate more often than not. In addition, all companies can benefit from a bottom-up approach at some point, because it requires less time and fewer resources before reporting can begin.

Starting with an OLTP Design

When you are designing a data warehouse, there is a strong chance that you will begin by reviewing an existing OLTP database. Although a data warehouse can be built on simple log files or XML data, it is much more likely that it will be built upon an OLTP database.

The standard OLTP database design has been defined for several decades. Many readers may be familiar with these patterns, but we do not expect everyone to have the same level of technical background, so let's take a brief look at these patterns before we move on.

The three basic table patterns in an OLTP database are one-to-one, one-to-many, and many-to-many relationship patterns. Figure 4-1 demonstrates these common design patterns.

Figure 4-1. *Standard OLTP design patterns*

In a one-to-one relationship pattern, a single table is divided along its columns into two tables. Although not unheard of, this pattern is somewhat unusual. Quite often, the goal with this pattern is to separate private information from public information or possibly to partition data onto a separate hard drive for performance reasons. In Figure 4-1, two employee information tables represent a one-to-one relationship. All of the data could have been stored in one employee table, but because Social Security information is considered private, it has been separated into an additional table. Note that the EmployeeId in either table is never repeated in this type of relationship pattern.

The one-to-many relationship pattern also seen in Figure 4-1 is by far the most common pattern in OLTP databases. Our example shows a one-to-many relationship between customers and sales. These tables demonstrate that one customer can have many sales, but an individual sale is associated with only one customer.

The many-to-many relationship pattern of Figure 4-1 demonstrates the process of how one student can attend many classes and one class can hold many students. Note that a third table records the relationship data by storing a copy of the two related tables' key values.

As we show later, these design patterns are also represented in a data warehouse, but for now let's look at an example of a typical OLTP database. It enables you to contrast these OLTP designs with that of a data warehouse design.

A Typical OLTP Database Design

The database we have created for our example consists of tables that record transactional data in the tables called Sales and SalesLineItems. Along with the transactional data are supporting tables. These tables describe the transactions that occur. Examples of these types of tables are the Employees, Stores, and Titles tables.

Normalized Tables

Standard OLTP databases are designed around the concept of normalization, and we have used the rules of normalization to design this example OLTP database.

In a normalized database, all columns contain a single unit of data, such as city or state, but never both city and state as a single entry. All rows contain a unique combination of values, and redundancy is eliminated wherever possible. For example, looking at Figure 4-2, you see that the Stores table has a link to the States table based on StateId.

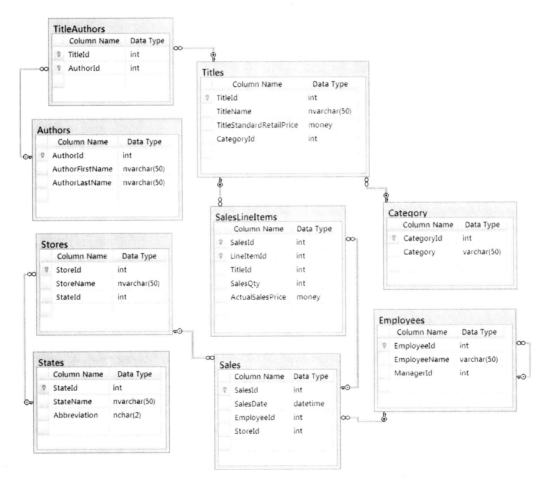

Figure 4-2. *A typical OLTP design*

We could move the state name right into the Stores table, but because both the state name and the state abbreviation are used, we have to move both columns into the Stores table. This means that both the abbreviation and the name redundantly show for each store. Normalization moves the data into their own table and uses an integer value to link the tables together. The integer values are repeated, but they represent less redundancy than repeating both the state name and abbreviation columns.

Table Relationships

Relational databases consist of a collection of related columns. Each set of columns forms a relation otherwise known as a *table*. In Figure 4-2, you can see a relationship line between Sales and Employees. Although the line only connects the tables together, it is not difficult to guess that the relationship is between the Sales.EmployeeId and Employees.EmployeeId columns.

Relationships between the tables are a vital part of any OLTP design, and when you are trying to understand a particular database, understanding these table relationships is also vital. Before we go any further, let's review the relationships between the tables in our example database in Figure 4-2.

Many-to-Many Tables

Let's take a look at the many-to-many relationships within our example. One is the relationship between sales and titles using three tables: Titles, SalesLineItems, and Sales. This relationship dictates that one title can be on many sales and one sale can have many titles. The SalesLineItems table contains both SalesId and TitleId, making this a bridge or junction table. These bridge tables are also called *associative entities*. Whatever you choose to call them, they provide the link between tables with a many-to-many relationship, like Sales and Titles.

Another example of a many-to-many relationship is the link documented by the TitleAuthors table. This table defines that one author can write many titles and one title can be written by many authors.

One-to-Many Tables

The Stores table has a one-to-many relationship with the Sales table. The relationship declares that one store can have many sales, but each sale is associated with only one store.

A similar one-to-many relationship is found between the Sales and Employees tables. This relationship describes that one sale is associated with a single employee, but one employee can be associated with many sales.

Parent–Child One-to-Many Tables

Now, let's take a look at an additional relationship defined in the Employees table. This relationship is bound to itself in the form of employees and managers. One manager can have many employees, but one employee has only one manager, at least as defined in this database.

A Managers table could have been created instead of defining the relationship in the table itself. In doing so, however, the relationship would have a single level between a manager and an employee. As it stands now, the relationship in the Employees table can take on many levels. By that we mean an employee could report to the manager, but that manager is also an employee who could report to another manager.

The relationship of employees to managers forms a jagged (sometimes called *ragged*) hierarchy. The chain between the parent and the child may consist of one level, two levels, or many levels. The fact that the number of levels is unknown is what exemplifies the pattern of parent–child relationships, not unlike how some human children may provide a parent with grandchildren, but others may not. Those grandchildren in turn may or may not have children of their own.

A Typical Data Warehouse Database Design

The design of the data warehouse is similar to the OLTP you just reviewed, but its focus is different. Instead of being concerned about normalization and the lack of redundancy, the focus is on report performance and simplicity. A data warehouse should provide your users with a simple, high-performance repository of report data. It should be easy to understand and consist of a minimal set of tables wherever possible.

To convert an OLTP design into an OLAP (data warehouse) design, start by identifying what reporting data is available. Using the bottom-up approach associated with Kimball's method, focus on a particular process, such as sales, and start building from there. It is important to try to make things very consistent so that additional processes can be added later, in what Kimball refers to as a *bus architecture*.

Measures

Measures are the aspects of your data warehouse that are reported on, such as if a client were to say, "I want a report on how many titles we have sold." After taking a look at the OLTP database in Figure 4-3, you know that the

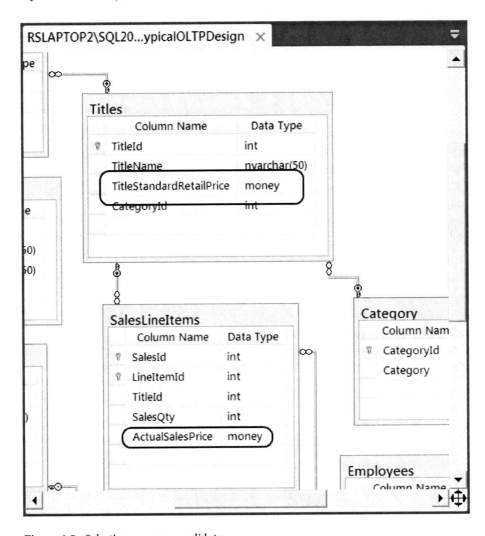

Figure 4-3. *Selecting measure candidates*

data related to the sales quantity will be required to answer this question. The *measure* provides information on a given process such as the process of selling something. The measure value, such as 15 items sold, for example, provides information about a specific event within that process.

Often when one question is asked, another question is suggested, such as, "How much was the standard retail price for that title?" Another question might be, "How much were the total sales for that title on that specific sale?" For the first question, the answer could be obtained by using the actual sales price, but for the second question, you would need to multiply the actual sales price by the sales quantity taken from specific dates to achieve a new value.

Looking at the Titles table, you see that there is a standard retail price (Figure 4-4). It would be tempting to assume that this was a measure as well. It could be used as a measure at some point in time by performing a calculation. But that might not be a mainstream calculation. Instead, it may be applicable to only a few reports.

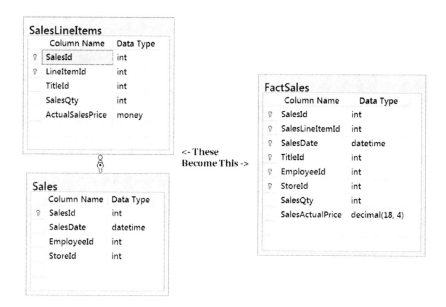

Figure 4-4. *A typical fact table*

To understand this, take a closer look at the StandardRetailPrice column. It defines information about the title and not about an actual sale. Therefore, although the ActualSalesPrice column describes something that occurred as part of a sales event, the StandardRetailPrice column describes an additional attribute of an individual title. In essence, this means that the standard retail price is more descriptive of a title, and not something that would be moved into the measures category directly. Any items that provide additional descriptions of a process are dimensional attributes and not measures.

Granularity

Measure values typically represent the lowest level of detail tracked by a given process. For example, you could store only a grand total for the sales of a given day by categories of products, but more likely you would instead record the individual sales event throughout the day and by the individual product itself and not solely by a product's category.

When you create a data warehouse, you can choose the level of detail you would like to report against. If you want to report only on the daily totals, you can do so by aggregating the values you stored in the OLTP data as they are turned into measures. Most of the time you will use the lowest level of detail you can obtain because it gives you the maximum amount of reporting options. Your chosen lowest level of detail defines the *granularity* of that measure.

The Fact Table

After you have defined your measures, you can begin designing your fact table. This fact table may look remarkably like the OLTP table where the measures were found, or it may combine data from multiple tables, which in turn denormalizes the data.

An example of this process would be collapsing the SalesLineItems table and the Sales table into one fact table. The SalesLineItems table defines a many-to-many relationship between sales and titles, but that relationship will still exist even if you collapse the SalesLineItems into the Sales table. When you do this, you end up with a single table with redundant values, as displayed in Figure 4-4. This denormalized design is typical in data warehouses. The SalesId will be repeated multiple times for each line item, but it is considered a small price to pay for simplification, and this pattern is representative of all fact tables.

Another common feature of the fact table is the simplification of datatypes. The actual sales price may have originally been recorded as a SQL Server money datatype, yet the fact table would represent this data as decimal (18, 4), as you see here (i.e., a total of 18 numbers with four of those numbers after the decimal point). The reason for this change is that SQL Server's money type is a custom datatype associated with Microsoft's SQL Server. Money is not an industry-standard datatype, but the decimal datatype is indeed an industry standard. The money datatype can be accurately represented by this decimal datatype; thus, it is logical to do so. Applications that use the data warehouse are more likely to work correctly using a standard datatype than a custom datatype.

In addition to the measures, the fact table contains IDs, or keys, that connect to dimension tables. These columns are referred to as *dimensional keys*. The dimensional keys may be listed at the front of the column list or placed at the end—it does not matter. We have chosen to represent them at the beginning of the column list, but it is simply a matter of preference.

Another preference is how you name a fact table. Some developers might choose the name Sales. Others would prefer the name SalesFacts. However, we chose the name FactSales for our fact table. Placing its designation as a fact table at the beginning of the table name will help organize our tables alphabetically. Our convention does not have much significance other than that it just makes it easier to group tables together in some applications. For instance, SQL Server's Management Studio will sort tables alphabetically in its Object Explorer window.

Dimensions

Once the fact table is defined, it is time to turn your attention to describing your measures with dimensions. At a minimum, each dimension table should contain a dimensional key column and a dimensional name column. The dimensional key column will typically be something such as ProductId or CustomerId. The name column provides a human-friendly description of that particular ID. Along with the name, this ID column may need additional descriptive data to allow for better organization and a clearer understanding, for example, the name of its category.

In Figure 4-5, we have created a Titles dimension table with four dimensional attributes (TitleId, TitleName, TitleStandardRetailPrice and TitleCategory). The TitleId is the dimensional key, but all four are dimensional attributes.

In Figure 4-5, note that the money datatype is now translated into a decimal for the same reasons we described with regard to the SalesActualPrice measure. We have also collapsed the data from two tables into one table by taking the category data from the Category table and moving into the TitleCategory column into the DimTitles table. In addition, the original Category table used a varchar datatype, but the new dimensional table uses an nvarchar, or Unicode variable character datatype. Unicode has now become the standard for most modern applications, and although it takes up twice the size in bytes to store this datatype, it is more consistent with the datatypes in the other columns within the database. Simplicity is often more important to the data warehouse design than absolute efficiency, and giving all columns consistent datatypes as well as consistent sizes is, well, simpler.

Figure 4-5. *A typical dimension table*

We realize that some may not agree with this decision, so let's examine this idea. It is true that performance might be enhanced using a smaller datatype, but if it can be determined that the Titles table will have only a few hundred rows, there will be little to no measurable decrease in performance. If, however, there are millions of titles, then of course you will want to change to the smaller datatype to increase performance.

Design your tables around the philosophy that simplicity is more important than pure efficiency and performance is more important than total simplicity. That is to say, if you can make it simple and not adversely affect performance, keep it simple! If, on the other hand, simplifying things decreases your performance to a noticeable degree, ignore your simplification efforts for this occurrence. Use common sense, and evaluate your needs on a case-by-case basis. It is possible to get too caught up in defending one style over another when often there is little impact either way.

Stars and Snowflakes

At one end of a lunch table you may hear the simplicity versus performance argument, and at the other the stars versus snowflake argument. Ignoring the fact that you need to find a more exciting place to eat lunch, let's examine the difference.

Star and snowflake designs reference a pattern form between dimension tables when compared to a fact table. A better name for them, however, would have been single-table dimensions and multitable dimensions; let us explain.

In Figure 4-6, you see a star design that forms a ring of dimension tables around a centralized fact table. In the star design there is only a single circle of dimension tables around the fact table of a data mart. Whether there are three, four or a hundred dimensions in the data mart, it makes no difference; as long as there is only a single table for each of these dimensions, this pattern still forms what is known as a star design.

Dimensions containing multiple tables per dimension are snowflake designs. Snowflakes form a circle of two or more tiers of dimensional tables around the fact table of a data mart. Whether it forms a circle of three, four or hundreds of tiers, it still is a snowflake design. Figure 4-6 outlines the pattern of a snowflake design compared to a star design.

If you are thinking, "Really? Is that all there is to it?" you are not alone. Randal has had many a student enroll in his classes specifically to learn the difference between a star and a snowflake design. How anticlimactic to discover the answer is so simple!

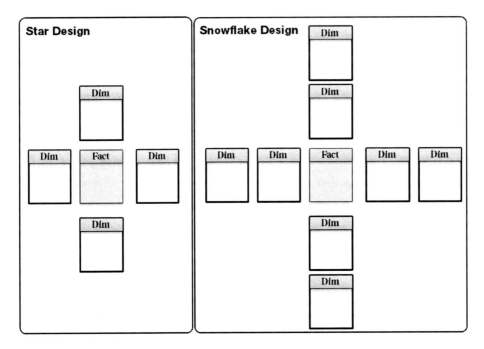

Figure 4-6. *Star and snowflake design patterns*

Keep in mind that when you are designing your tables, they may not necessarily display in a circular pattern as they do here, but it does not matter. The terms *snowflake* and *star* are simply descriptive words that portray how they are connected. Additionally, a single data warehouse can contain both design patterns.

Next, let's take a look at scenarios where one design pattern may be a better choice than the other.

Performance Considerations

In most cases, a set of tables can simply be collapsed into a single star dimension table. One argument in favor of the star design is that single tables are simpler to work with, even if they decrease storage efficiency by containing redundant values. The opposite argument is that by reducing redundancy, you increase storage efficiency; therefore, using separate tables by normalizing them is a better choice. Kimball and Inmon disagree on this subject as well, with Kimball siding with the single table star design and Inmon taking the snowflake stance.

Keep in mind that performance is relative to action, and the performance related to the action between the functionality of stars versus snowflakes is no exception. If you were to create a report based on star-designed tables, you would have fewer tables to connect to in your SQL joins. However, the redundancy of data in the tables means that there is more to transfer from the hard drive into memory before your query's results are assembled. The action of SQL Server finding and linking two tables decreases performance, but the hard drive I/O performance is lowered because of the reduction of redundancy data. From this single example, you can see that performance considerations are not as straightforward as they may appear at first.

In general, the simplified joins of a star design give the best performance for smaller tables (few columns and thousands of rows), whereas the snowflake design has better performance for larger tables (many columns and millions of rows).

Tip Remember that whether you choose to use a star or snowflake design, the data in your report is exactly the same. Even if you do not make the "best" choice, your design will still work.

In summary, it is more efficient to store data in a snowflake design, but it is more convenient to store it in a star design. Many data warehouses end up as a hybrid with some of the dimension tables designed in the star design pattern and other dimension tables designed in the snowflake design pattern. Nothing says you can't have both!

If you still can't decide which to use, follow this advice: when possible, use the star design for simplicity. If you come across a circumstance where you need to reduce redundancy, change the design to a snowflake. In the end both are simply tools that will enable you to get the job done, so choose the tool that is appropriate for the job.

Comparing Designs

Figure 4-7 compares the tables used in a star design and a snowflake design. If you design the Stores and States tables according to the star design philosophy, you would take the StateName and Abbreviation columns from the States table and collapse them into the Stores table (as shown in DimStores_StarVersion). Conversely, if you design the same two tables according to the snowflake philosophy, you leave it in two separate tables (as shown in the DimStores_SnowflakeVersion and the DimStates_SnowflakeVersion).

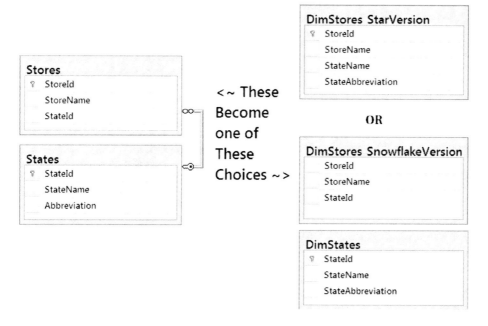

Figure 4-7. *Star versus snowflake designs*

Tip For learning purposes, we use a hybrid approach in our exercises, to provide experience with both. This means that some dimensions are designed as a snowflake and others as a star.

Notice that the Abbreviation from the States table has been changed to StateAbbreviation in the new examples. This is one example of how this change adds clarification and allows for a more workable and detailed table. Another change is that the StateId from the States table is no longer needed in the star design, as the need for that particular key is removed.

When you compare the original OLTP tables (Figure 4-8) to the OLAP tables in Figure 4-9, notice that they are similar but not exactly the same. The names are changed to reflect their usage, and some of the tables are combined into a single table. The FactSales table has a composite primary key on all the dimension key columns (with one exception, AuthorId). The FactTitleAuthors contains only two primary keys, AuthorId, and TitleId, forming a bridge table just as in the OLTP design.

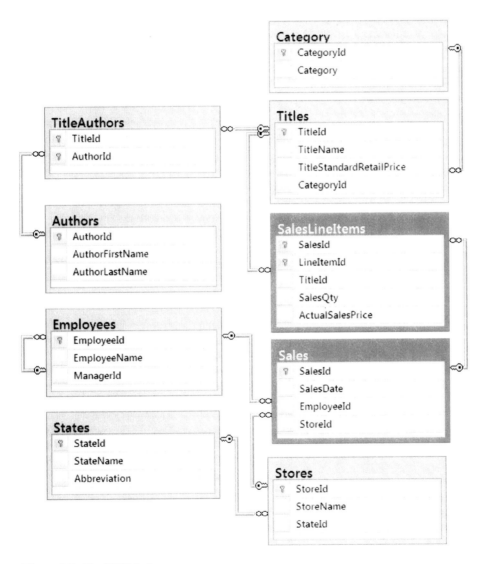

Figure 4-8. *The OLTP design*

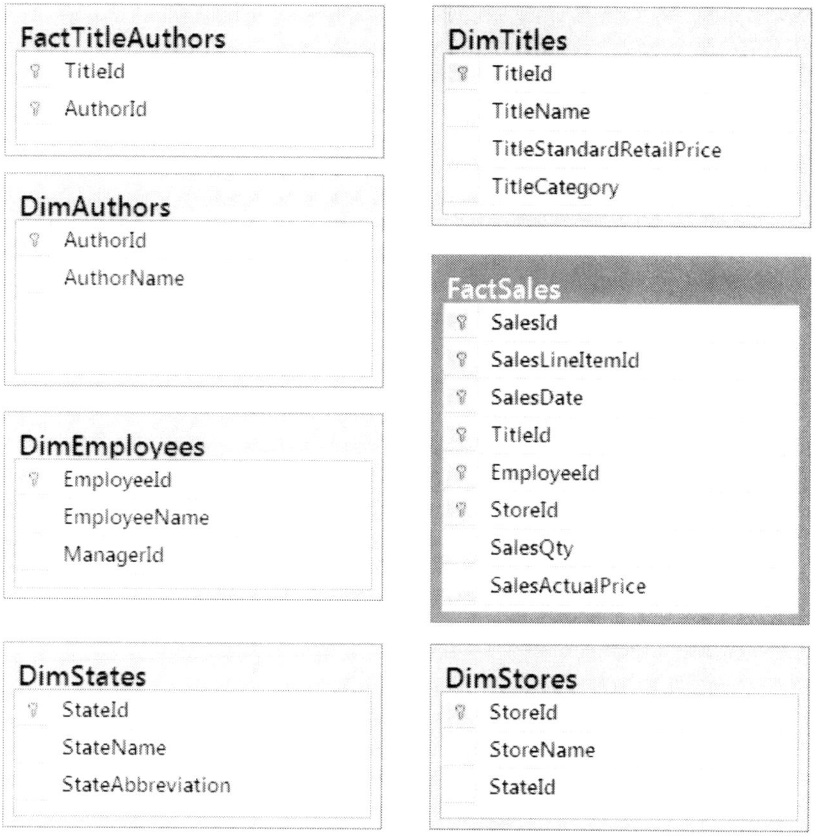

FactTitleAuthors
- ⚷ TitleId
- ⚷ AuthorId

DimAuthors
- ⚷ AuthorId
- AuthorName

DimEmployees
- ⚷ EmployeeId
- EmployeeName
- ManagerId

DimStates
- ⚷ StateId
- StateName
- StateAbbreviation

DimTitles
- ⚷ TitleId
- TitleName
- TitleStandardRetailPrice
- TitleCategory

FactSales
- ⚷ SalesId
- ⚷ SalesLineItemId
- ⚷ SalesDate
- ⚷ TitleId
- ⚷ EmployeeId
- ⚷ StoreId
- SalesQty
- SalesActualPrice

DimStores
- ⚷ StoreId
- StoreName
- StateId

Figure 4-9. *The OLAP design*

Foreign Keys

Foreign key relations exist between the tables, but the lack of lines connecting the columns indicate that we have not placed a foreign key constraint on the tables. This can cause confusion, so let's elaborate on this feature.

A foreign key *column* is a single column where the data represents values from another table. A foreign key *constraint* stops you from putting data in a foreign key column that does not exist in the original column you are referencing. Many developers who do not work with database development on a daily basis get the two terms confused. Most database administrators would shudder at the idea of not putting foreign key constraints in an OLTP database, yet it is not nearly as common to include them in an OLAP database.

The argument against putting them in the OLAP database is that the data has already been validated in the original OLTP database, so why validate it again? The argument for placing foreign key constraints in the OLAP database is that it is considered cheap insurance. It is a type of fail-safe. If someone imports data that is somehow incorrect, the foreign key constraints will catch it as an error. We prefer using foreign key constraints in both styles of databases, but we are not doing so for this example to provide additional contrast for learning purposes.

Missing Features

Foreign key constraints are a common feature of most databases. The example in Figure 4-9 is also missing some other common features. These features include several different types of dimensions, such as a time dimension and a parent–child dimension. Although it is common for data warehouses to have some, but not all possible features, it is good to understand these other design options so that you can include them when they are required. To help with that, let's look at common dimensional patterns.

Dimensional Patterns

Dimensional patterns are different ways of creating tables to hold your dimensional data. Using the right pattern for the right set of dimensional attributes is important. For example, if you incorrectly choose a standard dimensional pattern for dealing with a many-to-many design, your analysis server cubes will come up with incorrect data. The good news is that once you have seen the patterns, they are pretty easy to recognize.

Standard Dimensions

A standard dimension is a collection of one or more tables linked directly to the primary fact table (FactSales in Figure 4-9). The standard dimension is the one that you see most often, which is why it is called *standard*. Each standard dimension table should have a key column and a name column. In addition to those two columns, you can provide additional descriptive values that help further categorize the data.

In summary, in Figure 4-9, the fact table is the FactSales table. The DimTitles table is an example of a standard dimension, as are the snowflake tables, DimStores, and DimStates. These three tables represent two dimensions (Titles and Stores), and although one is in a star design and the other is in a snowflake design, they are both still considered standard dimensions.

Fact or Degenerate Dimensions

Fact dimensions (aka degenerate dimensions) have all their attributes stored in the fact table. They are most commonly referred to as *fact dimensions*, because *degenerate dimension* is not as descriptive and it sounds, well… degenerate." The names are synonymous, but for the purpose of this text, we refer to them as fact dimensions.

In a fact dimension with two dimensional attributes, both would be stored in the fact table. A classic example of this is the SalesId and the SalesLineItemId, as shown in Figure 4-10. These columns are not measures; they represent additional descriptions of the measures, and they do not link to any dimensional tables. This, by definition, makes them part of a fact dimension we call DimSales.

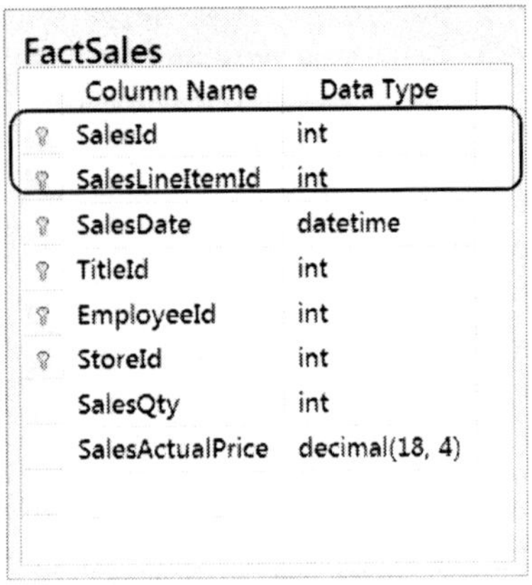

FactSales	
Column Name	Data Type
🔑 SalesId	int
🔑 SalesLineItemId	int
🔑 SalesDate	datetime
🔑 TitleId	int
🔑 EmployeeId	int
🔑 StoreId	int
SalesQty	int
SalesActualPrice	decimal(18, 4)

Figure 4-10. *Fact dimenisons*

We could create a DimSales table and put them both in it, but there is really not much point. Leaving the SalesId and SalesLineItemId in the fact table is more straightforward for reporting, and changing this would complicate the design. Besides, this new DimSales table will effectively have a one-to-one relationship between itself and the FactSales table.

Another fact dimension column is the SalesDate. From it, we can create a time dimension. You will almost always want a time dimension for every data warehouse. Evaluating how something changes over time is one of the most common types of reporting.

When you create time-based reports, chances are that you will want more than just a list of dates. It is likely you will want to categorize those dates into months, quarters, and years. Each of these items is an additional dimensional attribute that could be stored side by side with the date. In addition, you can extract all of these dimensional attributes from the one date by performing calculations on the data itself. Because all the dimensional attributes (either stored or calculated) are in the FactSales table, this means that, by definition, the time dimension is currently designed as a fact dimension.

Time Dimensions

Designing a time dimension as a fact dimension works but is not considered the best practice. Instead, you should create a date dimension table. A table called DimDates or DimTime is one of the most common features of every data warehouse.

At a minimum, a date table includes a date key and a date name, but it also includes other dimensional attributes such as the month, quarter, and year. These are basic date values and are easily calculated from individual dates in the fact table. But what if you want to include additional attributes such as holidays, corporate events, or fiscal weeks? These are not easy to calculate from a single date value. Creating a separate date dimension table that holds this information makes creating reports that include holidays or fiscal weeks, easy.

Consider this: Figure 4-11 shows a typical date dimension table linked to the fact table by a date ID. If you were to leave all of the additional attributes from the DimDates table in the fact table, it would dramatically increase the size of each row! Because the fact table commonly has thousands if not millions of rows, this has a big impact on the data warehouse.

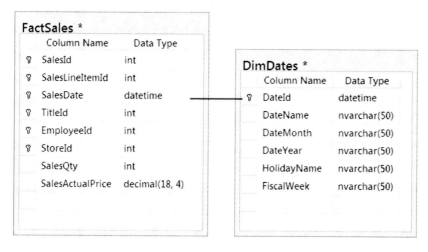

Figure 4-11. *Date or time dimension table*

Tracking Dates and Times

It should be noted that there have been debates about whether these tables should be called DimTime instead of DimDate. Arguments for calling your table DimDate and not DimTime usually revolve around the question, "What if I may want to have a separate table for tracking hours, minutes and seconds; wouldn't that table be called DimTimes?" It sounds like a good argument, but let us examine it further.

Currently, the lowest level of detail in the DimDates table of Figure 4-11 is an individual day. Therefore, we have 365 rows for each year of dates, or at least, for three out of four years we will. Ten years of dates can be stored in a table with less than 4,000 rows, which in database terms is not that large a table.

However, what would happen if we added in hours, minutes and seconds? The size of the table would swell to more than 31 million rows! This many rows are very likely to cause problems with performance. So, what are our options?

One option would be to create a separate time dimension table that held hours, minutes, and seconds columns. You would then link this table to the fact table just as you would the date dimension table. This can work, but there is a simpler option.

Using DateTime Keys

In this second option, leave the date and time data just as it was in the OLTP design and then link the DimDate table to the fact table based in this datetime column, instead of using an integer column (Figure 4-11).

You can derive hours, minutes, and seconds quite easily from a datetime column. Because in most cases hours, minutes, and seconds do not have additional descriptors associated with them, a simple datetime column is all you need. You can then include a datetime column in your DimDate table to provide connectivity between the dimension and fact tables (Figure 4-11).

You still have additional associations for holidays. Using this design, it is easy to store time dimension values down to the milliseconds without influencing the fact table size.

Having It All

There are times where having just a datetime column might not work for you. Consider requiring additional descriptors for hours, such as lunchtime or second shift. One way of handling these descriptors is to include them in the fact table along with the datetime column. You now have dates in a dimension table, times in the fact table, and an hour description in the fact table as well. This hour description column will have considerable redundancy, but it is better than having 31 million rows in your DimDate dimension table.

Of course, if you get too many descriptors, this still will not work, and you need to consider using a separate time dimension table, as shown in Figure 4-12.

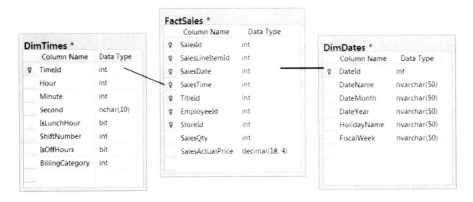

Figure 4-12. *Time and date dimension tables*

■ **Tip**   Once again, we recommend keeping your design simple! Most developers find that their reports do not need details about hours or minutes. If you do not regularly need this kind of information, just create a simple date dimension table and use a datetime key to connect it to the fact table. If a rare occasion comes up where you do need more than basic information, you can generate the required results with lookup tables and programming.

Using Foreign Key Constraints

In most dimensions, you want to use an integer value to connect a dimension table to a fact table. This is a practice recommended for almost all occasions, and Kimball defends this practice strongly. However, even Kimball agrees that for the time dimension, using a datetime key instead of an integer key is more practical for tracking periods of hours, minutes, and seconds. Still, there are issues you must consider.

In SQL Server, foreign key constraints compare values to determine whether the constraint is violated. If columns do not have matching values in both tables, an error occurs. Consider the values in Figure 4-13. The values are almost the same but vary by hours and minutes. Because of this, you cannot put a SQL foreign key constraint between these tables.

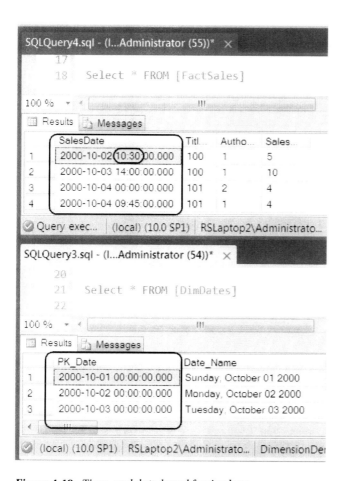

Figure 4-13. *Time- and date-based foreign keys*

You can, however, solve this problem programmatically by using validation code that will verify the values match at the level of days. In Figure 4-14 you can see an example of this.

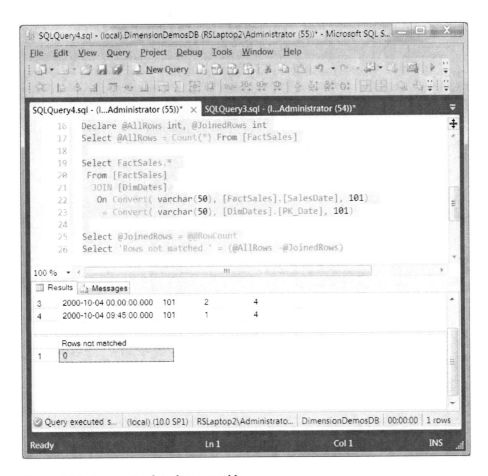

Figure 4-14. *Comparing dates between tables*

We discuss ETL programming further in Chapter 6, but for now let's move on to an example of how a date dimension table can be used in multiple ways.

Role-Playing Dimensions

Role-playing dimension tables are used repeatedly for slightly different purposes or, well, roles. So, the term *role-playing* comes from the fact that the same table plays many roles. The classic example of a role-playing dimension is a time dimension that links to a fact table multiple times. Figure 4-15 shows an example that can be found in Microsoft's AdventureWorks2008DW demonstration database.

This may seem to be too simplistic an explanation, but this is really all there is to it. It is possible to have role-playing dimensions that do not involve dates. An example might be geographic regions. But dates are by far the most common example.

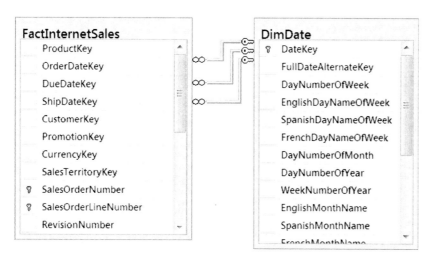

Figure 4-15. *A role-playing dimension*

Parent–Child Dimensions

Parent–child dimension tables look like their OLTP counterparts. The classic example of a parent–child dimension is an employee table where the relationship between employees and managers is mapped by associating a manager ID (parent) to an employee ID (child). To create a parent–child dimension table in the data warehouse, all you need to do is use the same design as you would in an OLTP environment, much as the one you see in Figure 4-16.

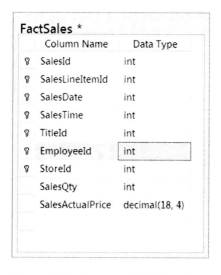

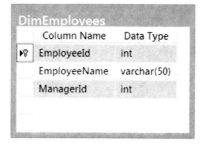

Figure 4-16. *A parent-child dimension*

Junk Dimensions

You may discover that you end up with a number of dimension tables that hold nothing more than the key column, name column and just a few rows. In a situation where there are a large number of these small tables, your data warehouse can become quite cluttered. One way to simplify things is to create a junk dimension. Although the name sounds a bit silly, a junk dimension can be quite useful.

With a junk dimension, you can bind a bunch of small dimensional tables into a single table. By taking the combination of possible values from these smaller tables and storing the combined values in a single table, you create a junk dimension table.

For example, if the values of one dimension table were "go to lunch" and "go home" and another dimension table had the values "yes" and "no," then the possible combinations would be as follows:

- Go to Lunch, Yes

- Go to Lunch, No

- Go Home, Yes

- Go Home, No

If you build a junk dimension table to hold these values, you will need an ID to link to the fact table and, in this case, two other columns to hold a combination of values. Figure 4-17 shows an example of taking two

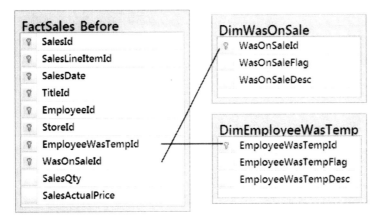

These two tables become one table

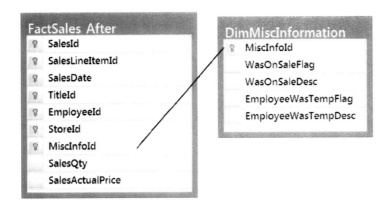

Figure 4-17. Converting to a junk dimension

dimensional tables and combining them into a single table. The table DimWasOnSale is combined with the table DimEmployeeWasTemp to form one junk dimension table, DimMiscInformation.

Another option could have been to move all columns in the fact table and create a fact dimension; however, that would add quite a lot of extra data to the fact table. With the current design, you end up with only a single integer linking to a particular combination of values.

Be careful not to go overboard on combining too many dimensions into a large junk dimension. We have seen this happen, and it does not work out well. It is very much like having a miscellaneous folder on your hard drive where you always have trouble finding what you are looking for because it is filled up with such a huge collection of junk.

Many-to-Many Dimensions

A many-to-many dimension is another common component in data warehouses. In these dimensions, you have a set of two tables connected by a bridge table. The bridge table defines the many-to-many relationship.

In Figure 4-18, our FactTitleAuthors bridge table links the DimAuthors and DimTitles together. In order to process the author's information, however, you have to go through the DimTitles table and its associative bridge table.

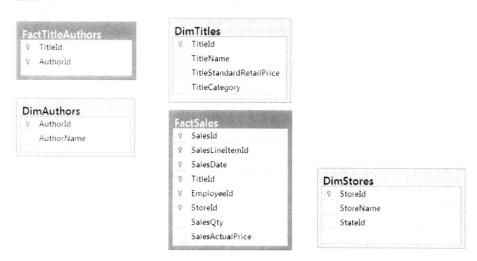

Figure 4-18. *A many-to-many dimension*

DimAuthors is not a standard dimension design because it does not link directly to the primary fact table, FactSales, and is instead linked to the bridge table, FactTitleAuthors.

Fact vs. Bridge Tables

Bridge tables are fact tables, just not the primary fact table of a data mart. A bridge table's purpose is to provide a connection between dimension tables, not store measures, and it is often referred to as a *factless fact table*.

On closer inspection, notice that all fact tables represent a many-to-many relationship between dimension tables; it is inherent in their design. Therefore, in our design (Figure 4-18), one title can sell in many stores, and one store can sell many titles. In this example, the FactSales table bridges DimTitles and DimStores and is the primary table of the data mart.

Still, this does not make both DimTitles and DimStores many-to-many dimensions; instead, they are just two regular dimensions connected to the primary fact table, FactSales. DimAuthors, on the other hand, does not connect directly to the primary fact table. Instead, it connects indirectly through DimTitles. This indirect connection is the hallmark of a many-to-many dimension.

Changing the Connection

It may seem confusing that the Authors dimension does not connect directly to the fact table with the measures, and you may be tempted to fix this by creating a direct connection. But if you try to do so, you will lose the correct many-to-many relationship between DimTitles and DimAuthors.

When you attach the DimAuthors table to the FactSales table, your reports will have issues. Initially this may seem like it works, but once you create an SSAS cube in the data warehouse, you will quickly spot the incorrect values in your cubes.

Wait a second! Didn't we say that all fact tables map a many-to-many relationship? What gives?

The problem is that there are two types of many-to-many relationships here, at least from the perspective of the primary fact table. We call them direct and indirect many-to-many relationships. Direct many-to-many relationships can be connected directly to the primary fact table, whereas indirect many-to-many relationships must be connected with a bridge table.

The distinction is based on measure granularity in the primary fact table, but to really understand what this means, we need to examine the two designs in more detail.

Direct Many-to-Many Relationships

In the direct design, one row in a dimension table is associated with another row in a different table, but it could be a different row for different events recorded in the primary fact table. A measured value in the fact table is associated with only one row of data from each of the dimension tables.

Let's look at an example. In Figure 4-19, we show the contents of two dimension tables, DimTitles and DimStores, and one fact table called FactSales. The relationship between DimTitles and DimStores is a many-to-many one, because sometimes a title is sold by one store, but on a different sales event, it sells in another store. This means that from the perspective of the sales event, there will be only one store ID for each title ID sold. After all, it makes sense that one part of the book will never be sold in one store and the other part of the book sold in another. So, although one title can sell in many stores and one store can sell many titles, there will never come a time where a single title has more than one store ID for a particular sales event. Therefore, from the position of a FactSales table, a row in the DimTitles table will map to one and only one row in the DimStores table for a given measure (such as the sales quantity), and each row in the fact table will have one title ID and one store ID (Figure 4-19).

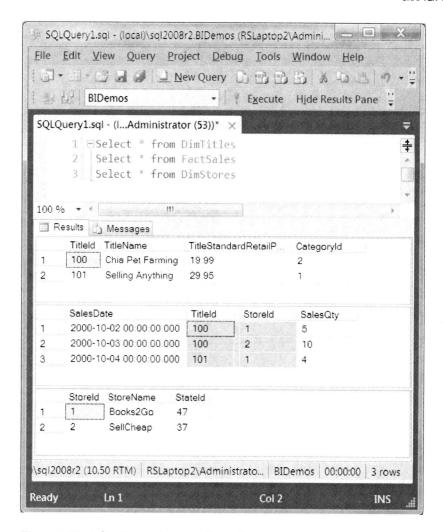

Figure 4-19. *A direct many-to-many dimension*

Indirect Many-to-Many Relationships

In an indirect many-to-many relationship, one row is associated with multiple rows of another table. A measured value in the primary fact table can be associated with one or more rows of data from each of the dimension tables. For example, one title may have only one author, whereas another has many authors. If a particular book has two authors, it will always have two authors regardless of the sales event in the fact table with which we are concerned. Therefore, from the position of a fact table's measures (such as sales quantity), sometimes a quantity will be associated with one author and other times with many authors!

If we try to track this in a fact table, we will need two rows for each title sold, as shown in Figure 4-20. Note, however, that this can lead to a problem known as *double counting*.

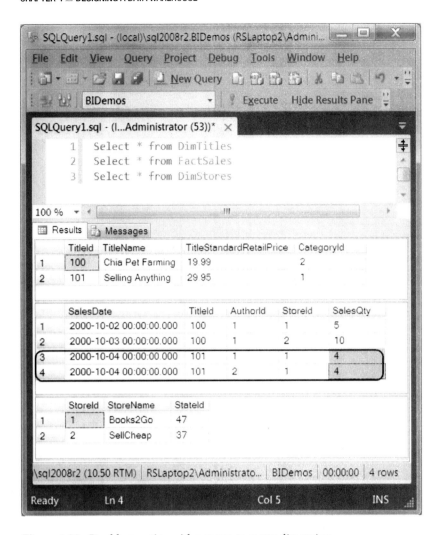

Figure 4-20. *Double counting with a many-to-many dimension*

On the surface, the fact that there are two rows for this one sales event does not seem too bad. When you make a report showing sales by authors, the data is even correct! There is a problem, however, if you make a report showing sales by titles alone. If you place the AuthorId in the primary fact table, you have to list the sales event multiple times, and title sales will aggregate once for each author. In software such as SSAS that performs aggregations of the sales quantity for a given day, title, and store for you, it will double-count the sale. Oops!

With a report like this, humans can figure out pretty quickly what the issue is. But computers? Well, computers are fast but not smart. You have to configure them around this issue because software, such as SSAS, for example, expects the many-to-many dimensions to use a bridge table.

When you move the AuthorId back out of the primary fact table and access it only indirectly through the bridge table, this problem is solved, and the sales event for that title will still be associated with all the authors who wrote that book.

The Takeaway

OK, no doubt we have made your head hurt with this one! The thing to remember is that many-to-many relationships connect to a fact table either directly or indirectly.

- In the *direct* many-to-many dimension, place both dimension keys from the two dimension tables in the central fact table.

- In the *indirect* many-to-many dimension, you need to have a separate, bridging fact table that is just between the two dimension tables.

- If you design them the wrong way, you will get incorrect values in your reports, so always check your values! If it's wrong, try the other way!

Conformed Dimensions

Conformed dimensions are not really dimensions in and of themselves; it is simply a phrase to describe dimensions that can be used from multiple data marts within a data warehouse. Basically, conformed dimension tables are utilized by many fact tables with each fact table based in a different subject.

For example, if we have a sales table and an inventory table, both of them can use the date dimension table. Since that is the case, the date dimension is a conformed dimension.

This is pretty straightforward, but it is important to be sure the granularity is appropriate. By that we mean, if the inventory is done monthly and the sales are tracked daily, you need a means of linking them based on their individual grain. A simple way of accomplishing this is by including both a DateID and a MonthID in the date table. You often see date dimension tables designed with two columns such as a MonthName column with a MonthID column as well. This can be further built upon by adding a YearID, aWeekID or whatever other increments that may be useful.

Other dimensions can be conformed as well. For example, consider the Titles dimension; in some databases, it is likely that inventory counts and sales will both be associated with titles and this, by definition, means that the titles dimension is also a conformed dimension.

If we continue with this example, the Authors dimension will not be a conformed dimension because it is unlikely that you will be taking inventory on authors.

Adding Surrogate Keys

It is a common practice to add artificial key columns. The idea is that instead of having values that naturally occur in the OLTP environment, you have artificial integer values that make up your data warehouse dimensional keys.

An example of this would be to add a new column to the DimAuthors table as you see in Figure 4-21. The new column, called AuthorKey, will connect to the fact table (not shown here) by replacing the AuthorId column with the new AuthorKey column. The artificial AuthorKey column is referred to as the *surrogate key*, whereas the original AuthorId column is known as the *natural key*.

Figure 4-21. Adding a surrogate key

It is considered best practice to do this on all tables with the possible exception of the time dimension table, as mentioned previously. The existence of surrogate keys helps when you are merging data into a data warehouse from many different OTLP databases as well as when you are tracking any changes to dimensional values, such as an author changing his or her name.

Slowly Changing Dimensions

Most of the time, dimensional values do not change over time; for example, it is unlikely that the names of the months will change any time soon. Some data does change, however, such as people names. We use the term *slowly changing dimension* (SCD) to describe such data.

When changes like these occur, you may want to track them for reporting purposes. For example, if the reports show sales connected with an old name, the author of those sales may not be given proper credit. It could be that whoever is reviewing the report knows only the new name of that author and has no way of knowing the previous name.

Slowly changing dimensions are certainly not a new concept. There have been several designs created over the years to help handle these types of situations. These designs are categorized into named types. The most common of these named types being used today are types I, II, and III.

Type I

The SCD type I, oddly enough, does not track changes at all. This is the normal state of a dimension table in which no additional columns or rows have been added for the purpose of tracking changes. If an author changes his or her name, simply record the new name and (using your best New York accent), forget about it. All reports now show the new name, and old ones will not be corrected. With type I you are not tracking historical changes, and perhaps that is preferred. The important thing here is that you made a conscious decision instead of letting it be the default behavior of your dimensions.

Type II

SCD type II is the complete opposite of type I. Type II tracks all changes rather than no changes at all. To do this, add additional columns and rows to the table specifically for the purpose of tracking the changes.

In Figure 4-22, the Authors table has been modified to include SCD columns. The modifications in this case included adding a column to indicate when a particular author's name was recorded using the start date column and when it ended using the end day column. In this way, every time an author changes his or her name, it can be recorded when the name change occurred and the length of time that it was applicable.

As shown in Figure 4-22, type II dimensions are distinguished by tracking the changes using additional rows. Each change adds a new row to the dimension table.

AuthorKey	AuthorId	AuthorName	StartDate	EndDate
1	1	Bob Smith	5/3/2000	NULL
2	2	Sue Jones	1/1/2000	1/1/2002
3	2	Sue Stevens	4/5/2002	6/2/2008
4	3	Tim Thomson	7/5/2002	NULL
5	2	Sue Jones	6/2/2008	NULL

Figure 4-22. *Adding type II slow changing dimension columns*

Type III

In the SCD type III, the process is simplified by tracking only the current and previous values. If an additional change occurs, the current value becomes a previous value, and the value before will be overwritten. Although not as popular as type II, it is simple to implement and may be appropriate on occasion. In Figure 4-23, the column AuthorPreviousName tracks the previous name, and the ChangeDate column tracks the date the change occurred.

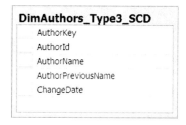

Figure 4-23. *A type III slow-changing dimension*

The distinguishing characteristic of a type III SCD is that changes are tracked by adding columns, not rows. So if you wanted to track the first, second, and third changes, you simply add columns for each version. You might even give each column a name such as AuthorNameVersion2, AuthorNameVersion3, and so on. The number of changes you track are determined by how many columns you add. Typically, though, it is most common in type III to track only one change, which would be the previously used version (AuthorPreviousName in this example).

Now that you have learned about the common dimensional design patterns, it is time to use your knowledge by performing the following exercise.

EXERCISE 4-1. REVIEWING THE DWPUBS DATA WAREHOUSE

In this exercise, you review a data warehouse design for the Pubs database using the design created by the authors.

Once again, there is not much to do in this exercise, but we did not see the point in having you fill in the spreadsheet line by line when reviewing and decided it would be just as enlightening and less tedious if we did it for you. However, in the "Learn by Doing" exercise of this chapter, you indeed get this type of practice.

Here are the steps to follow:

1. Open the `PubsBISolutionworksheet.xslx` file found in the downloadable book files.
2. Navigate to the data warehouse worksheet and compare the tables listed to the designs presented in this chapter.
3. Review the columns and datatypes defined in this worksheet comparing the original source datatypes to the destination datatypes. Figure 4-24 is an example of what this worksheet looks like, but the details will be more visible in the worksheet itself. (You create these tables in the next chapter, and details are provided there as well.)

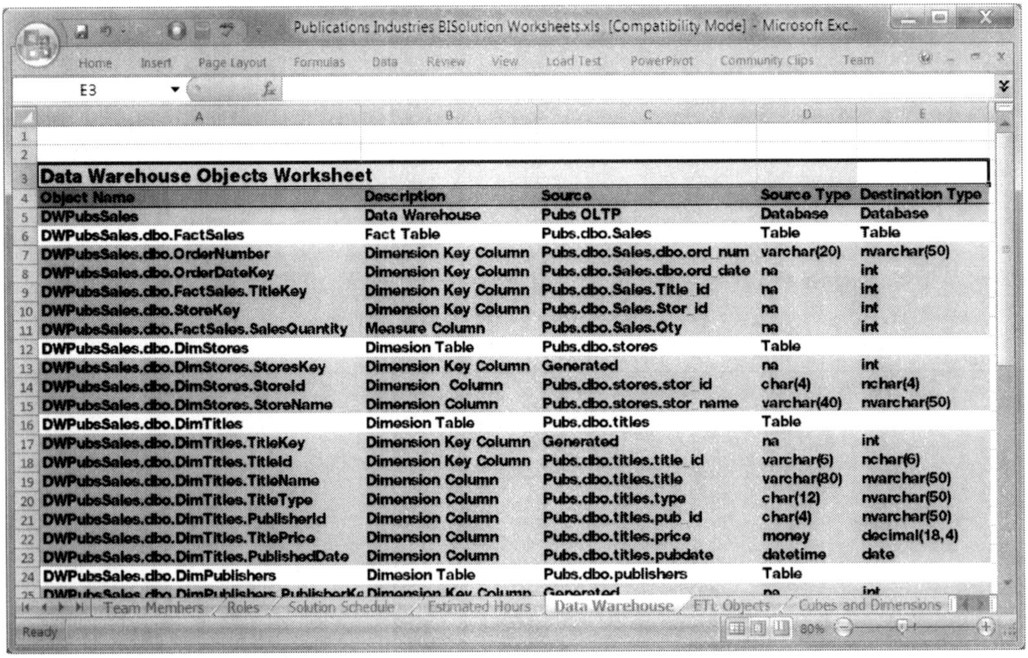

Data Warehouse Objects Worksheet

Object Name	Description	Source	Source Type	Destination Type
DWPubsSales	Data Warehouse	Pubs OLTP	Database	Database
DWPubsSales.dbo.FactSales	Fact Table	Pubs.dbo.Sales	Table	Table
DWPubsSales.dbo.OrderNumber	Dimension Key Column	Pubs.dbo.Sales.dbo.ord_num	varchar(20)	nvarchar(50)
DWPubsSales.dbo.OrderDateKey	Dimension Key Column	Pubs.dbo.Sales.dbo.ord_date	na	int
DWPubsSales.dbo.FactSales.TitleKey	Dimension Key Column	Pubs.dbo.Sales.Title_id	na	int
DWPubsSales.dbo.StoreKey	Dimension Key Column	Pubs.dbo.Sales.Stor_id	na	int
DWPubsSales.dbo.FactSales.SalesQuantity	Measure Column	Pubs.dbo.Sales.Qty	na	int
DWPubsSales.dbo.DimStores	Dimesion Table	Pubs.dbo.stores	Table	
DWPubsSales.dbo.DimStores.StoresKey	Dimension Key Column	Generated	na	int
DWPubsSales.dbo.DimStores.StoreId	Dimension Column	Pubs.dbo.stores.stor_id	char(4)	nchar(4)
DWPubsSales.dbo.DimStores.StoreName	Dimension Column	Pubs.dbo.stores.stor_name	varchar(40)	nvarchar(50)
DWPubsSales.dbo.DimTitles	Dimesion Table	Pubs.dbo.titles	Table	
DWPubsSales.dbo.DimTitles.TitleKey	Dimension Key Column	Generated	na	int
DWPubsSales.dbo.DimTitles.TitleId	Dimension Key Column	Pubs.dbo.titles.title_id	varchar(6)	nchar(6)
DWPubsSales.dbo.DimTitles.TitleName	Dimension Column	Pubs.dbo.titles.title	varchar(80)	nvarchar(50)
DWPubsSales.dbo.DimTitles.TitleType	Dimension Column	Pubs.dbo.titles.type	char(12)	nvarchar(50)
DWPubsSales.dbo.DimTitles.PublisherId	Dimension Column	Pubs.dbo.titles.pub_id	char(4)	nvarchar(50)
DWPubsSales.dbo.DimTitles.TitlePrice	Dimension Column	Pubs.dbo.titles.price	money	decimal(18,4)
DWPubsSales.dbo.DimTitles.PublishedDate	Dimension Column	Pubs.dbo.titles.pubdate	datetime	date
DWPubsSales.dbo.DimPublishers	Dimesion Table	Pubs.dbo.publishers	Table	

Figure 4-24. *The PubsBISolution data warehouse worksheet*

In this exercise, you reviewed the data warehouse design created by the authors. Now it's time to move on to creating the database using SQL Server 2012. As we do so, we revisit this spreadsheet in more detail (Figure 4-24).

Moving On

In this chapter, you learned techniques for designing a data warehouse. You also reviewed an example of an OLAP database created by the authors and compared it to its OLTP counterpart. Finally, you reviewed a design for a data warehouse built upon the Pubs OLTP database. Now it's time to implement that design by building a data warehouse in the next chapter.

LEARN BY DOING

In this "Learn by Doing" exercise, you perform the processes defined in this chapter using the Northwind database. We have included an outline of the steps you performed in this chapter and an example of how the authors handled them in two Word documents. These documents are found in the folder `C:\_BISolutionsBookFiles\_LearnByDoing\Chapter04Files`. Please see the `ReadMe.doc` file for detailed instructions.

What's Next?

We just gave you quite a bit of information on designing a data warehouse, but there is always more to tell. In this chapter, we focused only on core concepts that you can anticipate seeing on a regular basis. Because of this, you may be interested in researching more on the subject.

For more information, articles and videos on this subject can be found at `www.NorthwestTech.org/ProBISolutons`. Design tips posted by the Kimball Group can be found on its website at `www.kimballgroup.com/html/designtips.html`.

We also recommend the following books: *The Data Warehouse Toolkit* and *The Data Warehouse ETL Toolkit*, both by Ralph Kimball (Wiley).

CHAPTER 5

Creating a Data Warehouse

Plans are only good intentions unless they immediately degenerate into hard work.

—Peter F. Drucker

When all the planning is done, it is time to create the data warehouse. SQL Server makes this task quite simple and effortless.

In this chapter, we discuss how to use the SQL Server 2012 Management Studio (SSMS) application to create a data warehouse. We look at several techniques to accomplish this, including SQL code, a table designer, and a diagramming tool that lets you visually create all of the data warehouse objects. When you have completed this chapter, you will have a working data warehouse that is ready to be filled with data.

SQL Server Management Studio

SQL Server Management Studio is one of the most important applications of Microsoft's SQL Server 2012. It allows you to create and manage your databases as well as work with certain aspects of Microsoft's other BI servers (SSIS, SSAS, and SSRS).

Note   At the time of this writing, SQL Server Data Tools (SSDT) has been newly released for SQL 2012. SSDT adds many, but not all, of the features found in SQL Server Management Studio (SSMS) to the Visual Studio environment. It does not replace SSMS, and it does not replace the way databases are created. It does provide added convenience to Visual Studio, but because this book is about BI development rather than the new features of SQL 2012, we use SQL Server Management Studio for our examples. For more information about SSDT, visit the author's website at http://NorthwestTech.org/ProBISolutions/SSDTDemos.

To launch SQL Server Management Studio, navigate to Start ➤ All Programs ➤ Microsoft SQL Server 2012 menu item and click the title to expand the selection, as shown in Figure 5-1. Then, right-click the SSMS menu item and select Run as Administrator to open the SSMS application. It is important to run SQL Server BI applications as a Windows administrator to avoid permission issues.

Figure 5-1. *Opening SQL Server Management Studio with adminstrator rights*

Connecting to Servers

When SQL Server Management Studio first opens, you will be asked to connect to a SQL Server installation using the Connect to Server dialog window. You can connect to several different types of servers including SQL Server's database server, the Integration Services Server, the Analysis Cube Server and the Reporting Services Server. We work with all of these servers and make the connections to them in future chapters, but for now, we are focusing on the SQL Server database engine.

■ **Note**   When you first open SQL Server Management Studio, the Object Explorer window will not be populated with servers as shown in Figure 5-2 until after these connections have been made.

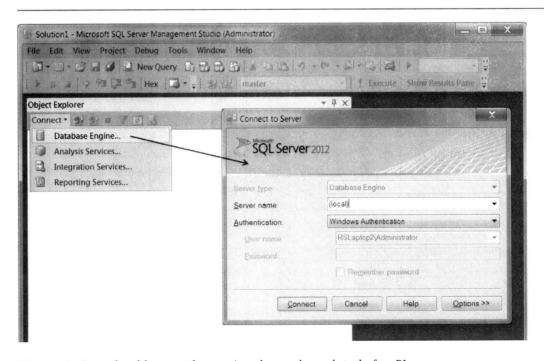

Figure 5-2. *Examples of the types of connections that can be made to the four BI servers*

To connect to a server, on the left of your SSMS window within the Object Explorer, click the Connect dropdown menu item and choose a server type. In Figure 5-2 we are selecting Database Engine. When the Connect to Server dialog window opens, enter the name of the server you want to connect to, as shown in Figure 5-2.

You can connect to more than one server at a time. In Figure 5-3, you can see that we have connected to all three of the BI servers from SQL Server Management Studio. When we did, Object Explorer indicated the type of server and the version number. Any server with a version number starting with 11 indicates SQL Server 2012.

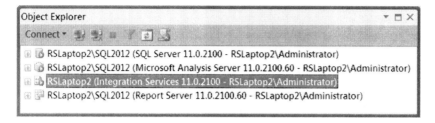

Figure 5-3. *Connecting to multiple servers at one time*

Note   The screen shots from Randal's computer are using a named instance called RSLaptop2\SQL2012. The exception of this is the Integration Services (SSIS) connection. SSIS does not allow for named instances. We talk more about named instances in just a moment.

Server Aliases

When connecting to one of the four BI servers on your machine, depending upon your configurations, you can use several different methods to connect; all of which mean the same thing. For instance, you can use localhost, (local), 127.0.0.1 or your computer name. You can even type a single period and SQL Server Management Studio understands to connect to your local server installation. For demonstration purposes, in Figure 5-4, we illustrate these four methods of connecting; each time it is to the SQL Server database engine.

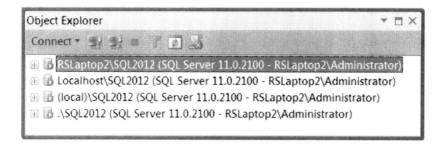

Figure 5-4. *Connecting to the database engine with various aliases*

It does not matter how you choose to connect, but it is important to be able to recognize each method. In this book, we use either *(local)* or *(local)\SQL2012* for most of our examples.

Use whichever works! Unfortunately, not all options work on all computers, so you will have to figure out which one is able to connect on your machine. For example; localhost allows Randal to connect to SSIS, but the alias (local) does not, even though they are supposedly equivalent. And Caryn connects in a different manner altogether, because her only installation of SQL Server is a named instance. We discuss how to handle named instances below.

To help you understand why some connection method's work and others do not, here is an overview of the differences:

- *Localhost*: Allows you to connect using the network protocol TCP/IP. As you may know, this is the protocol used throughout networks today. Localhost and the IP address 127.0.0.1 are equivalent to each other. They are both associated with TCP/IP, and each acts as an alias for your computer name. Using the name localhost to connect to your SQL Server works most of the time, but some configurations prevent this from working on all computers. Although there are ways to fix these issues, the easiest thing to do is try one of the other options.

- *(Local)*: Typing in the word **(local)** with parentheses will almost always connect when localhost does not. This name gives you access using an older protocol called NetBIOS. Microsoft originally used NetBIOS for networking and required NetBIOS on top of TCP/IP for many years. This requirement is no longer mandatory; however, legacy items remain, such as your ability to use (local) as an alias for your computer. Please note that (local) is the only computer alias that uses parentheses. This can confuse new developers into putting parentheses around aliases like *localhost*, leaving them unable to connect because *localhost*, with parentheses, has no significant meaning in NetBIOS or TCP/IP.

- *A period "."*: If, for some reason, you still cannot connect, another option is to type a single period as a server alias. This period symbol often works when no other selections will. It originates from an even older Microsoft protocol called *Named Pipes*. Named Pipes allows applications to talk to each other on a Windows machine and has been used extensively over the years.

- *Your computer name*: You can always just type in your computer name. Keep in mind that you can also connect to BI servers on a different computer by typing in its name. For example, let's say your company has a development server named DevServer; you could connect to it from your desk or laptop computer using its name, DevServer.

- *Computer Name or Alias\A_Named_Instance_on_Your_Server*: This allows you to connect to additional named instances installed on a single computer. An example of this would be (local)\SQLExpress or DevServer\SQLTestInstall. Keep in mind that you may only have named SQL Server instances installed on a computer. If this is the case, attempting to connect using the computer name is not enough. You must use the computer name or an alias such as (local) or localhost followed by the backslash "\" followed by your instance name—with no spaces in between.

WHAT IS UP WITH NAMED INSTANCES?

Named instances may be confusing, but they are very convenient. Here is a brief history of how and when they were added to SQL BI servers.

When SQL Server was first introduced by Microsoft, it could be installed only once per computer. In SQL 2000, this changed, and you could install the SQL database engine multiple times on the same machine.

To do this, each additional installation needed its own name. That is why it is called a *named instance*. Before SQL 2005, the first installation always installed without an additional identifier and was by definition the *default instance.* Only subsequent installations were given a unique name.

From 2005 on, the rules changed again, and you can now install SQL, SSAS, and SSRS with an additional identifier even on the first installation, without requiring a default instance. On your personal computer, if a default instance is not installed, you must qualify your SQL, SSAS, and SSRS servers with the full name; *Computer Name\named instance name.*

Another method of connecting is to use the Server Name dropdown menu. Select the < browse for more. . . > option, and then select your local machine or a network to connect to (Figure 5-5).

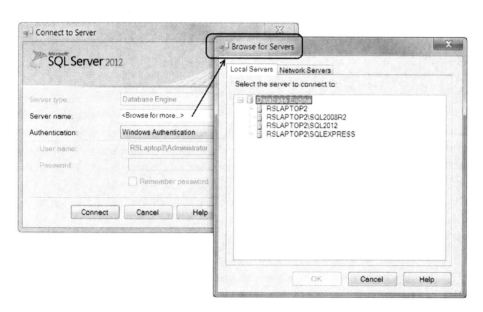

Figure 5-5. *Browsing for SQL Servers*

In Figure 5-5, you can see the local servers list on Randal's laptop. Selecting one of these creates a connection to the server as long as that database engine is running. Additionally, you can connect to remote SQL Servers as well, as long as it is running and remote connections to the other computer are allowed.

Configuration Manager

If you want to allow a remote connection to your server, you can enable this using the SQL Server Configuration Manager. You can also use this tool to make sure that an instance of SQL server is running on a given computer or to manage SQL Server, SSAS, SSIS, and SSRS start-up and service account settings.

Figure 5-6 lists SQL Server services and indicates that the named instance for (local)\SQLExpress is not currently running, but three other SQL servers are currently running. The SQLExpress named instance is set to start up manually, whereas the other three servers start automatically when the computer boots up. As it is now, you could not connect to the SQLExpress named instance from Management Studio with first starting the server by right-clicking its name and selecting Start from the context menu.

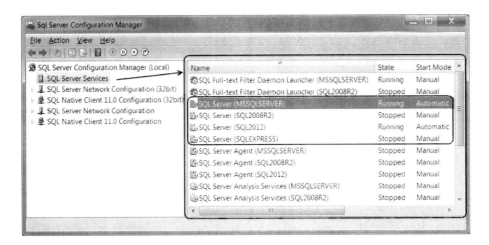

Figure 5-6. *Checking the SQL Servers with the Configuration Manager*

■ **Tip** In SQL Server Configuration Manager any server with the name (MSSQLSERVER) is the default instance and is accessed solely by the computer name or alias. No instance name is required. This is true for SQL Server, SSAS, and SSRS, but not for SSIS, which does not allow named instances.

To get to the SQL Server Configuration Manager, you need to navigate to the Start menu, select All Programs ➤ Microsoft SQL Server 2012 ➤ Configuration Tools, and right-click the SQL Server Configuration Manager menu option. From there, select the Run as Administrator option to launch the Configuration Manager.

When SQL Server Configuration Manager opens, you see a navigation tree on the left side of the screen. Selecting SQL Server Services shows you a window similar to the one in Figure 5-6. In this window, you can see which services are currently running and either stop or restart them by right-clicking the server name and selecting the Start or Restart option in the context menu. You can also access a number of server and start-up settings using the Properties option.

Clicking the SQL Server Network Configuration (32bit) node displays a window similar to Figure 5-7. Here you enable or disable the network protocols used for remote SQL connections. You will not be able to access the SQL BI servers remotely without TCP/IP enabled, and by default it is disabled.

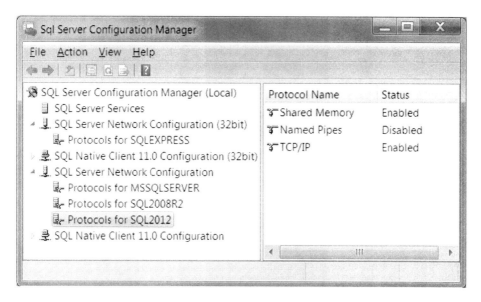

Figure 5-7. *Enabling TCP/IP*

To configure a network protocol, first expand the SQL Server Network Configuration node, and then click the "Protocols for [SQL Server instance name]" node for whichever instance you wish to configure.

If a protocol is disabled, you can right-click the protocol and then click Enable from the context menu to enable it. The status will change when the protocol is enabled, but you must restart the SQL Server service before applications can connect to the server. Additionally, Microsoft recommends restarting the SQL Server Browser service as well. Both of these services can be restarted under the SQL Server Services node (Figure 5-6).

Finally, you must use SQL Server Management Studio to allow remote access by checking the "Allow remote connections to this server" checkbox (Figure 5-8). This checkbox is found on the SQL Server Properties window under the Connection page.

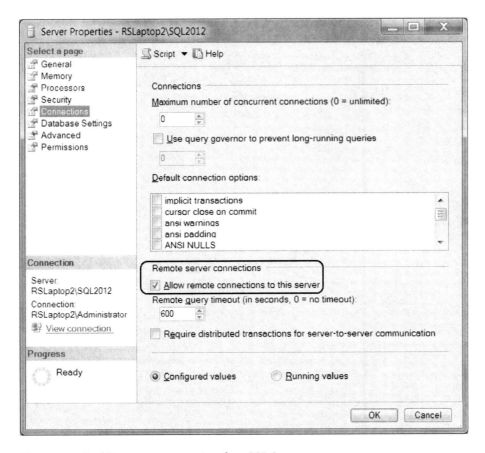

Figure 5-8. *Enabling remote connections for a SQL Server*

■ **Note** Our purpose is to allow you to become familiar with the network protocol settings, but there is no need to make changes at this time. No remote connections are necessary for the exercises in this book.

Management Studio Windows

SQL Server 2012 introduces a new SQL Server Management Studio user interface. The design may look different, but functionally it is very similar to previous versions, and it includes multiple windows. These windows can be repositioned and even moved outside Management Studio. This feature is great for use with multiple monitors. However, the majority of the time you will use only Object Explorer and a query window.

Object Explorer

Object Explorer is the first of the two main windows utilized in SQL Server Management Studio. In the default configuration, the Object Explorer window is displayed on the left side of the screen (Figure 5-9). Object Explorer allows you to view the various components of the database engine in a treeview format. By clicking the + next to each folder icon, you are able to expand the folder to access subcomponents.

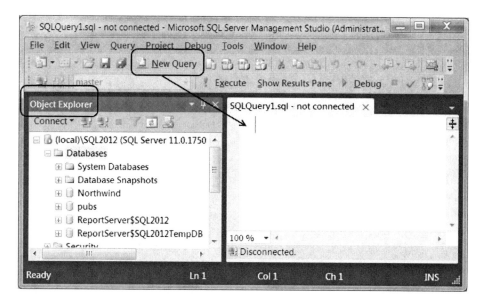

Figure 5-9. *The SQL Server Management Studio UI*

The Query Window

The second main window used on a regular basis is the query window. As indicated in Figure 5-9, there is a button on the toolbar called New Query which opens a new query window for you. This window is actually a set of windows, and just like modern web browsers, you can have many query windows open and each is accessed by its tab. Each tabbed query window has its own unique connection to a database engine.

By default each query window opens in the center of SQL Server Management Studio. Once opened, one or more SQL commands can be typed into a query window and then executed separately or collectively.

A query window must be connected in order to execute SQL code. In Figure 5-9, the disconnected status at the bottom of the screen is a clear indication that the query window is not yet connected. Disconnected queries are easily remedied by right-clicking anywhere on a query window, choosing the Connection option from the context menu, and selecting Connect.

It is important to note that the Object Explorer window represents a separate connection. You may be connected to a database engine in the Object Explorer, but you are not necessarily connected to a database engine in a query window. This can often be a source of confusion because developers expect one application to have only a single connection to a server at a time. Once you are able to get past the confusion, it is a very convenient feature nonetheless, since it allows you to work with multiple servers at once from a single application.

Changing the Query Window Focus

To focus on a particular database in Object Explorer, you need to expand the database folder and select a database from the expanded treeview. To focus on a particular database in the query window, use the *Available Databases* dropdown box (circled in Figure 5-10). This may be hard to spot because the name of the dropdown box is not shown, it can be repositioned on the toolbar, and the selected database may be different than expected. The Master database is typically displayed, but you can of course change this. In Figure 5-10, we have selected DWWeatherTracker instead, which causes the query window to be focused on that database.

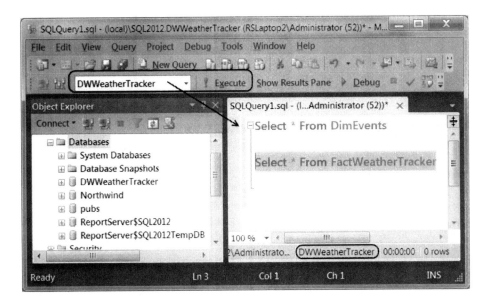

Figure 5-10. *Using the query window*

Although we are not showing it in Figure 5-10, you can also change the focus to a database by executing the command: USE DWWeatherTracker. Looking at the status panel at the bottom of the query window tells you which database is currently in focus.

Executing a Query

After you have added code to the query window, you are then able to execute that code by clicking the "! Execute" button, as shown in Figure 5-10. You can also execute a query using the keystrokes Alt + X or Ctrl + E.

■ **Important**   We do *not* recommended using the Debug button (with the green arrow) to execute your code. A debugging session can start other processes and complicate what you are working on. If you are a .NET programmer, you know that this is contrary to how you normally run code in Visual Studio in C# or VB.NET applications. Nevertheless, it is important to remember that while working in SQL Server Management Studio, execute your code with the "! Execute" button rather than debugging it.

In many database applications, a query window may hold only a single statement or batch of statements at a time, but this is not true in SQL Server Management Studio. Instead, you can type in hundreds of lines of SQL code and execute them independently, as a batch, or as several batches by selecting whichever statements you want to run.

If no statements are highlighted, all the statements in the query window run sequentially. In Figure 5-10 you see two SQL statements have been typed into the query window, but only one statement has been selected. In this example, only the highlighted statement will be executed when "! Execute" is clicked.

■ **Tip**   The term *SQL batch* is used to describe one or more SQL statements that are submitted to the database engine as a unit. Some statements, such as the CREATE PROCEDURE statement, must be the first statement in a batch, but most statements can be submitted in any order. If you want to create multiple procedures in one query window, you can use the batch separator keyword GO to divide the SQL statements into multiple batches. You may also use the GO keyword between any of the individual statements, but this is more of a stylistic choice than a programmatic one and is not necessary for your statements to execute correctly. Examples of this are shown throughout the code samples of this book.

SQL code can be submitted to the database engine from other applications as well as SQL Server Management Studio. Simple examples of this are executing code from SSIS and SSRS. We investigate both of these in later chapters of this book.

Creating Data Warehouse Database

With SQL Server Management Studio, you can easily create a database by right-clicking the database icon in Object Explorer, as shown in Figure 5-11. This launches the New Database dialog window. When the dialog window opens, there is a selection of pages on the left of the dialog window, and on the right are associated text boxes and grids. To create a database, select the general page and provide a name for the database such as DWPubSales; then indicate the owner of the database as the system administrator account, SA.

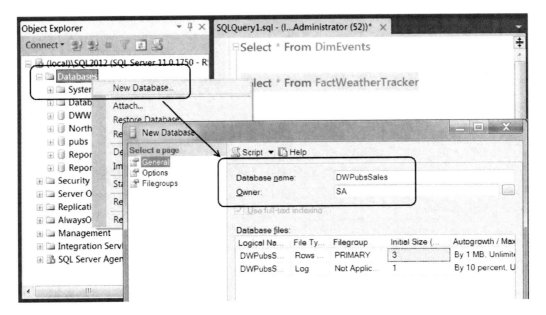

Figure 5-11. *Creating the data warehouse database*

■ **Note** Next, we discuss a number of advanced options in the database creation process, but you do not need to use these options for the exercises of this book.

Setting the Database Owner

Most of the time, the database owner setting has little impact on working with the database. An example of when it is necessary, however, is when using SQL Server's diagram features. If there is no owner assigned to the database or the one that was assigned is no longer valid, the diagramming tools will not work.

The database owner can map to either a Windows login account or a SQL Server login account. Both are similar, but a SQL Server login's name and password are passed onto the network as plain text. And although this network traffic can be encrypted, by default it is not. Because of this, many SQL administrators have stopped using SQL logins for most occasions and just use Windows logins instead. Thus, the standard choice for a database owner is an existing Windows login.

There are times, however, where using a SQL login for the database owner makes sense. For example, if you created a database using your personal login account as the owner of the database and then backed it up to send to a co-worker's computer, the database owner may not be valid on your co-worker's computer. When your co-worker restores the database on his or her own machine, if your personal login account does not exist on his or her machine, the database ownership is broken.

At first, this would probably go unnoticed, but if your co-worker tried to create a database diagram, which requires the use of valid owner, your co-worker will get the error message shown in Figure 5-12.

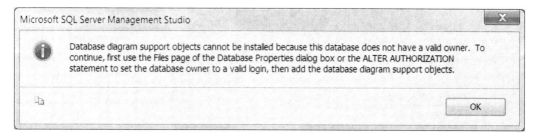

Figure 5-12. *This database is missing a valid owner*

Perhaps the easiest way to resolve this issue is to assign a valid owner using a SQL login that you know will be on all SQL Server installations. (The SA login was used in Figure 5-7.) The SA login is built into Microsoft SQL Server as a system administrator and works very nicely for this occasion. Although it is disabled by default, it still exists on each and every installation of SQL Server and by its mere existence can be used as a valid owner. When you create your database using the new database dialog window, the SA account is a simple and effective choice that can safely be used for nonproduction computers. On a production machine, you normally use a Windows domain account.

■ **Note**   The authors, and the legal department, are really hoping that you are not using a production server to perform the exercises in this book, so the SA login should work just fine throughout the text! Please let the lawyers sleep at night and use only nonproduction servers. They get grouchy when they don't get their sleep!

Setting the Database Size

During this creation process, many decisions must be made; such as where the files for the database should go or what their initial size will be. In the examples from this book, you can safely leave these at the default settings because the samples are very small. In real life, however, you will probably want to put your data warehouse on its own separate hard drive and try to match the initial size as close as possible to its expected size.

Every database you create on SQL Server will have at least two files with which to be concerned. They are the data file and the log file. The data file holds all the data from your tables, and the log file records any changes to the data file. All changes are recorded to the log file before they are changed in the data file. A synchronization process, known as a *checkpoint,* occurs on a regular basis to record all the changes from the log file into the data file. This is performed automatically.

When estimating the size of the data warehouse, the two files should be adjusted according to their function. The data file needs to be large enough to hold all your data, but the log file can be quite small. In an OLTP database, the transaction log is expected to be receiving continual transaction entries, but the log file has much less work to do in an OLAP data warehouse. The log file records data entries while importing data from the OLTP source to the OLAP destination, but reporting statements are not logged. Microsoft recommends an OLTP database log file be set to 25% of the size of its associated data file. In an OLAP data warehouse, you can usually get by with a lot less: approximately 10% of the data file. So, if you made your data file 100 MB, the log file will then be 10 MB.

Listing 5-1 shows the code for creating a typical OLAP database. We have placed the files on a different hard drive to maximize performance and started with the D drive, so the Windows operating system is on a separate hard drive as well.

Listing 5-1. Creating a New Database

```
/* (Note the following code is expected to error out and is only a demo) */
USE [master]GO
CREATE DATABASE [DWWeatherTracker] ON PRIMARY
( NAME = N'DWWeatherTracker'
, FILENAME = N'D:\_BISolutions\DWWeatherTracker.mdf' -- On the D:\ hard drive
, SIZE = 10MB
, MAXSIZE = 1GB
, FILEGROWTH = 10MB )
LOG ON
( NAME = N'DWWeatherTracker_log'
, FILENAME = N'F:\_BISolutions\DWWeatherTracker_log.LDF' -- On the F:\ hard drive
, SIZE = 1MB
, MAXSIZE = 1GB
, FILEGROWTH = 10MB)
GO
EXEC [DWWeatherTracker].dbo.sp_changedbowner @loginame = N'SA', @map = false
GO
ALTER DATABASE [DWWeatherTracker] SET RECOVERY BULK_LOGGED
GO
```

Estimating the exact size of the data warehouse files can be tricky. You do not have to be exact, however, just close. SQL Server databases have the capacity to grow in size automatically as more data is added, but it is considered best practice not to rely on this feature. An easy way to estimate the data warehouse size is by looking at the source databases or source files to identify what is going to be imported and how much space it currently takes up.

If your source is a SQL Server OLTP database, the process is greatly simplified by a number of reports included in SQL Server Management Studio. To access these reports, right-click an object in the treeview of Object Explorer and look for the Reports menu option. From there you see a list of standard reports, a selection for custom reports and a list of currently used reports. Upon selecting a particular report, such as the one shown in Figure 5-13, a report is generated for you. You can estimate the size of your data warehouse tables by

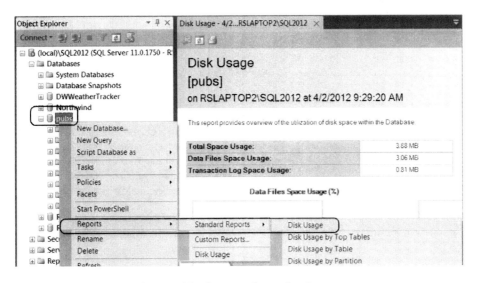

Figure 5-13. *Estimating the size of the data warehouse database*

determining which OLTP tables you intend to use as the source for your dimension and fact tables and then noting their current size.

■ **Note**   SQL Server Management Studio reports are rendered using a miniature version of the Report Server engine and do not require that SSRS be installed or configured on your machine. For information on setting up custom reports, search the Web for "Customer Reports Management Studio."

This is a simple effective way to estimate the size of your data warehouse, but it will not work for all occasions. For instance, you may have to pull data from sources other than SQL Server, such as a set of flat files or an Oracle server. In those cases, you have to estimate the database size the old-fashioned way: by estimating the size of each row in a table and multiplying it by the number of rows in that table. You can find more information on estimating the size of a database on Microsoft's website at http://msdn.microsoft.com/en-us/library/ms187445.aspx.

Setting the Recovery Model

After you have selected the name, owner, and size of your data warehouse, you may want to move to the options page and change the recovery model. The recovery model controls the behavior of a database transaction log file. Setting this correctly can increase your data warehouse loading performance.

The three settings are Full, Simple, and Bulk-Logged. Each functions slightly differently. In both full and simple modes, the log file will record every row that you add to the table. The difference between these two modes is that in simple mode, the transaction log file is periodically cleared automatically. When the database is set to full mode, the log file continues to fill up until you back up the log file. In an OLTP database, it is best practice to put the database in full mode and do regular backups on the transaction log. This makes sense, given that the purpose of an OLTP database is to process transactions. Therefore, making sure that you have a backup of these transactions is important. In an OLAP database, however, the focus is not on transactions but on analysis of data. Because of this, using full mode is not necessary because you do not need to make backups of the individual transactions.

A SQL Server transaction will occur whenever you insert, update or delete data from a table. Consequently, you will still end up with transactional events each time you import to the data warehouse. In both simple and full mode, if you import a million rows, the log records a million insert transactions. Therefore, regardless of whether you use the simple or the full, the log file records each and every insert.

Using the full mode increases the size of the log files substantially because it is not automatically cleared for you but instead relies on regular backups. Using the simple mode, which has automatic truncations, will keep the log file size small, although it still has the extra overhead of tracking each and every insert. This brings us to the third option: bulk-logged.

The bulk-logged option may be a more logical choice for an OLAP database. Figure 5-14 shows this selection. With this setting, large imports of data are recorded with the minimal amount of transaction log entries, which provides additional performance during data warehouse loading and minimal impact on the size of the transaction log. You will, however, still need to perform regular backups on the log file to clear it out. Also, there will still be times when bulk-logged is not the best choice, such as when you have slow-changing columns that you could lose data from, as we discuss in a moment.

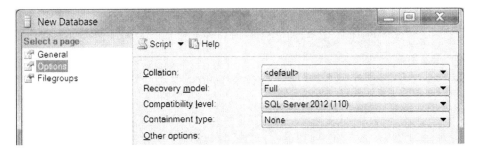

Figure 5-14. *Setting the recovery model*

Performing Database Backups

Backing up the transaction log is quite simple, and you can do it with a single line of SQL code. You cannot back up the log, however, unless you have already backed up the whole database at least once. You must do this even if you are not going to use the database backup, but you can reuse the same backup file multiple times by clearing its contents using the INIT option. Listing 5-2 shows a full database backup, followed by a transaction log backup to the same file.

Listing 5-2. Backing Up the Log File

```
BACKUP DATABASE [DWWeatherTracker] TO DISK = N'C:\_BISolutions\DWWeatherTracker.bak'
GO
BACKUP LOG [DWWeatherTracker] TO DISK = N'C:\_BISolutions\DWWeatherTracker.bak' WITH INIT
```

When you back up the transaction log, it clears out entries in the log file, freeing up space for more log entries. Performing regular backups of the transaction log keeps the transaction log file from growing on your hard drive. But, remember that the backup file(.bak) increases in size unless you clear this file each time you perform a backup. Initializing the backup file clears it, allowing you to maintain a smaller backup file. The WITH INIT command triggers initialization (Listing 5-2).

Shrinking Log Files

Occasionally, when you do a very large import, the transaction log file may temporarily need to expand. SQL Server does this for you automatically, but it does not automatically shrink the file back to its original sizes afterwards. You manually shrink the file by first backing up the transaction log to clear out its contents, and then using the shrink file option to reduce its size. The code in Listing 5-3 does just that.

Listing 5-3. Shrinking the Log File

```
USE [DWWeatherTracker]
GO
BACKUP LOG [DWWeatherTracker] TO DISK = N'C:\_BISolutions\DWWeatherTracker.bak' WITH INIT
GO
DBCC SHRINKFILE (N'DWWeatherTracker_log' , 0, TRUNCATEONLY)
GO
```

■ **Note**   In previous versions of SQL, Microsoft included a TRUNCATE_ONLY option for clearing the log file entries without having to back them up. But, this caused problems that could introduce corruption into your database if not handled properly. Therefore, since SQL 2008, Microsoft removed the ability to clear log file entries with the

TRUNCATE_ONLY command, requiring log file entries to be backed up before truncation. Oddly, Management Studio offers a TRUNCATEONLY option for the DBCC SHRINKFILE command, but it is only applicable to data files. We are able to use it in Listing 5-3 because it does not cause corruption even when shrinking a transaction log, and it does not require data files to be backed up before truncation.

Keeping Data Warehouse Backups

As any database administrator (DBA) will attest, database backups are vital to every business. Still, you may not need to keep a series of database backups for your data warehouse to the same degree as expected of an OLTP database. In an OLTP database, a daily cycle of full and transactional backups is recommended. Data warehouse databases, however, may be updated only once a night or once a week and do not require repeated backups during the middle of the day or even daily. In point of fact, it is sometimes argued that because you can rebuild the entire data warehouse from the OLTP database—which is continually backed up—you do not need a backup of the data warehouse. If the data warehouse ever crashes, you can always rebuild it and reload.

This approach makes the DBAs very nervous, sometimes for good reason. Many OLTP databases are not designed to track changing dimensional values over time. When an update or deletion occurs, previous data in a column and table is lost. On the other hand, in a data warehouse this data is often tracked in some, if not all, dimensions. If your data warehouse tracks changes to dimensional values and the OLTP database does not, you need to retain your backups.

This is an important point, so let's take a look at an example. Suppose you have a table that is tracking changes to product prices where the standard retail price of a product can change as time goes on. In the data warehouse, you could add dimension columns to track these changes. (Tracking columns are referred to as *slow-changing* dimension columns, as discussed in Chapter 4.)

In the example in Figure 5-15, you see two different table designs: an OLTP table called Products without any slow-changing dimension columns and an OLAP table called DimProducts. The DimProducts table includes four slow-changing dimension columns (RecordStartDate, RecordEndDate, IsCurrent and Productkey).

Figure 5-15. *Tables with and without slow-changing dimension columns*

If the standard retail price is overridden in the Products table, the original price will not have been lost, but rather it will be recorded in the DimProducts table during the ETL process. In a perfect world, the slow-changing dimension columns would have been included in the Products table and tracked in the OLTP environment. There are times, however, where you will have no control over the design of the OLTP environment but may have control over the OLAP environment. And if that is the case, you will be left implementing slow-changing dimension tracking similar to this example (Figure 5-15).

Were the data warehouse database to crash, you would not be able to reload it from the current OLTP database, but rather, it must be reloaded from the backups. In this scenario, you need to maintain a set of data warehouse backups.

Using the Filegroups Option

The Filegroups page of the New Database window allows you to create and configure filegroups for your data warehouse (Figure 5-16). Using filegroups can increase performance, but doing so adds complexity. With filegroups, you can control the placement of your fact and dimension tables in a specific set of one or more files. These files, in turn, are located on separate hard drives allowing the database engine to read and write from several hard drives at once. With this design, not only will you benefit from ETL performance but also from data retrieval performance.

Figure 5-16. *Creating filegroups*

Every database contains a filegroup called *primary*, and a master data file (.MDF) within it. All the internal system tables are mapped to the primary filegroup and are therefore placed in the .MDF file.

To create an additional filegroup, just click the Add button for each filegroup you want to create. As an example, you might create one filegroup to store your fact tables, another filegroup to store one set of dimension tables, and yet a third to store other dimension tables.

At first filegroups have no files associated with them. To use the filegroup, you must go back to the General page (Figure 5-16), click the Add button to create a new file, type in a logical file name, select the file group you want to use, define a file path, and type in a physical file name (Figure 5-17). Other options are available as well, but these are the common ones.

■ **Note**   The dropdown box showing the filegroups (Figure 5-17) "automagically" appears only when you click the file group cell; it can be confusing because it is set to Primary by default with no dropdown option in sight!

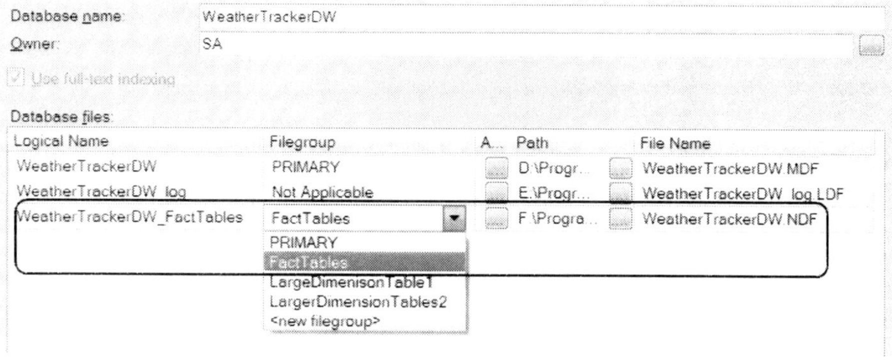

Figure 5-17. Adding files to a filegroup on the General page

When creating the new file, be sure to change the Path property of that file to point to a particular hard drive, as shown in Figure 5-17. The whole point of using filegroups is to be able to dictate which hard drive to use to store a particular table's data. If you go to the trouble of creating filegroups and files without distributing the files on multiple hard drives, you have just added administrative overhead with no benefit.

In addition, you should name your file with an `.NDF` extension rather than an `.MDF` or `.LDF` extension. All files beyond the original `.MDF` file are given the extension of `.NDF`. The explanation of the file names are as follows: the `.MDF` file is your main data file, the `.LDF` is your log file, and any `.NDF` files are every other additional data file. If you add six additional files, they will all have `.NDF` as their extension.

■ **Note** These extensions are not case sensitive; they are capitalized here for clarity.

Once you have the filegroups and files created, you can add a table in a particular filegroup by using the syntax shown in Listing 5-4. Although you cannot determine which file a table goes into, you can determine which file group a table belongs to. This has the advantage of allowing you to have multiple files in a single file group. If each file in the same filegroup is on a different hard drive, your table will span multiple hard drives.

Listing 5-4. Creating Tables in a Particular Filegroup

```
/* (Note the following code is expected to error out and is only a demo) */

CREATE TABLE [dbo].[FactWeather]
([Date] [datetime] NOT NULL,
 [EventKey] [int] NOT NULL,
 [MaxTempF] [int] NOT NULL,
 [MinTempF] [int] NOT NULL,
 CONSTRAINT [PK_FactWeathers] PRIMARY KEY CLUSTERED ( [Date] ASC,[EventKey] ASC )
) ON [FactTables] -- Name of the File Group not the file!
```

■ **Note** We cover filegroups in this chapter because the Filegroups page is part of the New Database creation window, but you most likely will not need to use them unless you are working with very large data warehouses. Be aware that filegroups can give increased performance when reading and writing to multiple hard drives but do not provide any fault tolerance to hard drive failure. If one drive fails, you have to replace the failed drive and restore your most recent backup or at least refill the Data Warehouse with all of the OLTP data. Although this does not represent a total tragedy, you may still want to avoid this by using RAID technology, which provides similar performance, simpler administration, and automatic recovery from a hard drive failure. It is also easier from a BI designer perspective, because the network team will set it up for you most of the time!

EXERCISE 5-1. CREATING THE PUBLICATION INDUSTRIES DATABASE

In this exercise, you create a database for the Publication Industries data warehouse.

Important: You practice administrator-level tasks in this book, so you need administrator-level privileges. The easiest way to achieve this is to right-click the menu item, select Run as Administrator, and then answer Yes to access administrator-level privileges while running this program. In Windows 7 and Vista, just logging in with an administrator account is not enough. For more information, search the Web for "Windows 7 True Administrator and User Access Control."

1. Open SQL Server Management Studio 2012. (You can do so by clicking the Start button and navigating to All Programs ➤ Microsoft SQL Server 2012 ➤ SQL Server Management Studio. Right-click SQL Server Management Studio 2012 and click the Run as Administrator menu item. If the UAC message box appears asking, "Do you want the following program to make changes to this computer?" click Yes [or Continue depending upon your operating system] to accept this request.)

2. When Management Studio opens, choose to connect to the database engine by selecting this option in the Server Type dropdown box. Then click the Connect button to connect to the database engine (Figure 5-2).

3. Use the Query Window to create a database using the code in Listing 5-5.

Tip: The files for this exercise, as well as all of the exercises throughout this book, are available in the downloadable book content.

Listing 5-5. Creating the DWPubsSales Data Warehouse Database

```
USE [master]
GO
IF EXISTS (SELECT name FROM sys.databases WHERE name=N'DWPubsSales')
  BEGIN
    -- Close connections to the DWPubsSales database
    ALTER DATABASE [DWPubsSales] SET SINGLE_USER WITH ROLLBACK IMMEDIATE
    DROP DATABASE [DWPubsSales]
  END

GO
CREATE DATABASE [DWPubsSales] ON PRIMARY
( NAME=N'DWPubsSales'
, FILENAME=N'C:\_BISolutions\PublicationsIndustries\DWPubsSales.mdf'
, SIZE=10MB
, MAXSIZE=1GB
, FILEGROWTH=10MB )
LOG ON
```

```
( NAME = N'DWPubsSales_log'
, FILENAME = N'C:\_BISolutions\PublicationsIndustries\DWPubsSales_log.LDF'
, SIZE = 1MB
, MAXSIZE = 1GB
, FILEGROWTH = 10MB)
GO
EXEC [DWPubsSales].dbo.sp_changedbowner @loginame = N'SA', @map = false
GO
ALTER DATABASE [DWPubsSales] SET RECOVERY BULK_LOGGED
GO
```

If you are typing the SQL code, highlight and execute the code by clicking the "! Execute" button.

In this exercise, you created the database that is used as the Publication Industries sales data warehouse. Next you need to fill it with fact and dimension tables.

Creating Tables

As you saw from Listing 5-5, creating a database with SQL code is pretty straightforward. However, many developers prefer to use the New Database dialog windows instead.

Creating a table is just as easy, and as with creating a database, you can use either code or designer windows. SQL Server Management Studio has two tools for designing tables. The first one we look at is the table designer.

■ **Note** Different versions of SQL Server may not include all the designer tools mentioned here. In this book, we assume that you are using SQL Server's developer edition, which includes all SQL Server's designer tools. For more information, see SQL Books Online.

Using the Table Designer

SQL Server Management Studio's table designer provides a graphical way to create tables by using a designer window along with a property window. To create a new table with the table designer, start by using Object Explorer to select the database in which you want to create the table, and then expand the treeview until you see the Tables folder. From there, right-click the Tables folder to get a context menu that allows you to add a new table (Figure 5-18).

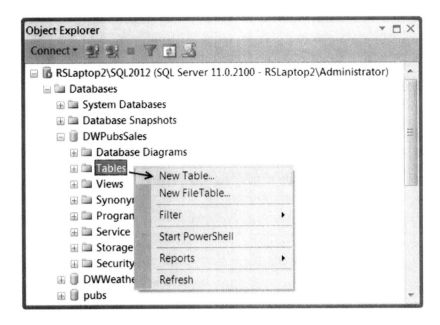

Figure 5-18. *Launching the table desiner*

By default, the table designer window opens in the center of SQL Server Management Studio, and the Properties window displays to the right of it (Figure 5-19). The Properties window does not always appear unless you press the F4 key or use Management Studio's View ➤ Properties Window menu item (Figure 5-19).

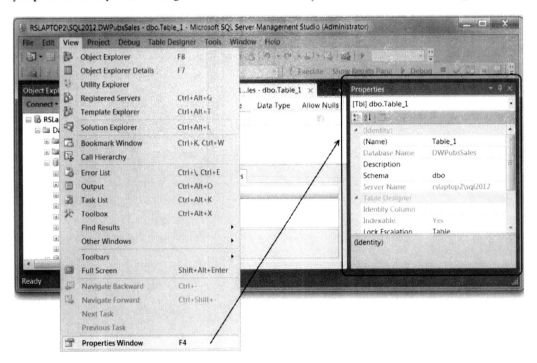

Figure 5-19. *Displaying the Properties window*

■ **Note** The window in SQL Management Studio can be repositioned, so your screen may not look the same as shown in the figure. You can reset the layout to its default settings at any time using the Window ➤ Reset Windows Layout menu item.

When you open the table designer initially, there will be no columns specified. Therefore, your first job is to type in the column name, choose a data type, and then choose whether to allow nulls. Additional settings can be added as well using the Properties window. For example, if the first column was a customer ID column and you decided to use the autonumbering feature in SQL Server known as the identity setting, you could use the Properties window to set the identity column on the table, as shown in Figure 5-20.

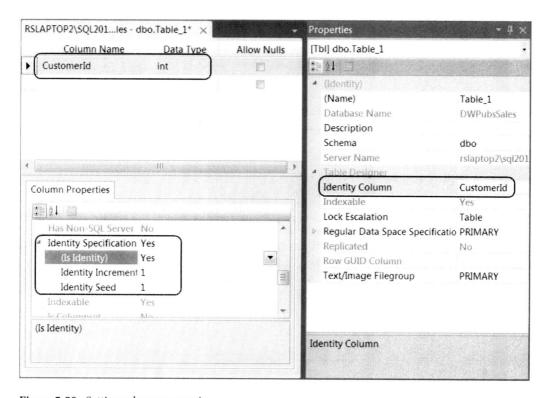

Figure 5-20. *Setting column properties*

You can easily add a primary key to the table by right-clicking a column and selecting the Set Primary Key option from the context menu. If you want to create a composite primary key constraint on a table, this is also easily accomplished. To do so, you must select both columns at once by holding down the Control button as you click each column. After that, you can right-click to get the context menu and select the Set Primary Key option to create a composite primary key constraint (Figure 5-21).

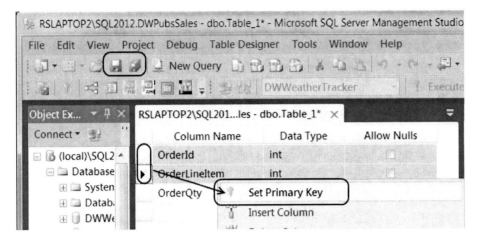

Figure 5-21. *Setting a composite primary key constraint*

Once you have designed the table to your satisfaction, save your work by using the File menu and choosing the Save Table option. You can also click the button on the toolbar that looks like a floppy disk (Figure 5-21). Upon saving, you will be prompted for a table name, unless if you have configured a new name in the Properties window already (Figure 5-20).

Generating SQL Scripts

The SQL Server Management Studio's designer tool invisibly creates SQL code for you each time you create or modify a table. You can access this code by using the Generate Change Script menu option, as shown in Figure 5-22.

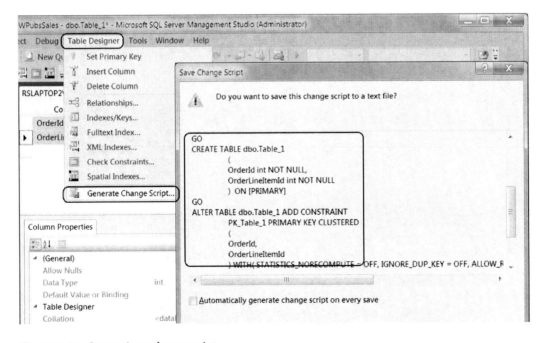

Figure 5-22. *Generating a change script*

Be aware that menus in Management Studio are context sensitive, meaning that the table designer window will not be available unless the table designer is open. It also means that if there are no changes since the change script was last run, this menu item is grayed out, because there are no changes recorded.

After you have saved your table, you can continue to work on it and save it again, or you can close the designer window by clicking the small *x* on the table designer tab (Figure 5-21). Each time you save and make a new change, a new script is created.

Changing an Existing Table

If you wish to work on an existing table you can reopen the table designer by right-clicking the table in Object Explorer and selecting Design from the context menu. When the table designer opens, you can modify the table's current columns, or add a column by clicking the blank row at the bottom of the column list. Alternately, you can add a column by right-clicking existing columns and choosing the Insert Column option from the context menu.

In the table designer, changes to a table sometimes force it to be re-created behind the scenes. When this happens, SQL Server Management Studio will do the following:

1. Create a temporary table.

2. Import data from an original table into the temporary table.

3. Delete the original table.

4. Rename the temporary table to the original table name.

This re-creation process could take a long time if the table had a lot of data. Because of this, Microsoft's default setting does not allow for operations that force a table re-creation. You can, however, change the SQL Server Management Studio options to allow this to occur. To do so, go to the Tools ➤ Options menu item. An Options dialog window will open, as shown in Figure 5-23. You need to navigate down to the Designers section in the treeview of this dialog window and then expand the section before being able to see the Tables and Database Designers page. Navigate to this page and look for the "Prevent saving changes that require table re-creation" checkbox. Uncheck this box, and then click OK.

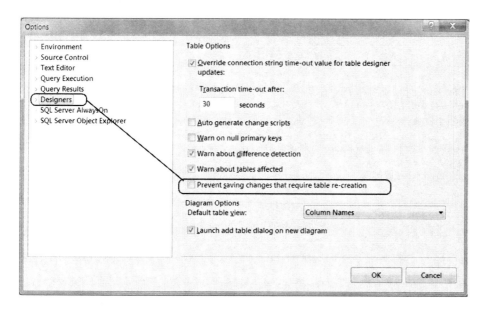

Figure 5-23. Allowing table re-creation in the designers page

Once you have unchecked this option, you will be able to make changes in the table designer that force a re-creation event to occur. You will not notice the difference overtly, except that certain items you may have wanted to change previously, such as the primary key settings, will now work; whereas they did not before.

Using the Diagramming Tool

Although creating a table with the table designer is easy, you may prefer another method even more: using the diagramming tool. To create a database diagram, right-click the diagram folder in Object Explorer and use the context menu to create a new diagram, as shown in Figure 5-24.

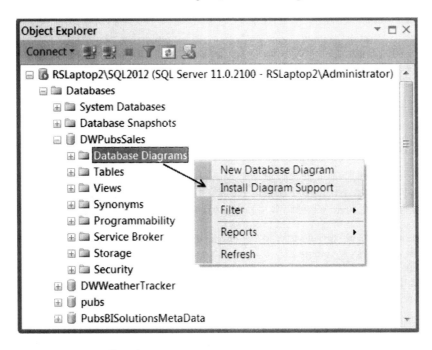

Figure 5-24. *Installing diagram support*

The SQL Server database diagram provides a schematic of a database's tables and their relationships, which can be quite helpful in understanding a database's design. In addition, the diagramming tool also allows you to create, edit, and delete tables. Each database can have one or more diagrams. Diagrams are stored within database system tables, which can be backed up along with other database objects and data.

SQL Server database diagrams have a couple of useful features you should know about. Firstly, you can show all or only some of the tables in the database. This is convenient because you can make diagrams specific to a particular subject matter, which is a big plus when you have a data warehouse with several data marts inside. Secondly, multiple database diagrams can reference the same table, which is advantageous when you have conformed dimensions that are shared between different data marts.

To create a database diagram, you must install some supporting objects within the database. If you have not installed the support objects before you attempt to make a database diagram, you will be prompted to install them.

■ **Tip** Although you can select the New Database Diagram option in the context menu before the Install Diagram Support option, no matter what selection you choose, you have to install the supporting objects (which makes you wonder why they even bothered including a menu option). As such, a message-box may appear informing you that it needs to create the supporting object when you click the New Database Diagram option first. Simply close the message box when it appears by clicking the Yes button.

Each time you create a new database diagram, an Add Table dialog box appears, as shown in Figure 5-25. This allows you to select which table to add to the new diagram. You can select one or all of the tables in the dialog box. And, you can only select tables that exist in the same database as the diagram.

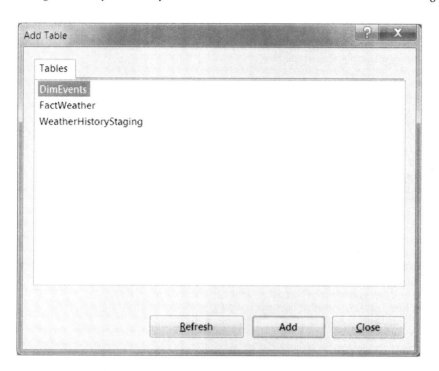

Figure 5-25. *Adding existing tables to a diagram*

Creating New Tables with the Diagram Tool

If there are no tables in the database, the Add Table dialog box appears empty. This is not a problem per se, but it is often disconcerting. If you have no tables in the database, yet want to use a diagramming tool to create them, just close the empty Add Table dialog box and then create a new table (Figure 5-26). Remember that if the diagram does not include any existing tables, you have the equivalent of a blank sheet of paper. Once again, this may initially be confusing, but it is perfectly normal.

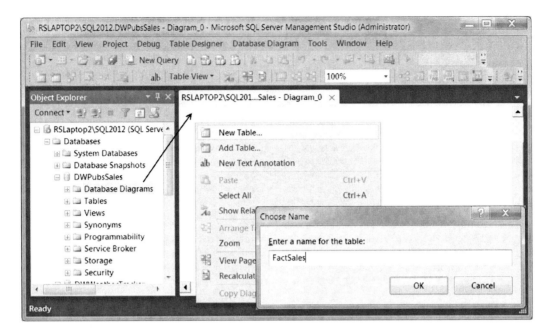

Figure 5-26. *Adding a table to the diagram*

Whether you have tables in the diagram already or the diagram is completely blank makes no difference to your ability to add a new table. You can easily add a new table to the diagram by right-clicking a blank spot and selecting the New Table option from the context menu, as shown in Figure 5-26.

Unlike the table designer, which asks you for the name of the table after you save it, the diagramming tool asks you for the name of the table straightaway, as shown in Figure 5-26. In addition, you can change the name later by modifying the properties of the property sheet as you did with the table designer.

A square representing your table is presented on the diagram immediately after you click OK in the Choose Name dialog box. At that point, you can add columns just as you did in the table designer. You can set the data type and the null special specification as well. You can also set primary keys by clicking to select the column or columns if you are using a composite primary key and select the Set Primary Key from the context menu, as shown in Figure 5-27.

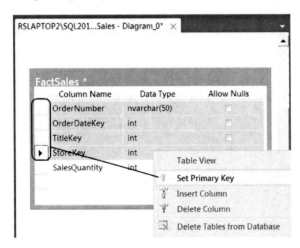

Figure 5-27. *Adding a primary key constraint in the diagram tool*

A classic way to use a diagramming tool is to document your table designs using the Excel spreadsheet and then use the worksheet to create the new tables, much as a carpenter uses a blueprint to build a house. As you can see in Figure 5-28, we created a new fact table while having both the diagramming tool and an Excel spreadsheet open.

Figure 5-28. *Using a BI solution plan as a blueprint*

Creating Foreign Keys with the Diagramming Tool

Once you have more than one table in the diagram, you can easily create foreign key constraints between them. You can accomplish this by dragging and dropping between the columns where you want to place a foreign key constraint, as shown in Figure 5-29.

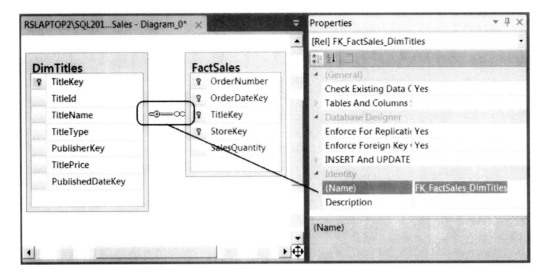

Figure 5-29. *Tables with a foreign key constraint*

Here are the steps:

1. Click the foreign key or child column.

2. With the mouse button still depressed, drag the mouse cursor to the primary key or parent column.

3. Release the mouse button.

We have seen many struggle with this method; therefore, it may take some practice before you become comfortable with it. Perhaps one reason for the difficulty is that the diagram tool does not place the foreign key constraint line directly next to the columns being used but instead shows the connection between the tables themselves. And although you can move the connecting line toward the columns manually to make it look more presentable, it is not done automatically. Once you get used to it, you will find this technique is not as confusing as it may seem at first. And as long as the columns are named similarly, you will easily be able to figure out which columns are interconnected.

Using the Query Window

You can also use code to create your tables. For some developers, this is the most straightforward way to do so. Listing 5-6 shows the syntax for creating a table with code.

Listing 5-6. Creating the Fact Table with SQL Code

```
CREATE TABLE [dbo].[FactSales](
  [OrderNumber] [nvarchar](50) NOT NULL,
  [OrderDateKey] [int] NOT NULL,
```

```
    [TitleKey] [int] NOT NULL,
    [StoreKey] [int] NOT NULL,
    [SalesQuantity] [int] NOT NULL,
    CONSTRAINT [PK_FactSales] PRIMARY KEY CLUSTERED
      ( [OrderNumber] ASC,[OrderDateKey] ASC, [TitleKey] ASC, [StoreKey] ASC )
)
GO
```

By typing this code into a query window and clicking the "! Execute" button, the table is created with the composite primary key on the first four columns. After you create all the tables in a database, you can create foreign key constraints between the tables using code similar to that shown in Listing 5-7.

Listing 5-7. Adding Foreign Key Constraints

```
ALTER TABLE [dbo].[FactSales] WITH CHECK ADD CONSTRAINT [FK_FactSales_DimStores]
FOREIGN KEY([StoreKey]) REFERENCES [dbo].[DimStores] ([Storekey])
GO

ALTER TABLE [dbo].[FactSales] WITH CHECK ADD CONSTRAINT [FK_FactSales_DimTitles]
FOREIGN KEY([TitleKey]) REFERENCES [dbo].[DimTitles] ([TitleKey])
GO
```

Note that in Listing 5-7 there are two foreign key constraint commands. If a table has multiple foreign key relationships, as most fact tables do, you need to execute a separate command for each of them.

At this point, you have seen three ways to create tables: using SQL code, using the table designer and using the diagramming tool. All three ways accomplish the same thing.

Now, after so many pages of theory, it is time to do another exercise!

EXERCISE 5-2. CREATING TABLES AND FOREIGN KEY CONSTRAINTS

In this exercise, you create the tables and foreign key constraints in the Publication Industries data warehouse. You can choose to use the SQL code presented here, the table designer or the diagramming tools to accomplish your goal.

The files for this exercise, as well as all the exercises throughout this book, are available in the downloadable book content.

Creating the Tables

The first thing we need to do is create the tables. Let's do that now.

1. If it is not open already, open SQL Server Management Studio (see Exercise 5-1 for more details).

2. Decide on a method for creating the tables and begin creating them utilizing the table names, column names, data types, nullablity, and primary keys represented in Listing 5-8. You can find a copy of this SQL script in the downloadable book files.

Listing 5-8. Creating the DWPubsSales Tables

```
USE [DWPubsSales]
GO

/****** Create the Dimension Tables ******/
```

```sql
CREATE TABLE [dbo].[DimStores](
  [StoreKey] [int] NOT NULL PRIMARY KEY Identity,
  [StoreId] [nchar](4) NOT NULL,
  [StoreName] [nvarchar](50) NOT NULL
)
GO

CREATE TABLE [dbo].[DimPublishers](
  [PublisherKey] [int] NOT NULL PRIMARY KEY Identity,
  [PublisherId] [nchar](4) NOT NULL,
  [PublisherName] [nvarchar](50) NOT NULL
)
GO

CREATE TABLE [dbo].[DimAuthors](
  [AuthorKey] [int] NOT NULL PRIMARY KEY Identity,
  [AuthorId] [nchar](11) NOT NULL,
  [AuthorName] [nvarchar](100) NOT NULL,
  [AuthorState] [nchar](2) NOT NULL
)
GO

CREATE TABLE [dbo].[DimTitles](
  [TitleKey] [int] NOT NULL PRIMARY KEY Identity,
  [TitleId] [nvarchar](6) NOT NULL,
  [TitleName] [nvarchar](100) NOT NULL,
  [TitleType] [nvarchar](50) NOT NULL,
  [PublisherKey] [int] NOT NULL,
  [TitlePrice] [decimal](18, 4) NOT NULL,
  [PublishedDateKey] [int] NOT NULL
)
GO

/****** Create the Fact Tables ******/

CREATE TABLE [dbo].[FactTitlesAuthors](
  [TitleKey] [int] NOT NULL,
  [AuthorKey] [int] NOT NULL,
  [AuthorOrder] [int] NOT NULL,
CONSTRAINT [PK_FactTitlesAuthors] PRIMARY KEY CLUSTERED
    ( [TitleKey] ASC, [AuthorKey] ASC )
)
GO

CREATE TABLE [dbo].[FactSales](
  [OrderNumber] [nvarchar](50) NOT NULL,
  [OrderDateKey] [int] NOT NULL,
  [TitleKey] [int] NOT NULL,
  [StoreKey] [int] NOT NULL,
  [SalesQuantity] [int] NOT NULL,
CONSTRAINT [PK_FactSales] PRIMARY KEY CLUSTERED
    ( [OrderNumber] ASC,[OrderDateKey] ASC, [TitleKey] ASC, [StoreKey] ASC )
)
GO
```

3. If you are typing the SQL code, highlight and execute the code by clicking the "! Execute" button. If you choose to use the table designer or the diagramming tool, make sure that you use the Save button to create the tables in the database.

Adding Foreign Key Constraints

Now that the tables are created, we want to add foreign key constraints between the tables.

1. Using either SQL code or the diagramming tool, create foreign key constraints as described by the code in Listing 5-9.

Listing 5-9. Creating the DWPubsSales Foreign Key Constraints

```
/****** Add Foreign Keys ******/
ALTER TABLE [dbo].[DimTitles] WITH CHECK ADD CONSTRAINT [FK_DimTitles_DimPublishers]
FOREIGN KEY([PublisherKey]) REFERENCES [dbo].[DimPublishers] ([PublisherKey])
GO

ALTER TABLE [dbo].[FactTitlesAuthors] WITH CHECK ADD CONSTRAINT
[FK_FactTitlesAuthors_DimAuthors]
FOREIGN KEY([AuthorKey]) REFERENCES [dbo].[DimAuthors] ([AuthorKey])
GO

ALTER TABLE [dbo].[FactTitlesAuthors] WITH CHECK ADD CONSTRAINT
[FK_FactTitlesAuthors_DimTitles]
FOREIGN KEY([TitleKey]) REFERENCES [dbo].[DimTitles] ([TitleKey])
GO

ALTER TABLE [dbo].[FactSales] WITH CHECK ADD CONSTRAINT [FK_FactSales_DimStores]
FOREIGN KEY([StoreKey]) REFERENCES [dbo].[DimStores] ([Storekey])
GO

ALTER TABLE [dbo].[FactSales] WITH CHECK ADD CONSTRAINT [FK_FactSales_DimTitles]
FOREIGN KEY([TitleKey]) REFERENCES [dbo].[DimTitles] ([TitleKey])
GO
```

2. If you are typing the SQL code, highlight and execute the code by clicking the "! Execute" button. If you choose to use the table designer or the diagramming tool, make sure you use the Save button to create the tables in the database.

Verifying Your Tables

Once the tables are made and the foreign keys created, it is best to check your work by comparing it to the design document, the Publication Industries BI solution worksheet you reviewed in Chapter 3.

1. Open the file C:_BISolutions\PublicationsIndustries\ BISolutionWorksheets.xlsx in Microsoft Excel, and verify that you have created your data warehouse accurately (Figure 5-28).

In this exercise, you created the data warehouse tables and foreign key constraints. Currently these tables are empty. In Chapters 6 through 8 we create an ETL process to fill them.

Creating a Date Dimension Table

One common practice in data warehousing is the creation of a date dimension table. We discussed the use of these tables in Chapter 4 but have not discussed how to create them. There are a number of ways to accomplish this, but using SQL code is probably the most common way to do so.

As a standard practice, the creation of a date dimension table includes a surrogate key in the form of an integer value, a natural date key in the form of a datetime value and other descriptive attributes such as month, quarter, and year. Listing 5-10 shows an example of a date dimension table being created with this design.

Listing 5-10. Creating the DimDates Table

```
-- We should create a date dimension table in the database
CREATE TABLE dbo.DimDates (
[DateKey] int NOT NULL PRIMARY KEY IDENTITY
, [Date] datetime NOT NULL
, [DateName] nVarchar(50)
, [Month] int NOT NULL
, [MonthName] nVarchar(50) NOT NULL
, [Quarter] int NOT NULL
, [QuarterName] nVarchar(50) NOT NULL
, [Year] int NOT NULL
, [YearName] nVarchar(50) NOT NULL
)
```

Once the table is created, fill it with dimensional values. You can do so during the ETL process or immediately after creating the table. This is different from the other dimensional tables, because the data for a data dimension table is not imported from the OTLP database and instead is programmatically generated.

The simplest way to accomplish this is to create a SQL WHILE loop and specify the range of dates that are to be placed inside the table. Listing 5-11 shows an example of a transact SQL statement that accomplishes this goal.

Once the data dimension table is filled with data, it can be referenced from both fact and dimensional tables. Therefore, you create foreign key constraints to all the tables that reference this new table.

In the DWPubsSales example, we must specifically use dates that were appropriate to Microsoft's Pubs database. These dates included sales records from the 1990s. If the date table was going to be utilized by multiple data marts or data warehouses, you would include a much broader range of dates.

Listing 5-11. Filling the DimDates Table

```
-- Because the date table has no associated source table we can fill the data
-- using a SQL script.

-- Create variables to hold the start and end date
DECLARE @StartDate datetime = '01/01/1990'
DECLARE @EndDate datetime = '01/01/1995'

-- Use a while loop to add dates to the table
DECLARE @DateInProcess datetime
SET @DateInProcess = @StartDate

WHILE @DateInProcess <= @EndDate
 BEGIN
 -- Add a row into the date dimension table for this date
 INSERT INTO DimDates (
   [Date]
 , [DateName]
 , [Month]
```

```
, [MonthName]
, [Quarter]
, [QuarterName]
, [Year]
, [YearName]
)
VALUES (
  -- [Date]
   @DateInProcess
  -- [DateName]
, Convert(varchar(50), @DateInProcess, 110)+', '
  + DateName( weekday, @DateInProcess )
  -- [Month]
, Month( @DateInProcess )
  -- [MonthName]
, Cast( Year(@DateInProcess) as nVarchar(4) )+' - '
  + DateName( month, @DateInProcess )
  -- [Quarter]
, DateName( quarter, @DateInProcess )
  -- [QuarterName]
, Cast( Year(@DateInProcess) as nVarchar(4) )+' - '
  + 'Q'+DateName( quarter, @DateInProcess )
  -- [Year]
, Year(@DateInProcess)
  -- [YearName]
, Cast( Year(@DateInProcess) as nVarchar(4) )
)

-- Add a day and loop again
SET @DateInProcess=DateAdd(d, 1, @DateInProcess)
END

-- Check the table SELECT Top 10 * FROM DimDates
```

In the next exercise, you create this date table using the code in Listing 5-10. Then, you create the foreign key constraints from the DimTitles and FactSales tables.

EXERCISE 5-3. CREATE A DATE DIMENSION

In this exercise, you create a date dimension table in the Publication Industries data warehouse. You can choose to use either the SQL code presented here, the table designer or the diagramming tools to accomplish your goal. (In Chapter 7 you fill the table with data using the code in Listing 5-11.)

Tip: The code files for this exercise, as well as all of the exercises throughout this book, are available in the downloadable book content.

Create the DimDates Table

You first task is to create a table to hold date dimension data.

1. If it is not open already, open SQL Server Management Studio; see Exercise 5-1 for more details.

2. Decide on a method for creating the DimDates table, and begin creating them utilizing the design represented in both Listing 5-10 and Figure 5-30.

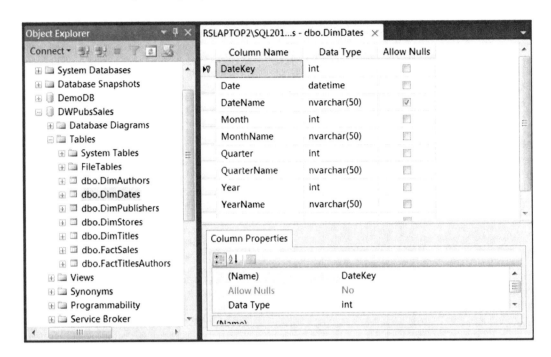

Figure 5-30. *The DimDates table*

3. If you are typing the SQL code, highlight and execute the code by clicking the "! Execute" button. If you choose to use the table designer or the diagramming tool, make sure you use the Save button to create the tables in the database.

Adding Foreign Key Constraints

We have added a new table to the data warehouse database, but now we want to connect this new table to the existing ones using foreign key constraints.

1. Add the foreign key constraints to the DimDates table using the code in Listing 5-12.

Listing 5-12. Creating the DWPubsSales Foreign Key Constraints

```
USE [DWPubsSales]
GO

ALTER TABLE [dbo].[FactSales] WITH CHECK ADD CONSTRAINT [FK_FactSales_DimDates]
FOREIGN KEY([OrderDateKey])
REFERENCES [dbo].[DimDates] ([DateKey])
GO

ALTER TABLE [dbo].[DimTitles] WITH CHECK ADD CONSTRAINT [FK_DimTitles_DimDates]
```

```
FOREIGN KEY([PublishedDateKey])
REFERENCES [dbo].[DimDates] ([DateKey])
GO
```

In this exercise, you created a date dimension table in the data warehouse and added foreign key constraints to the tables that reference it. We will fill the table with data during the ETL process, so for now we leave it empty.

Getting Organized

Excellent, we now have the database and all the tables, and we can get started with the ETL process. Before we do, however, we need to organize our hard work by performing a few simple tasks.

Backing Up the Data Warehouse

We have already talked about how it is sometimes necessary to back up a data warehouse database. Well, this is one of those times! The reason to do so now is to have a copy of the data warehouse in its empty state so that you can hand it over to your testers and other developers for review. These team members can easily restore the backup and do their work simultaneously with yours. Also, a database backup is a simple way to save your progress thus far.

The backup and restore process is easy. Listing 5-13 exhibits code that does both.

Listing 5-13. Backing Up and Restoring a Database

```
BACKUP DATABASE [DWPubsSales]
TO DISK =
N'C:\_BISolutions\PublicationsIndustries\DWPubsSales\DWPubsSales_BeforeETL.bak'
GO

RESTORE DATABASE [DWPubsSales]
FROM DISK =
N'C:\_BISolutions\PublicationsIndustries\DWPubsSales\DWPubsSales_BeforeETL.bak'
WITH REPLACE
Go
```

▨ **Important** The SQL backup statement will not create folders if they do not exist. Consequently, before you execute the backup command, it is important to create any folders indicated in the backup path. If you try running the code in Listing 5-13 without the folder, you can expect an error.

Scripting the Database

Another way that you can preserve your work is by using the database scripting tool. SQL Server has long provided a tool for scripting the objects in the database, and SQL 2012 is no exception. You can launch the scripting tool by right-clicking the database in Object Explorer and selecting Tasks ➤ Generate Scripts from the context menu. The selection launches the Generate and Publish Scripts Wizard you see in Figure 5-31.

▨ **Note** You may notice that there is also a Script Database menu item, but this option only generates code to create the database and not all the tables within it.

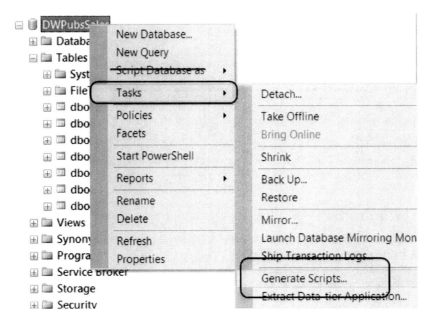

Figure 5-31. *Generating SQL scripts for your database and objects*

When the wizard appears, walking through the steps enables the generation of SQL code that can be used to quickly re-create your data warehouse. The first page of the wizard displays a list of the wizard pages on the left of its window (Figure 5-32).

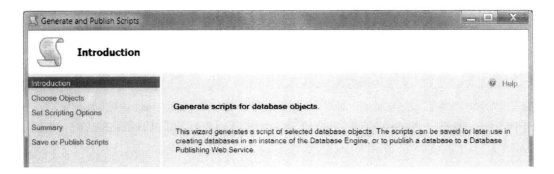

Figure 5-32. *The first page of the Generate and Publish Scripts Wizard*

The second page of the wizard allows you to choose the items you script (Figure 5-33). For our purposes, we want all of the database objects, which is the default choice.

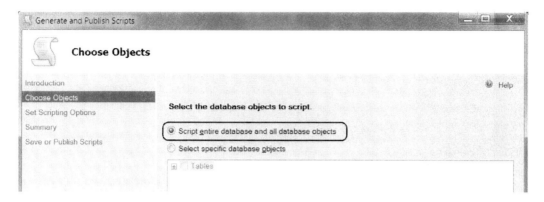

Figure 5-33. *Selecting which database objects to script*

The third page allows you to choose where your script is saved. The two most common choices are to save to a file, which is the default, and to save to a New Query window (Figure 5-34). This second option is convenient because you will most likely want to review the code and make changes before you save it as one of your BI solution files.

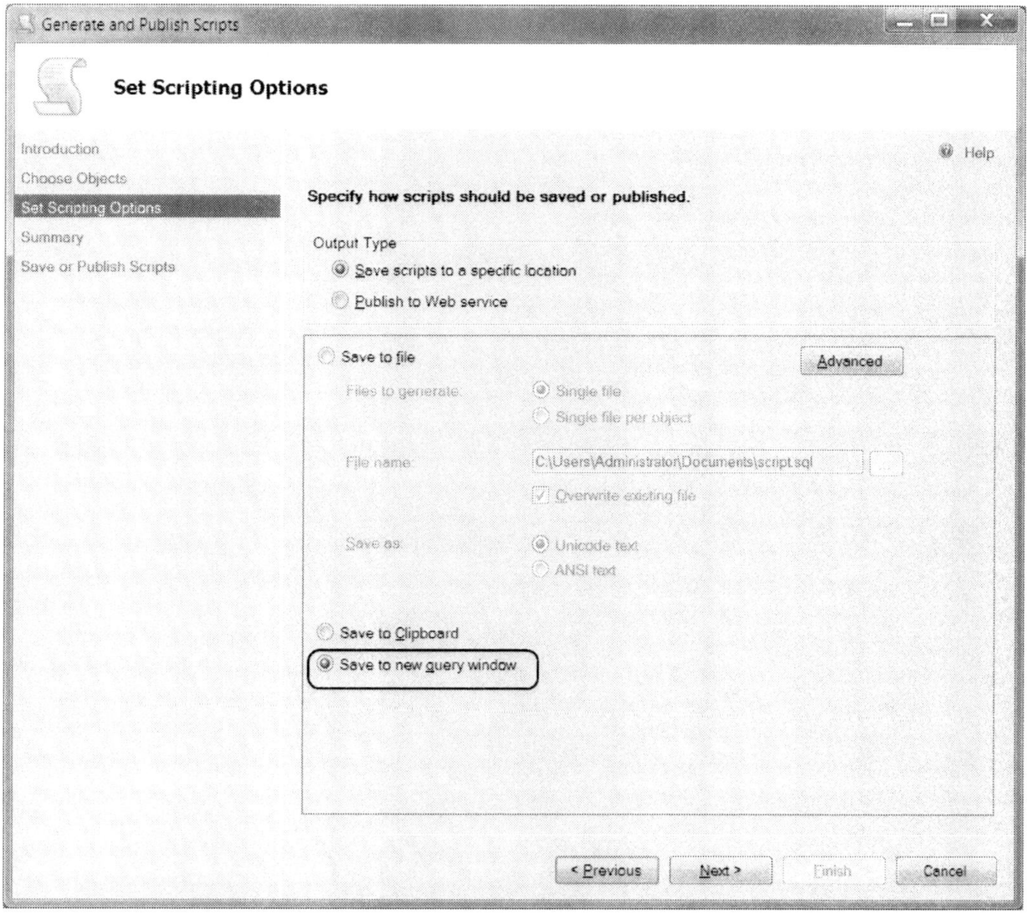

Figure 5-34. *Selecting the scripting options*

The next page provides you with a summary of your choices (Figure 5-35).

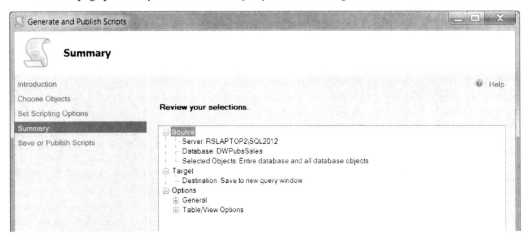

Figure 5-35. *Reviewing the Generate and Publish Scripts summary page*

When you navigate to the last page, it starts the scripting process and indicates the successful completion or failure of each item scripted (Figure 5-36).

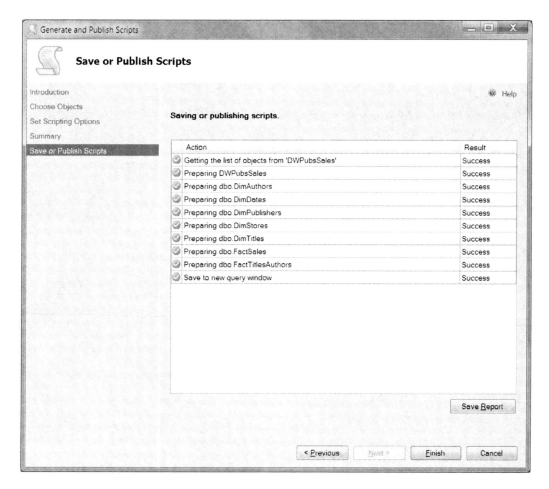

Figure 5-36. *Saving and publishing the SQL Scripts*

Having a script file in addition to a database backup is practical for a couple of reasons. It is useful for training new team members, familiar with SQL code, about the database design. Another advantage is that you can include the script along with your business intelligence projects in a Visual Studio solution. Although this is also true of a backup file, you cannot open a backup directly in a Visual Studio solution. With a script file, you can simply double-click the file in Solution Explorer, and its contents conveniently display within Visual Studio.

■ **Tip** The scripting tool generates working code but often adds more settings and features than you need in your script. We use this tool to create the basic outline of the script and then modify it to remove any superfluous code and add our own comments.

Organizing Your Files with Visual Studio

As you saw in Chapter 2, using Visual Studio to manage your BI solution files is practical because most of the BI servers already store their files in Visual Studio projects. The exception to this rule is the SQL Server database engine. This can be done in Visual Studio, but the means of accomplishing this can be quite obscure, especially because SQL Server Management Studio has its own type of solutions for code files. Sadly, SQL Management Studio and Visual Studio solutions are not aligned. This means you cannot take a Visual Studio project and open it in SQL Server Management Studio, and vice versa. To work around this dilemma, you can use a solution folder in Visual Studio and add your SQL files to it.

■ **Note** Visual Studio 2010 finally has a true SQL Server Database project type when you install the SQL Server Data Tools (SSDT) plug-in. We use SQL Server Management Studio instead, because it is simpler to use and applicable to all versions of SQL Server. We have included information about SSDT at http://NorthwestTech.org/SSDTDemos.

One nice feature of Visual Studio is the ability to add logical folders to a solution. This helps you organize files that are not part of a standard Visual Studio project, such as SSIS or SSAS, but are still part of your overall BI solution.

In Chapter 3, Exercise 3-4, you created a blank Visual Studio Solution and then added a solution folder to it called SolutionDocuments. You then placed the planning documents you created in Chapter 3 into that folder. This organized your planning documents within Visual Studio. Your SQL scripts and backup files can also be added in a Visual Studio solution in a similar manner (Figure 5-37). Let's see how this is done in the following exercise.

Figure 5-37. *Organizing files with solution folders*

EXERCISE 5-4. ADDING DATABASE FILES TO VISUAL STUDIO

In this exercise, you add scripts and backup files to the Visual Studio Solution you created in Chapter 3. You start by creating the files and a subfolder to hold the files. Then add these files to a logical Visual Studio folder, as shown in Figure 5-37.

Important: You are practicing administrator-level tasks in this book, so you need administrator-level privileges. The easiest way to achieve this is to remember to always right-click a menu item, select Run as Administrator, and then answer Yes to access administrator-level privileges while running this program. In Windows 7 and Vista, logging in with an administrator account is not enough. For more information, search the Web on the keywords "Windows 7 True Administrator and User Access Control."

Create a Database Backup and Restore Script

Your first task is to create an operating system folder on your hard drive to hold your scripts and backup files.

1. Create a subfolder to hold the script. To do this, use Windows Explorer and navigate to C:_BISolutions\PublicationsIndustries; then right-click the PublicationsIndustries folder and select the New ➤ Folder option from the context menu. The new folder is created as a subfolder of the PublicationsIndustries folder.

2. Rename the new folder **DWPubsSales**. To do this, right-click the new folder and select Rename from the context menu. Type in the name **DWPubsSales** when prompted (Figure 5-38).

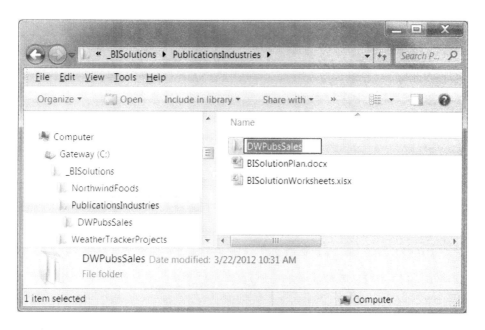

Figure 5-38. *The current files and subfolders in the _BISolutions folder*

Create a Database Backup and Restore Script

Now we need to place files within the subfolder. Let's start by making a backup and restore script and then use it to create a SQL backup file.

1. If it is not open already, open SQL Server Management Studio; see Exercise 5-1 for more details.

2. Using the code in Listing 5-14 to create the backup and restore script in a new query window.

Tip: If you want to use our script file, use the File ➤ Open ➤ File menu in SQL Server Management Studio to open the file C:_BookFiles\Chapter05Files\Listing 5-14. Backup and Restore DWPubsSales. sql"

Listing 5-14. Backing Up and Restoring the DWPubsSales Database

```
/**********************************************
1) Make a copy of the empty database
before starting the ETL process
**********************************************/
BACKUP DATABASE [DWPubsSales]
TO DISK =
N'C:\_BISolutions\PublicationsIndustries\DWPubsSales\DWPubsSales_BeforeETL.bak'
GO

/**********************************************
2) Send the file to other team members
and tell them they can restore the database
with this code...
**********************************************/

-- Check to see if they already have a database with that name…
IF EXISTS (SELECT name FROM sys.databases WHERE name=N'DWPubsSales')
  BEGIN
  -- If they do, they need to close connections to the DWPubsSales database, with this code!
    ALTER DATABASE [DWPubsSales] SET SINGLE_USER WITH ROLLBACK IMMEDIATE
  END

-- Now they can restore the empty database...
USE Master
RESTORE DATABASE [DWPubsSales]
FROM DISK =
N'C:\_BISolutions\PublicationsIndustries\DWPubsSales\DWPubsSales_BeforeETL.bak'
WITH REPLACE
GO
```

3. Save the script file into the new DWPubsSales folder as the Backing up and restoring database.sql file (Figure 5-39). To do this, use SQL Server Management Studio and click the File ➤ Save ➤ current file name ➤ As . . . menu item.

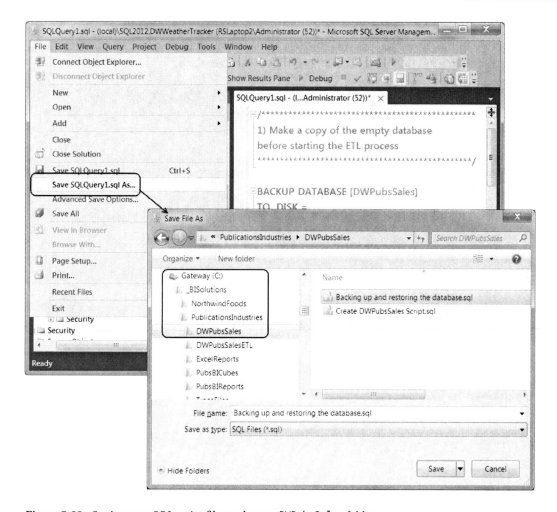

Figure 5-39. *Saving your SQL script files to the new DWPubsSales folder*

Create a Database Backup

Now that you have the backup and restore script, you need to use it to back up your database.

1. Execute the BACKUP DATABASE statement to create a database backup file in the DWPubsSales folder. To do this, highlight the BACKUP DATABASE statement in the Query Window, as shown in Figure 5-40, and click the "! Execute" button.

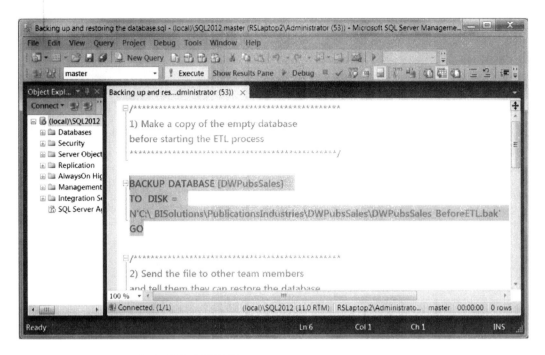

Figure 5-40. *Executing the BACKUP DATABASE statement*

2. Test the restore process by restoring the database. You can do this by highlighting all the code under the IF statement, the USE statement and the RESTORE DATABASE statement, as shown in Figure 5-41.

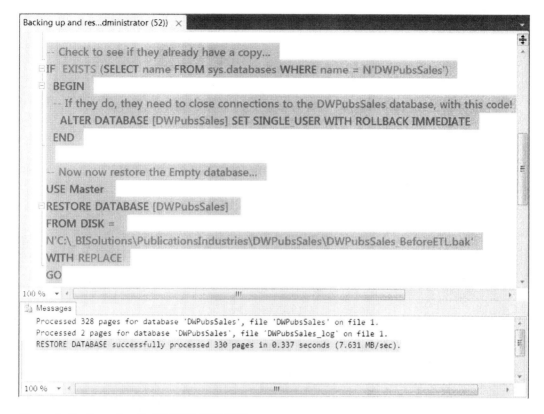

Figure 5-41. Executing the RESTORE DATABASE statement

Create a Database Script File

You now have a tested backup and restore script as well as a database backup file in the DWPubsSales folder. Let's add a complete script for creating the database. We have created a custom script for this process, so you just need to review the code and add it to the folder.

1. Open the author's version of the script file and review its contents. To do this, use the File ➤ Open ➤ File menu in SQL Server Management Studio and open the file C:_BookFiles\Chapter05Files\Create DWPubsSales Script.sql.

2. Review the contents of the file and then save it to the DWPubsSales folder. To do so use SQL Server Management Studio and click the File ➤ Save ➤ current file name ➤ As . . . menu item.

Add the Files to Visual Studio

We now have three files in the operating system folder, the backup and restore script, the backup file and the database creation script. Let's add them to the Publications Industries Visual Studio solution you created in Chapter 3.

1. Open Visual Studio 2010. (You can do so by clicking the Start button, navigating to All Programs ➤ Microsoft Visual Studio 2010, and right-click Microsoft Visual Studio 2010 to see an additional context menu [Figure 2-7]. In this new menu, click the Run as Administrator menu item. If the UAC message box appears asking "Do you want the following program to make changes to this computer?" click Yes [or Continue depending upon your operating system] to accept this request.)

To add these files to the Visual Studio solution you made for Publication Industries in Chapter 3, follow these steps:

2. When Visual Studio opens, open the solution you created in Chapter 3. You can do so by using the File ➤ Open-Project/Solution menu item. A dialog window will open allowing you to select your solution file.

3. Navigate to the SLN file `C:\_BISolutions\PublicationsIndustries\PublicationsIndustries.sln`, select it, and then click the Open button at the bottom of the dialog window.

4. Right-click the solution icon in the Solution Explorer window and select Add ➤ New Solution Folder from the context menu (Figure 5-42).

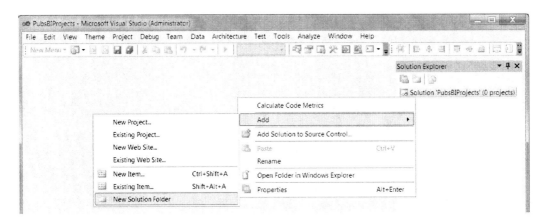

Figure 5-42. *Adding a new solution folder using Solution Explorer*

5. Rename the new solution folder (we called ours DWPubsSales in Figure 5-43).

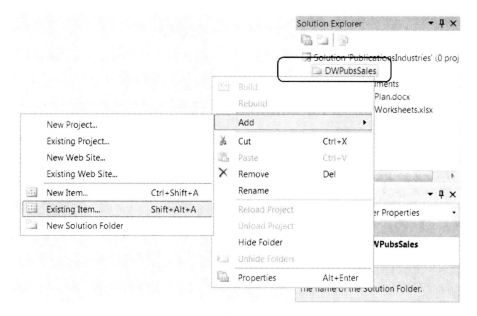

Figure 5-43. *Adding existing files to the new solution folder*

6. Right-click the new folder and select Add ➤ Existing Item from the context menu.

7. Select the files you want to add (in our example those are the backup and script files).

When this is completed, the files are displayed in Solution Explorer alongside the solution documents (Figure 5-44). Visual Studio tries to open the files added in this manner. If it cannot interpret the type of file it is, such as the backup file, it shows a hex representation of the file. You can close any windows, tabs, or applications that appear.

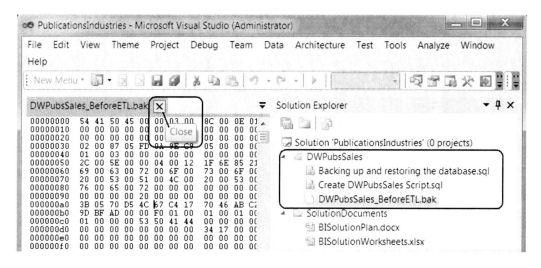

Figure 5-44. *Adding folders and files to the solution folders*

In this exercise, you opened a preexisting Visual Studio solution and added various solution files. Currently there are no projects within the solution, but we change that shortly by adding an SSIS project to it in Chapter 6.

Moving On

In this chapter, you saw how to implement the creation of a data warehouse by creating a database and tables using SQL Server 2012. We examined various options that you can use to create your database, as well as three possible ways to create the tables: using SQL code, the table designer, and the database diagramming tool.

Thus far, we have covered three steps of our eight-step outline. In Chapter 3, we walked through the interview process. In Chapter 4, we looked at data warehouse designs and planned the solution. And in this chapter, we have completed the process of creating the data warehouse. As you can see in Figure 5-45, we are now ready to move on to step 4, the ETL process.

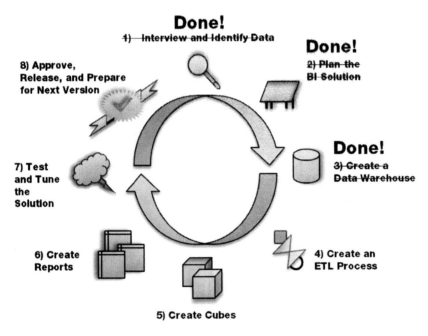

Figure 5-45. *Progressing through the BI solution steps*

In Chapter 6, we examine commonly used SQL code for the ETL process. Then, we follow this up in Chapters 7 and 8 by showing how this code is used in SQL Server integration Services to create your ETL process. Until then, we recommend you practice creating a data warehouse based on the Northwind database in this "Learn by Doing" exercise.

LEARN BY DOING

In this "Learn by Doing" exercise, you perform the process defined in this chapter using the Northwind database. We have included an outline of the steps you performed in this chapter and an example of how the authors handled them in two Word documents. These documents are found in the folder C:_BISolutionsBookFiles_LearnByDoing\Chapter05Files. Please see the ReadMe.doc file for detailed instructions.

What's Next?

Using the SQL Server database engine effectively is a complex task. We have included only a minimum of what you need to be effective when creating a BI solution. If you want a deeper understanding, you will find that many books have been written about the subject; some of them are excellent for database administrators, whereas others are more for general knowledge.

Most BI developers are not the actual database administrators and do not necessarily need to have a great degree of knowledge about the database engine itself. Still, most BI developers can benefit from having a more complete understanding of this subject. For this reason, we recommend the following beginning book on SQL administration: *Beginning SQL Server 2008 Administration* by Grant Fritchey and Robert Walters (Apress, 2009).

CHAPTER 6

■ ■ ■

ETL Processing with SQL

The universe is transformation; our life is what our thoughts make it.

—Marcus Aurelius

The ETL process, and projects associated with it, involve extracting vital information from outside sources, transforming data into clean, consistent, and usable data, and loading it into the data warehouse. This process is vital to the success of your BI solution. It is what makes the difference between a professional, functional data warehouse versus one that is messy, insufficient, and unusable. The ETL process is also one of the longest and most challenging steps in developing a BI solution.

In this chapter, we explain how to perform the ETL tasks required for your BI solution. Our ultimate goal in both this chapter and the next is to demonstrate a technique where you use a combination of SQL programming and SQL Server Integration Service (SSIS) to create a professional ETL project that will be a cornerstone of your BI solution. We cover common SQL programming techniques used to identify issues and provide resolutions associated with the ETL process. And we show how code for these SQL techniques can be placed into views and stored procedures to be used for ETL processing.

This chapter is a prelude to Chapter 7 where we delve into how SSIS and the SQL programming techniques learned in this chapter are combined. Let's begin now by taking a look at the overall process.

Performing the ETL Programming

Figure 6-1 outlines the typical steps involved in creating an ETL process using a combination of SQL Server and SSIS. Note that the process includes deciding between filling up a table completely with fresh data, loading it incrementally (as explained in the following section), and updating any changes from the original source.

These steps are followed by locating the data to extract and examining its contents for validity, conformity, and completeness.

When you have verified that the data available meets your needs, it is likely that you still may need to manipulate it to some degree to fit your destination tables. This manipulation can come in the form of renaming columns or converting the original data types to their destination data types. Once all of these preparations have been completed, you can load the data into your data warehouse tables and begin the process again for each data warehouse table you need to fill.

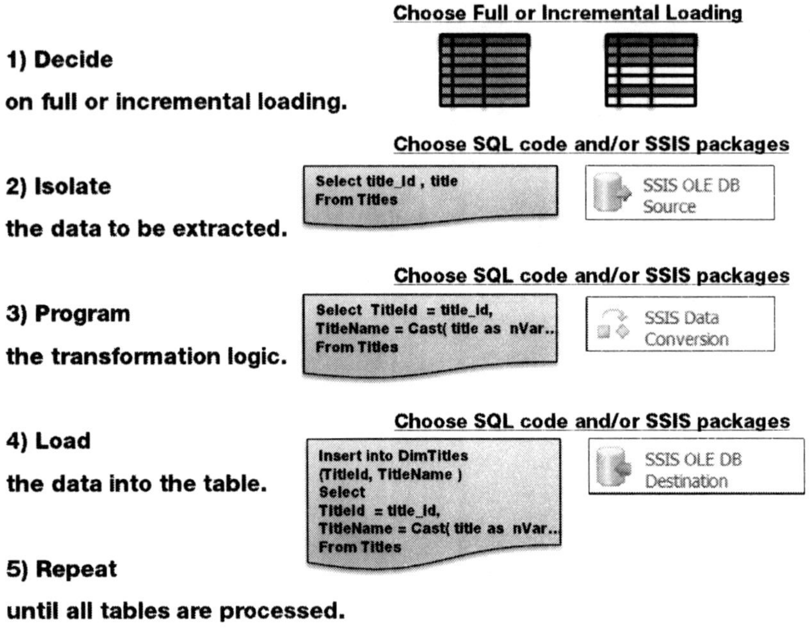

Choose Full or Incremental Loading

1) Decide

on full or incremental loading.

Choose SQL code and/or SSIS packages

2) Isolate

the data to be extracted.

```
Select title_Id , title
From Titles
```

SSIS OLE DB Source

Choose SQL code and/or SSIS packages

3) Program

the transformation logic.

```
Select TitleId = title_id,
TitleName = Cast( title as nVar..
From Titles
```

SSIS Data Conversion

Choose SQL code and/or SSIS packages

4) Load

the data into the table.

```
Insert into DimTitles
(TitleId, TitleName )
Select
TitleId = title_id,
TitleName = Cast( title as nVar..
From Titles
```

SSIS OLE DB Destination

5) Repeat

until all tables are processed.

Figure 6-1. *The ETL process with SQL Server and SSIS*

Most of these steps can be completed using either SQL programming statements or SSIS tasks, and we examine both in this book. You will likely understand the role of the SSIS tasks more thoroughly if we start by examining the SQL statements that they represent. For that reason, let's examine the code necessary to complete these steps using SQL programming statements.

Deciding on Full or Incremental Loading

Tables in the data warehouse can be either cleared out and refilled or loaded incrementally. Clearing out the tables and then completely refilling them is known as a *flush and fill* technique. This technique is the simplest way to implement an ETL process, but it does not work well with large tables. When dimension tables are small and have only a few thousand rows, the flush and fill technique works quite well. Large tables, such as the fact table, for example, may have millions of rows. Therefore, the time it takes to completely clear the table out and then refill it with fresh data may be excessive. In those cases, filling up only data that has changed in the original source is a much more efficient choice.

To use the flush and fill technique, clear the tables in the data warehouse of a SQL Server database using either the DELETE command or the TRUNCATE command. When using the delete command, rows are deleted from the table one by one. Accordingly, if there are 1,000 rows in a table, the delete will be processed 1,000 times. This happens very quickly, but it will still take more time than a simple truncation. The TRUNCATE command de-allocates data pages that internally store the data in SQL Server. These invisible data pages are then free to be reused for other objects in a SQL Server database. Truncation represents the quickest way to clear out a SQL table, but you are not allowed to truncate a table if there are foreign key constraints associated with that table.

Listing 6-1 is an example of what your SQL code looks like using the DELETE command.

▓ **Note**   It may help to have the listing files open in SQL Management Studio as we discuss them. All of the SQL code we are discussing in this book is provided for you as part of the downloadable book files. The files for this chapter are found in the `C:\_BookFiles\Chapter06Files` folder.

Listing 6-1. Deleting Data from the Data Warehouse Tables Using the Delete Command

```
Delete From dbo.FactSales
Delete From dbo.FactTitlesAuthors
Delete From dbo.DimTitles
Delete From dbo.DimPublishers
Delete From dbo.DimStores
Delete From dbo.DimAuthors
```

If you choose to use the truncation statement here, your code must include statements that drop the foreign key relationships before truncation. Listing 6-2 is an example of what your SQL code looks like using the TRUNCATE command.

Listing 6-2. Truncating the Table Data and Resetting the Identity Values

```
/****** Drop Foreign Key s ******/
Alter Table [dbo].[DimTitles] Drop Constraint [FK_DimTitles_DimPublishers]
Alter Table [dbo].[FactTitlesAuthors] Drop Constraint [FK_FactTitlesAuthors_DimAuthors]
Alter Table [dbo].[FactTitlesAuthors] Drop Constraint [FK_FactTitlesAuthors_DimTitles]
Alter Table [dbo].[FactSales] Drop Constraint [FK_FactSales_DimStores]
Alter Table [dbo].[FactSales] Drop Constraint [FK_FactSales_DimTitles]
Go

/****** Clear all tables and reset their Identity Auto Number ******/
Truncate Table dbo.FactSales
Truncate Table dbo.FactTitlesAuthors
Truncate Table dbo.DimTitles
Truncate Table dbo.DimPublishers
Truncate Table dbo.DimStores
Truncate Table dbo.DimAuthors
Go

/****** Add Foreign Keys ******/
Alter Table [dbo].[DimTitles] With Check Add Constraint [FK_DimTitles_DimPublishers]
Foreign Key ([PublisherKey]) References [dbo].[DimPublishers] ([PublisherKey])

Alter Table [dbo].[FactTitlesAuthors] With Check Add Constraint [FK_FactTitlesAuthors_DimAuthors]
Foreign Key ([AuthorKey]) References [dbo].[DimAuthors] ([AuthorKey])

Alter Table [dbo].[FactTitlesAuthors] With Check Add Constraint [FK_FactTitlesAuthors_DimTitles]
Foreign Key ([TitleKey]) References [dbo].[DimTitles] ([TitleKey])

Alter Table [dbo].[FactSales] With Check Add Constraint [FK_FactSales_DimStores]
Foreign Key ([StoreKey]) References [dbo].[DimStores] ([Storekey])

Alter Table [dbo].[FactSales] With Check Add Constraint [FK_FactSales_DimTitles]
Foreign Key ([TitleKey]) References [dbo].[DimTitles] ([TitleKey])
Go
```

An additional benefit of truncation over deletion is that if you have a table using the identity option to create integer key values, truncation will automatically reset the numbering scheme to its original value (typically 1). Deletion, on the other hand, will not reset the number; therefore, when you insert a new row, the new integer value will continue from where the previous insertions left off before deletion. Normally this is not what you want, because the numbering will no longer start from 1, which may be confusing.

Listings 6-1 and 6-2 are examples of code used in the flush and fill process. But, if you choose to do an incremental load, do not clear the tables first. Instead, compare the values between the source and destination tables. Then either add rows to the destination tables where new rows are found in the source, update rows in the destination that are changed in the source or delete rows from the destination that are removed in the source.

In the example in Listing 6-3, a Customer's OLTP table contains a flag column called RowStatus. Each time a row in this OLTP table is added, updated or marked for deletion, a flag is set indicating the operation. The flags are examined to determine which data in the Customers table needs to be synchronized in the DimCustomers table. Then an INSERT, UPDATE or DELETE takes place depending on the flag found in the table.

Tip We have included this code in the Chapter06 folder of the downloadable files if you would like to test it.

Listing 6-3. Synchronizing Values Between Tables

```
Use TEMPDB
Go
-- Step #1. Make two demo tables
Create Table Customers
( CustomerId int
, CustomerName varchar(50)
, RowStatus Char(1) check(RowStatus in ('i','u','d') ) )
Go
Create Table DimCustomers
( CustomerId int
, CustomerName varchar(50) )
Go

-- Step #2. Add some starting data
Insert into Customers (CustomerId, CustomerName, RowStatus )
Values(1, 'Bob Smith', 'i')
Go
Insert into Customers (CustomerId, CustomerName, RowStatus )
Values(2, 'Sue Jones', 'i')
Go

-- Step #3. Verify that the tables are not synchronized
Select * from Customers
Select * from DimCustomers
Go

-- Step #4 Synchronize the tables with this code
BEGIN TRANSACTION
Insert into DimCustomers
(CustomerId, CustomerName)
  Select CustomerId, CustomerName
  From Customers
```

```
  Where RowStatus is NOT null
    AND RowStatus='i'
-- Synchronize Updates
Update DimCustomers
  Set DimCustomers.CustomerName = Customers.CustomerName
  From DimCustomers
  JOIN Customers
    On DimCustomers.CustomerId = Customers.CustomerId
    AND RowStatus = 'u'
-- Synchronize Deletes
Delete DimRows
  From DimCustomers as DimRows
  JOIN Customers
  On DimRows.CustomerId = Customers.CustomerId
  AND RowStatus = 'd'
-- After we import data to the dim table
-- we must reset the flags to null!
Update Customers Set RowStatus = null
COMMIT TRANSACTION

-- Step #5. Test that both tables now contain the same rows
Select * from Customers
Select * from DimCustomers
Go

-- Step #6. Test the Updates and Delete options
Update Customers
Set
  CustomerName = 'Robert Smith'
, RowStatus = 'u'
Where CustomerId = 1
Go
Update Customers
Set
  CustomerName = 'deleted'
, RowStatus = 'd'
Where CustomerId = 2
Go

-- Step #7. Verify that the tables are not synchronized
Select * from Customers
Select * from DimCustomers
Go

-- Step #8. Synchronize the tables with the same code as before
BEGIN TRANSACTION
Insert into DimCustomers
(CustomerId, CustomerName)
  Select CustomerId, CustomerName
  From Customers
  Where RowStatus is NOT null
    AND RowStatus = 'i'
-- Synchronize Updates
```

```
Update DimCustomers
  Set DimCustomers.CustomerName = Customers.CustomerName
  From DimCustomers
  JOIN Customers
    On DimCustomers.CustomerId = Customers.CustomerId
    AND RowStatus = 'u'
-- Synchronize Deletes
Delete DimRows
  From DimCustomers as DimRows
  JOIN Customers
    On  DimRows.CustomerId = Customers.CustomerId
    AND RowStatus = 'd'
-- After we import data to the dim table
-- we must reset the flags to null!
Update Customers Set RowStatus = null
COMMIT TRANSACTION

-- Step #9. Test that both tables contain the same rows
Select * from Customers
Select * from DimCustomers
Go

-- Step #10. Setup an ETL process that will run the Synchronization code
```

As you can see, creating SQL code to accomplish incremental loading can be quite complex. The good news is that you will not need to do this for most tables. Many tables are too small to benefit from the incremental approach, and in those cases, you should try to keep your ETL processing as simple as possible and stick with the flush and fill technique. For example, all the tables in our three demo databases have small amounts of data; consequently, this book focuses on the flush and fill technique for all the tables.

■ **Note**   The problem with the approach we just demonstrated is that the OLTP table needs to have a tracking column. Since SQL 2005, Microsoft has introduced the SQL Merge command that performs these same comparison tasks without a tracking column. We use SQL statements in Listing 6-3 as an example of an original method that will work with most database software, but remember that there is more than one way to hook a fish. Although we don't want to confuse readers by introducing multiple ways to solve the same tasks, we have created a web page detailing a number of historic and modern approaches to this task. For more information, visit www.NorthwestTech.org/ProBISolutions/ETLProcessing.

Isolating the Data to Be Extracted

We now need to examine the data needed for the ETL process. Selecting all the data from the table you are working on is a good start. You can do this by launching a query window, typing in a simple SELECT statement, and executing it to get the results. We begin the process with a statement such as the one shown in Listing 6-4.

Listing 6-4. Selecting All the Data from the Source Table

```
Select * from [Pubs].[dbo].[Titles]
```

Formatting Your Code

Often the code you use to isolate the data is later used in the ETL process. Because of this, you may want to take time to make your code look professional. One way of making your code more professional is to format the ETL code. For example, although using a star symbol to indicate all columns implicitly is acceptable practice for ad hoc queries, a better practice is to explicitly list columns individually as we have in Listing 6-5.

Listing 6-5. Explicitly Listing the Columns

```
Select
    [title_id]
  , [title]
  , [type]
  , [pub_id]
  , [price]
  , [advance]
  , [royalty]
  , [ytd_sales]
  , [notes]
  , [pubdate]
From [Pubs].[dbo].[Titles]
```

You may notice that we are using square brackets around column and table names. This is optional; however, this convention is also considered a best practice.

One additional convention is identifying a table using its full name, or at least most of it. The standard parts of a table's name are < ServerName > . < DatabaseName > . < SchemaName > . < TableName >. It is common to use the last three parts of the fully qualified name, but you seldom use the server name part. Doing so indicates that you want to access a table on a remote server. Although this can be advantageous, it necessitates that SQL Server be configured to use linked servers, something that is not commonly done on a production server. Without a linked server, including the server name as part of the full name of the table generates an error. You therefore use the three-part name most of the time.

Identifying the Transformation Logic

Identifying which transformation is needed and then programming the transformation logic is the portion of the ETL process that usually takes the longest. Let's take a look at an example to understand what is involved. In Figure 6-2, you see two tables: the Titles OLTP database table and the DimTitles OLAP data warehouse table. Let's compare the two tables:

- The DimTitles table has fewer columns.

- The column names are different between the tables.

- The data types are different on some columns.

- There is now a surrogate key that can be used for foreign key references.

- Nullability has been changed in many columns.

- Some values have been cleansed and made more readable.

titles

	Column Name	Data Type	Allow Nulls
🔑	title_id	tid:varchar(6)	☐
	title	varchar(80)	☐
	type	char(12)	☐
	pub_id	char(4)	☑
	price	money	☑
	advance	money	☑
	royalty	int	☑
	ytd_sales	int	☑
	notes	varchar(200)	☑
	pubdate	datetime	☐
			☐

DimTitles

	Column Name	Data Type	Allow Nulls
🔑	TitleKey	int	☐
	TitleId	nvarchar(6)	☐
	TitleName	nvarchar(50)	☐
	TitleType	nvarchar(50)	☐
	PublisherKey	int	☐
	TitlePrice	decimal(18, 4)	☐
	PublishedDate	datetime	☐
			☐

Figure 6-2. *Comparing the Titles table to DimTitles*

All these differences make up a list of transformations that must be addressed as the data moves from the Titles table to the DimTitles table.

■ **Note** The following pages have a lot of SQL programming code. If you are not a SQL programmer, many examples will seem obscure and perhaps even difficult to read. We have endeavored to keep the examples simple to alleviate confusion; however, the ETL process is complex and most examples can be simplified only so much. Consequently, consider our listings as general examples of how a programmer could create an ETL process. Not every ETL process will be coded the same way, and they may not be this simplistic. If you happen to find these samples too difficult, keep in mind that you may never be asked to create the SQL code on your own. Nevertheless, you may be expected to understand what some of this SQL code does. Therefore, we recommend that you focus on the explanation of each process instead of the details on how the code is written.

Programming Your Transformation Logic

You need to create code and programming structures to transform any data that requires it. This transformation code can use SQL or application code, such as C#, or a combination of both.

This code is generated for you using tools such as SSIS. However, automatically generated code is not efficient code, so you may have to optimize it yourself. For that matter, sometimes you even need to fix it before you can use it in production.

The more you work with ETL processing, the more you will find that having a thorough understanding of the code that performs the transformations will help you effectively create SSIS packages. Using tools that help you create code and knowing how to optimize that code are two aspects of ETL processing that go hand in hand. Let's take a look at some common programming techniques that are simple to implement but still provide a great deal of benefit for your efforts.

Reducing the Data

It is unlikely that you will need to extract every column from the original table; therefore, you can simply leave out the columns you do not want from the select clause. This simple procedure represents the first task in optimizing your ETL process.

Listing 6-6 shows code requesting only the columns needed to fill up the DimTitles table. If you compare the listed columns to the possible Titles table's columns shown in Figure 6-2, you see that this represents about a 30% reduction over selecting all the columns by using a query such as SELECT * FROM Titles.

Listing 6-6. Selecting Only Data Required for the Destination Table

```
Select
    [title_id]
  , [title]
  , [type]
  , [pub_id]
  , [price]
  , [pubdate]
From [Pubs].[dbo].[Titles]
```

Using Column Aliases

You may want to change the name of your source columns to match the names of your destination columns by using a column alias. This makes it easier for others to see the correlation between your sources and destinations, as well as aids in the SSIS configuration (covered in the next chapter). SQL Server allows you to create column aliases in two formats; the first is [column name] AS [alias], and the second is [alias] = [column name]. Both of these accomplish the same thing, and both are shown in Listing 6-7.

Listing 6-7. Using Different Styles of Column Aliases

```
-- Older style column aliases: [column name] as [alias]
Select
    [title_id] as [TitleId]
  , [title] as [TitleName]
  , [type] as [TitleType]
  --, [pub_id] Will be replaced with a PublisherKey
  , [price] as [TitlePrice]
  , [pubdate] as [PublishedDate]
From [Pubs].[dbo].[Titles]

-- Newer style column aliases: [alias] = [column name]
Select
    [TitleId] = [title_id]
  , [TitleName] = [title]
  , [TitleType] = [type]
  --, [pub_id] Will be replaced with a PublisherKey
  , [TitlePrice] = [price]
  , [PublishedDate] = [pubdate]
From [Pubs].[dbo].[Titles]
```

▓ **Tip** The first style is an older one that many programmers, including the authors of this book, were introduced to when we first started SQL programming. Although it is familiar and simple, it has one basic problem: the aliases are more difficult to spot when reading the code. The newer style aligns all the aliases on the left side of the column listings and provides for easier reading and troubleshooting. Perhaps this is why Microsoft documentation typically uses the left-hand alias style. But whatever the reason may be, to provide consistency, the newer style is what we use throughout this book.

Converting the Data Types

Another transformation is data type conversion between the source and destination tables. Listing 6-8 shows SQL code that converts the data types as the data is selected from the source table. This is a very simple and fast way of performing these types of transformations.

Listing 6-8. Using the Function to Convert Data Types

```
Select
    [TitleId]=Cast( [title_id] as nvarchar(6) )
  , [TitleName]=Cast( [title] as nvarchar(50) )
  , [TitleType]=Cast( [type] as nvarchar(50) )
--, [pub_id] Will be replaced with a PublisherKey
  , [TitlePrice]=Cast( [price] as decimal(18, 4) )
  , [PublishedDate]=[pubdate] -- has the same data type in both tables
From [Pubs].[dbo].[Titles]
```

If you have programmed in SQL before, you may have noted that this listing utilizes the CAST() function to change existing data types from the source table. The cast function is one of the oldest and simplest functions in Microsoft SQL Server. It has a sister function called CONVERT() that can also perform these conversions. The CONVERT() function has additional options that are not available in CAST(). One example of this is how the CONVERT() function can be used to change the format of the date. Figure 6-3 shows an example.

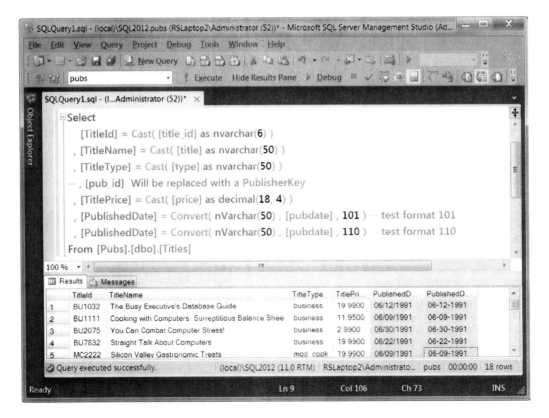

Figure 6-3. *Using the CONVERT() function with additional formatting options*

As you can see, using the CONVERT() function allows you to change datetime data types into a more readable character format, but because we are not interested in storing the dates in a character format, it is irrelevant in our current ETL process. It is nice to know that it exists, however, and that it provides additional options beyond what the CAST() function provides. A word of caution: keep in mind that CAST() is a standard ANSI function and CONVERT() is not. As a result, if you need to pull source data from tables in an Oracle or MySQL database, CAST() is still the better option.

Looking Up Surrogate Key Values

In Chapter 5, we mentioned adding surrogate keys to your dimension tables as a best practice. These artificial columns allow you to merge data from different sources and record data changes more effectively. Therefore, it is no surprise that we use surrogate keys in addition to natural keys in all our dimension tables in this book.

This necessitates that the foreign key relationships between our data warehouse tables are based on the surrogate keys rather than original natural columns. These are different from the relationships defined in the source database. As a result, we need to identify a way to look up the new surrogate key value based on the natural key value in the original tables.

To help you more thoroughly understand this issue, take a look at Figure 6-4. In this figure you can see that the original link between the titles and publishers table in the OLTP database was defined in the pub_id column. This is no longer true in the data warehouse where the link is provided based on the PublisherKey column.

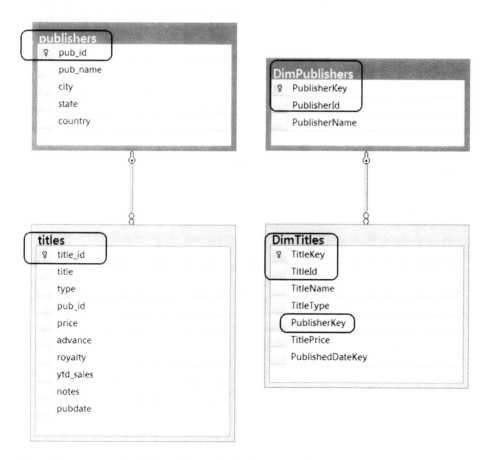

Figure 6-4. *Comparing tables with and without surrogate keys*

To fill the DimTitles table with correct PublisherKey data, we first need to look up the publisher's new surrogate key value by referencing the original relationship between publisher IDs that formed the original natural keys. A simple way to accomplish this is by creating a SQL join that queries the natural key columns in the source and destination tables. Listing 6-9 shows an example.

Listing 6-9. Referencing the Natural Keys to Find the Surrogate Key Value

```
Select
    [TitleId]=Cast( [title_id] as nvarchar(6) )
  , [TitleName]=Cast( [title] as nvarchar(50) )
  , [TitleType]=Cast( [type] as nvarchar(50) )
  , [PublisherKey]=[DWPubsSales].[dbo].[DimPublishers].[PublisherKey]
  , [TitlePrice]=Cast( [price] as decimal(18, 4) )
  , [PublishedDate]=[pubdate]
From [Pubs].[dbo].[Titles]
Join [DWPubsSales].[dbo].[DimPublishers]
  On [Pubs].[dbo].[Titles].[pub_id]=[DWPubsSales].[dbo].[DimPublishers].[PublisherId]
```

For the publisher key data to be available, you have to fill the DimPublishers table first. Otherwise, the surrogate key that is autogenerated will not be in existence. In Listing 6-10 we have provided a simple insert statement that accomplishes this.

Listing 6-10. Inserting Values into the DimPublishers Table

```
Insert Into [DWPubsSales].[dbo].[DimPublishers]
( [PublisherId], [PublisherName] )
Select
    [PublisherId]=Cast( [pub_id] as nchar(4) )
  , [PublisherName]=Cast( [pub_name] as nvarchar(50) )
From [pubs].[dbo].[publishers]
```

Provide Conformity

One additional common ETL task is the process of conforming data to be more readable or more consistent. There are two simple ways to accomplish this using SQL code; the first is to use a SELECT-CASE statement, and the second involves using a lookup table.

A SELECT-CASE statement is created by adding the CASE clause to any statement. For example, in Listing 6-11, we extract the values of the TitleType column in the WHEN clause and match the patterns of "business," "mod_cook," "popular_comp" and so on. We do this because the current values do not read well. They should have been capitalized and fully spelled out to read as "Modern Cooking" and "Popular Computing." Because the abbreviations are not particularly legible, a transformation that conforms the data into a more readable format is in order.

To do this, examine a particular expression and then compare it to the pattern that you want to match. When the result of the expression, defined in the WHEN clause, matches a particular pattern, the output of the select statement is transformed into the value of the expression found in the THEN clause. Listing 6-11 shows how to accomplish this type of transformation using the SQL SELECT-CASE statement.

Listing 6-11. Conforming Values with a SELECT-CASE Statement

```
Select
    [TitleId]=Cast( [title_id] as nvarchar(6) )
  , [TitleName]=Cast( [title] as nvarchar(50) )
```

```
    , [TitleType]=Case Cast( [type] as nvarchar(50) )
        When 'business' Then 'Business'
        When 'mod_cook' Then 'Modern Cooking'
        When 'popular_comp' Then 'Popular Computing'
        When 'psychology' Then 'Psychology'
        When 'trad_cook' Then 'Traditional Cooking'
        When 'UNDECIDED' Then 'Undecided'
    End
    , [PublisherKey]=[DWPubsSales].[dbo].[DimPublishers].[PublisherKey]
    , [TitlePrice]=Cast( [price] as decimal(18, 4) )
    , [PublishedDate]=[pubdate]
From [Pubs].[dbo].[Titles]
Join [DWPubsSales].[dbo].[DimPublishers]
    On [Pubs].[dbo].[Titles].[pub_id]=[DWPubsSales].[dbo].[DimPublishers].[PublisherId]
```

When this code runs, values such as "mod_cook" are converted into the value of "Modern Cooking," and so on. The select statement is applied to every row in the table and takes place for every value in the column.

This method is simple and effective, but there is a downside. If you need to apply the same transformation to other tables, you need to repeat the same SQL code for each table. And if one select statement uses capitalized values, but another uses lowercase values, your tables will have multiple versions of the same data (i.e., Modern Cooking vs. modern cooking). These title type examples are unlikely to be used in any other dimension table, so it is very unlikely that this would become a problem, but it is something to keep in mind with naming conventions.

If you are going to use the same transformation data for more than one table, you may want to use the second option. The second option provides these types of transformations by creating a lookup table and then comparing the values in the lookup table to the values of the original table. When the match is found, extract the conformed value from the lookup table, replacing the original value.

The advantage of this option is that it can be reused multiple times without repeatedly having to define the list of conformed values. For example, this method is appropriate when you have a lookup table that holds two-letter state abbreviations in some table cells and the state's full name in others. Any time you want to convert the two-letter abbreviation to the state's full name in one or more tables, you could reference a lookup table to accomplish that goal. Because state names are likely to appear in many different tables within a given data warehouse, a lookup table is a better choice compared to using a Select-Case statement to conform your data.

Listing 6-12 shows a lookup table being created and then filled with original and transformed values via the lookup table.

Listing 6-12. Conforming Values with a Lookup Table

```
-- Create the lookup table
Create table [TitleTypeLookup] (
    [TitleTypeKey] int Primary Key Identity
  , [OriginalTitleType] nvarchar(50)
  , [CleanTitleType] nvarchar(50)
)

-- Add the original and transformed data
Insert into [TitleTypeLookup]
  ( [OriginalTitleType] , [CleanTitleType] )
Select
    [OriginalTitleType]=[Type]
  , [CleanTitleType]=Case Cast( [type] as nvarchar(50) )
      When 'business' Then 'Business'
      When 'mod_cook' Then 'Modern Cooking'
      When 'popular_comp' Then 'Popular Computing'
```

```
        When 'psychology' Then 'Psychology'
        When 'trad_cook' Then 'Traditional Cooking'
        When 'UNDECIDED' Then 'Undecided'
    End
From [Pubs].[dbo].[Titles]
Group By [Type] -- get distinct values

-- Combine the data from the lookup table and the original table
Select
    [TitleId]=Cast( [title_id] as nvarchar(6) )
  , [TitleName]=Cast( [title] as nvarchar(50) )
  , [TitleType]=[CleanTitleType]
  , [PublisherKey]=[DWPubsSales].[dbo].[DimPublishers].[PublisherKey]
  , [TitlePrice]=Cast( [price] as decimal(18, 4) )
  , [PublishedDate]=[pubdate]
From [Pubs].[dbo].[Titles]
Join [DWPubsSales].[dbo].[DimPublishers]
  On [Pubs].[dbo].[Titles].[pub_id]=[DWPubsSales].[dbo].[DimPublishers].[PublisherId]
Join [DWPubsSales].[dbo].[TitleTypeLookup]
  On [Pubs].[dbo].[Titles].[type]=[DWPubsSales].[dbo].[TitleTypeLookup].[OriginalTitleType]
```

■ **Note**   Rather than using the DISTINCT keyword, you may find it advantageous to use a SQL GROUP BY statement that can give you different results than the DISTINCT keyword will when working with some functions, such as ROW_NUMBER(). In the example in Listing 6-12, there is no change in results, but it is a good technique to keep in mind as part of your ETL toolkit.

Generate Date Data

There are times when the ETL process will not just copy and transform existing data but instead will generate entirely new data. For example, many data warehouses have a table holding a sequential list of dates which are generated using an INSERT statement nested inside a programmatic loop.

In Chapter 5, we created a DimDates table, but did not fill it with data. As we will see in Chapter 8, Exercise 8-1, the SQL code in Listing 6-13 can be used during the ETL process to fill the DimDates table.

Listing 6-13. Filling the DimDates Table

```
-- Because the date table has no associated source table we can fill the data
-- using a SQL script.

-- Create variables to hold the start and end date
DECLARE @StartDate datetime = '01/01/1990'
DECLARE @EndDate datetime = '12/31/1995'

-- Use a while loop to add dates to the table
DECLARE @DateInProcess datetime
SET @DateInProcess=@StartDate

WHILE @DateInProcess <= @EndDate
 BEGIN
 -- Add a row into the date dimension table for this date
INSERT INTO DimDates (
```

```
    [Date]
  , [DateName]
  , [Month]
  , [MonthName]
  , [Quarter]
  , [QuarterName]
  , [Year]
  , [YearName]
  )
  VALUES (
  -- [Date]
    @DateInProcess
  -- [DateName]
   , Convert(varchar(50), @DateInProcess, 110)+', '
     + DateName( weekday, @DateInProcess )
  -- [Month]
, Month( @DateInProcess )
  -- [MonthName]
, Cast( Year(@DateInProcess) as varchar(4) )+' - '
  + DateName( month, @DateInProcess )
  -- [Quarter]
, DateName( quarter, @DateInProcess )
  -- [QuarterName]
, Cast( Year(@DateInProcess) as varchar(4) )+' - '
  + 'Q'+DateName( quarter, @DateInProcess )
  -- [Year]
, Year(@DateInProcess)
  -- [YearName]
, Cast( Year(@DateInProcess) as Char(4) )
)

  -- Add a day and loop again
  SET @DateInProcess=DateAdd(d, 1, @DateInProcess)
  END

-- Check the table SELECT Top 10 * FROM DimDates
```

Note that we used SQL variables to indicate the range of our dates as starting on 1/1/1990 and ending on 12/31/1995. This may seem odd, but if you look at the dates in the Pubs database, you will see that all of the dates are within this range.

Dealing with Nulls

Null values represent a special challenge because there are many different opinions on how to handle them in a data warehouse environment. Because of their very nature, nulls present you with multiple options, and it's up to you to sort out which approach works best in a given case. Nulls are most often considered to be an unknown value, but what does the term *unknown* really mean? Does it mean that it is unknowable and could never be known? Does it mean that is unknown at the moment but will soon be available? Does it mean that a value just does not apply in this particular instance? Any of these can be true: the value may not be known, may be missing, or may not be applicable.

Because of the ambiguity of the term *null*, a decision has to be made as to which interpretation is most accurate, or you may face endless arguments over the validity of your reports. Let's break each option down to help with this decision process.

Nulls in a Fact Table

Null values in the fact table can be either dimensional keys or measured values. When the null is a measured value, we recommend that you keep the null value indicating that it is either unknown or simply does not exist yet. This is because most database products can properly aggregate null values into totals and subtotals. If you substitute values such as a zero for null, the calculations may become incorrect.

To understand this issue, take a look at Listing 6-14. It shows an example of the difference when using the aggregate function AVERAGE().

Listing 6-14. How Nulls Work with Aggregate Functions

```
-- Using a null you get the correct answer of 4
Select Avg([Amt]) From (
  Select [Amt] = 2
    Union
  Select [Amt] = 6
    Union
  Select [Amt] = null
) as aDemoTableWithNull

-- Using a zero you get the incorrect answer of 2
Select Avg([Amt]) From (
  Select [Amt] = 2
    Union
  Select [Amt] = 6
    Union
  Select [Amt] = 0
) as aDemoTableWithZero
```

In cases where the fact table's null value is a dimensional key, we suggest that you create an artificial key value that can further describe the meaning of the null.

Be aware that you may need to provide more than one artificial key for each interpretation of the null values. Programmers often want to group null data together and assign implicit meanings to these nulls. Although this approach is useful for determining how many orders have not shipped, there is a downside to this grouping. What happens when the implicit meaning is not correct? For example, if a package was shipped but the shipping date was not recorded, the package may be sent twice.

To alleviate these issues, provide ways to replace null values with a set of descriptive dimensional attributes. You can then use a SELECT-CASE lookup transformation, similar to the one we just discussed in the "Provide Conformity" section of this chapter, to exchange the null values for these more descriptive dimensional values.

Here is an example of a simple null lookup: if you have a sales record in your fact table that does not have a ShipperId associated with it, for no other reason than one has not been chosen yet, you could use a ShipperId whose name column had a value of "Shipper not selected," as shown in Figure 6-5.

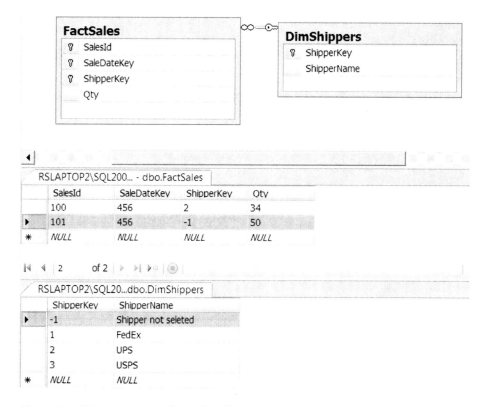

Figure 6-5. *Using a surrogate key with null values*

■ **Note**　We use negative numbers for the null related dimensional keys, because SSAS does not handle using a zero as a dimension key well. But even if it did, the advantage to using negative numbers is that you can specify multiple interpretations by marking each with a different number such as −1, −2, −3, and so on, and easily distinguish them from the shippers that have positive ID numbers.

Nulls in a Dimension Table

We do not recommend leaving null values in dimensional tables. Instead, they can be either excluded or transformed.

The simplest technique to handle null values is to disallow them in the data warehouse. The idea is that these null values can still be reported against the OLTP environment when necessary but are too ambiguous to be used in the data warehouse. For specific occasions where disallowing nulls is appropriate, it is easily handled, as shown in Listing 6-15.

Listing 6-15. Excluding Nulls with the Where Clause

```
Select
    [TitleId]=Cast( [title_id] as nvarchar(6) )
  , [TitleName]=Cast( [title] as nvarchar(50) )
  , [TitleType]=Cast( [type] as nvarchar(50) )
```

```
    , [PublisherKey] = [DWPubsSales].[dbo].[DimPublishers].[PublisherKey]
    , [TitlePrice] = Cast( [price] as decimal(18, 4) )
    , [PublishedDate] = [pubdate]
From [Pubs].[dbo].[Titles]
Join [DWPubsSales].[dbo].[DimPublishers]
  On [Pubs].[dbo].[Titles].[pub_id] = [DWPubsSales].[dbo].[DimPublishers].[PublisherId]
Where [Pubs].[dbo].[Titles].[Title_Id] Is Not Null
And [Pubs].[dbo].[Titles].[Title] Is Not Null
And [Pubs].[dbo].[Titles].[Type] Is Not Null
And [Pubs].[dbo].[Titles].[Price] Is Not Null
And [Pubs].[dbo].[Titles].[PubDate] Is Not Null
```

This approach works for some tables; however, it has the potential to exclude much of the data that you need for your BI solution. Whenever possible, it is more accurate to transform nulls into descriptive values. One way to accomplish this is by using the ISNULL() function. In Listing 6-16 the ISNULL() function converts null values into the word *Unknown*. The value of Unknown is more descriptive than the use of null, although admittedly not by much.

Listing 6-16. Converting Null Values with the ISNULL() Function

```
Select
    [TitleId] = Cast( isNull( [title_id], -1 ) as nvarchar(6) )
    , [TitleName] = Cast( isNull( [title], 'Unknown' ) as nvarchar(50) )
    , [TitleType] = Cast( isNull( [type], 'Unknown' ) as nvarchar(50) )
    , [PublisherKey] = [DWPubsSales].[dbo].[DimPublishers].[PublisherKey]
    , [TitlePrice] = Cast( isNull( [price], -1 ) as decimal(18, 4) )
    , [PublishedDate] = isNull( [pubdate], '01/01/1900' )
From [Pubs].[dbo].[Titles]
Join [DWPubsSales].[dbo].[DimPublishers]
  On [Pubs].[dbo].[Titles].[pub_id] = [DWPubsSales].[dbo].[DimPublishers].[PublisherId]
```

Nulls in columns that consist of noncharacter data types, such as integers and datetime, will not accept the string value of Unknown. Consequently, you can use the ISNULL() function to convert the published date into something like 01/01/1900. The year 1900 is a starting point for many date data types and is unlikely to mark a real event, at least in this particular database. This means that it has no relevance to an actual publication date and can be used as an indicator for missing data. Keep in mind, this date will not be appropriate when the data includes this particular date as a legitimate value.

A Null Lookup Table

Although the previous techniques for dealing with nulls are used in many data warehouses, another option is to reference lookup values in either a dedicated lookup table or existing dimensional tables.

Lookup tables (also known as *domain tables*) are additional tables that are not part of the dimension model but instead have descriptive values that can be referenced from the dimensional tables. Figure 6-6 shows the contents of a typical lookup table.

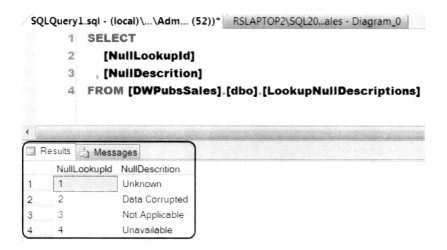

Figure 6-6. *A typical lookup table*

For the lookup table to function properly, you have to determine the meaning of each null found in the source data and transform the null value into a lookup ID that will cross-reference the lookup table.

The dilemma with a dedicated lookup table is that each dimension with null values now forms a snowflake design. Although this is not a problem per se, it violates the "keep it simple" rule.

Often you will find that a lookup table can be collapsed into your dimension tables. For example, when a date column in the fact or dimension table has a null value, you can use a dimension table such as the DimDates table as a lookup table, as long as it includes rows that contain lookup data. These lookup rows contain descriptive values that represent your null interpretations. Listing 6-17 shows an example of adding two lookup rows to the DimDates table.

Listing 6-17. Adding Additional Lookup Values to the DimDates Table

```
Set Identity_Insert [DimDates] On
INSERT INTO [DWPubsSales].[dbo].[DimDates] (
   [DateKey] -- This is normally added automatically
 , [Date]
 , [DateName]
 , [Month]
 , [MonthName]
 , [Quarter]
 , [QuarterName]
 , [Year]
 , [YearName]
 )
VALUES
( -1 -- This will be the Primary key for the first lookup value
 , '01/01/1900'
 , 'Unknown Day'
 , -1
 , 'Unknown Month'
 , -1
 , 'Unknown Quarter'
 , -1
```

```
, 'Unknown Year'
)
, -- add a second row
( -2 -- This will be the Primary key for the second lookup value
, '02/01/1900'
, 'Corrupt Day'
, -2
, 'Corrupt Month'
, -2
, 'Corrupt Quarter'
, -2
, 'Corrupt Year'
)
Set Identity_Insert [DimDates] Off
```

■ **Note**   We now have lookups for two interpretations of null values, unknown and corrupt. Remember that you can create as many null definitions as you want, but you will also have to create additional programming logic to distinguish between each case. Determining the meaning of a null value is not an easy task, nor is it within the scope of this book. A temporary simplistic approach is to use one interpretation for all your nulls, such as Unknown, until you have time to create a programmatic resolution.

In Chapter 5, you created the DimDates table with an IDENTITY option on the DateKey column. This option automatically inserts numeric values every time a new row is added. In Listing 6-13 we added rows of dates to the DimDates table but did not include null lookup values.

By default the IDENTITY option prevents values from being inserted into the column manually. However, you can manually insert a value if you enable SQL's IDENTITY_INSERT option, by using the SET IDENTITY_INSERT < table name > ON SQL command. Once that is done, additional lookup values can be added to the dimension table, as shown in Figure 6-7.

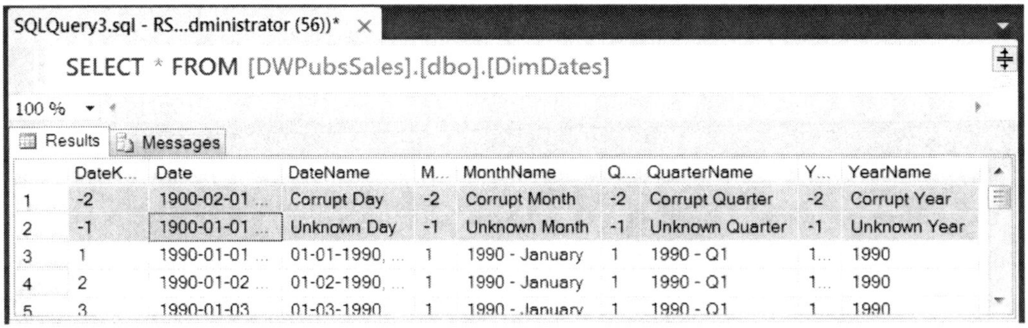

Figure 6-7. *Additional lookup dates have been added.*

Once you add the new lookup rows, you can use them in your other tables. Listing 6-18 shows code that will do this very thing. It utilizes an outer join to include every value in the DimTitles, even when no matching date is found in the DimDates table. On every row where a matching date is not found, the outer join forces a null value to automatically be inserted into the results. Therefore, we can use the ISNULL() function to convert the new null value into -1, which we then cross-reference to the Unknown date in the DimDates table.

Listing 6-18. Cross-Referencing the DimDates Table's Dimensional Keys

```
Select
    [TitleId]=Cast( isNull( [title_id], -1 ) as nvarchar(6) )
  , [TitleName]=Cast( isNull( [title], 'Unknown' ) as nvarchar(50) )
  , [TitleType]=Cast( isNull( [type], 'Unknown' ) as nvarchar(50) )
  , [PublisherKey]=[DWPubsSales].[dbo].[DimPublishers].[PublisherKey]
  , [TitlePrice]=Cast( isNull( [price], -1 ) as decimal(18, 4) )

  , [PublishedDateKey]=isNull( [DWPubsSales].[dbo].[DimDates].[DateKey], -1 )

From [Pubs].[dbo].[Titles]
Join [DWPubsSales].[dbo].[DimPublishers]
  On [Pubs].[dbo].[Titles].[pub_id]=[DWPubsSales].[dbo].[DimPublishers].[PublisherId]
Left Join [DWPubsSales].[dbo].[DimDates] -- The "Left" keeps dates not found in DimDates
  On [Pubs].[dbo].[Titles].[pubdate]=[DWPubsSales].[dbo].[DimDates].[Date]
```

▦ **Note**   The problem with each of these solutions is your ability to interpret the meaning of each null value. If your source data does not provide you any way to identify the meaning, the null values will have to be assigned as an Unknown. This is not particularly satisfying, but there is no magic that can fix these issues. You must work to improve the source data before you can resolve the interpretation of nulls in your data warehouse.

The SQL Query Designer

As you can see, SQL code can become complex very quickly. To make the programming process easier, you can utilize the SQL Query Designer that comes as part of SQL Management Studio or the one that comes with Integration Services (SSIS). Both these tools are nearly identical and create SQL code using a GUI interface.

The Query Designer that is part of SQL Management Studio makes it easy for you to create multiple statements in a single SQL script file, whereas the one that is part of the SSIS development environment does not. Therefore, it is likely that you will find Management Studio's Query Designer to be the more powerful of the two options.

To start SQL Server Management Studio's Query Designer, open a new query window and simultaneously press the Ctrl + Shift + Q keys. You can also access the designer through the Query menu at the top of Management Studio or by right-clicking a blank area of a query window and selecting the Design Query in Editor option from the context menu. We recommend using this last option. Usually, we start by adding a comment in the query window to identify what we are trying to accomplish, then creating a new line below the comment and right-clicking the blank spot to access the context menu (Figure 6-8).

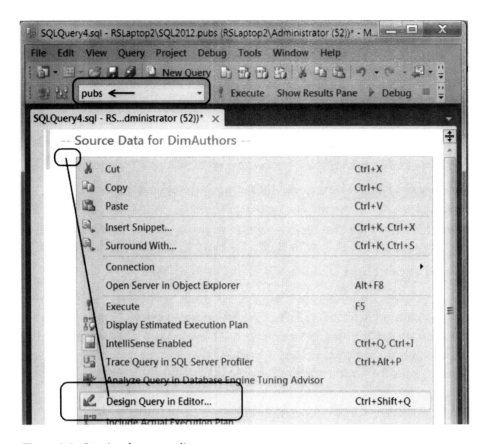

Figure 6-8. *Starting the query editor*

When the Query Designer launches, it will present the Add Table dialog window and a list of database tables. Selecting one or more tables in this window adds the table names to the FROM clause of the new SQL SELECT statement you are creating. For example, if you select the Authors table and click the Add button, the Authors table will be added to the query window (Figure 6-9).

■ **Important** When the Query Designer is open, you cannot choose which database to use. If you do not see the tables you were expecting, close the Query Designer, change the focus to the correct database, and then reopen the Query Designer. You can use the database selector, indicated with a circle and an arrow in Figure 6-8, to change the database focus.

If you want your query to include more than one table, you can hold down the Ctrl button, select multiple tables, and then click the Add button. This feature allows you to create complex queries, such as SQL joins. After you have selected all the tables you want to use, click the Close button to close the Add Tables dialog window.

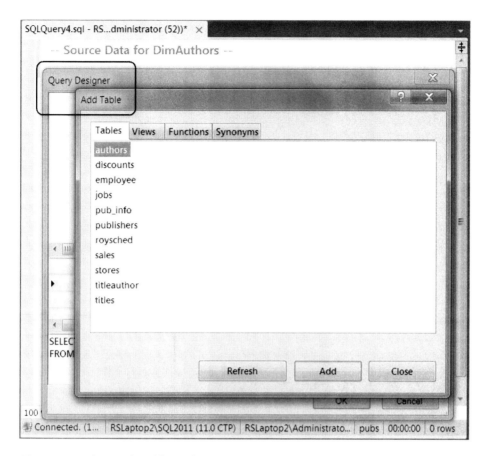

Figure 6-9. *Selecting the tables in the Query Designer*

When the Add Tables dialog window closes, your tables will be represented at the top of the Query Designer window, and the SQL code to query the table will be at the bottom (Figure 6-10).

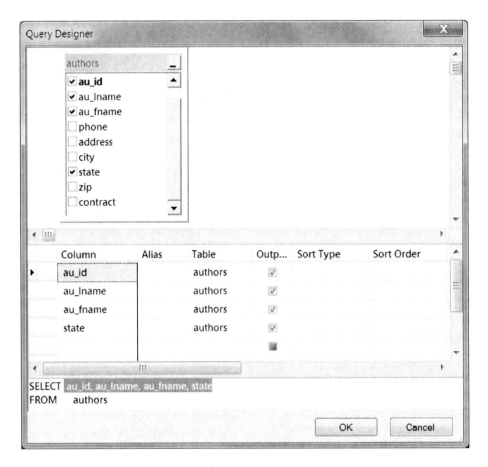

Figure 6-10. *Selecting the columns in the Query Designer*

We need to extract data from the original Author's source table. Therefore, we write a SQL SELECT statement to retrieve this data. If you check the checkboxes on each column you want to use, the Query Designer will add them to the SELECT statement you are building (Figure 6-10).

The Authors and DimAuthors tables contain several differences. As a result, we need to apply a number of transformations to the basic query. These include combining data from multiple columns into one column of data, renaming the existing columns, and converting column data types.

Common transformations are concatenating two columns to form a single name and converting source data into a different data type. In our example, we want to combine an author's first and last name into a single column. Keeping the names separate in an OLTP database is common, but it is unlikely to be of use in our sales data warehouse, that is, unless you think that reports will be made that will aggregate the measures on a person's first or last name.

In the DWPubsSales data warehouse, we are going to assume that a sales report detailing the sales quantity based on an author's last name is not needed. Therefore, we are going to combine an authors' first and last names to keep the reporting process as simple as possible. To combine this data, we can add a simple concatenation to the query, as shown in Figure 6-11.

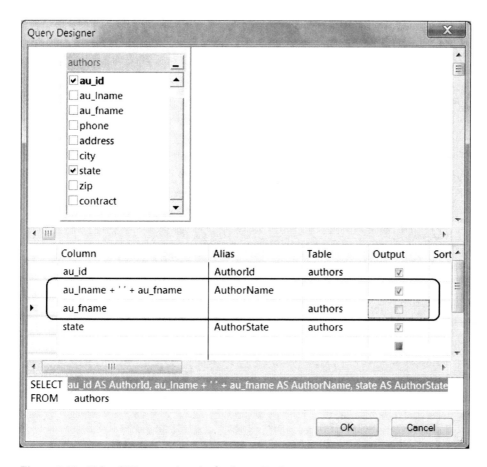

Figure 6-11. *Using SQL expressions in the Query Designer*

Renaming columns is about the simplest transformation you can do. This transformation is accomplished by adding the alias to the Alias column in the Query Designer (Figure 6-11).

As we have seen previously, you change data types with either the CAST() or CONVERT() function, and both of these can be typed right into the query window to form another SQL expression.

There is no reason why you cannot combine both the concatenation and added casting transformations as shown in Figure 6-12. Unfortunately, the Query Designer tool is not designed for this and will misunderstand what you are trying to do even though you are using perfectly legal SQL syntax. As you click away from the Column cell where your expression is typed, you will receive an error message stating that the conversion may be unnecessary (Figure 6-12). You can ignore this error message as long as you are sure that the SQL syntax is correct and appropriate to what you are trying to accomplish, but at least it will change your code, in spite of having to work around this error message.

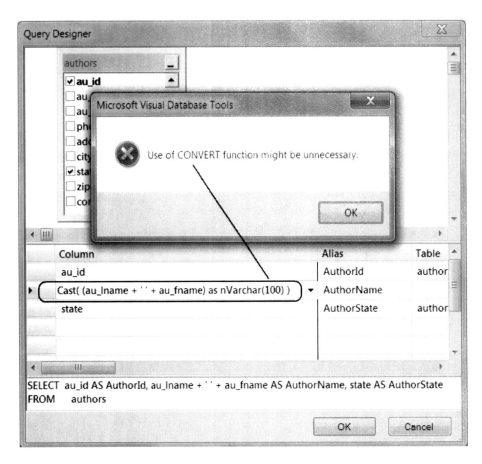

Figure 6-12. *A misleading error from the Query Designer*

Your formatted code will now look something like the code in Listing 6-19. Notice that an N has been added to the data conversion. This N indicates that the space between the two single quotes should be considered Unicode. Strictly speaking, there is nothing wrong with this; however, it may be confusing to someone working with this code. Using the N symbol is well documented and can easily be researched to determine the meaning, but it is still commonly misunderstood by developers new to SQL programming.

Listing 6-19. An Example of Code That the Query Designer Will Change Without Your Consent

```
SELECT au_id AS AuthorId, CAST( (au_fname+N' '+au_lname) AS nVarchar(100)) AS AuthorName, state
AS AuthorState
FROM authors
```

Another issue with the Query Designer is the way it formats your code. As we mentioned at the beginning of this chapter, formatting your code consistently is the mark of a professional. If you try to format the code in the Query Designer window, however, it will ignore what you type and change the code to what it thinks is better. Yes, the Query Designer does have an ego.

You can always clean up the code and the conversion code after you close the Query Designer, but if you forget this feature, it can be frustrating. Figure 6-13 shows how we reformatted the code created using the Query Designer. We have also added a data conversion to the state column after the Query Designer closes to avoid seeing the error in Figure 6-12 again.

■ **Note** To solve the issues of code conversion and reformatting, we recommend using the Query Designer to type out the basic syntax for your SQL statements and then closing the tool and completing the statement by hand. Using this combination will provide you with a quick and easy way to write SQL code.

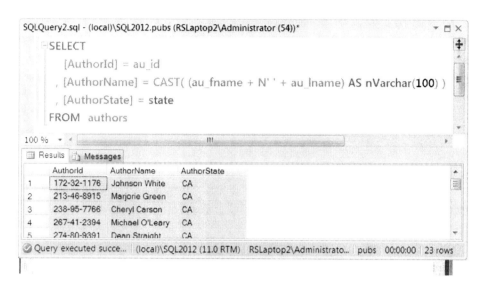

Figure 6-13. *Reformatting and updating your query code*

Updating Your BI Documentation

This chapter has focused on developing SQL code for your ETL process. The general consensus seems to be that most developers dislike creating documentation. And yet, documentation is an important aspect of completing any solution, and that documentation must be upgraded as progress is made toward the completion of the solution.

In this book, we have tried to keep the documentation requirements to a minimum. So far, we have simplified this process by using an Excel spreadsheet to document the solution objects. Because this is currently our only form of documentation, we need to keep it up-to-date by recording our transformation logic in the spreadsheet.

Figure 6-14 shows an example of adding the transformations required to the data warehouse objects worksheet. Doing this provides two distinct advantages. First, it gives you the opportunity to let other developers know what kind of transformations have occurred without having to read complex SQL statements. And second, it provides you with a way of verifying that you have completed coding all the transformations in your list.

Pubs.dbo.titles.pubdate	datetime	int	lookup in Dim table
Pubs.dbo.publishers	Table	Table	
Generated	na	int	na
Pubs.dbo.publishers.pub_id	char(4)	nchar(4)	Cast to nchar(4)
Pubs.dbo.publishers.pub_name	varchar(40)	nvarchar(50)	Cast to nvarchar(50)
Pubs.dbo.titleauthor	Table	Table	
Pubs.dbo.titleauthor.title_id	varchar(6)	int	na
Pubs.dbo.titleauthor.au_id	varchar(11)	int	na
Pubs.dbo.titleauthor.au_ord	tinyint	int	Cast to Int
Pubs.dbo.authors	Table	Table	
Generated	na	int	na
Pubs.dbo.authors.au_id	varchar(11)	nchar(11)	Cast to nvarchar(50)
Pubs.dbo.authors.au_fname	varchar(40)	nvarchar(100)	Cast(Fname + Lname) to nvarchar(100)
Pubs.dbo.authors.au_lname	varchar(20)	na	na
Pubs.dbo.authors.state	char(2)	nchar(2)	Cast to nchar(2)
na		na	Table
na		na	int

Figure 6-14. Documenting your ETL transformations

■ **Note** When creating your ETL code, updating your documents to define the transformations required may be the most convenient place to start. Although we do not believe that this is a replacement for formal documentation, it can allow developers to track their updates and provide you with accurate information about those changes. Later this document can be turned into formal documentation. We show you examples of this in Chapter 19.

Building an ETL Script

As you work with the ETL process, it is a good idea to organize all the ETL code you generate. This can easily be done by opening up SQL Server Management Studio to create a new query, typing in the necessary code, and saving the code as a SQL script file. The advantages to this are that you can later add script to your SSIS project as a miscellaneous file, and it can also be reused in the future for other projects that are similar. In the next exercise, you do just that.

EXERCISE 6-1. CREATING AN ETL SCRIPT

In this exercise, you create the ETL code needed to fill the Publication Industries data warehouse. You can choose either to type the SQL code presented here or to use the Query Designer tool to accomplish your goal. We recommend using both.

Important: You are practicing administrator-level tasks in this book, so you need administrator-level privileges. The easiest way to achieve this when opening a program is to remember to always right-click a menu item, select Run as Administrator, and then answer Yes to access administrator-level privileges while running this program. In Windows 7 and Vista, logging in with an administrator account is not enough. For more information, search the Web on the keywords "Windows 7 True Administrator and User Access Control."

1. Open Excel from the Start menu using the Run as Administrator option.

2. Locate and open the following Excel spreadsheet:
    C:_BookFiles\Chapter06Files\BISolutionWorksheets.xlsx. You can do so by
    clicking the File tab in Microsoft Excel and choosing the Open option from the menu.

3. We have added a column called Transformations to the spreadsheet you used in Chapter 5. It shows the transformations we need. Review the transformations defined on the data warehouse worksheet.

4. Open SQL Server Management Studio 2012. You can do so by clicking the Start button and navigating to All Programs ➤ Microsoft SQL Server ➤ SQL Server Management Studio. Right-click SQL Server Management Studio 2012 and click the Run as Administrator menu item. If the UAC message box appears asking, "Do you want the following program to make changes to this computer?" click Yes (or Continue depending upon your operating system) to accept this request.

5. When SQL Server Management Studio opens, choose to connect to the database engine by selecting this option in the Server Type dropdown box. Then click the Connect button to connect to the database engine. (For more information on connecting to your database, see Chapter 5.)

6. Decide on a method for creating the SQL Script either by clicking the New Query button and typing the code by hand, using the Query Designer, or both. Then, type the following code in Listing 6-20 to create the script.

Tip: This code can be found in the file C:_ BookFiles\Chapter06Files from the downloadable book files. We recommend that you type the following code yourself, but it is nice to know that you don't have to.

Listing 6-20. The ELT Script for DWPubsSales

```
/*******************************************************************************
The code in this file is used to create an ETL process for the
Publication Industries data warehouse.

INPORTANT: You must run the "Creating the Publication Industries Data Warehouse.sql"
file before you can use this code. This file is in the Chapter05 folder.
*******************************************************************************/

-- Step 1) Code used to Clear tables (Will be used with SSIS Execute SQL Tasks)
Use DWPubsSales

-- 1a) Drop Foreign Keys
Alter Table [dbo].[DimTitles] Drop Constraint [FK_DimTitles_DimPublishers]
Alter Table [dbo].[FactTitlesAuthors] Drop Constraint [FK_FactTitlesAuthors_DimAuthors]
Alter Table [dbo].[FactTitlesAuthors] Drop Constraint [FK_FactTitlesAuthors_DimTitles]
Alter Table [dbo].[FactSales] Drop Constraint [FK_FactSales_DimStores]
Alter Table [dbo].[FactSales] Drop Constraint [FK_FactSales_DimTitles]
Alter Table [dbo].[FactSales] Drop Constraint [FK_FactSales_DimDates]
Alter Table [dbo].[DimTitles] Drop Constraint [FK_DimTitles_DimDates]
-- You will add Foreign Keys back (At the End of the ETL Process)
Go

--1b) Clear all tables data warehouse tables and reset their Identity Auto Number
Truncate Table dbo.FactSales
Truncate Table dbo.FactTitlesAuthors
Truncate Table dbo.DimTitles
Truncate Table dbo.DimPublishers
Truncate Table dbo.DimStores
Truncate Table dbo.DimAuthors
```

```
Truncate Table dbo.DimDates
Go

-- Step 2) Code used to fill tables (Will be used with SSIS Data Flow Tasks)

-- 2a) Get source data from pubs.dbo.authors and
-- insert into DimAuthors
Select
  [AuthorId]=Cast( au_id as nChar(11) )
, [AuthorName]=Cast( ( au_fname+' '+au_lname ) as nVarchar(100) )
, [AuthorState]=Cast( state as nChar(2) )
From pubs.dbo.authors
Go

-- 2b) Get source data from pubs.dbo.stores and
-- insert into DimStores
Select
  [StoreId]=Cast( stor_id as nChar(4) )
, [StoreName]=Cast( stor_name as nVarchar(50) )
From pubs.dbo.stores
Go

-- 2c) Get source data from pubs.dbo.publishers and
-- insert into DimPublishers
Select
  [PublisherId]=Cast( pub_id as nChar(4) )
, [PublisherName]=Cast( pub_name as nVarchar(50) )
From pubs.dbo.publishers
Go

-- 2d) Create values for DimDates as needed.

-- Create variables to hold the start and end date

Declare @StartDate datetime = '01/01/1990'
Declare @EndDate datetime = '01/01/1995'

-- Use a while loop to add dates to the table
Declare @DateInProcess datetime
Set @DateInProcess=@StartDate

While @DateInProcess<= @EndDate
 Begin
 -- Add a row into the date dimension table for this date
 Insert Into DimDates
 ( [Date], [DateName], [Month], [MonthName], [Quarter], [QuarterName], [Year],
   [YearName] )
 Values (
  @DateInProcess -- [Date]
 , DateName( weekday, @DateInProcess ) -- [DateName]
 , Month( @DateInProcess ) -- [Month]
 , DateName( month, @DateInProcess ) -- [MonthName]
 , DateName( quarter, @DateInProcess ) -- [Quarter]
 , 'Q'+DateName( quarter, @DateInProcess )+' - '
     + Cast( Year( @DateInProcess) as nVarchar(50) ) -- [QuarterName]
 , Year( @DateInProcess )
```

```
    , Cast( Year( @DateInProcess ) as nVarchar(50) ) -- [Year]
    )
  -- Add a day and loop again
  Set @DateInProcess =DateAdd( d, 1, @DateInProcess )
  End

-- 2e) Add additional lookup values to DimDates
Set Identity_Insert [DWPubsSales].[dbo].[DimDates] On
Go

Insert Into [DWPubsSales].[dbo].[DimDates]
  ( [DateKey]
  , [Date]
  , [DateName]
  , [Month]
  , [MonthName]
  , [Quarter]
  , [QuarterName]
  , [Year], [YearName] )
Select
    [DateKey] = -1
  , [Date] =Cast( '01/01/1900' as nVarchar(50) )
  , [DateName] =Cast( 'Unknown Day' as nVarchar(50) )
  , [Month] = -1
  , [MonthName] =Cast( 'Unknown Month' as nVarchar(50) )
  , [Quarter] = -1
  , [QuarterName] =Cast( 'Unknown Quarter' as nVarchar(50) )
  , [Year] = -1
  , [YearName] =Cast( 'Unknown Year' as nVarchar(50) )
  Union
  Select
    [DateKey] = -2
  , [Date] =Cast( '01/01/1900' as nVarchar(50) )
  , [DateName] =Cast( 'Corrupt Day' as nVarchar(50) )
  , [Month] = -2
  , [MonthName] =Cast( 'Corrupt Month' as nVarchar(50) )
  , [Quarter] = -2
  , [QuarterName] =Cast( 'Corrupt Quarter' as nVarchar(50) )
  , [Year] = -2
  , [YearName] =Cast( 'Corrupt Year' as nVarchar(50) )
Go

Set Identity_Insert [DWPubsSales].[dbo].[DimDates] Off
Go

-- 2f) Get source data from pubs.dbo.titles and
-- insert into DimTitles
Select
    [TitleId] =Cast( isNull( [title_id], -1 ) as nvarchar(6) )
  , [TitleName] =Cast( isNull( [title], 'Unknown' ) as nvarchar(100) )
  , [TitleType] =Case Cast( isNull( [type], 'Unknown' ) as nvarchar(50) )
      When 'business' Then N'Business'
      When 'mod_cook' Then N'Modern Cooking'
```

```
      When 'popular_comp' Then N'Popular Computing'
      When 'psychology' Then N'Psychology'
      When 'trad_cook' Then N'Traditional Cooking'
      When 'UNDECIDED' Then N'Undecided'
    End
  , [PublisherKey]=[DWPubsSales].[dbo].[DimPublishers].[PublisherKey]
  , [TitlePrice]=Cast( isNull( [price], -1 ) as decimal(18, 4) )
  , [PublishedDateKey]=isNull( [DWPubsSales].[dbo].[DimDates].[DateKey], -1 )
From [Pubs].[dbo].[Titles]
Join [DWPubsSales].[dbo].[DimPublishers]
  On [Pubs].[dbo].[Titles].[pub_id]=[DWPubsSales].[dbo].[DimPublishers].[PublisherId]
Left Join [DWPubsSales].[dbo].[DimDates] -- The "Left" keeps dates not found in DimDates
  On [Pubs].[dbo].[Titles].[pubdate]=[DWPubsSales].[dbo].[DimDates].[Date]
Go

-- 2g) Get source data from pubs.dbo.titleauthor and
-- insert into FactTitlesAuthors
Select
  [TitleKey]=DimTitles.TitleKey
--, title_id
, [AuthorKey]=DimAuthors.AuthorKey
--, au_id
, [AuthorOrder]=au_ord
From pubs.dbo.titleauthor
JOIN DWPubsSales.dbo.DimTitles
  On pubs.dbo.titleauthor.Title_id=DWPubsSales.dbo.DimTitles.TitleId
JOIN DWPubsSales.dbo.DimAuthors
  On pubs.dbo.titleauthor.Au_id=DWPubsSales.dbo.DimAuthors.AuthorId

-- 2h)Get source data from pubs.dbo.Sales and
-- insert into FactSales
Select
  [OrderNumber]=Cast( ord_num as nVarchar(50) )
, [OrderDateKey]=DateKey
--, title_id
, [TitleKey]=DimTitles.TitleKey
--, stor_id
, [StoreKey]=DimStores.StoreKey
, [SalesQuantity]=qty
From pubs.dbo.sales
JOIN DWPubsSales.dbo.DimDates
  On pubs.dbo.sales.ord_date=DWPubsSales.dbo.DimDates.date
JOIN DWPubsSales.dbo.DimTitles
  On pubs.dbo.sales.Title_id=DWPubsSales.dbo.DimTitles.TitleId
JOIN DWPubsSales.dbo.DimStores
  On pubs.dbo.sales.Stor_id=DWPubsSales.dbo.DimStores.StoreId

-- Step 3) Add Foreign Key s back (Will be used with SSIS Execute SQL Tasks)
Alter Table [dbo].[DimTitles] With Check
Add Constraint [FK_DimTitles_DimPublishers]
Foreign Key ( [PublisherKey] ) References [dbo].[DimPublishers] ( [PublisherKey] )
```

```
Alter Table [dbo].[FactTitlesAuthors] With Check
Add Constraint [FK_FactTitlesAuthors_DimAuthors]
Foreign Key ( [AuthorKey] ) References [dbo].[DimAuthors] ( [AuthorKey] )

Alter Table [dbo].[FactTitlesAuthors] With Check
Add Constraint [FK_FactTitlesAuthors_DimTitles]
Foreign Key ( [TitleKey] ) References [dbo].[DimTitles] ( [TitleKey] )

Alter Table [dbo].[FactSales] With Check
Add Constraint [FK_FactSales_DimStores]
Foreign Key ( [StoreKey] ) References [dbo].[DimStores] ( [Storekey] )

Alter Table [dbo].[FactSales] With Check
Add Constraint [FK_FactSales_DimTitles]
Foreign Key ( [TitleKey] ) References [dbo].[DimTitles] ( [TitleKey] )

Alter Table [dbo].[FactSales] With Check
Add Constraint [FK_FactSales_DimDates]
Foreign Key ( [OrderDateKey] ) References [dbo].[DimDates] ( [DateKey] )

Alter Table [dbo].[DimTitles] With Check
Add Constraint [FK_DimTitles_DimDates]
Foreign Key ( [PublishedDateKey] ) References [dbo].[DimDates] ( [DateKey] )
```

7. When you are done, if you found any transformations that were missed, go back and finish them by adding the necessary SQL code to your script file. Once that is done, save your script file to: C:_BISolutions\PublicationsIndustries\ PublicationIndustriesETLCode.sql.

In this exercise, you created the code that gathers data for the data warehouse objects. Most of these tables are still empty because we have defined only the source of the data, but we start filling them up using a combination of the SQL statements you have just created along with SSIS tasks in the next chapter. Before we do that, however, we need to look at three common ways of abstracting your SQL code: using views, stored procedures, and user-defined functions.

Working in the Abstract

It is common knowledge in the programming industry that building software by defining multiple layers can give you greater flexibility and lower maintenance costs. One of the classic ways of doing this is by providing an abstraction layer between the objects that contain data and the software that uses that data. In Microsoft SQL Server, this can be done with three simple database objects: views, stored procedures and functions.

The concept of abstraction basically means designing your software so that the underlying objects are always used indirectly. In the case of a database table, you do not directly connect your SSIS applications to the table itself, but instead utilize one of these abstraction objects.

The advantages of this design include the following:

- It can mask the complexity of programming statements by binding these statements to a named database object.

- Underlying data structures can undergo changes as a normal part of maintenance, but the abstraction objects will hide these changes.

- Permissions can be given to the abstraction objects and not to the underlying data structures, thus protecting the data structure from misuse.

- The same data structure can be represented multiple ways without having to create duplicate data structures.

You can probably think of more advantages than those we have listed, but these are enough to make one consider utilizing them in your ETL solutions. In fact, Microsoft has long recommended it as a best practice to do so.

Views

One of the most common tools used in a database is a SQL view. A SQL view consists of a named select statement that is saved internally in the database. The view is used as if it were a table. Indeed, it is sometimes called a *virtual table*. This is a misnomer, however, because it gives the impression that the table and the view are more similar than they truly are.

One of the biggest differences between a table and a view is that table data has a physical representation on your hard drive, whereas the view is just a saved select statement. Yet views still act as if they were tables, and you can select against the view exactly as you would against a table.

Although there are some restrictions about which SQL SELECT clauses are allowed in a view, they are still useful for many ETL processing tasks. For example, the code in Listing 6-21 creates a view around a SQL statement that extracts data required for a sales fact table. Note that the view transforms the column names using column aliases. It also transforms data found in multiple tables into a single logical table using SQL JOIN statements.

Listing 6-21. Creating a View for ETL Processing

```
Create View vEtlFactSalesData
as

Select
  [OrderNumber]=ord_num
, [OrderDateKey]=DateKey
, [TitleKey]=DimTitles.TitleKey
, [StoreKey]=DimStores.StoreKey
, [SalesQuantity]=qty
From pubs.dbo.sales
JOIN DWPubsSales.dbo.DimDates
  On pubs.dbo.sales.ord_date=DWPubsSales.dbo.DimDates.date
JOIN DWPubsSales.dbo.DimTitles
  On pubs.dbo.sales.Title_id=DWPubsSales.dbo.DimTitles.TitleId
JOIN DWPubsSales.dbo.DimStores
  On pubs.dbo.sales.Stor_id=DWPubsSales.dbo.DimStores.StoreId
```

Using column aliases and SQL Joins in a query may be rather simple for experienced SQL programmers, but it is considered advanced by most novices. Once the view is created, however, the complexity of this query is masked by the view. You can then query the view using a simple SQL statement, such as the one in Listing 6-22, that is understood by everyone.

Listing 6-22. Querying the View

```
Select * from vEtlFactSalesData
```

> ■ **Note** SSIS has a data source view object that can be used in conjunction with your SSIS packages. This may make the concept of views seem redundant, but in practice, you will find that the SSIS data source views are somewhat limited since they can only be used within an SSIS project, whereas SQL views can be used by any application that connects to SQL Server.

The view is a very simple and effective tool, but it does have some disadvantages. You cannot define parameters on a view. Parameters allow you to pass in specific arguments to get back differing results. This is a useful technique when incrementally loading a data warehouse. For example, the code in Listing 6- 23 filters out data that was not added on today's date. Because SQL views cannot contain parameters, this query could not be saved inside a Create View statement.

Listing 6-23. A Select Statement with a Parameter

```
Select
  [OrderNumber] = ord_num
, [OrderDateKey] = DateKey
, [TitleKey] = DimTitles.TitleKey
, [StoreKey] = DimStores.StoreKey
, [SalesQuantity] = qty
From pubs.dbo.sales
JOIN DWPubsSales.dbo.DimDates
  On pubs.dbo.sales.ord_date = DWPubsSales.dbo.DimDates.date
JOIN DWPubsSales.dbo.DimTitles
  On pubs.dbo.sales.Title_id = DWPubsSales.dbo.DimTitles.TitleId
JOIN DWPubsSales.dbo.DimStores
  On pubs.dbo.sales.Stor_id = DWPubsSales.dbo.DimStores.StoreId

Where pubs.dbo.sales.ord_date = @TodaysDate - This will not work in a View!
```

Stored Procedures

Like views, stored procedures consist of a named set of SQL statements. Unlike views, stored procedures can contain multiple statements, work with variables, and process transaction statements. Stored procedures can therefore be quite complex and contain hundreds of lines of SQL code, but they can also be as simple as a saved select statement. For example, Listing 6-24 holds the same select statement as the one in the view created in Listing 6-22. Notice that the syntax looks almost the same. The only difference is the command CREATE PROCEDURE in place of CREATE VIEW.

Listing 6-24. Creating a Stored Procedure for ETL Processing

```
Create Procedure pEtlFactSalesData
as

Select
  [OrderNumber] = ord_num
, [OrderDateKey] = DateKey
, [TitleKey] = DimTitles.TitleKey
, [StoreKey] = DimStores.StoreKey
, [SalesQuantity] = qty
From pubs.dbo.sales
```

```
JOIN DWPubsSales.dbo.DimDates
  On pubs.dbo.sales.ord_date=DWPubsSales.dbo.DimDates.date
JOIN DWPubsSales.dbo.DimTitles
  On pubs.dbo.sales.Title_id=DWPubsSales.dbo.DimTitles.TitleId
JOIN DWPubsSales.dbo.DimStores
  On pubs.dbo.sales.Stor_id=DWPubsSales.dbo.DimStores.StoreId
Go
```

Once you create a stored procedure, you run its code by executing it. The code to execute a stored procedure is in Listing 6-25.

Listing 6-25. Executing a Stored Procedure

```
Execute pEtlFactSalesData
```

If a stored procedure contains a SQL select statement, the selected results are returned to the client software. Client software includes SQL Server Management Studio, but as we show in Chapter 8, SSIS also acts as client software when it processes the stored procedure results.

As mentioned previously, one advantage of using stored procedures is their ability to process parameters. Parameters can be used to filter results and modify output or supplied transactional data. For example, you might filter a select statement specifically to return results that were associated with the current day's sales, as shown in Listing 6-26.

Listing 6-26. Altering the Stored Procedure to Use a Parameter

```
Alter Procedure pEtlFactSalesData

( @OrderDate datetime )
as

Select
  [OrderNumber]=ord_num
, [OrderDateKey]=DateKey
, [TitleKey]=DimTitles.TitleKey
, [StoreKey]=DimStores.StoreKey
, [SalesQuantity]=qty
From pubs.dbo.sales
JOIN DWPubsSales.dbo.DimDates
  On pubs.dbo.sales.ord_date=DWPubsSales.dbo.DimDates.date
JOIN DWPubsSales.dbo.DimTitles
  On pubs.dbo.sales.Title_id=DWPubsSales.dbo.DimTitles.TitleId
JOIN DWPubsSales.dbo.DimStores
  On pubs.dbo.sales.Stor_id=DWPubsSales.dbo.DimStores.StoreId
Where pubs.dbo.sales.ord_date=@OrderDate
```

Once a stored procedure is created, you can execute it by supplying the current date, using a built-in function such as the SQL Server GETDATE() function. If you were interested in incrementally loading your fact tables with only data that came in on a particular day, a stored procedure similar to this could be useful. Listing 6-27 shows an example of executing a stored procedure with a parameter.

Listing 6-27. Executing a Stored Procedure with a Parameter

```
Declare @TodaysDate datetime
Set @TodaysDate=Cast( GetDate() as datetime )
Execute pEtlFactSalesData

  @OrderDate=@TodaysDate
```

▪ **Note**   Whenever you are working with parameters like @OrderDate, list the name of the parameter on the left side of the assignment operator = and the argument @TodaysDate on the right side. This may seem backwards for beginner programmers who are used to reading an expression from left to right, (1 + 1 = 2), but in programming the code is reversed: (2 = 1 + 1).

Like all things in programming, stored procedures have both good and bad qualities. For the most part, their good qualities far outweigh their bad ones. Indeed, most programmers would be hard-pressed to find a strong negative aspect to using stored procedures in your ETL processing. But if you look hard enough, you will discover that stored procedures, when used in conjunction with SSIS packages, require special configurations: not difficult, just special. We look at how stored procedures are configured in SSIS in Chapter 7.

Stored procedures cannot be used as an expression. In other words, you cannot integrate a stored procedure call within a SQL select statement and have the stored procedure execute for each individual row. This is yet another aspect of stored procedures that is not all bad per se, but notable. One way around this issue is the use of user-defined functions (UDFs).

User-Defined Functions

Like views and stored procedures, UDFs consist of a named set of SQL statements. Unlike views or stored procedures, they can be used as an expression. For example, the GETDATE() function will evaluate into the current date much as the expression 5 + 6 will evaluate into 11, such as in the SQL statement in Figure 6-15.

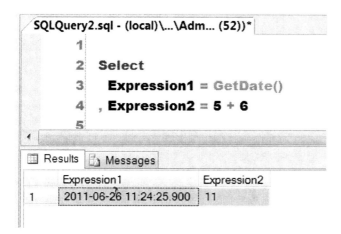

Figure 6-15. *Using a function as an expression*

The syntax for creating a UDF is quite similar to that of a stored procedure. First, list the name of the UDF, followed by a list of any parameters that you want to use and then the return type of the function. The return type of a stored procedure is always an implied integer, but UDFs, on the other hand, can return either a single value or a table of values. Listing 6-28 shows an example of this.

Listing 6-28. Creating a User-Defined Function

```
Create Function fEtlTransformStateToLongName

  ( @StateAbbreviation nChar(2) )
  Returns nVarchar(50)
As

  Begin
    Return
    ( Select Case @StateAbbreviation
        When 'CA' Then 'California'
        When 'OR' Then 'Oregon'
        When 'WA' Then 'Washington'
    End )
  End
```

The code in Listing 6-28 creates a UDF that returns the full name of the state whenever a two-letter abbreviation for the state is given. For this example, we have added only three states, but it goes without saying that you would normally include all the states, regions, or territories that your data warehouse requires. You can test your code to verify that it works by executing select statements similar to the three test queries in Figure 6-16.

Figure 6-16. *Working with user-defined functions*

Moving On

We could easily devote more time to ETL programming, but we would like to take some time to demonstrate how these statements are utilized in conjunction with the SSIS. So, let's move on from here and create an SSIS project for the ETL processing needed in the DWPubsSales data warehouse. We use the SQL statements you created in this chapter to do so. Chapter 7 shows you how easy it is to combine SQL programming statements and SSIS tasks to create effective ETL projects.

LEARN BY DOING

In this "Learn by Doing" exercise, the process defined in this chapter is performed using the Northwind database. We have included an outline of the steps you performed in this chapter and an example of how the authors handled them in two Word documents. These documents are found in the folder `C:\_BISolutionsBookFiles\_LearnByDoing\Chapter06Files`. Please see the `ReadMe.doc` file for detailed instructions.

What's Next?

The purpose of this book is to describe the process of creating a professional BI solution in conjunction with SQL Server 2012. We have provided simple SQL code examples that allow readers to get an idea of how this code can be used as part of the ETL process. If you are a beginning SQL programmer, you may want to devote some time to improving your SQL skills. We suggest you check out a free tutorial on the W3 Schools website at `http://www.w3schools.com/sql/default.asp.` A book that we have found to be both fun and informative on SQL programming is *Beginning Microsoft SQL Server 2008 Programming* by Robert Vieira (Wrox, 2009).

CHAPTER 7

■ ■ ■

Beginning the ETL Process with SSIS

I have this hope that there is a better way. Higher-level tools that actually let you see the structure of the software more clearly will be of tremendous value.

—Guido van Rossum

In the previous chapter, we discussed how to extract data from a source table and transform that data using SQL programming statements. This is a very traditional approach and works well for many professionals, particularly SQL programmers, but can feel cumbersome and tedious to work with.

To be successful with the SQL programming approach, you must thoroughly understand the code you are working with and visualize each action the code performs. The greater the amounts of ETL processing you do the more difficult it is to visualize and coordinate the programmatic flow of your SQL statements and the more difficult it is to explain it to others on the team.

In this chapter, we begin creating the ETL process with Microsoft's SQL Server Integration Services (SSIS). SSIS can help simplify your ETL process by providing the following:

- A visual overview of the ETL process

- Visual and programmatic control of a sequence of ETL tasks

- A way to collectively schedule automations

- The ability to individually execute tasks and track processing

- Access to any data source that supports OLE DB or ODBC connections

SSIS is a platform that can perform various tasks pertaining to data collection and automating processes. It features tools that can be extremely helpful in performing the ETL process. Although the ETL process is its main function, it has several other features as well, such as the ability to run .NET code, interact with the file system, and access various operation system resources.

For the purpose of this book, we focus specifically on the common tasks needed for developing BI solutions. Our goal is to show how easy SSIS is to use in conjunction with the Microsoft SQL Server database engine and to make the ETL process as pain-free as possible.

Starting Your SSIS Project

To begin creating the SSIS ETL process, you can either create a new SSIS project in its own Visual Studio solution or add a new SSIS project to an existing solution. This second option is useful when you want to make a Visual Studio solution that is a collection of all of your BI projects.

In Chapter 5, you used an existing Visual Studio solution to hold your BI solution documents and SQL scripts that you created for the data warehouse. So far, we have added only documents and scripts to the solution, but now we need to add an SSIS project. To do so, we start by opening the solution. Figure 7-1 illustrates the menu items that allow you to open an existing solution from Visual Studio.

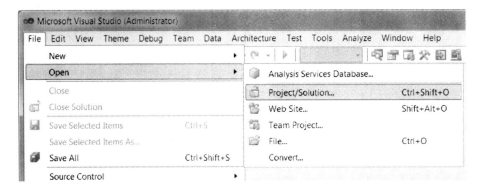

Figure 7-1. *Opening an existing Visual Studio solution*

From here, we navigate to the folder containing our files and then select the solution file (with .sln as the extension), as shown in Figure 7-2.

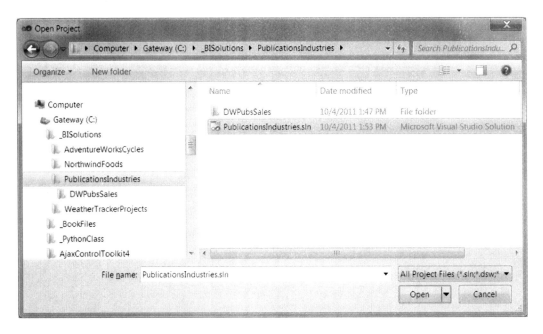

Figure 7-2. *Locating the existing PublicationsIndustries solution*

Adding a Project to an Existing Solution

After the Visual Studio solution opens, we can add the new project using the context menu in Solution Explorer. Another option is to add a new project to an open solution from the File menu, but be careful not to choose File ➤ New ➤ Project option. Instead, choose File ➤ **Add** ➤ New Project (Figure 7-3).

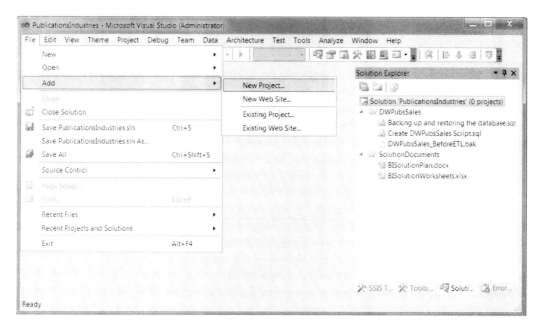

Figure 7-3. *Adding a new SSIS project to the solution*

Once the new project has been added, choose between the different templates available within Visual Studio, as shown in Figure 7-4. The template options include the BI servers: Analysis Services, Reporting Services, and Integration Services.

■ **Note**   The terms *server* and *service* are often interchangeable in Microsoft terminology. Additionally, when creating new projects, the icons do not necessarily represent a specific category (Figure 7-4).

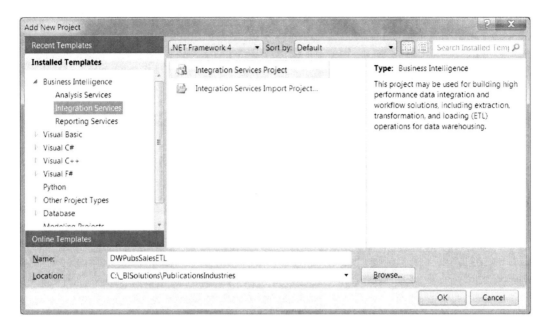

Figure 7-4. *Selecting a project template for the DWPubsSalesETL project*

We are using the Integration Services Project template. This template creates an integration service project with one starter SSIS package. An SSIS project contains a collection of one or more SSIS packages, and each package contains the programming instructions for your ETL process.

Be sure to give an appropriate name to your new integration service project. An appropriate name describes its purpose. In Figure 7-4, we have chosen to call it DWPubsSalesETL to indicate that its mission is to do the ETL processing for the DWPubsSales data warehouse.

Renaming Your SSIS Package

When you first create a new SSIS project, Visual Studio immediately changes the layout by displaying the SSIS package editor on the left of your screen, the Getting Started (SSIS) help window in the middle of your screen, and the SSIS Toolbox on the right of your screen (Figure 7-5).

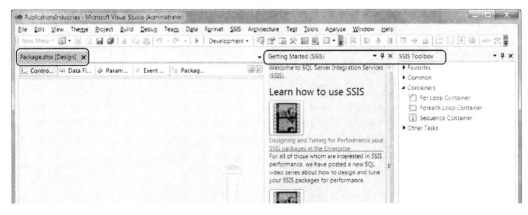

Figure 7-5. *Visual Studio default look when creating a new SSIS project*

Your SSIS project includes a default package called Package.dtsx. As always, you want to rename the default package to something more appropriate. To do this, begin by displaying Solution Explorer again and right-click the package to access the Rename option in the context menu (Figure 7-6). Use the keyboard shortcut Ctrl+Alt+L to display Solution Explorer, or select the Solution Explorer menu item under Visual Studio's View menu at the top of your screen.

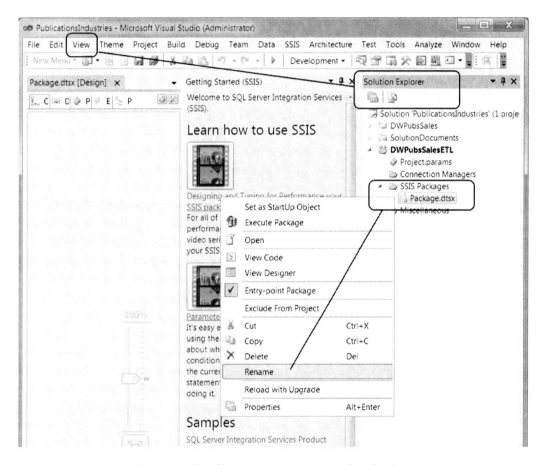

Figure 7-6. *Renaming the Package.dtsx file to ETLProcessForDWPubsSales.dtsx*

■ **Note** In the previous version of SSIS, if the package was renamed, Visual Studio would ask if you wanted to rename the package object as well. The package object refers to the internal name of the package, which can be different from the external file name. Selecting Yes to rename the package object changes an entry in the underlying XML code that makes up an SSIS package file so that the internal and external names are the same.

As you can see, getting started with SSIS is not very difficult, and it becomes easier with practice. Let's start that practice now by completing this chapter's first exercise.

EXERCISE 7-1. CREATING AN SSIS PROJECT

In this exercise, you create an SSIS project to perform the ETL processing required to fill the Publication Industries data warehouse, DWPubsSales.

1. Open Visual Studio by clicking the Windows Start button and selecting All Programs ➤ Microsoft SQL Server 2012 ➤ Data Tools.

Important: You are practicing administrator-level tasks in this book; therefore, you need administrator-level privileges. The easiest way to achieve this is to remember to always right-click a menu item, select Run as Administrator, and then answer Yes to access administrator-level privileges. In Windows 7 and Vista, logging in with an administrator account is not enough. For more information, search the Web on the keywords "Windows 7 True Administrator and User Access Control."

2. Open the solution you have been working on in Chapters 3 and 5. You can do so by using the File ➤ Open Project/Solution menu item, as shown in Figure 7-1. A dialog opens.

3. Navigate to the SLN file and select C:_BISolutions\PublicationsIndustries\PublicationsIndustries.sln.

4. Click the Open button at the bottom of the dialog (Figure 7-2). This opens the solution.

Add a Project to the Current Solution

1. Add a new project to the solution by selecting File ➤ Add ➤ New Project from the main menu at the top of Visual Studio (Figure 7-3). An Add New Project dialog appears (Figure 7-4).

2. Select the Business Intelligence template category on the left of the dialog. Then select the Integration Services Project option from the center of the dialog (Figure 7-4).

3. Name the Project DWPubsSalesETL at the bottom of the dialog and click the OK button to add the new project to your current solution (Figure 7-4).

4. Click the View ➤ Solution Explorer menu item to display Solution Explorer (Figure 7-6). Note that an SSIS package has been created for you.

5. Rename the Package.dtsx file to ETLProcessForDWPubsSales.dtsx (Figure 7-6).

6. Save your work using File ➤ Save All on the Visual Studio main menu.

7. Close the Getting Started window to provide more development space on your screen (Figure 7-6).

Your Visual Studio solution should now look like Figure 7-7. You need this solution for the next exercise, so leave it open for now.

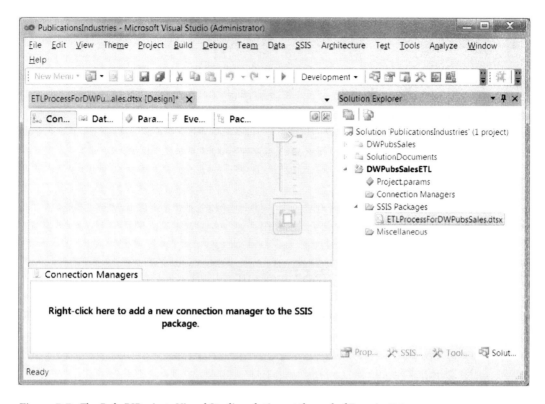

Figure 7-7. *The PubsBIProjects Visual Studio solution at the end of Exercise 7-1*

In this exercise, you created a new SSIS project and added it to a current solution. Next we configure our SSIS package. We begin by outlining the ETL process using SSIS sequence containers and tasks.

The Anatomy of an SSIS Package

Once we have made and then renamed the package, it is important to become familiar with the SSIS design environment. We have listed the most frequently used components and labeled them with a letter that corresponds to the letters shown in Figure 7-8:

  A. *Package Designer*: A collection of design tabs that visually represent underlying XML code

  B. *Control Flow tab*: A designer surface used to collect SSIS tasks and containers

  C. *Data Flow tab*: A designer surface used to collect SSIS data sources, transformation tasks, and data destinations

  D. *Parameters tab*: A collection of parameter used to configure your package

  E. *Event Handlers tab*: A designer surface used to hold SSIS tasks and containers that are executed when an event occurs

  F. *Package Explorer tab*: A tree view of all the objects in the current SSIS package

G. *Zoom control*: A slider control that changes the display magnification

H. *Designer surface*: The location where various SSIS tasks and containers are placed

I. *Connection Managers tray*: A list view of SSIS connection objects

J. *Properties window*: A configuration tool allowing you to set the properties of individual SSIS objects

K. *SSIS Toolbox*: A list of SSIS tasks and containers

L. *Visual Studio Toolbox*: Used in previous versions of SSIS to hold tasks and containers

M. *Solution Explorer*: A list of files and projects

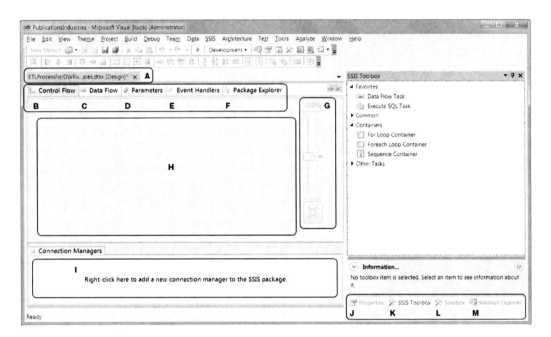

Figure 7-8. *Components of the SSIS design environment*

Let's take a closer look at a few key items from our list.

The Control Flow Tab

The Control Flow tab lets you configure the basic tasks that make up your ETL process. To work with the Control Flow tab, add sequence containers and control flow tasks from the SSIS Toolbox to the designer surface. The SSIS Toolbox offers many items to choose from, but only a select few are used frequently (Figure 7-9).

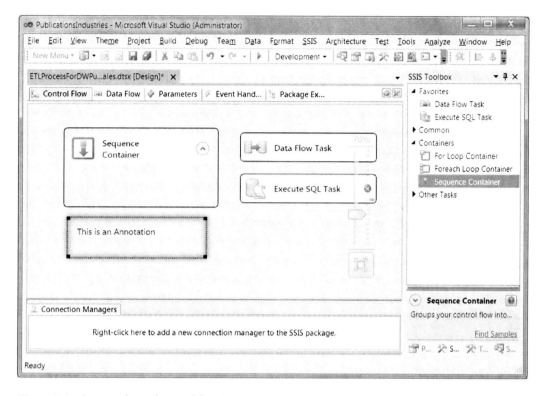

Figure 7-9. *Commonly used control flow items*

The following list describes each of the most frequently used control flow items:

- *Annotations*: Text descriptions that can be added to your SSIS package to provide additional clarity. (To create an annotation in an SSIS package, click anywhere on the designer surface within the Control Flow tab and select Add Annotation from the context menu.)

- *Data Flow Task*: Handles the importing and exporting of data.

- *Execute SQL Task*: Allows SQL commands to be sent to a database server.

- *Sequence container*: Allows tasks to be grouped into a single unit of work.

■ **Note**　For more information on additional tasks, please see "What's Next?" for recommended reading at the end of this chapter.

The Data Flow Tab

When working in SSIS, you soon discover that the data flow tasks are the most common tasks used on the Control Flow designer surface. Perhaps that is why data flows are the only tasks that have their own designer surface.

Data flows are made up of other subtasks. In this way, data flows are similar to control flows in that they both contain various SSIS tasks; however, data flow tasks are specialized for transferring data from one location to another.

A data flow task can be edited either by double-clicking the task while working in the Control Flow tab or by navigating to the Data Flow tab at the top of the designer window.

Figure 7-10 illustrates the three categories of data flow tasks that are used to configure a data flow: Sources, Transformations, and Destinations.

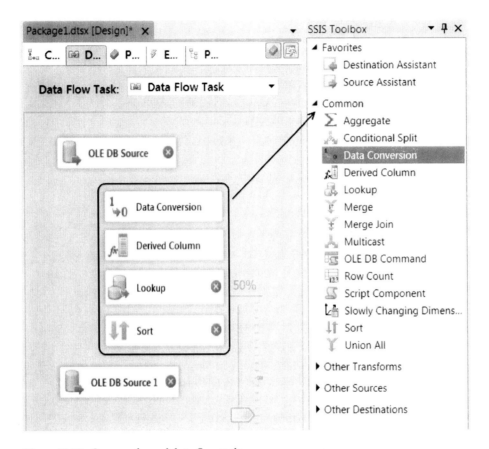

Figure 7-10. *Commonly used data flow tasks*

In Figure 7-10, we have outlined four transformation tasks that you regularly see in SISS data flow tasks. These transformations are found under the Common collection in the SSIS Toolbox. Many other transformations exist, but these are commonly used in ETL processing. These four transformations perform the same actions as the SQL programming statements detailed in Chapter 6, just not as efficiently.

Each data flow needs one or more data sources, zero or more transformations, and one or more destinations. Both a data source and a data destination are required for data to "flow" from one place to another, but SSIS transformations are not required. Let's review each of these categories of tasks.

Data Sources

Data sources collect data from a variety of locations but usually pull data from a database table by executing a SQL select statement. These select statements are either generated for you by SSIS or manually added to the data flow source editor, as shown in Figure 7-11. For example, setting the data access mode (shown in Figure 7-11) to Table or View and choosing a table or view by name automatically creates a SQL statement for you. This is a simple way to configure a data source but is usually not the most efficient. Writing your own SQL code, as shown in the Figure 7-11, is a much better practice.

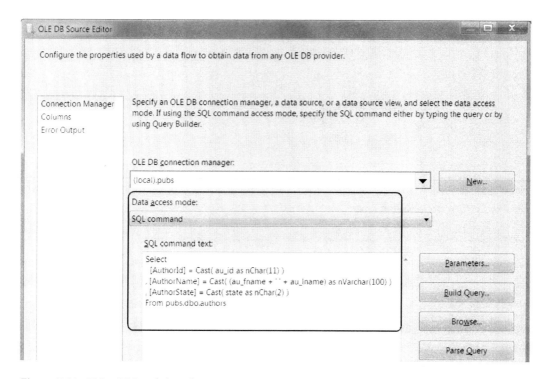

Figure 7-11. *Using SQL code in a data source*

Data Transformations

Data transformations are a group of tasks that perform transformations such as concatenations, lookups, data conversions, sorting, and column aliasing, all of which can be performed using SQL code as well. We recommend using a combination of both SQL programming and SSIS tasks to perform ETL transformations. To do this, place SQL programming code that includes these transformations inside your data source tasks (Figure 7-11). All of the transformations are performed when your SQL query is executed. You are then able to send the transformed data directly to the data destinations.

This process may not always work. If, for example, your data source is pulling directly from a flat file, such as a comma-delimited data file (also known as a .csv file), you are not able to perform SQL programming statements directly on the file. Instead, you can use the SSIS data flow transformation tasks to accomplish the same thing (Figure 7-12). This is helpful when your business needs require that you pull data directly from a file.

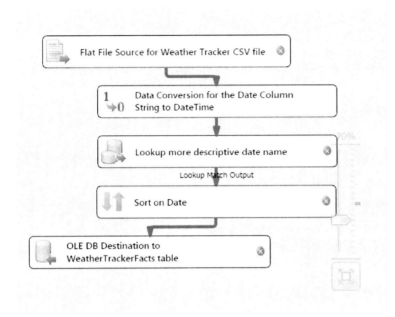

Figure 7-12. *Using data transformation tasks*

CONSIDER STAGING TABLES

If your data source is a file, you may find it advantageous to load the data from the file into a temporary staging table in the database. You can perform ETL transformations using SQL code from there.

This additional step may seem like extra work, but if there are enough transformations involved, you will find the process faster than performing transformations directly through SSIS. This is because the database engine can process transformations on large datasets quicker than SSIS can. While SSIS is fast, especially compared to its predecessors, it is still not as fast as a dedicated relational database engine for processing data.

Another advantage to this technique is that it is easier to find team members who understand SQL programming statements than it is to find SSIS experts. Using a combination of SQL statements and data flows provide a fast and easy way to include ETL programming code within a Visual Studio solution.

Data Destinations

Data flow destination tasks allow you to place data in files, in tables, or even in an in-memory dataset. There are quite a few data flow destinations that come with SSIS at first install, and you can download others from the Microsoft website. The installed items are listed at the bottom of the SSIS Toolbox (Figure 7-13). If they are not enough to meet your needs, you can create custom destinations.

A data flow destination must have an input from another data flow component, and it must be connected before you start editing it. This is important because the input provides the metadata describing the list of source columns that are mapped to columns in the destination. Without this metadata, the data flow destination cannot be configured correctly.

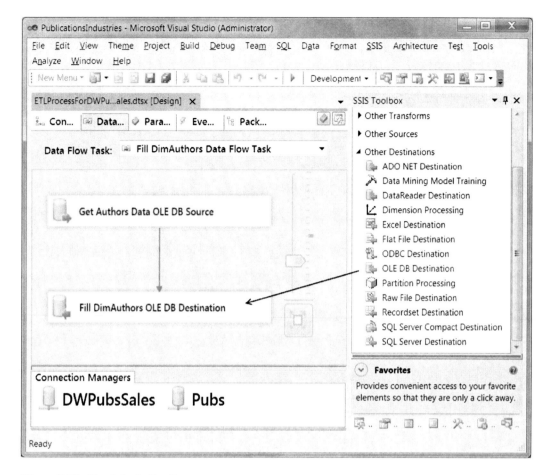

Figure 7-13. *The data destinations*

Data flow destinations can have an error output containing errors that occur when writing data to the destination data store. Errors can occur for a number of reasons, but one example is attempting to insert duplicate data into a primary key column.

Using Sequence Containers

Another common element of an SSIS package is the *sequence container*. Its main purpose is to group control flow tasks. These containers have additional helpful features within the sequence container. These features include the following:

- Expand and collapse arrows on the upper-right side of the sequence container allow you to show or hide tasks within it.

- Properties within a sequence container affect the same property setting on the tasks inside the container.

- Disabling a sequence container disables the tasks inside it.

- Configuring a sequence container as part of an SSIS transaction allows you to commit or roll back the tasks inside.

Although both the Control Flow tab and the Data Flow tab look similar, sequence containers are available only on the Control Flow tab and not on the Data Flow tab. This means you can group control flow tasks only within them, but it is really not much of a limitation since data flow tasks are by their very nature a way of grouping SSIS tasks.

Figure 7-14 shows an example of a typical control flow design. You will note that similar tasks are grouped together to form a set of those actions, such as preparing the data warehouse tables to be filled. Other sequence containers may have different actions, such as filling up dimension tables or filling up fact tables.

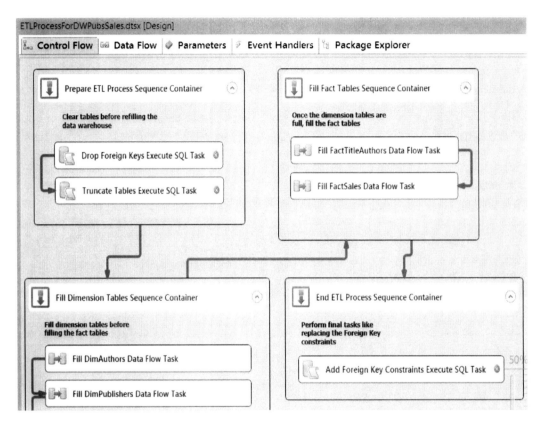

Figure 7-14. *Grouping control flow tasks with the sequence container*

In cases where you are using the flush-and-fill technique to fill up your data warehouse tables, you need a task that clears the tables before they are refilled. In Chapter 6, we discussed using the SQL truncate command for this process. You must drop the foreign key constraints before you begin truncation. This is accomplished by adding the code that you created in Chapter 6 to an Execute SQL task.

While the code to drop the foreign key constraints and the code to truncate the tables could be placed in one control flow task, it provides a better visual to create two tasks—one to drop the foreign key constraints and one to do the truncation. Since both of these tasks are part of a single process, it is logical to group them. In SSIS, a sequence container is specifically designed for this job. In Figure 7-14 you can see that we have created two Execute SQL Task items and placed them inside a sequence container called Prepare ETL Process Sequence Container.

> ■ **Note** It is important to give your data flows and your sequence containers unique names that identify their purpose. The naming convention we have chosen to give our SSIS tasks is the task description followed by the task type. For example, to fill up an authors table using a data flow task, we would name the task Fill DimAuthors Data Flow Task. This is a very long name, but it is self-explanatory, and you seldom—if ever—have to type it in code elsewhere.

Using Precedence Constraint Arrows

In some cases, you must perform a particular operation before another can begin. You can connect different control flow tasks using a precedence constraint arrow. These are the arrows shown in Figure 7-14.

Precedence constraint arrows control when a task runs in relationship to another task. When two tasks do not have a control flow arrow between them, the tasks run simultaneously. For example, looking closer, you see that the two data flow tasks in Figure 7-15 do not have a precedence constraint arrow between them. Therefore, they execute at the same time.

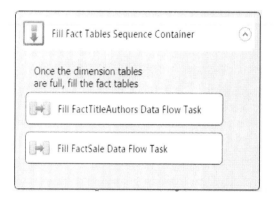

Figure 7-15. *Control flow tasks without precedence constraint arrows*

When you initially drag an SSIS task to the control flow surface, you may not see the precedence constraint arrow, just as you do not see one in Figure 7-15, but one magically appears once you click a task. When the arrow appears, you can click the arrow and then drag it to the other control flow task that you want to connect.

> ■ **Note** Rules, rules, and more rules…! You cannot drag a precedence constraint arrow between tasks inside different sequence containers. Instead, you must connect the containers. Containers can connect to other containers, and containers can connect to individual tasks, but individual tasks can only connect to other individual tasks.

It is possible to create more than one precedence constraint arrow per task. After creating the first precedence constraint arrow, click the initial task again. This causes another arrow to appear that you can then connect to other tasks, thereby performing complex control flow scenarios (Figure 7-16).

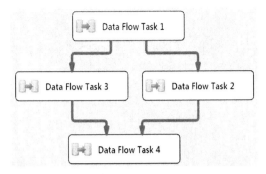

Figure 7-16. *Control flow tasks can have multiple precedence constraints arrows.*

Notice that Data Flow Task 1 has two precedence constraints attached to it. Once this task completes, control passes to both Data Flow Task 2 and Data Flow Task 3.When either one of these finishes, control passes to Data Flow Task 4, but this can also be configured so that both tasks must complete before control is passed. Precedence constraints can be configured to allow the execution control to flow between tasks based on success, failure, or completion. This configuration is important because it allows you to apply conditional logic to your design. For example, if we were to add a task that sends us an email when a portion of the package fails, we could do so by adding a Send Mail Task, renaming it, and connecting a precedence constraint arrow from the existing task container to the new send mail task (Figure 7-17). We could then configure it to execute the send mail task only upon the condition that the previous task failed.

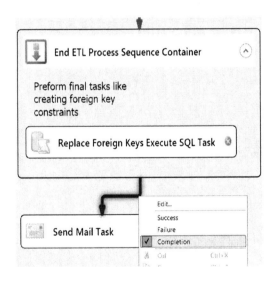

Figure 7-17. *Setting the Constraint operation value*

■ **Note**   Although it cannot be seen in the black-and-white images of this book, the precedence constraints are color-coded to indicate their status. The green arrows indicate success, red indicates failure, and blue indicates completion.

Precedence constraint arrows can be configured using the context menu, as shown in Figure 7-17. Or you can right-click the arrow and select Edit from the context menu to access the Precedence Constraint Editor (Figure 7-18).

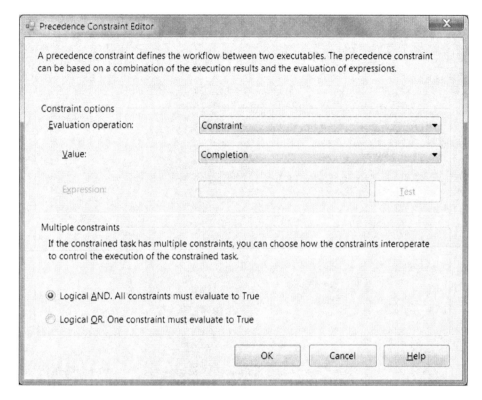

Figure 7-18. *The Precedence Constraint Editor dialog*

In the Precedence Constraint Editor, you can set the arrow to configure expression values as well as constraint operation values, such as Success, Completion, and Failure.

Expressions values always evaluate a true or false Boolean value, but they can be combined with the constraint values to give you a rather elaborate logic. Figure 7-19 shows an example of using both expressions and constraints.

The Evaluation Operation dropdown box allows you to set the precedence constraint to consider the following combinations:

- Constraint

- Expression

- Constraint and Expression

- Constraint or Expression

These combinations let you fine-tune the constraint process. In Figure 7-19, if the expression value evaluates the @RowCount variable as equal to zero or the constraint evaluates that the connected task has failed its execution, then at least one of these conditions is met, and SSIS passes control onto the next task for processing.

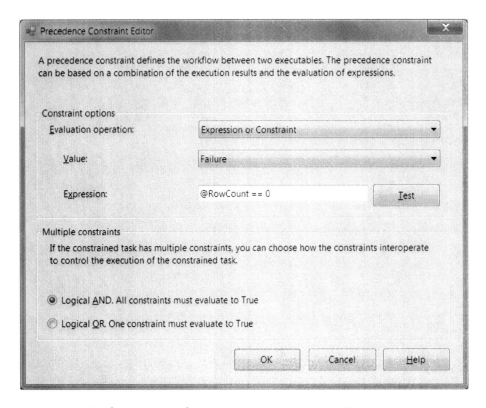

Figure 7-19. *Configuring a precedence constraint to use an expression*

Figure 7-19 also shows the "Multiple constraints" area containing radio buttons that allow you to configure a constraint to work in conjunction with other constraints on the same task. When a control flow task contains two or more constraints, as in Figure 7-16, you can determine whether just one or all constraints are required to be true before moving to the next task.

■ **Note**　You may notice that the expression in Figure 7-19 uses syntax similar to C#. At other times, you may see syntax that is more like SQL or perhaps Visual Basic. This is not C# but rather SSIS's own expression language. Microsoft has published updates to this language for each of the last four versions of SQL Server.

You can find more about this language by searching the Web using the keywords "SSIS Expression Reference." Here is an example page that was available at the time of this writing:

http://msdn.microsoft.com/en-us/library/ms141232.aspx.

SSIS Variables

In Figure 7-19 we use a custom SSIS variable called @RowCount. The intended purpose of this variable is to determine whether there are zero rows of data found in a given table. If this is true, a particular action is performed or avoided.

An SSIS variable can be used to temporarily hold results or configuration data in random-access memory (RAM). Variables can be added to your package by using the Variables window. This window does not normally appear in a new package but can be opened using the Variables submenu item found in the SSIS menu, as shown in Figure 7-20.

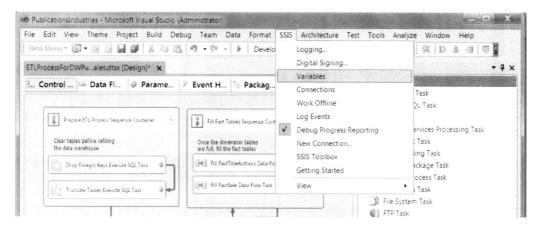

Figure 7-20. *Displaying the SSIS Variables window*

> ■ **Note**   The SSIS menu options change depending on the element of the UI you have as your current focus. If you do not see the Variables menu item, try clicking the Control Flow designer surface and look under the SSIS menu once more.

SSIS has many premade variables available, but when you first open the Variables window, you will not see them. And although we do not use them in this chapter, you can display the premade system variables by clicking the fourth button on the Variables window's toolbar, circled in Figure 7-21, and checking the "Show system variables" checkbox.

Additional variables can be added by clicking the first button on the Variables window toolbar. The other buttons on the toolbar control features such as deleting a variable or moving variables between scopes.

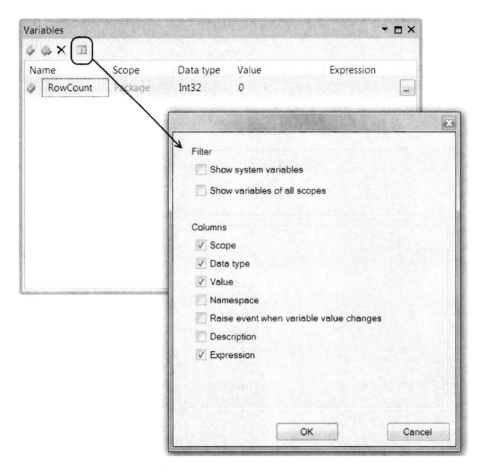

Figure 7-21. *The SSIS Variables window*

Note The Variables submenu items do not show under Visual Studio's SSIS menu if you are not currently focused on the SSIS package. You can change the focus to the package by clicking the package file in Solution Explorer or the package designer window.

When you create a new variable, you must give it a name, define its data type, and optionally set an initial value.

The scope of the variable identifies which tasks have access to the variable. If the variable is scoped at the package level, which is the default, then all the tasks in the package have access to it. If, on the other hand, it is scoped at an individual data flow task, then only that data flow task has access to it.

The scope of the variable defaults to the package, but the variable scope can be changed using the second button of the Variables window toolbar. Clicking this button displays a new dialog that includes a tree view of all the containers and tasks in your SSIS package. By selecting one of the tree-view items and clicking the dialog window's OK button, you can change the scope of the variable.

One oddity in SSIS that you should be aware of is that the name of the variable as it appears in the Variables window is addressed differently in other parts of SSIS. For example, a variable named RowCount must be accessed as @RowCount from a Precedence Constraint Editor dialog window, but as User::RowCount from an SSIS Script task. That's all part of the fun of working with SSIS!

Outlining Your ETL Process

If you are new to creating SSIS projects, you may be overwhelmed with all the information we have shown you. Rest assured that many developers feel this way when they first start working with SSIS but soon find that it is not as difficult as it appears.

The best way to keep from being overwhelmed is by outlining the activities you need to perform in your ETL process. You can outline these activities by adding SSIS tasks to the control flow surface, grouping them together with sequence containers, and connecting them with precedence constraints. Don't forget that you can add annotations to the control flow surface to give more specific information about each task.

This practice of outlining the tasks can be very helpful. And we recommend doing so before configuring each task. You will get a feel for this as you perform this next exercise.

EXERCISE 7-2. FILLING THE DATA WAREHOUSE

In this exercise, we outline the steps needed to perform the ETL processing. The goal is to fill the DWPubsSales data warehouse using the SSIS project that was created in Exercise 7-1. Figure 7-20 is an example of what your SSIS package should look like after having completed this exercise, and it can be used as a reference for each of the following steps.

1. If the solution from the previous exercise is closed, open Visual Studio by clicking the Windows Start button and selecting All Programs ➤ Microsoft SQL Server 2012 ➤ Data Tools. Then follow the instructions from Exercise 7-1 to open C:_BISolutions\PublicationsIndustries\PublicationsIndustries.sln.

Remember to right-click the menu item, select Run as Administrator, and answer Yes to close the UAC dialog!

Add Sequence Containers

1. Select the Control Flow table of the ETLProcessForDWPubsSales.dtsx SSIS package you created in Exercise 7-1. Your screen should look like Figure 7-7.

2. On the Control Flow tab, drag four Sequence Containers to the control flow surface from the SSIS Toolbox, rename them by right-clicking the container, and select the Rename context menu item. When prompted, rename them as listed here (Figure 7-22):

   • Prepare ETL Process Sequence Container

   • Fill Dimension Tables Sequence Container

   • Fill Fact Tables Sequence Containers

   • End ETL Process Sequence Containers

Adding Annotations

3. Right-click the empty space in each of the sequence containers, and select Add Annotation from the context menu.

4. Type in the annotations as listed here (Figure 7-22):

   • Clear tables before refilling the data warehouse

   • Fill dimension tables before filling the fact tables

   • Once the dimension tables are full, fill the fact tables

   • Perform final tasks like replacing the foreign key constraints

5. Move the annotation to the upper-left corner of each sequence container, as indicated in Figure 7-22.

Your SSIS package should now look similar to that shown in Figure 7-22.

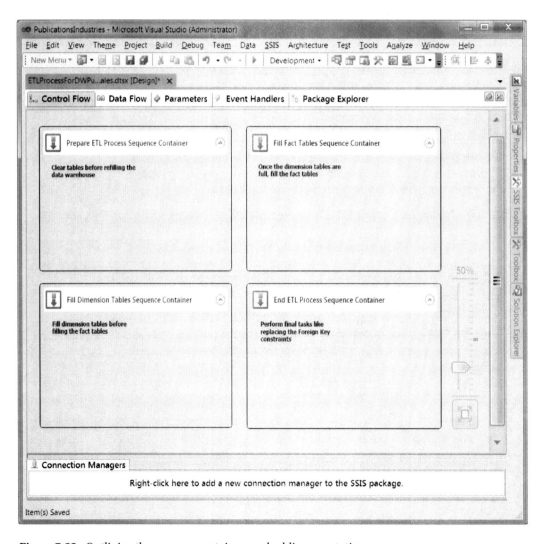

Figure 7-22. *Outlining the sequence containers and adding annotations*

Add Tasks to the Sequence Containers

6. Drag and drop two Execute SQL Tasks into the Prepare ETL Process Sequence Container, and rename them to the following:

   - Drop Foreign Keys Execute SQL Task

   - Truncate Tables Execute SQL Task

7. Drag and drop four Data Flow Tasks into the Fill Dimension Tables Sequence Container, and rename them to the following:

   - Fill DimAuthors Data Flow Task

   - Fill DimStores Data Flow Task

   - Fill DimPublishers Data Flow Task

   - Fill DimTitles Data Flow Task

8. Drag and drop one Execute SQL Task into the Fill Dimension Tables Sequence Container, and rename it to the following:

   - Fill DimDates Execute SQL Task

9. Drag and drop one Execute SQL Task into the Fill Dimension Tables Sequence Container, and rename it to the following:

   - Add Null Date Lookup Values Execute SQL Task

10. Drag and drop two Data Flow Tasks into the Fill Fact Tables Sequence Container, and rename them to the following:

    - Fill FactTitlesAuthors Data Flow Task

    - Fill FactSales Data Flow Task

11. Drag and drop one Execute SQL Task into the End ETL Process Tasks Sequence Containers, and rename them to the following:

    - Add Foreign Key Constraints Execute SQL Task

12. Because DimTitles uses data from the DimDates table, we need to load it after the DimDates table is filled, moving the Fill DimTitles Data Flow Task to the bottom of the Fill Dimension Tables Sequence Container.

When you are done, your SSIS package should now look similar to Figure 7-23.

Note: A red dot may appear on some tasks indicating that the task is not configured. We remedy this later in the chapter.

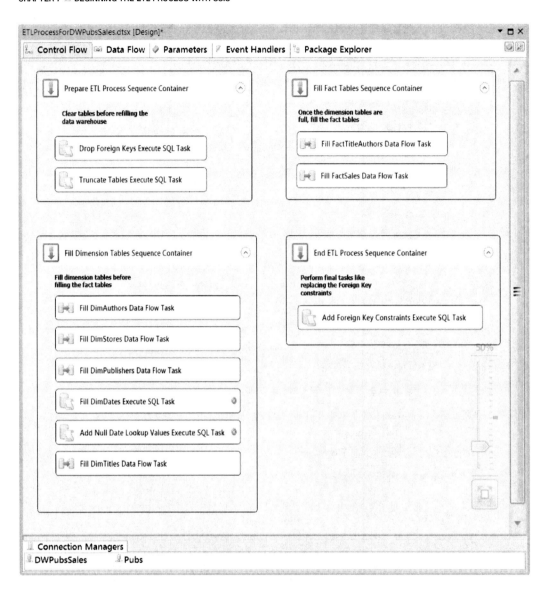

Figure 7-23. *Outlining the tasks within the sequence containers*

Add Precedence Constraints

1. Click the Drop Foreign Keys Execute SQL Task, and a precedence constraint arrow appears. Connect this arrow to the Truncate Tables Execute SQL Task. Continue this for steps 14–22.

2. Add a constraint arrow to the Prepare ETL Process Sequence Container, and connect it to the Fill Dimension Tables Sequence Container.

3. Add a constraint arrow to the Fill DimAuthors Data Flow Task, and connect it to the Fill DimStores Data Flow Task.

4. Add a constraint arrow to the Fill DimStores Data Flow Task, and connect it to the Fill DimPublishers Data Flow Task.

5. Add a constraint arrow to the Fill DimPublishers Data Flow Task, and connect it to the Fill DimDates Execute SQL Task.

6. Add a constraint arrow to the Fill DimDates Execute SQL Task, and connect it to the Add Null Date Lookup Values Execute SQL Task.

7. Add a constraint arrow to the Add Null Date Lookup Values Execute SQL Task, and connect it to the Fill DimTitles Data Flow Task.

8. Add a constraint arrow to the Fill Dimension Tables Sequence Container, and connect it to the Fill Fact Tables Sequence Container.

9. Add a constraint arrow to the Fill FactTitlesAuthors Data Flow Task, and connect it to the Fill FactSales Data Flow Task.

10. Add a constraint arrow to the Fill Fact Tables Sequence Container, and connect it to the End ETL Process Tasks Sequence Container.

11. Move the precedence constraint arrows around by clicking them and dragging them to a different position. Once the package looks similar to the one illustrated in Figure 7-24, save your SSIS package using File ➤ Save from the Visual Studio main menu.

Your SSIS package should now look similar to Figure 7-24.

Note: A red dot may appear on some tasks indicating that the task is not configured. As noted, we resolve this later in the chapter.

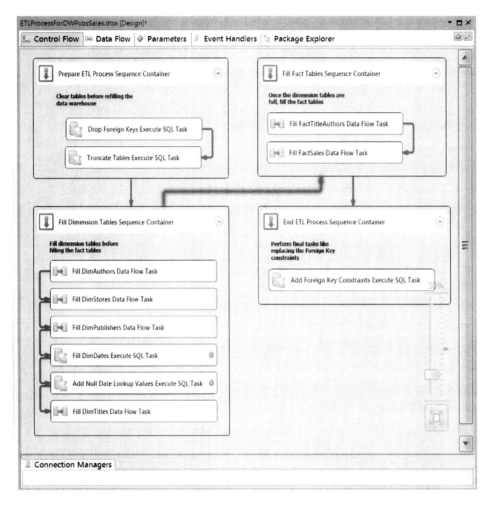

Figure 7-24. *The SSIS package at the end of Exercise 7-2*

In this exercise, you began configuring your SSIS package by outlining the various steps required utilizing sequence containers and SSIS tasks. Currently these tasks do not do anything because they have not been fully configured. In addition, the package will not work until you create one or more connections to the source and destination objects. We look at how this is done in the next section.

Data Connections

After you have outlined your ETL process on the Control Flow tab, the next step is to create connections so that you can configure each task. Creating connections is easy; all you have to do is go down to the bottom of the screen, find the Connection Managers tray, and add a connection by right-clicking the background of the Connection Managers surface area and selecting New Connection from the context menu (Figure 7-25). This action brings up the Add SSIS Connection Manager dialog. From this dialog, you are able to select from several connection managers.

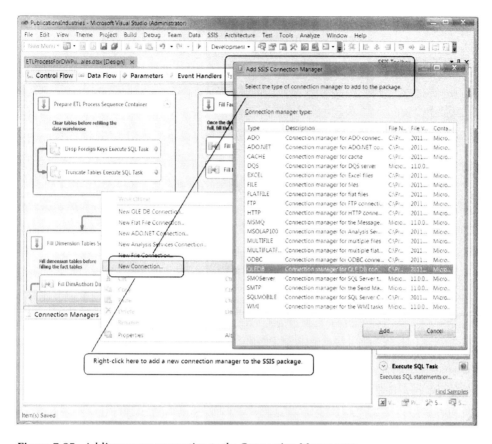

Figure 7-25. *Adding a new connection to the Connection Manager tray*

Connection managers are defined by their connection type. There are about 20 different connection types available; some are installed by default, and others can be downloaded. Of all of these choices, you will frequently use the following connection types:

- The file connection manager
- The OLE DB connection manager
- The ADO.NET connection manager

Selecting one of these options is equivalent to selecting New Connection and choosing the provider from the dialog you saw in Figure 7-25. Microsoft conveniently placed these choices at the top of the context menu.

The File Connection Manager

The file connection manager allows you to work with both files and folders. This handy little connection manager can perform operations such as creating new files and folders as well as accessing data from existing files.

In the business intelligence world, it is common to import data from files into staging tables and then distribute it to the various tables in the data warehouse. Therefore, it is very likely that you will use these connections when working with SSIS.

An example scenario would be to have a number of branch offices that upload files to a corporate folder. You could process these files one by one into a staging table by combining the file connection manager with an SSIS For-Each container, which allows you to loop through multiple files one by one and apply processing tasks to each of them in turn.

The OLE DB Connection Manager

The OLE DB connection manager is the most common connection manager you will use, since it allows you to connect to anything that has an OLE DB provider. That is a very wide range of connection options when you consider that anything from an Excel spreadsheet to an old mainframe database will have OLE DB providers available.

The OLE DB connection manager is designed to be flexible when working with data types that differ from the originating and destination data sources, and because of this built-in flexibility, the OLE DB connection manager is the one that gives you the least amount of trouble when it comes to data conversion error.

◼ **Note**   SSIS has a picky attitude toward data conversions, which can be quite frustrating. These issues are less likely to occur when using OLE DB connections. This is an important consideration when choosing to use OLE DB connections in your SSIS packages!

As usual, a gain in flexibility comes at a cost. The cost in this case is a decrease in speed. As a result, for raw performance (when your project requires speed), you probably want to use a more specialized connection manager such as the ADO.NET connection manager, for example. And for ease of use and compatibility (for comparatively smaller projects), use the OLE DB connection manager.

If you are working with a mere several thousand rows, we recommend using an OLE DB connection manager. With only thousands of rows (as opposed to millions), you will notice very little difference in performance and will appreciate the flexibility and ease of use of the OLE DB connection manager.

The ADO.NET Connection Manager

The ADO.NET connection manager is preconfigured to use a .NET provider for accessing data sources, and depending upon the size and type of data you are working with, it may give increased performance over the generic OLE DB connection manager. The performance increase when used with the newer versions of SQL Server can be substantial, but with older or non-Microsoft databases, there are little to no performance gains, no matter how much data is involved.

The ADO.NET connection manager is limited to the types of connections it can make, specifically in comparison to the OLE DB connection manager, but it connects to all versions of Microsoft SQL Server.

The data types used in an ADO.NET connection manager are much more specific to Microsoft's .NET data types. As you work with them, you begin to see that the list of types looks very different than the more generic OLE DB data types, which are based on an ANSI standard and not the .NET standard. This is important because SSIS creates metadata to describe all of its source, destination, and transformation components. If the metadata of your SSIS task does not identify a compatible data type, your task will either process with a warning or fail to process at all.

◼ **Note**   We recommend using the OLE DB connection manager whenever possible since it provides the greatest flexibility and the least amount of problems, which is especially important when you are first learning to use a complex tool like SSIS. To be clear, there is nothing wrong with the more precise .NET data types or the ADO.NET

---

connection managers. But, keep in mind that they are more difficult to work with. Microsoft's website provides a great deal of information on this topic. For more information, search for the topic of "SSIS data types" at http://msdn.microsoft.com.

Configuring a Connection

Each time you add a new connection, you need to configure it. Once you select a connection manager and click the Add button, a configuration dialog appears (Figure 7-26).

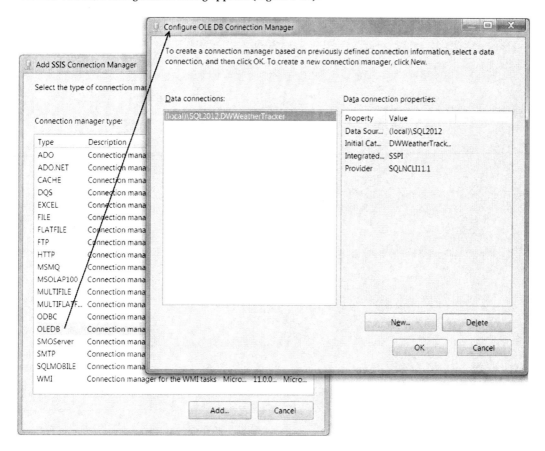

Figure 7-26. *Each connection manager type has its own configuration manager*

This new dialog is dependent on which type of connection manager you selected, and each connection manager type has its own configuration dialog. If you select an ADO.NET connection manager, for example, you are presented with the Configure ADO.NET Connection Manager dialog. If, instead, you choose an OLE DB connection manager, you will be presented with the OLE DB version of this dialog.

In both the OLE DB and ADO.NET configuration dialogs, Visual Studio remembers any previous connections you have created either in this project or in past projects. Therefore, if you have connected to a database in a previous project, these connections are still available. For example, since we previously created a connection to the DWWeatherTracker database in Chapter 2, this connection is displayed in Figure 7-26.

If you have not created a connection to a given database prior to this, you will not see a connection available and will need to click the New button to create one.

Once you click the New button, yet another dialog appears! In this one, you start your configuration by typing in the name of the database server. If you are connecting to the server on your computer, you can type in **localhost**. And as covered in previous chapters, using *(local)* with parentheses or simply putting a period (.) can work, if localhost does not.

■ **Tip**   If you are using a named instance of SQL Server, you have to configure the name as localhost\<MyNamedInstance>. For more information on connecting to your local server, see Chapter 5.

After choosing the name of the server, you choose the database you want to connect to. To determine which database a connection manager uses, you can type in the database name or can use the dropdown button beneath the "Select or enter a database name" label, as shown in Figure 7-27.

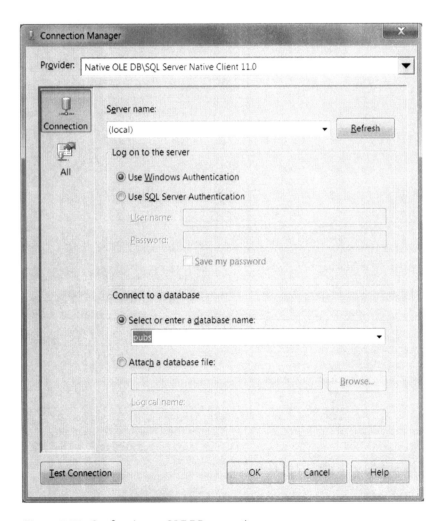

Figure 7-27. *Configuring an OLE DB connection*

With both the server and the database selected, clicking the Test Connection button tells you whether your connection is successful. When the connection works correctly, click OK to close the dialog and create a new connection manager in the Connection Manager tray.

In our example, we are importing data from Pubs to our DWPubsSales data warehouse. Therefore, we need a connection to both databases. Each connection manager provides a connection to only a single database at a time, so we must create and configure a separate connection manager for each database.

After you have created your connections, you can edit and review their properties using the Properties window (Figure 7-28). This is convenient since you often create a package on one computer but move it to another later. When you do so, you can adjust the connection for use on the new computer by clicking the ConnectionString property to launch the same dialog shown in Figure 7-27.

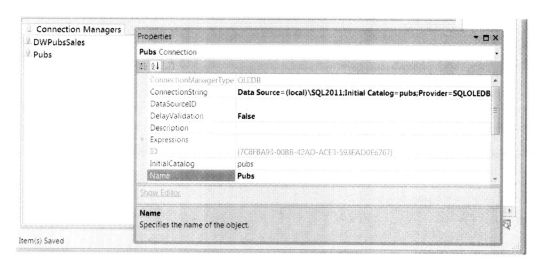

Figure 7-28. *Reviewing the connection manager properties*

Although it is possible to have multiple connections to the same database, you typically use only one connection manager per database for all of your SSIS tasks. This allows you to reconfigure one connection manager while affecting all of the SSIS tasks using that connection.

■ **Tip**   After you create a connection manager, you can reconfigure it using either its property dialog or the Visual Studio property window, but be aware that some settings appear only in the Visual Studio property window and not in the dialog. And for other tasks the opposite is true.

Execute SQL Tasks

An Execute SQL task allows you to run SQL code on a connected database. For example, to use an Execute SQL task designed to drop foreign key constraints, you configure it to connect to the proper database by selecting the appropriate connection manager and then add SQL code to drop the constraints.

Editing Your Execute SQL Task

You add code to an Execute SQL Task Editor window by right-clicking the task and choosing Edit from the pop-up context menu (Figure 7-29).

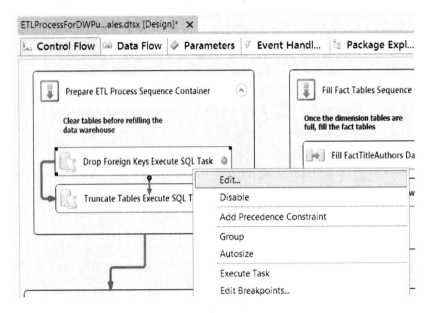

Figure 7-29. *Editing an Execute SQL task*

When the editing dialog displays, you will see a number of properties that can be configured; however, the two you must configure are the Connection and SQL Statement properties.

In SSIS, editing windows are usually divided into *pages*. These pages are listed on the left side of the dialog. In an Execute SQL task, configure the connection by first selecting the General page and then clicking the Connection setting. A dropdown box appears allowing you to select an existing connection manager or even to create a new one (Figure 7-30).

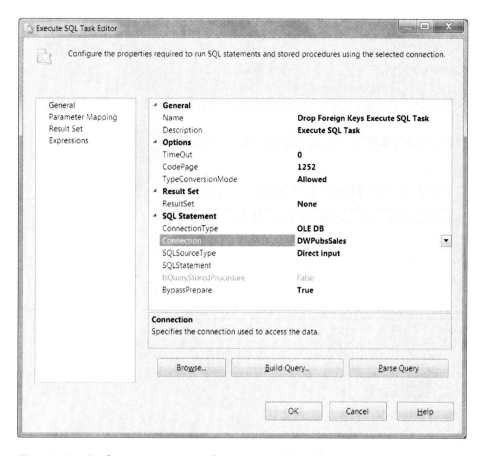

Figure 7-30. Configuring a connection for an Execute SQL task

The connections available in the dropdown box are context sensitive to the connection type selected in the ConnectionType property. If the connection type is not set to OLE DB but instead is set to another connection type such as ADO.NET, any OLE DB connection managers that have been created will not appear in the dropdown selection. Switching the ConnectionType property to OLE DB allows them to appear (Figure 7-30).

To add SQL code, you need to configure the SQL Statement property. After clicking this property, you will see that an ellipsis button appears. When you click the ellipsis button, a dialog appears where you can add your SQL code (Figure 7-31). Hidden buttons like this one are becoming commonplace in Microsoft applications. If you do not immediately see a way to configure a property, try clicking it to see whether it contains a hidden button.

■ **Tip**   In Microsoft's most recent user interfaces, dropdown boxes, buttons, or ellipsis icons do not appear until you click a property setting. This newer, sexier interface can cause some problems when trying to figure out how a given property is configured. Just remember; when in doubt, click the property!

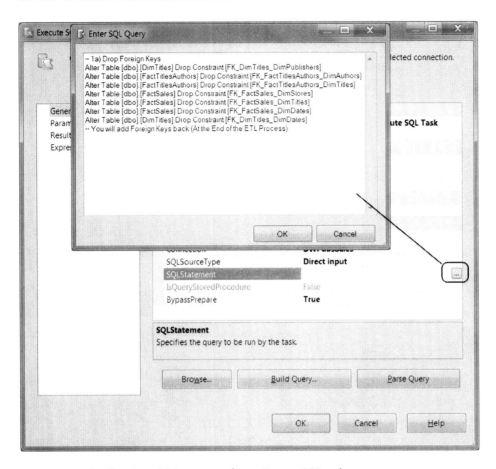

Figure 7-31. *Configuring a SQL statement for an Execute SQL task*

SQL code can be typed into the dialog, but a better way is to create the ETL code beforehand with SQL Management Studio (as we did in Chapter 6) and then copy and paste the code into the SSIS window. One advantage to this method is that you can test and correct your code before adding it to your Execute SQL Task.

Once you have selected a connection and added your SQL code, you can close the Execute SQL Task dialog window and test your work by executing it.

Executing Your Execute SQL Tasks

SSIS packages consist of XML code that describes various tasks. Each task placed on the designer surface is a collection of programming instructions in an XML format. When you configure a task, you are filling in attributes and elements of the XML programming code, as shown in Figure 7-32.

Like any other programming language, the code by itself does not do anything unless software exists that can read the programming instructions and perform the action accordingly. This type of software is often referred to as a *runtime environment* or *runtime engine*. Visual Studio includes an SSIS debugging engine that launches the SSIS runtime environment. After creating your package, you will no longer need Visual Studio to run your SSIS code, since it can run on any computer that has the SSIS runtime installed on it, even if Visual Studio is not installed.

```
ETLProcessForDWPubsSales.dtsx [XML]*                                              ▼ ◘ ×
  289   DTS:DTSID="{A420BBD3-58A0-4F0A-A446-A54C66873427}"                          ⊕
  290   DTS:ExecutableType="Microsoft.SqlServer.Dts.Tasks.ExecuteSQLTask.ExecuteSQLTask
  291   DTS:LocaleID="-1"
  292   DTS:ObjectName="Drop Foreign Keys Execute SQL Task"
  293   DTS:TaskContact="Execute SQL Task; Microsoft Corporation; SQL Server &quot;Dena
  294   DTS:ThreadHint="0">
  295   <DTS:Variables />
  296   <DTS:ObjectData>
  297     <SQLTask:SqlTaskData
  298       SQLTask:Connection="{7EDB9C04-008B-4C9D-B3F2-32F78285B45F}"
  299       SQLTask:SqlStatementSource="Alter Table [dbo].[DimTitles] Drop Constraint [
  300   </DTS:ObjectData>
  301   /DTS:Executable>
  302   )TS:Executable
  303   DTS:refId="Package\Prepare ETL Process Sequence Container\Truncate Tables Execu
  304   DTS:CreationName="Microsoft.SqlServer.Dts.Tasks.ExecuteSQLTask.ExecuteSQLTask,
100 %  ◄                                                                              ►
```

Figure 7-32. *The XML code behind the Execute SQL task*

Still, for a developer running the code within, Visual Studio is convenient. Right-click a task and select Execute Task from the context menu, as shown in Figure 7-33, to launch Visual Studio's debugging engine, which in turn executes the code associated with the SSIS task you selected using the SSIS runtime environment.

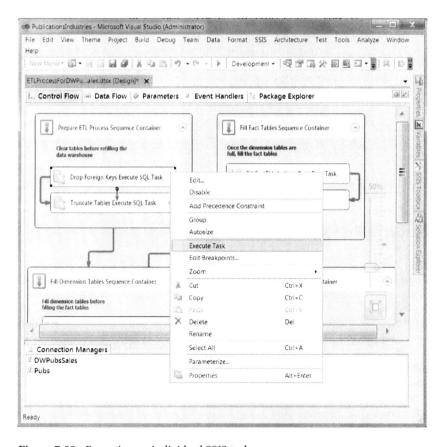

Figure 7-33. *Executing an individual SSIS task*

As each task executes, the task displays a yellow wheel and then either a red *X* or green check mark (Figure 7-34). The yellow status means that the underlying code within the task is currently running in the debugging engine. When the task icon turns green, it means that all the code has stopped running and that the execution of that code was successful. When the task icon turns red, it means that the execution of the code was unsuccessful.

No matter the outcome, the debugging engine does not automatically shut down but instead continues running in the background. This is evidenced by the word *(Running)* within parentheses being displayed at the top of the Visual Studio window (Figure 7-34).

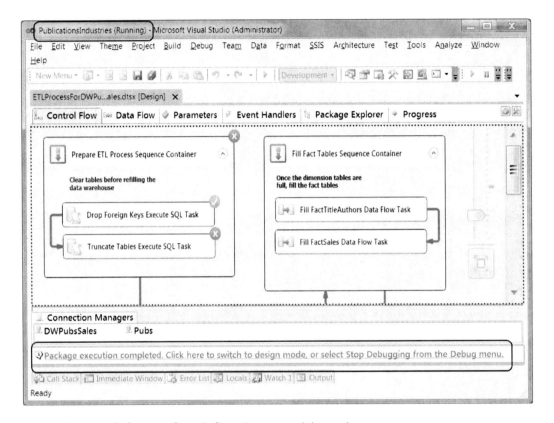

Figure 7-34. *A task changes color to indicate its status as it is running*

You cannot edit the SSIS package while the debugger is running. Therefore, once all of the tasks have completed, stop the debugger by selecting the hyperlinked message that we circled in Figure 7-34 or by selecting the Stop Debugging option from the Visual Studio Debug menu. Once the SSIS code stops running, you can continue to edit your SSIS package.

You may have noticed that there is an indicator on the sequence container as well as the individual tasks (Figure 7-34). You can debug more than one task at a time within Visual Studio, either by executing the SSIS package file as a whole or by executing all of the tasks within a sequence container.

To execute a set of tasks within a sequence container, right-click the container and choose Execute Container from the context menu. To execute all the tasks within a particular SSIS package, select the package file in Solution Explorer, right-click it, and then choose Execute Package from the context menu, as shown in Figure 7-35.

Executing individual tasks as you configure them allows you to test and troubleshoot errors immediately. If you have a collection of tasks that must work together, you will want to test them collectively as well.

Figure 7-35. *Executing the entire SSIS package*

We recommend that you test individual SSIS tasks as you go so that any error can be resolved sooner rather than later. This approach is not always practical, however, when certain tasks are contingent on others running or when the database must be in a certain state before your SSIS code executes, such as having the foreign keys dropped before table data can be truncated. In cases like these, you need to pay close attention to the logical order of your ETL tasks and create a strategy that is appropriate to what you are trying to accomplish.

Be sure to test the tasks both singularly and collectively whenever possible, and you will resolve many issues that developers have when creating, deploying, and executing SSIS packages.

■ **Note** We recommend that you run the entire package twice. This is because your testing process may have inadvertently set the state of your database objects to something other than normal. For example, some tables may be filled, but others may not be. Since most packages eventually are set up by a SQL administrator to run automatically at night, nobody wants to find out that a package scheduled to run at 2 a.m. has failed. Running the package successfully the first time may give you a false positive, while running it a second time will be similar to how it was scheduled.

The Progress/Execution Results Tabs

The Visual Studio debugger tracks whether each task completes successfully on both the Progress and Execution Results tabs. It may seem odd, but both of these tabs are the same. The tab title changes to Progress while it is running, as shown in Figure 7-36, but changes back the Execution Results (Figure 7-37) when it stops.

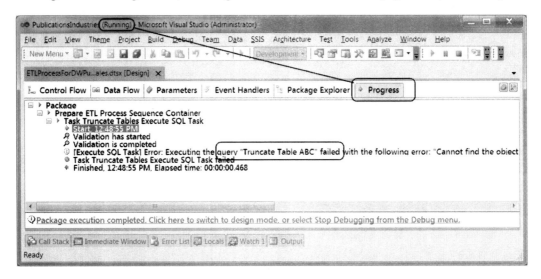

Figure 7-36. *Viewing the Progress tab*

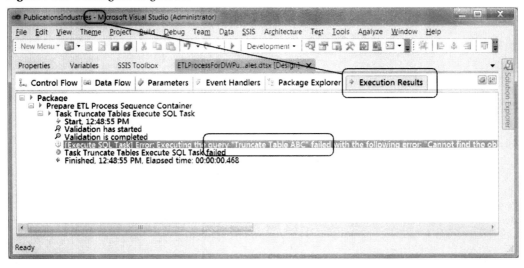

Figure 7-37. *Viewing the Execution Results tab*

When all the tasks are executed successfully, you see "100 percent complete" in the breakdown of each task. If the task fails, you are notified in the same window with a circled red "!" symbol and an associated error message. You use this information to troubleshoot the cause of your problem.

Sometimes the information can be hard to read, so you can get a better look at the text by opening the Output window using the View ➤ Output Window menu option of Visual Studio. (There is also a table at the bottom of Visual Studio while SSIS is in debug mode.)

To see an example of this, you could try running the task that drops a foreign key constraint once and then attempt to rerun that same task again. SQL Server raises an error when you try to drop foreign key constraints that do not exist. This error is then caught by the SSIS runtime engine and sent to Visual Studio's Output window. That is the error you see in Figure 7-38.

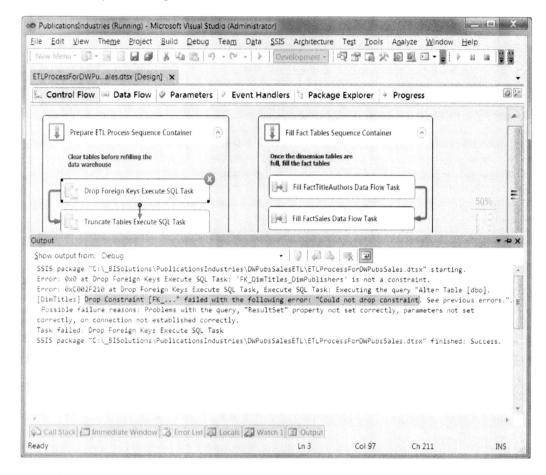

Figure 7-38. *Reading error messages on the Output window*

The information about the error is not always useful and, if not read carefully, can even be misleading. In Figure 7-38, note that the output message states "connection not established correctly," but if you look carefully, you find that the output text uses the word *possible* to describe the reason for the failure.

When we made this screenshot, the connection was working fine. The problem was that we had already run the "Drop the Foreign Key Constraints" task, and when the task ran a second time, the error occurred because those foreign key constraints no longer existed.

The error message also mentions a problem with the query, but we know there is no problem with it because we tested the code in SQL Server Management Studio. In the end, you must use common sense to troubleshoot the cause of these errors and not necessarily rely on information from the Output window alone.

You have seen how to create a new SSIS project, outline the ETL process in the package, create connections, and configure an SSIS task. You will soon put this knowledge to work by creating some connections and configuring the two Execute SQL tasks in your ETLProcessForDWPubsSales.dtsx package during Exercise 7-3. Before we start the exercise, let's make sure the current state of the database is ready for the ETL process. You can do this by resetting the database.

Resetting Your Destination Database

Whenever you test your ETL process, it is important to verify that the destination database (in this case the DWPubsSales data warehouse) is currently in a preload state, as it would be if the SSIS package was scheduled to run automatically. Over the years, we have seen a number of errors stemming from this issue.

To understand this better, let's go back to an example mentioned earlier. Consider a database with tables and foreign key constraints and an SSIS package that drops the foreign key constraints, clears out the tables, and replaces the foreign key constraints—in that order. Everything should run smoothly if the database is in its preload state (where the foreign key constraints exist before the execution of the SSIS package begins). If, while testing your work, you ran an Execute SQL Task that was configured to drop the foreign key constraints, the database would no longer be in its normal, preload state.

At this point, if you try to execute the entire SSIS package, the task that drops these constraints fails because the foreign key constraints are now missing! In fact, that is how we forced the error in Figure 7-38 to appear.

You can resolve this problem by replacing the foreign key constraints and rerunning the SSIS package again, but occasionally these issues are less obvious and you find troubleshooting is taking an exorbitant amount of time. When this happens, resetting the database to its preload state is the quickest way to resolve these issues.

There are two common ways to reset a database: running a SQL script or restoring a database backup.

In Chapter 6, we wrote a SQL script that creates the DWPubsSales database and all of the objects inside of it. Running this SQL script resets the database to its normal, empty state: resetting the database. The downside of this approach is that the script will not normally place data into the tables. Therefore, if your SSIS package is expecting at least some data to be present, you have to find a way to add it. In that case, the script can be modified to insert data, but you might want to consider restoring a copy of the database from a SQL backup instead.

To create a SQL backup, you can open a SQL Query window in SQL Server Management Studio, as we did in Chapter 6, and execute code similar to Listing 7-1.

Listing 7-1. Creating a Backup Copy of a Data Warehouse for the Developer Team

```
BACKUP DATABASE [DWPubsSales]
TO DISK=N'C:\_BISolutions\Publications Industries\DevBackup.bak'
WITH INIT -- Initialize the backup file by clearing its contents
```

This code copies all of the tables, constraints, views, stored procedures, and other objects in the database along with all of the data in the tables and indexes. A common scenario is for a SQL administrator to make a backup for the developer team to use in their development and testing process and then to place it on a network share that the development team has access to. They then either restore the backup on a development server, let each developer restore a copy to their local computers, or both.

Listing 7-2 is an example of the code used to restore a database.

Listing 7-2. Restoring a Backup Copy of a Data Warehouse for the Developer Team

```
RESTORE DATABASE [DWPubsSales]
FROM DISK=N'C:\_BISolutions\Publications Industries\DevBackup.bak'
WITH REPLACE -- replace the existing DB files as needed
```

■ **Tip** Being able to back up and restore a SQL database is a very useful skill. We recommend you do some more research on this subject when you can. A place to start is by performing an Internet search on the keywords "TSQL backup command."

In the next exercise, you start the process of configuring and executing the various tasks that make up the SSIS package that fills the DWPubsSales data warehouse. If at any point you find that you are receiving errors that are difficult to resolve and you suspect this may be because your database is out of sync with the current execution of the package objects, keep the option of resetting the data warehouse database in mind!

EXERCISE 7-3. CONFIGURING AND TESTING TASKS

In this exercise, you continue to work on the SSIS package outlined in Exercise 7-2.

In the previous exercise, you outlined all of the tasks and organized them within sequence containers. Now you focus on creating two SSIS connection managers, configuring three Execute SQL tasks, and testing that they run successfully. You also add an existing SQL code file containing all the SQL code you need to complete this exercise to the project.

Important: Before you begin, remember that in Chapter 6, Exercise 6-1, you created a SQL script file that contained ETL code and saved it to C:_BISolutions\PublicationsIndustries\PublicationIndustriesETLCode.sql. You need this file in that location for the current exercise. If you do not have this file for some reason, you can use the one that comes with the downloadable book files. You will find it at C:_BookFiles\Chapter07Files\ PublicationIndustriesETLCode.sql.

1. If the solution from the previous exercise is closed, open Visual Studio by clicking the Windows Start button and selecting All Programs ➤ Microsoft SQL Server 2012 ➤ Data Tools. Then follow the instructions from Exercise 7-1 to open C:_BISolutions\PublicationsIndustries\PublicationsIndustries.sln.

Remember to right-click the menu item, select Run as Administrator, and answer Yes to close the UAC dialog!

Create Your Connections

Before you start to configure your SSIS tasks, you should create its connections. You need two connections: one to the Pubs database and another to the DWPubsSales database. Let's set those up now!

1. At the bottom of the SSIS package designer, locate the Connection Managers tab; then right-click its background, and select New OLE DB Connection from the context menu. The page is similar to what you already saw in Figure 7-25. The OLE DB Connection Manager dialog appears (Figure 7-26).

2. In the OLE DB Connection Manager dialog, click the New button to create a new connection. The Connection Manager dialog appears (Figure 7-27).

3. In the Connection Manager dialog, type in your SQL Server's name, typically the word *(local)* will suffice unless you are using a named instance of SQL server (Figure 7-27). (For more information on connecting to your local server or a named instance of your server, see Chapter 5.)

4. On this same dialog, select or type **Pubs** into the "Select or enter a database name" textbox (Figure 7-27).

5. Repeat this process to create a new OLE DB connection to the DWPubsSales database.

6. Rename the connections managers to match the naming conventions used in Figure 7-39.

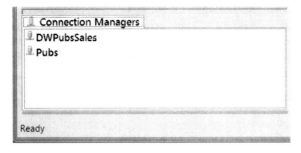

Figure 7-39. *Two connection managers have been configured and renamed*

Adding an ETL Code File to Your Project

OK! The connections are in place; now it is time to configure the Execute SQL tasks. Let's start with the task that drops the foreign keys from the DWPubsSales database at the beginning of the ETL process, then configure the task that clears the tables out, and finally configure the task that replaces the foreign keys at the end of the ETL process. You will use your Chapter 6 script file for these tasks, so there is no need to re-create the same code. Instead, you add this SQL script file into your SSIS project and then copy and paste the sections of code you need for each task.

7. Right-click the DWPubsSalesETL project icon in Solution Explorer, and select the Add ➤ Existing Item option from the context menu (Figure 7-40). An Add Existing Item dialog appears.

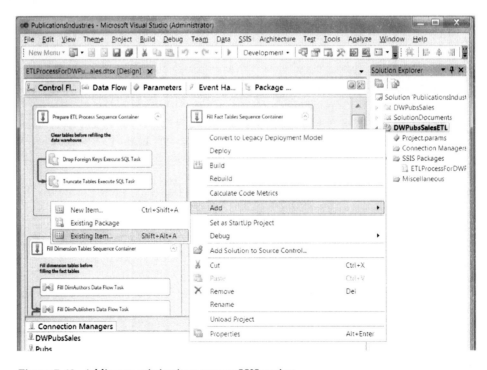

Figure 7-40. *Adding an existing item to your SSIS project*

8. Locate the C:_BISolutions\PublicationsIndustries\PublicationIndustriesETLCode.sql file; highlight it with your mouse, and then click the Add button at the bottom of the dialog (Figure 7-41). The new file is placed in a Miscellaneous folder as part of the SSIS project.

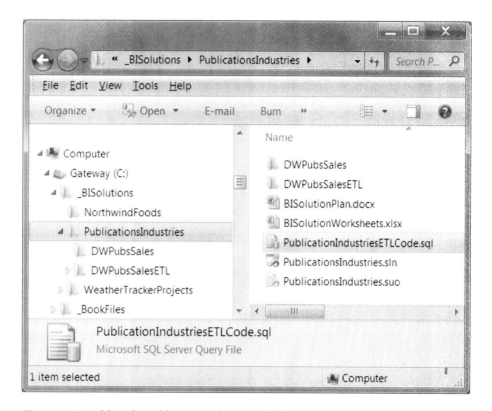

Figure 7-41. *Adding the Publication Industries ETL Code.sql file to your SSIS project*

9. Once the SQL code file has been added to your project, right-click the file and select the Open With option from the context menu. This opens the Open With dialog, as shown in Figure 7-42.

10. In the Open With dialog, note the choice of editing tools avaliable.

11. Select Transact-SQL Editor, and click the OK button.

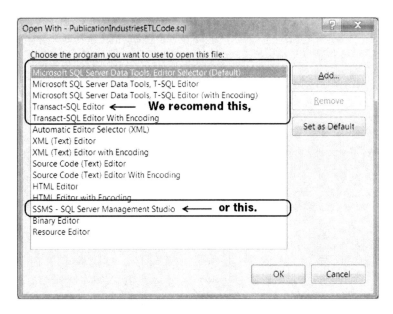

Figure 7-42. *Selecting an editor to open the SQL file in Visual Studio*

Important: Depending on your Visual Studio configuration, you may not have all these options, and even if you do, some may not work. For example, opening an existing SQL code file with the Microsoft SQL Server Data Tool options does not work on some computers, and instead of seeing the code shown in Figure 7-43, a black screen appears! To avoid this issue, we recommend using the Transact-SQL Editor in this book instead of the more modern Data Tools options because this works on most computers. If you still have problems, you can always open the file in SQL Server Management Studio instead. That option is not as convenient, but it always works.

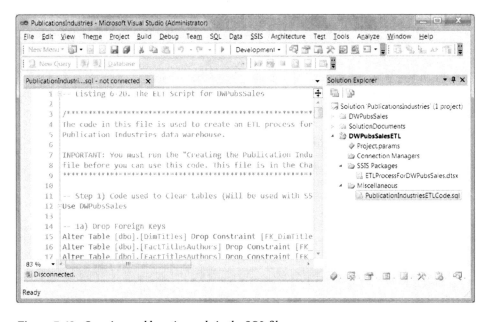

Figure 7-43. *Opening and locating code in the SQL file*

Configure the Drop Foreign Keys Execute SQL Task

With the SQL script file added and displayed in Visual Studio, it is time to use that code to configure your Execute SQL Tasks. For each task, locate code within the file, copy the code by right-clicking it, choose Copy from the context menu, and paste it into the appropriate SSIS task.

1. Locate the code labeled "1a)" and verify that its purpose is to drop the foreign key constraints. The code you are looking for should look like Listing 7-3.

Listing 7-3. Code That Drops the Foreign Key Constraints

```
-- 1a) Drop Foreign Keys
Alter Table [dbo].[DimTitles] Drop Constraint [FK_DimTitles_DimPublishers]
Alter Table [dbo].[FactTitlesAuthors] Drop Constraint [FK_FactTitlesAuthors_
DimAuthors]
Alter Table [dbo].[FactTitlesAuthors] Drop Constraint [FK_FactTitlesAuthors_
DimTitles]
Alter Table [dbo].[FactSales] Drop Constraint [FK_FactSales_DimStores]
Alter Table [dbo].[FactSales] Drop Constraint [FK_FactSales_DimTitles]
Alter Table [dbo].[FactSales] Drop Constraint [FK_FactSales_DimDates]
Alter Table [dbo].[DimTitles] Drop Constraint [FK_DimTitles_DimDates]
```

2. Highlight this code, right-click it, and choose Copy from the context menu. In a moment you will paste it into the Execute SQL Task called Drop Foreign Keys Execute SQL Task.

3. Edit the Drop Foreign Keys Execute SQL Task by right-clicking the task and selecting Edit from the context menu.

4. Select the Connection property. A configuration dropdown box appears.

5. Configure the Connection property by selecting the option DWPubsSales from the dropdown box, as shown in Figure 7-28.

6. Select the SQL Statement property. An ellipsis button appears.

7. Configure the SQL Statement property by clicking the ellipsis button and pasting the code from Listing 7-3 that you copied earlier into the Enter SQL Query dialog that appears (Figure 7-29).

8. Verify that there are no SQL Go statements your code. SQL Go statements cause the Execute SQL Task to fail.

9. Click the OK button to close the Enter SQL Query dialog.

10. Click the OK button to close the Execute SQL Task editor.

Configure the Truncate Tables Execute SQL Task

1. Locate the code labeled "1b)" and verify that its purpose is to truncate the tables' data. The code you're looking for should look like Listing 7-4.

Listing 7-4. Code That Clears the Tables of Data

```
--1b) Clear all tables data warehouse tables and reset their Identity Auto
Number
Truncate Table dbo.FactSales
Truncate Table dbo.FactTitlesAuthors
Truncate Table dbo.DimTitles
Truncate Table dbo.DimPublishers
Truncate Table dbo.DimStores
Truncate Table dbo.DimAuthors
```

2. Highlight this code, right-click it, and choose Copy from the context menu.

3. Edit the Truncate Tables Execute SQL Task by right-clicking the task and selecting Edit from the context menu.

4. Select the Connection property. A dropdown box appears.

5. Configure the Connection property by selecting the option of DWPubsSales from the dropdown box, as shown in Figure 7-28.

6. Select the SQL Statement property. An ellipsis button appears.

7. Configure the SQL Statement property by clicking the ellipsis button and pasting the code from Listing 7-4 that you copied earlier into the Enter SQL Query dialog that appears.

8. Remove any SQL Go statements from your code.

9. Click the OK button to close the Enter SQL Query dialog.

10. Click the OK button to close the Execute SQL Task editor.

Configure the Add Foreign Key Execute SQL Task

1. Locate the code labeled "Step 3)" and verify that its purpose is to replace the foreign key constraints that were dropped earlier. The code you're looking for should look like Listing 7-5.

Listing 7-5. Code That Replaces the Foreign Key Constraints

```
-- Step 3) Add Foreign Key s back (Will be used with SSIS Execute SQL Tasks)
Alter Table [dbo].[DimTitles] With Check Add Constraint [FK_DimTitles_
DimPublishers]
Foreign Key ([PublisherKey]) References [dbo].[DimPublishers] ([PublisherKey])

Alter Table [dbo].[FactTitlesAuthors] With Check Add Constraint [FK_
FactTitlesAuthors_DimAuthors]
Foreign Key ([AuthorKey]) References [dbo].[DimAuthors] ([AuthorKey])

Alter Table [dbo].[FactTitlesAuthors] With Check Add Constraint [FK_
FactTitlesAuthors_DimTitles]
Foreign Key ([TitleKey]) References [dbo].[DimTitles] ([TitleKey])

Alter Table [dbo].[FactSales] With Check Add Constraint [FK_FactSales_
DimStores]
Foreign Key ([StoreKey]) References [dbo].[DimStores] ([Storekey])
```

```
Alter Table [dbo].[FactSales] With Check Add Constraint [FK_FactSales_
DimTitles]
Foreign Key ([TitleKey]) References [dbo].[DimTitles] ([TitleKey])

Alter Table [dbo].[FactSales] With Check Add Constraint [FK_FactSales_DimDates]
Foreign Key ([OrderDateKey]) References [dbo].[DimDates] ([DateKey])

Alter Table [dbo].[DimTitles] With Check Add Constraint [FK_DimTitles_DimDates]
Foreign Key ([PublishedDateKey]) References [dbo].[DimDates] ([DateKey])
```

2. Highlight this code, right-click it, and choose Copy from the context menu.

3. Edit the Add Foreign Key Execute SQL Task by right-clicking the task and selecting Edit from the context menu.

4. Select the Connection property. A dropdown box appears.

5. Configure the Connection property by selecting the option of DWPubsSales from the dropdown box, as shown in Figure 7-28.

6. Select the SQL Statement property. An ellipsis button appears.

7. Configure the SQL Statement property by clicking the ellipsis button and pasting the code from Listing 7-5 that you copied earlier into the Enter SQL Query dialog that appears.

8. Remove any SQL Go statements from your code.

9. Click the OK button to close the Enter SQL Query dialog.

10. Click the OK button to close the Execute SQL Task editor.

Reset the Database Objects Before Testing Your ETL Process

With the three Execute SQL tasks configured, it is important to test that they work correctly. But, before we do so, let's make sure that the database is in its normal empty state by running the SQL code that re-creates the database.

1. Locate the Create DWPubsSales Script.sql file in Visual Studio or in SQL Server Management Studio. You should find this file under the DWPubsSales solution folder you created in Exercise 5-4.

2. Open this file and execute its code. You can do this by right-clicking an area in the SQL code window to bring up the context menu and select Execute SQL. You may remember doing this back in Chapter 2 (Figure 2-15).

3. When the Connect to Database Engine dialog box appears, type in your server name, and click the Connect button.

4. In a few seconds, you receive a message stating "The DWPubsSales data warehouse is created."

Testing Your Tasks

Now we test the Execute SQL tasks you just configured.

1. Right-click your SSIS task called Drop Foreign Keys Execute SQL Task, and select Execute Task from the context menu (Figure 7-33).

2. Verify that the tasks runs successfully, and if not, troubleshoot why (Figure 7-38). Remember, if you need to change settings in the task, you must first stop the debugging engine.

3. Right-click your SSIS task called Truncate Tables Execute SQL Task, and select Execute Task from the context menu.

4. Verify that the tasks runs successfully, and if not, troubleshoot why.

5. Right-click your SSIS task called Add Foreign Keys Execute SQL Task, and select Execute Task from the context menu.

6. Verify that the tasks runs successfully, and if not, troubleshoot why.

7. Stop the debugger using the Debug menu or by clicking the link at the bottom of the screen (Figure 7-36).

In this exercise, you created two new SSIS connection managers and configured three Execute SQL tasks to utilize them. Each task was tested to verify proper execution, and then the database objects such as tables and foreign key constraints were re-created by executing a SQL script so that you could continue working with the database and other SSIS tasks.

Moving On

In this chapter, we began the ETL process with SSIS. We worked with various data connections and began configuring SQL tasks to properly execute. But our ETL process is just getting started. In Chapter 8 we continue working on our ETL process.

LEARN BY DOING

In this "Learn by Doing" exercise, create an ETL process similar to the one defined in this chapter. This time you need to use the Northwind database. We have included an outline of the steps you performed in this chapter and an example of how the authors handled them in two Word documents. You can find these documents in the folder C:_BISolutionsBookFiles_LearnByDoing\Chapter07Files. Please see the ReadMe. doc for detailed instructions.

What's Next?

SSIS can be a very exciting tool to work with for some and a completely frustrating tool for others. If you found that you liked working with SSIS, we suggest the following book: *Microsoft SQL Server 2008 Integration Services Unleashed*, by Kirk Haselden (Sams, ISBN-10: 0672330326).

CHAPTER 8

Concluding the ETL Process with SSIS

Sometimes good things fall apart so better things can fall together.

—Marilyn Monroe

Thus far, the ETL process has taken us through a lot of decision making. Data has had to be cleaned up and handled with care to ensure it is accurate and that it can be worked with. In Chapter 7, we created an SSIS package and set up our data connections. We also configured and tested the Execute SQL tasks.

In this chapter, we continue the SSIS process by configuring data flows and data destinations that enable us to load our data warehouse, and we handle errors and round out this subject by executing the entire package.

Data Flows

Data Flows are the most complex of the SSIS tasks, because they are made up of a composite of many subcomponents. There are three different types of data flow subcomponents: sources, transformations, and destinations. Most often sources pull data from tables and views, but you can also extract data from multiple file types, and even SSAS cubes. Transformations modify, summarize, and clean data. Destinations load data into tables, cubes, files, or in-memory datasets that can be used by other tasks within the SSIS package itself.

Data flows are configured using the Data Flow tab. You can access this tab at the top of the package designer window or right-click any data flow task and select Edit from the context menu. Either way, the user interface navigates to the Data Flow tab (Figure 8-1).

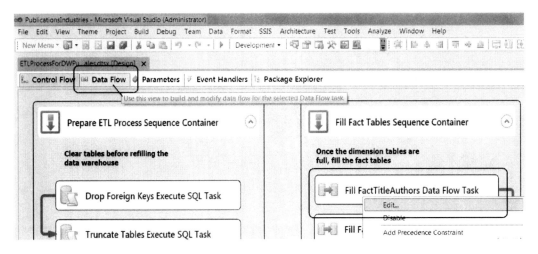

Figure 8-1. *Editing a data flow task*

All the data flows in your SSIS package are edited from one tab. The dropdown box at the top of the Data Flow tab allows you to select which data flow task you would like to work with, as shown in Figure 8-2.

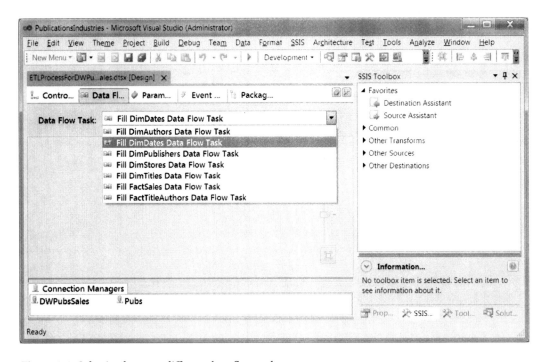

Figure 8-2. *Selecting between different data flow tasks*

Outlining a Data Flow Task

Data flow tasks require at least one data source component and one data destination component. Often you just need one of each, but you can have more than one data source and more than one data destination per data flow.

To outline your Data Flow task steps, begin by placing data source and data destination components from the Toolbox onto the designer surface (Figure 8-3). At this point, even though it is possible to connect them with arrows between each of the components, do not do so! Adding arrows before configuring the components causes problems. The arrows of the data flow do not work the same as the arrows on the Control Flow tab. Control flow arrows represent a precedent constraint and control the order of execution.

Data flow arrows represent a data flow path and pass metadata from the component at the arrow's origin to the component at the end of the arrow. If arrows are connected between components before the source component is configured, there will be no metadata to pass on to the end component. Therefore, connect the arrows only *after* each source component is configured.

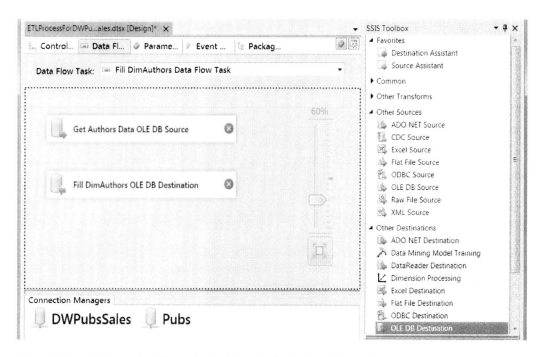

Figure 8-3. *Outlining a data flow task should not include a data flow path*

Configuring the Data Source

A data source component is the heart of each data flow task. You can add a data source component to the data flow by clicking the SSIS Toolbox and locating either the Source Assistant or one of the data source component lists under the Other Sources category. Selecting the Source Assistant will launch a wizard that helps you select one of the other sources. Therefore, you will not need to use it when you already know which type of component you want.

You should always choose the data source and destination components to match the connection manager objects in your SSIS package. We are using OLE DB connections in our SSIS package; therefore, we want to use an OLE DB Source component.

When you click a component, you will see data path arrows displayed, but remember that until the component is properly configured, you cannot use them to connect to other components. A data source component is configured by right-clicking the component and choosing Edit from the context menu.

■ **Note** In Figures 8-3 and 8-4, the two data path arrows represent a blue data flow path and a red error output path. Previous versions of SQL use green for the data flow rather than blue, but the function is the same. For more information, see "Data Flows: Data Flow Paths" later in this chapter.

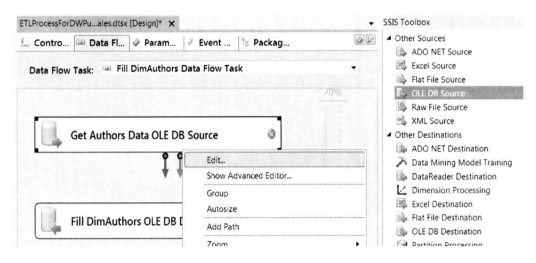

Figure 8-4. Editing a data source

The OLE DB Source Editor

Selecting Edit from the context menu shown in Figure 8-4 launches the OLE DB Source Editor window—a window that you will become quite familiar with if you work with SSIS for any length of time!

The OLE DB Source Editor window consists of three separate pages:

- The Connection Manager page
- The Columns page
- The Errors Output page

The Connection Manager Page

On the Connection Manager page, select which connection manager object to use from the "OLE DB connection manager" dropdown box (Figure 8-5). If you have not previously created a connection manager for the SSIS package, the dropdown box will not offer any selections. The connection mangers are typically created first, but if you have not done so yet, click the New button and create one.

■ **Tip** One connection manager can be used by many different data flow components. It is a common mistake for developers to create a new connection for each separate component using the New button.

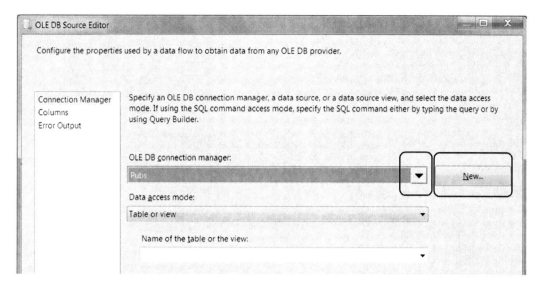

Figure 8-5. *Configuring the Data Source connection manager*

After you have chosen your connection, you will want to choose your data access mode. As you can see in Figure 8-6, you can select from four separate options in an OLE DB connection manager:

- Table or view
- Table name or view name variable
- SQL command
- SQL command from a variable

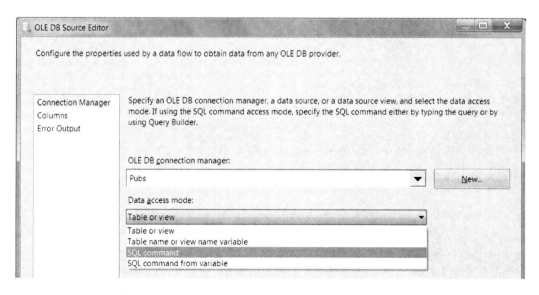

Figure 8-6. *Configuring the data source data access mode*

The OLE DB Source Connection Manager Page

The "Table or view" option shown in Figure 8-7 allows you to select a table or view from the data source. The table name and view name variables do the same thing, as long as you place the name of the table or view in an SSIS variable. When you use the "Table or view" option, you are presented with a list of tables and views to select from, based on your chosen connection manager.

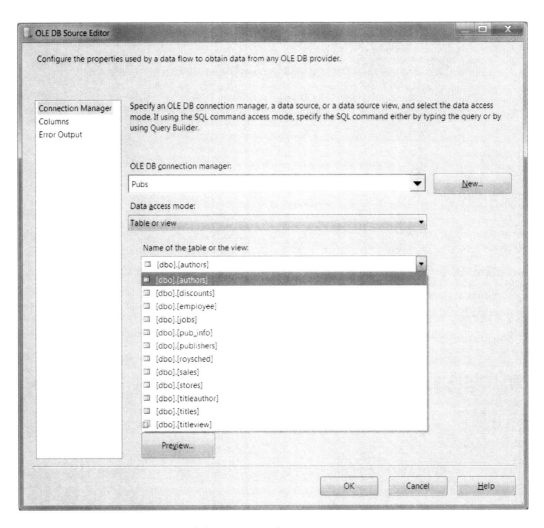

Figure 8-7. Using the "Table or view" data access mode

■ **Note**   Using either one of these options is similar to using SELECT * FROM < Some table>, bringing all of the columns from that table even if they will never be of use. Therefore, always try to use either a SQL view that restricts the data you get from a table or the "SQL command" data access mode and include only the required columns in your SQL code.

SQL Command and SQL Command from a Variable

The "SQL command" choice allows you to type or copy and paste a SQL statement into the command window (Figure 8-8). As you might have already concluded, the SQL command from a variable does the same thing, as long as you place your SQL code into an SSIS variable first. You can use either of these data access modes to execute stored procedures as well.

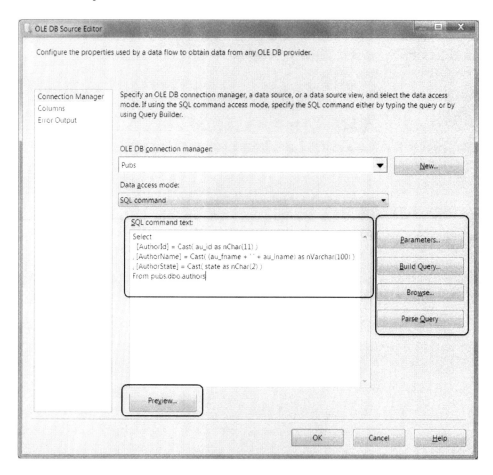

Figure 8-8. *Using the SQL command data access mode*

■ **Tip**　The recommended practice is to use SQL Server Management Studio to create and test your SQL code. The steps involved include writing the select statement, wrapping the statement into a SQL stored procedure, testing that it works as expected, and then using the stored procedure's name in the SQL command text window. The stored procedure would include most, if not all, the ETL transformations required and provides your source component with clean transformed data. All the actual ETL transformation processing happens within the database engine as the stored procedure is executed. This is both faster and less prone to errors. You may remember that in Chapter 6 we created all the select statements we needed, but to keep things simple, we did not put them into stored procedures. In a production environment, you should consider using stored procedures instead.

Let's discuss the purpose of several useful buttons in the OLE DB source editor window, as indicated in Figure 8-8:

- *Parameters button*: Allows you enter a parameterized query in the query text using the question mark symbol (?) for OLE DB sources or the at (@)sign for ADO.NET sources

- *Build Query button*: Allows you to build a SQL with the Query Builder dialog window, similar to the one in SQL Server Management Studio (also discussed in Chapter 6)

- *Browse button*: Allows you to look for a file that contains the SQL code you want to run

- *Parse Query button*: Checks the syntax of the SQL code and verifies if the objects exist in the connected database

- *Preview button*: Runs the SQL code and shows you the results

■ **Note** Some of these buttons do not become available until code is typed in the window.

The Columns Manager Page

The Columns Manager page allows you to filter out any columns that were part of the input but no longer wanted as part of the output. Unchecking the checkbox next to a column name will remove it from the output of your data source.

Figure 8-9 shows all of the columns checked. Remember that we explicitly selected the SQL query, and typically we would use all of them. If we chose to use a table or view name, however, instead of a SQL query, we could uncheck any columns we did not need.

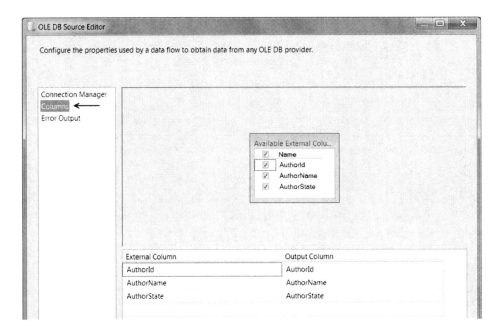

Figure 8-9. *Viewing the Columns page of the OLE DB data source editor*

■ **Tip** Sometimes, the data source component will not properly register the internal XML code unless you click the Columns page. Therefore, we recommend you always click the Columns page even if you are not going to make any additional configurations. It is simple to do and helps avoid this occasional bug.

The Error Output Page

The Error Output page allows you to redirect data that produces an error to a separate error output path. The idea is that if an error of some type occurs, you could route the rows causing these errors to a file on your hard drive or perhaps a SQL table that was made to hold incorrect or inconsistent data. Figure 8-10 shows this configuration page.

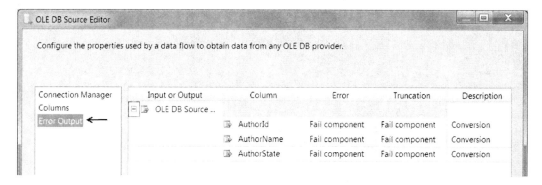

Figure 8-10. Viewing the Error Output page of the OLE DB Source Editor window

To understand error outputs better, you need to first understand what an error output path is. To understand that, you need to understand what a data flow path is. So, without further ado, let's take a look at each of these in order.

Data Flow Paths

To advance our discussion of data flow paths, let's start with a short review. When you place a data source component on the surface of the Data Flow designer surface, you will notice that both a blue line and a red line appear (Figure 8-4). These lines represent data flow and error flow paths, respectively. Data flow paths look similar to control flow precedence constraints, but they behave very differently.

Precedence constraints contain conditional logic that defines the programmatic flow of the tasks in an SSIS package. Each data flow path originates in a source component and ends in a destination component and does not have a way to conditionally define program flow. Instead, data flow paths contain metadata from the original data source component and any subsequent transformation components that come after. The idea is to pass all the metadata from the source component and all intermediate components on to the final destination component so that the final destination component can be properly configured. To see this metadata, double-click the data flow path arrow, and a dialog window will appear like the one shown in Figure 8-11.

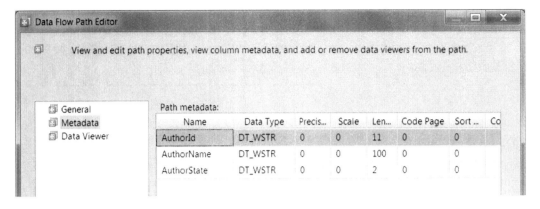

Figure 8-11. *The Data Flow Path Editor window*

If you try to connect the arrows before the source component is configured, there will be no metadata to pass on to the destination component. This causes errors that are tedious to resolve—not difficult, just tedious. We recommend avoiding this practice and connecting the arrows only after properly configuring each beginning-point component. Be sure that you do not open the destination component editor until after the source component metadata is available.

■ **Important**   The most common problem we see when working with data flows is connecting the blue or red arrows during the outlining process. If, for some reason, you connect an arrow between two components before the source component is configured, simply delete the destination component and try again. It is possible to fix the issue without deleting components, but most of the time, deleting is the simplest or fastest way to resolve the issue.

Error Outputs Paths

Error flows have the ability to route data to a separate location in case of an error. The error output path arrow looks similar to the data flow path arrow, but it is red in color, instead of blue (Figure 8-12). To use this feature, first add an additional destination component, such as the additional OLE DB destination component shown in Figure 8-12, and then connect the error output path arrow to the new destination component. The destination component will usually be a flat file or a SQL table.

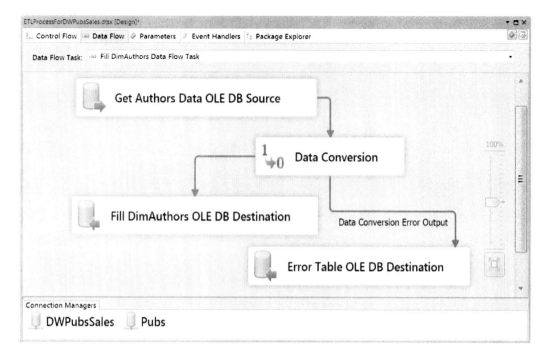

Figure 8-12. *Adding an error output path to the data flow*

Connect the error flow path from a transformation task to a destination component by dragging and dropping the red arrow just as you would when working with a data flow path output. Once you connect the arrow, the Configure Error Output window automatically appears (Figure 8-13).

In the Configure Error Output window (Figure 8-13), you can configure SSIS to redirect the row causing the error, ignore the error, or fail the source component. The default setting is Fail, but leaving it set to default defeats the purpose of the error output, because no rows would be redirected to the log file or table.

In Figure 8-13, an arrow points to the Selected Cells dropdown box at the bottom of the dialog window. Its purpose is to allow you to highlight multiple cells within the dialog window and set all of their values at once. This is an optional feature, and it can be safely ignored until you have many outputs that you want to configure collectively. (The only reason to point out this option is because it is so prominent on the dialog window that it seems more significant than it is.)

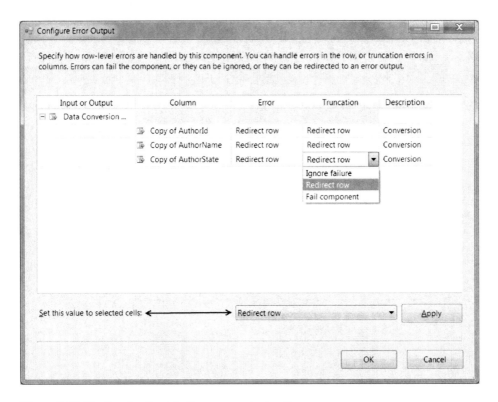

Figure 8-13. *Configuring the data flow error output path*

■ **Note**   Error outputs are a common feature in production ETL processing, but we keep things simple by not using error outputs in our exercises.

Configuring the Data Destination

Data destinations insert data into database tables, files, spreadsheets, and so on. The OLE DB Destination Editor has three pages: Connection Manager, Mappings, and Error Output. You can see these listed on the left of Figure 8-14.

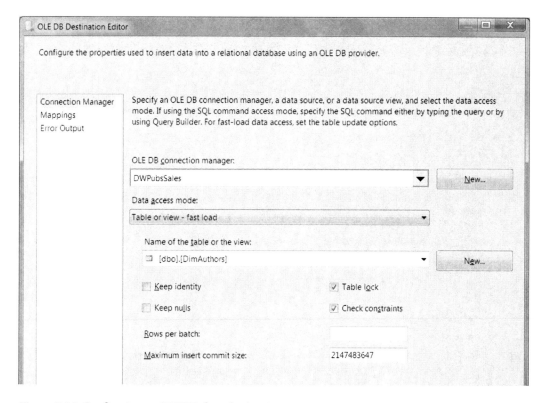

Figure 8-14. *Configuring an OLE DB data destination*

Each data destination component can look slightly different depending on which type of destination component you use. The SSIS Toolbox has several data destination components to choose from, but you are likely to use the OLE DB destination the majority of the time. Therefore, we focus on the settings of this type of data destination.

■ **Important**   If you are using an OLE DB source task, you should choose the OLE DB destination versus the SQL Server destination, even when you are importing data into a SQL Server database. Although the SQL Server destination works, it may expect different metadata types than the OLE DB destination provides. If you try to use an OLE DB data source with a SQL Server destination, you can anticipate errors unless you have transformed all the data types appropriately.

The Connection Manager Page

On the Connection Manager page, set the connection to point to your destination table and decide what access mode you are going to use (Figure 8-15).

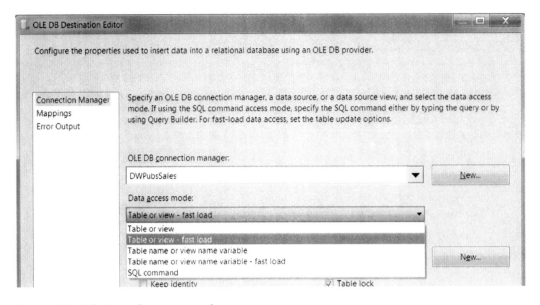

Figure 8-15. *Selecting a data access mode*

To select your data connection, use the "OLE DB connection manager" dropdown box, shown in Figure 8-15, and select the OLE DB connection object that has already been created in the SSIS package. If, for some reason, you have not created an OLE DB connection before you get to this point, clicking the New button allows you to create one.

After you select your connection, configure the data access mode you want by using the "Data access mode" dropdown box (Figure 8-15). An OLE DB destination has five data access modes to choose from:

- *Table or view*: Insert values into a new or existing table or view.

- *Table or view - fast load*: Bulk insert into a new or existing table or view.

- *Table name or view name variable*: Same as the "Table or view" option, but you pass the table or view name through an SSIS variable

- *Table name or view name variable - fast load*: Same as the "Table or view - fast load" option, but you pass the table or view name through a SSIS variable.

- *SQL command*: You use a SQL query to describe the table and column names.

You will use the "Table or view - fast load" option for most occasions, because it provides additional configuration options as well as it is easy to use.

■ **Tip**   Fast load uses SQL Server bulk insert statements. The other option uses regular SQL Server insert statements. Since SQL Server 2005 and later, BULK INSERT enforces new, stricter data validation. The two most common issues occur when Unicode text data is not used or when the decimal data type is not used for floating-point numbers. You can find more information in SQL Books Online, but since we converted all of our examples to Unicode and decimal data types, this should not be an issue for our exercises.

Mappings Page

The Mappings page allows you to dictate which input columns are mapped to which destination output columns. SSIS automatically maps the input and destinations for you if the names of both source and destination columns are the same. If you use column aliases SQL code, as we did, the column names will be the same. If, on the other hand, the column names are not the same, you have to map them manually.

Mapping the input and destination columns is an easy process. Just drag and drop the available input columns in the table to destination columns. Another option is to use the dropdown box to select an input column and an associated destination column (Figure 8-16).

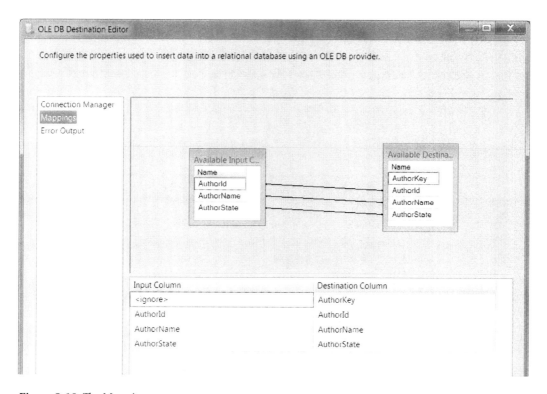

Figure 8-16. *The Mappings page*

You do not have to map input columns to all destination columns. If you have more input columns than you have destination columns, set the input column to < ignore>, thus excluding columns from the output.

■ **Tip** Sometimes the XML code in the SSIS package will not be properly written unless you click the Mappings page. Because of this, we recommend you always click the Mappings page even if you are not going to make any additional configurations.

Error Output Page

Just like the data sources and transformation tasks, the Error Output page allows you to redirect data that causes an error to a file or SQL table. This can be convenient in some cases, but typically you want your errors to be taken care of before you get to this stage of the data flow.

■ **Note** The destination components were not specifically set up to handle errors but rather to insert clean data into a destination. All duplications, conversions, foreign key values, and so on, need to be configured before you get to the destination!

Now that you have seen how to configure a data flow, it is time to put that knowledge into practice. Let's do so in the next exercise.

EXERCISE 8-1. COMPLETING THE SSIS PACKAGE

In this exercise, you complete the SSIS package by configuring all of the remaining tasks and then testing them to verify that the entire SSIS package can execute as a whole. Use the same code file from Exercise 7-3 of Chapter 7 for all of SQL code required in this exercise.

Tip: If you have any problems with this exercise, don't worry, we have you covered. You will still be able to complete the exercises in the rest of the book using a prefilled version of the DWPubsSales database. You will find this file in the `C:\_BookFfiles\Chapter07Files\FilledDWPubsSalesDB` folder.

Reset the Database Objects Before Testing Your ETL Process

Before we begin configuring and testing our tasks, let's make sure that the database is in its normal empty state by running the SQL code that re-creates the database.

1. Locate the Create DWPubsSales Script.sql file in Visual Studio or SQL Server Management Studio (Figure 8-17). You should find this file under the DWPubsSales solution folder you created in Exercise 5-4.

Figure 8-17. *Locating the create DWPubsSales script*

2. Open this file and execute its code in either Visual Studio or SQL Server
   Management Studio to re-create the database.

You can do this by right-clicking an area in the SQL code window to bring up the context menu and select
Execute SQL. If the Connect to Database Engine dialog box appears, type in your server name and click the
Connect button. In a few seconds, you should receive a message stating "The DWPubsSales data warehouse
is now created."

Fill DimAuthors Data Flow Task

Now that you located the ETL code, you need to use it in the task that fills DimAuthors. This is the same
process we used in Chapter 7.

1. Open the PublicationIndustriesETLCode.sql SQL script file in Visual Studio or SQL
   Server Management Studio (Figure 8-17) and locate the code that is commented as
   `-- Step 2 Code used to fill tables`.

2. Navigate to the Data Flow tab and select Fill DimAuthors Data Flow Task from the
   Data Flow Task dropdown box.

3. Add an OLE DB Data Source to the data flow surface. Rename it to Get Authors Data
   OLE DB Source.

4. Locate the comment `-- Step 2a) Get source data from pubs.dbo.`
   `authors` in the SQL code file and review the code beneath it. The code should look
   like Listing 8-1.

Listing 8-1. *SQL Code from pubs.dbo.authors*

```
Select
  [AuthorId]=Cast( au_id as nChar(11) )
```

```
, [AuthorName] = Cast( (au_fname + ' ' + au_lname) as nVarchar(100) )
, [AuthorState] = Cast( state as nChar(2) )
From pubs.dbo.authors
```

5. Add this code to the SQL command text code area of the OLE DB data source as shown in Figure 8-8.

6. Click the Preview button to verify that the query worked successfully.

7. Click the Columns page to force the XML to be written properly in the .dtsx file.

8. Close the OLE DB Source Editor window by clicking the OK button.

9. Add an OLE DB Data Destination to the data flow. Rename it to Fill DimAuthors OLE DB Destination.

10. Connect the data flow path from the source to the destination.

11. Edit the OLE DB data destination as shown in Figure 8-14.

12. Click the Mappings page to force the XML to be written properly in the .dtsx file and verify that the mappings look like Figure 8-16.

13. Verify that the data flow looks like Figure 8-18.

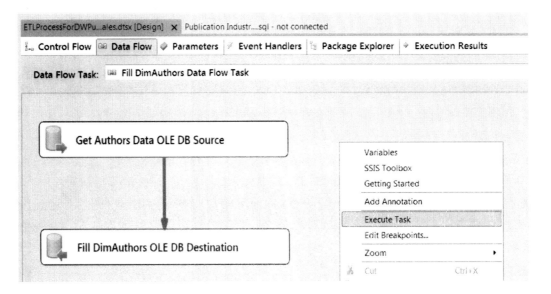

Figure 8-18. *Testing the Fill DimAuthors data flow task*

Important: You do not need to configure anything more than the connection and the table name. Leave all the checkboxes on their default settings as shown. Nor should you need to do anything on the mappings page since our code uses column aliases that match the column names in the data warehouse tables. Just verify that the mappings are correct and move on. In fact, we will save some trees by not showing screenshots for the other tables, since the configurations are all the same (with the exception of the table and column names, of course).

14. Right click the data flow design surface and choose Execute Task from the context menu. Visual Studio launches this data flow task with its debugging engine (Figure 8-18).

15. When the task is complete, stop Visual Studio's debugger and verify that the task ran successfully; or, if there are errors, troubleshoot and try again. You can stop the debugger using the Debug menu or by clicking the link at the bottom of the screen.

16. Open SQL Server Management Studio and verify that DimAuthors has data in it.

Tip: Common errors include primary and foreign key violations. If those happen to you, try running the Execute SQL tasks that you created in Exercise 7-3 to clear the table data and drop the foreign key constraints. You can run both by executing the Prepare ETL Process sequence container.

Fill DimStores Data Flow Task

Congratulations, DimAuthors has data! Let's do the same for DimStores.

1. Navigate to the Data Flow tab and select the Fill DimStores Data Flow Task from the Data Flow Task dropdown box.

2. Add an OLE DB Data Source to the data flow surface. Rename it to Get Stores Data OLE DB Destination.

3. Locate the comment -- 2b) Get source data from pubs.dbo.stores in the SQL code file and review the code beneath it. The code should look like Listing 8-2.

Listing 8-2. Source Data from pubs.dbo.stores

```
Select
  [StoreId]=Cast( stor_id as nChar(4) )
, [StoreName]=Cast( stor_name as nVarchar(50) )
From pubs.dbo.stores
```

4. Add this code to the SQL command text code area of the OLE DB data source (similar to Figure 8-8).

5. Click the Preview button to verify that the query worked successfully.

6. Click the Columns page to force the XML to be written properly in the .dtsx file.

7. Close the OLE DB Source Editor window by clicking the OK button.

8. Add an OLE DB Data Destination to the data flow surface. Rename it to Fill DimStores OLE DB Destination.

9. Connect the data flow path from the source to the destination.

10. Edit the OLE DB data destination so that it imports data to the DimStores table (similar to Figure 8-14).

11. Click the Mappings page to force the XML to be written properly in the .dtsx file and verify that the mappings look logically correct.

12. Right click the data flow design surface and choose Execute Task from the context menu (similar to Figure 8-18).

13. When the task is complete, stop Visual Studio's debugger and verify that the task ran successfully or troubleshoot any errors and try again.

14. Open SQL Server Management Studio and verify that DimStores has data in it.

Configure the Fill DimPublishers Data Flow Task

DimPublishers is next on the list. Let's tackle it now.

1. Navigate to the Data Flow tab and select the Fill DimPublishers Data Flow Task from the Data Flow Task dropdown box.

2. Add an OLE DB Data Source to the data flow surface. Rename it to Get Publishers Data OLE DB Source.

3. Locate the comment -- 2c) Get source data from pubs.dbo.publishers in the SQL code file and review the code beneath it. The code should look like Listing 8-3.

Listing 8-3. SQL Code from pubs.dbo.publishers

```
Select
  [PublisherId]=Cast( pub_id as nChar(4) )
, [PublisherName]=Cast( pub_name as nVarchar(50) )
From pubs.dbo.publishers
```

4. Add this code to the SQL command text code area of the OLE DB data source (similar to Figure 8-8).

5. Click the Preview button to verify that the query worked successfully.

6. Click the Columns page to force the XML to be written properly in the .dtsx file.

7. Close the OLE DB Source Editor window by clicking the OK button.

8. Add an OLE DB Data Destination to the data flow surface. Rename it to Fill DimPublishers OLE DB Destination.

9. Connect the data flow path from the data source to the data destination.

10. Edit the OLE DB data destination so that it imports data to the DimPublishers table (similar to Figure 8-14)

11. Click the Mappings page to force the XML to be written properly in the .dtsx file and verify that the mappings look logically correct.

12. Right click the data flow design surface and choose Execute Task from the context menu (similar to Figure 8-18).

13. When the task is complete, stop Visual Studio's debugger and verify that the task ran successfully or troubleshoot any errors and try again.

14. Open SQL Server Management Studio and verify that DimPublishers has data in it.

Configure the Fill DimDates Execute SQL Task

We should to fill up the DimDates table before we fill the DimTitles table. This is because of the foreign key relationship between the two tables. This is not strictly necessary since we have dropped the foreign key constraints in the database, but filling the tables in the same order as you would as if the foreign key constraints were still there documents the logical progression of the ETL process.

1. Locate the comment `-- 2d) Create values for DimDates as needed` in the SQL code file and review the code beneath it. The code should look like Listing 8-4.

    Listing 8-4. Code That Fills the DimDates Table

    ```
    -- 2d) Create values for DimDates as needed.

    -- Create variables to hold the start and end date
    Declare @StartDate datetime = '01/01/1990'
    Declare @EndDate datetime = '01/01/1995'

    -- Use a while loop to add dates to the table
    Declare @DateInProcess datetime
    Set @DateInProcess = @StartDate

    While @DateInProcess <= @EndDate
      Begin
      -- Add a row into the date dimension table for this date
      Insert Into DimDates
      ( [Date], [DateName], [Month], [MonthName], [Quarter], [QuarterName], [Year],
      [YearName] )
      Values (
      @DateInProcess -- [Date]
      , DateName( weekday, @DateInProcess ) -- [DateName]
      , Month( @DateInProcess ) -- [Month]
      , DateName( month, @DateInProcess ) -- [MonthName]
      , DateName( quarter, @DateInProcess ) -- [Quarter]
      , 'Q'+DateName( quarter, @DateInProcess )+' - '+Cast( Year(@DateInProcess) as
      nVarchar(50) ) -- [QuarterName]
      , Year( @DateInProcess )
      , Cast( Year(@DateInProcess ) as nVarchar(50) ) -- [YearName]
      )
      -- Add a day and loop again
      Set @DateInProcess = DateAdd(d, 1, @DateInProcess)
      End
    ```

2. Highlight this code, right-click it, and choose Copy from the context menu.

3. Navigate to the Control Flow tab and select the Fill DimDates Execute SQL Task.

4. Right-click this task and select Edit from the context menu.

5. Edit the Fill DimDates Execute SQL Task by right-clicking the task and selecting Edit from the context menu.

6. Select the Connection property. A dropdown box appears in the dialog window that will let you configure this property.

7. Configure the Connection property by selecting the option DWPubsSales from the dropdown box, as shown in Figure 8-19.

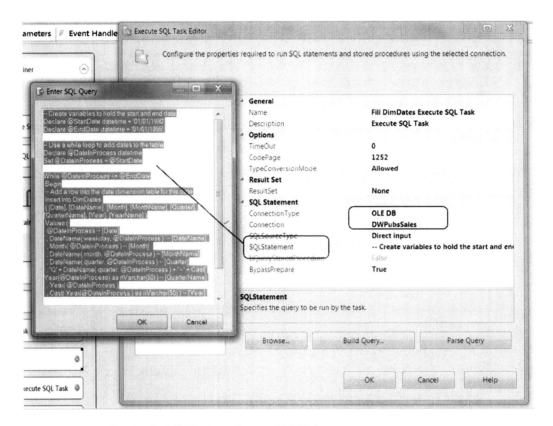

Figure 8-19. *Configuring the Fill DimDates Execute SQL Task*

8. Select the SQL Statement property. An ellipsis button appears in the dialog window that will let you configure this property.

9. Configure the SQL Statement property by clicking the ellipsis button and pasting the code from Listing 8-4 (that you copied earlier) into the Enter SQL Query dialog window that appears (Figure 8-19).

10. Remove any SQL Go statements from your code. SQL Go statements will cause the Execute SQL Task to fail.

11. Click the OK button to close the Enter SQL Query dialog window.

12. Click the OK button to close the Execute SQL Task editor.

13. Right-click the Fill DimDates Execute SQL Task and choose Execute Task from the context menu (Figure 8-20).

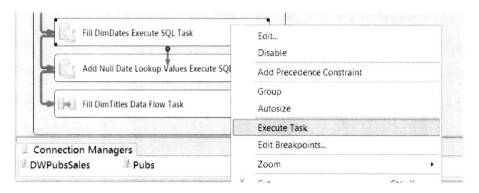

Figure 8-20. *Executing the Fill DimDate Execute SQL Task*

14. When the task is complete, stop Visual Studio's debugger and verify that the task ran successfully; if there are errors, troubleshoot and try again.

15. Open SQL Server Management Studio and verify that DimDates has data in it.

Configure the Add Null Date Lookup Values Execute SQL Task

Now we need to add a couple of dates used to for null value lookups to the DimDates table. This was an option we discussed in Chapter 6 but have waited until now to implement it.

1. Locate the comment `-- 2e) Add additional lookup values to DimDates` in the SQL code file and review the code beneath it. The code should look like Listing 8-5.

Listing 8-5. Code That Adds New Dates to the DimDates Table

```
Set Identity_Insert [DWPubsSales].[dbo].[DimDates] On
Insert Into [DWPubsSales].[dbo].[DimDates]
 ( [DateKey]
 , [Date]
 , [DateName]
 , [Month]
 , [MonthName]
 , [Quarter]
 , [QuarterName]
 , [Year], [YearName] )
Select
[DateKey] = -1
 , [Date] = Cast('01/01/1900' as nVarchar(50) )
 , [DateName] = Cast('Unknown Day' as nVarchar(50) )
 , [Month] = -1
 , [MonthName] = Cast('Unknown Month' as nVarchar(50) )
 , [Quarter] = -1
 , [QuarterName] = Cast('Unknown Quarter' as nVarchar(50) )
 , [Year] = -1
```

```
, [YearName]=Cast('Unknown Year' as nVarchar(50) )
Union
Select
[DateKey]=-2
, [Date]=Cast('01/01/1900' as nVarchar(50) )
, [DateName]=Cast('Corrupt Day' as nVarchar(50) )
, [Month]=-2
, [MonthName]=Cast('Corrupt Month' as nVarchar(50) )
, [Quarter]=-2
, [QuarterName]=Cast('Corrupt Quarter' as nVarchar(50) )
, [Year]=-2
, [YearName]=Cast('Corrupt Year' as nVarchar(50) )
Set Identity_Insert [DWPubsSales].[dbo].[DimDates] Off
```

2. Highlight this code, right-click it, and choose Copy from the context menu.

3. Navigate to the Control Flow tab and select the Add Null Date Lookup Values Execute SQL Task.

4. Edit the Add Null Date Lookup Values Execute SQL Task by right-clicking the task and selecting Edit from the context menu.

5. Select the Connection property. A dropdown box appears in the dialog window that will let you configure this property.

6. Configure the Connection property by selecting the option DWPubsSales from the dropdown box (Figure 8-21).

Figure 8-21. *Configuring the Add Null Dates Lookup Values Execute SQL Task*

7. Select the SQL Statement property. An ellipsis button appears in the dialog window that will let you configure this property.

8. Configure the SQL Statement property by clicking the ellipsis button and pasting the code from Listing 8-5 that you copied earlier into the Enter SQL Query dialog window that appears (Figure 8-21).

9. Verify that there are no SQL Go statements your code. SQL Go statements will cause the Execute SQL Task to fail.

10. Click the OK button to close the Enter SQL Query dialog window.

11. Click the OK button to close the Execute SQL Task editor.

12. Right click then Add Null Date Lookup Values Execute SQL Task and choose Execute Task from the context menu (similar to Figure 8-20). Visual Studio will launch this Execute SQL task with its debugging engine.

13. When the task is complete, stop Visual Studio's debugger and verify that the task ran successfully; if there are errors, troubleshoot and try again.

14. Open SQL Server Management Studio and verify that DimDates has data in it. Figure 8-22 displays what the data should look like.

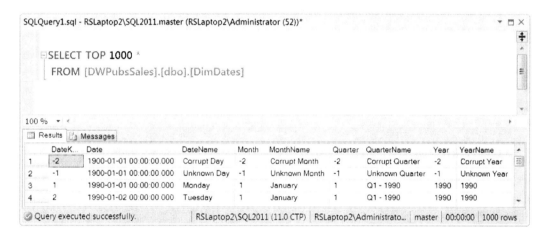

Figure 8-22. *Verifing the DimDates data*

Configure the Fill DimTitles Data Flow Task

The last dimension table we need to fill is DimTitles. So…

1. Navigate to the Data Flow tab and select the Fill DimTitles Data Flow Task from the Data Flow Task dropdown box.

2. Add an OLE DB Data Source to the data flow surface. Rename it to Get Titles Data OLE DB Destination.

3. Locate the comment `-- 2e) Get source data from pubs.dbo.titles` in the SQL code file and review the code beneath it. The code should look like Listing 8-6.

Listing 8-6. SQL Code for pubs.dbo.titles

```
Select
    [TitleId]=Cast( isNull( [title_id], -1 ) as nvarchar(6) )
  , [TitleName]=Cast( isNull( [title], 'Unknown' ) as nvarchar(50) )
  , [TitleType]=Cast( isNull( [type], 'Unknown' ) as nvarchar(50) )
```

```
  , [PublisherKey]=[DWPubsSales].[dbo].[DimPublishers].[PublisherKey]
  , [TitlePrice]=Cast( isNull( [price], -1 ) as decimal(18, 4) )
  , [PublishedDateKey]=isNull( [DWPubsSales].[dbo].[DimDates].[DateKey], -1)
From [Pubs].[dbo].[Titles]
Join [DWPubsSales].[dbo].[DimPublishers]
 On [Pubs].[dbo].[Titles].[pub_id]=[DWPubsSales].[dbo].[DimPublishers].
 [PublisherId]
Left Join [DWPubsSales].[dbo].[DimDates] -- The "Left" keeps dates not found in
DimDates
 On [Pubs].[dbo].[Titles].[pubdate]=[DWPubsSales].[dbo].[DimDates].[Date]
```

4. Edit the OLE DB Data Source as shown in Figure 8-23. (We are using the DWPubsSales connection, but reference the original Pubs.dbo.titles table in our code.)

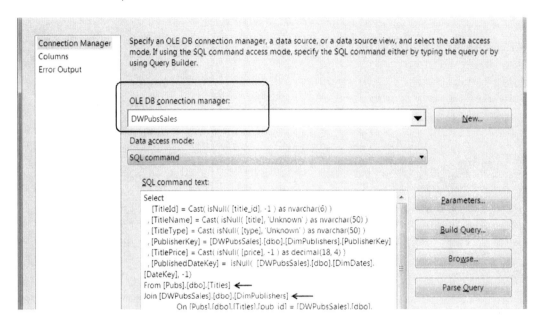

Figure 8-23. The OLE DB source for the Fill DimTitles Data Flow Task

5. Click the Preview button to verify that the query works successfully.

6. Click the Columns page to force the XML to be written properly in the .dtsx file.

7. Close the OLE DB Source Editor window by clicking the OK button.

8. Add an OLE DB Data Destination to the Data Flow surface. Rename it to Fill DimTitles OLE DB Destination.

9. Connect the data flow path from the source to the destination.

10. Edit the OLE DB data destination so that it imports data to the DimTitles table (similar to Figure 8-14)

11. Click the Mappings page to force the XML to be written properly in the .dtsx file and verify that the mappings look logically correct.

12. Right click the Data Flow design surface and choose Execute Task from the context menu. Visual Studio will launch this data flow task with its debugging engine.

13. When the task is complete, stop Visual Studio's debugger and verify that the task ran successfully; if there are errors, troubleshoot and try again.

14. Open SQL Server Management Studio and verify that DimTitles has data in it.

Configure the Fill FactTitlesAuthors Data Flow Task

With all the dimension tables filled, it is time to fill the fact tables. It does not matter which one you begin with, but let's start with the FactTitlesAuthors table.

1. Navigate to the Data Flow tab and select the Fill FactTitlesAuthors Data Flow Task from the Data Flow Task dropdown box.

2. Add an OLE DB Data Source to the data flow surface. Rename it to Get TitleAuthors Data OLE DB Destination.

3. Locate the comment -- 2f) Get source data from pubs.dbo.titleauthor in the SQL code file and review the code beneath it. The code should look like Listing 8-7.

Listing 8-7. SQL Code from pubs.dbo.titleauthor

```
Select
 [TitleKey]=DimTitles.TitleKey
--, title_id
,[AuthorKey]=DimAuthors.AuthorKey
--, au_id
,[AuthorOrder]=au_ord
From pubs.dbo.titleauthor
JOIN DWPubsSales.dbo.DimTitles
 On pubs.dbo.titleauthor.Title_id=DWPubsSales.dbo.DimTitles.TitleId
JOIN DWPubsSales.dbo.DimAuthors
 On pubs.dbo.titleauthor.Au_id=DWPubsSales.dbo.DimAuthors.AuthorId
```

4. Edit the OLE DB data source to use the DWPubsSales connection (similar to Figure 8-23).

5. Click the Preview button to verify that the query worked successfully.

6. Click the Columns page to force the XML to be written properly in the .dtsx file.

7. Close the OLE DB Source Editor window by clicking the OK button.

8. Add an OLE DB Data Destination to the Data Flow surface. Rename it to Fill FactTitlesAuthors OLE DB Destination.

9. Connect the Data Flow Path from the source to the destination.

10. Edit the OLE DB data destination so that it imports data to the FactTitlesAuthors table (similar to Figure 8-14).

11. Click the Mappings page to force the XML to be written properly in the .dtsx file and verify that the mappings look logically correct.

12. Right click the data flow design surface and choose Execute Task from the context menu. Visual Studio will launch this data flow task with its debugging engine.

13. When the task is complete, stop Visual Studio's debugger and verify that the task ran successfully; if there are errors, troubleshoot and try again.

14. Open SQL Server Management Studio and verify that DimFactTitlesAuthors has data in it.

Configure the Fill FactSales Data Flow Task

Our last table to fill is the FactSales table.

1. Navigate to the Data Flow tab and select the Fill FactSales Data Flow Task from the Data Flow Task dropdown box.

2. Add an OLE DB Data Source to the Data Flow surface. Rename it to Get Sales Data OLE DB Destination.

3. Locate the comment `-- 2g) Get source data from pubs.dbo.sales` in the SQL code file and review the code beneath it. The code should look like Listing 8-8.

Listing 8-8. SQL Code from pubs.dbo.sales

```
Select
 [OrderNumber]=Cast(ord_num as nVarchar(50))
,[OrderDateKey]=DateKey
--, title_id
,[TitleKey]=DimTitles.TitleKey
--, stor_id
,[StoreKey]=DimStores.StoreKey
,[SalesQuantity]=qty
From pubs.dbo.sales
JOIN DWPubsSales.dbo.DimDates
 On pubs.dbo.sales.ord_date=DWPubsSales.dbo.DimDates.date
JOIN DWPubsSales.dbo.DimTitles
 On pubs.dbo.sales.Title_id=DWPubsSales.dbo.DimTitles.TitleId
JOIN DWPubsSales.dbo.DimStores
 On pubs.dbo.sales.Stor_id=DWPubsSales.dbo.DimStores.StoreId
```

4. Edit the OLE DB data source to connect to the DWPubsSales connection.

5. Click the Preview button to verify that the query worked successfully.

6. Click the Columns page to force the XML to be written properly in the .dtsx file.

7. Close the OLE DB Source Editor window by clicking the OK button.

8. Add an OLE DB Data Destination to the Data Flow surface. Rename it to Fill FactSales OLE DB Destination.

9. Connect the Data Flow Path from the source to the destination.

10. Edit the OLE DB data destination so that it imports data to the FactSales table (similar to Figure 8-14).

11. Click the Mappings page to force the XML to be written properly in the .dtsx file and verify that the mappings look logically correct.

12. Right-click the data flow design surface and choose Execute Task from the context menu. Visual Studio will launch this data flow task with its debugging engine.

13. When the task complete, stop Visual Studio's debugger and verify that the task ran successfully or troubleshoot any errors and try again.

14. Open SQL Server Management Studio and verify that FactSales has data in it.

Configure the Add Foreign Key Constraints Execute SQL Task

One last task to go! We need to configure an Execute SQL task that replaces the foreign key constraints we dropped that the beginning of the process.

1. Locate the comment `-- Step 3) Add Foreign Keys` back in the SQL code file and review the code beneath it. The code should look like Listing 8-9.

   Listing 8-9. Code That Adds the Foreign Keys Back to the Database

   ```
   -- Step 3) Add Foreign Keys back (Will be used with SSIS Execute SQL Tasks)
   Alter Table [dbo].[DimTitles] With Check Add Constraint
   [FK_DimTitles_DimPublishers]
   Foreign Key ([PublisherKey]) References [dbo].[DimPublishers] ([PublisherKey])

   Alter Table [dbo].[FactTitlesAuthors] With Check Add Constraint
   [FK_FactTitlesAuthors_DimAuthors]
   Foreign Key ([AuthorKey]) References [dbo].[DimAuthors] ([AuthorKey])

   Alter Table [dbo].[FactTitlesAuthors] With Check Add Constraint
   [FK_FactTitlesAuthors_DimTitles]
   Foreign Key ([TitleKey]) References [dbo].[DimTitles] ([TitleKey])

   Alter Table [dbo].[FactSales] With Check Add Constraint [FK_FactSales_DimStores]
   Foreign Key ([StoreKey]) References [dbo].[DimStores] ([Storekey])

   Alter Table [dbo].[FactSales] With Check Add Constraint [FK_FactSales_DimTitles]
   Foreign Key ([TitleKey]) References [dbo].[DimTitles] ([TitleKey])

   Alter Table [dbo].[FactSales] With Check Add Constraint [FK_FactSales_DimDates]
   Foreign Key ([OrderDateKey]) References [dbo].[DimDates] ([DateKey])

   Alter Table [dbo].[DimTitles] With Check Add Constraint [FK_DimTitles_DimDates]
   Foreign Key ([PublishedDateKey]) References [dbo].[DimDates] ([DateKey])
   ```

2. Highlight this code, right-click it, and choose Copy from the context menu.

3. Navigate to the Control Flow tab and select the Add Null Date Lookup Values Execute SQL Task.

4. Edit the Add Null Date Lookup Values Execute SQL Task by right-clicking the task and selecting Edit from the context menu.

5. Select the Connection property. A dropdown box appears in the dialog window that will let you configure this property.

6. Configure the Connection property by selecting the option DWPubsSales from the dropdown box (similar to Figure 8-21).

7. Select the SQL Statement property. An ellipsis button appears in the dialog window that will let you configure this property.

8. Configure the SQL Statement property by clicking the ellipsis button and pasting the code from Listing 8-9 (that you copied earlier) into the Enter SQL Query dialog window that appears (Figure 8-21).

9. Verify that there are no SQL Go statements in your code. SQL Go statements will cause the Execute SQL Task to fail.

10. Click the OK button to close the Enter SQL Query dialog window.

11. Click the OK button to close the Execute SQL Task editor window.

12. When the task is complete, stop Visual Studio's debugger and verify that the task ran successfully; if there are errors, troubleshoot and try again.

In this exercise, you configured the SSIS tasks that filled the dimension and fact tables of the DWPubsSales data warehouse. You also tested each tasks as you went. Now we need to test that the entire SSIS package runs successfully as a whole.

Executing the Entire Package

When you have completed configuring and testing all of the tasks in your SSIS package, you need to test the entire package as a whole. This vital step is often missed by developers, and they only find out later that the package does not work as a unit—usually while giving a presentation showing how well it works!

The best way to reset the database back to its normal, preload state, is to use the SQL code script and then right-click the package in Solution Explorer to access the context menu (Figure 8-24). Clicking the Execute Package option will start the package for the beginning of the first task in the chain of precedence constraints.

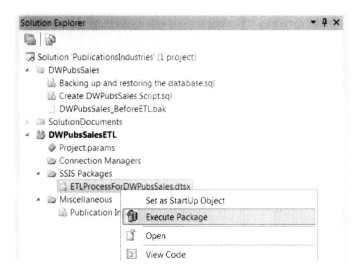

Figure 8-24. *Executing the entire SSIS package*

As each task completes, a green check mark indicates the success of each until all the tasks and sequences containers are done. When the package has completed, it should look like Figure 8-25.

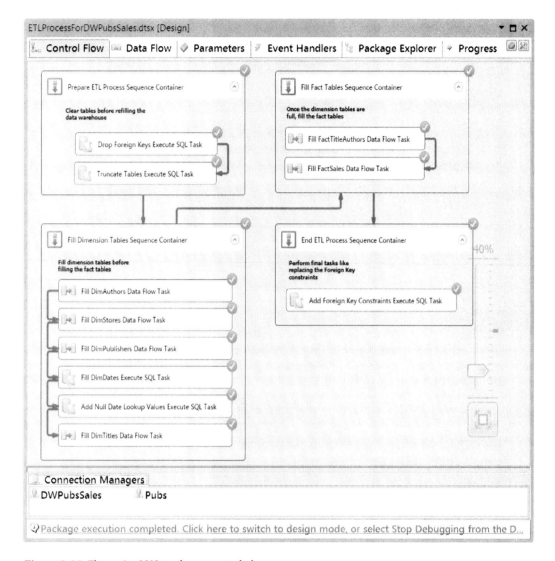

Figure 8-25. *The entire SSIS package succeeded*

Let's finish up this chapter by resetting the database one last time and execute the package in the next exercise.

EXERCISE 8-2. TESTING YOUR SSIS PACKAGE

In this exercise, you will test that the entire package works as expected by resetting the database and executing the package.

Reset the Database Objects Before Testing Your ETL Process

1. Locate the Create DWPubsSales Script.sql file in Visual Studio. You should find this file under the DWPubsSales solution folder you created in Exercise 5-4.

2. Open this file and execute its code. You can do this by right-clicking an empty area in the SQL code window to bring up the context menu and select Execute SQL. You may remember doing this back in Chapter 2 (Figure 2-15).

3. A Connect to Database Engine dialog box will appear. Type in you server name and click the Connect button.

4. In a few seconds, you should receive a message stating "The DWPubsSales data warehouse is now created."

Testing Your SSIS Package

Now it is time to execute the entire SSIS package to verify that it runs as a unit.

1. In Solution Explorer, right-click your SSIS package called ETLProcessForDWPubsSales.dtsx and select Execute Task from the context menu, as shown in Figure 8-24. Visual Studio will launch this Execute SQL task with its debugging engine.

2. When the task is complete, stop Visual Studio's debugger and verify that the task ran successfully or troubleshoot any errors and try again. It should look like Figure 8-25.

3. Open SQL Server Management Studio and verify that all the tables have data in them. You can use the code from Listing 8-10.

Listing 8-10. SQL Code to Verify DWPubSales' Table Data

```
Select Top 100 * from dbo.DimAuthors
Select Top 100 * from dbo.DimStores
Select Top 100 * from dbo.DimPublishers
Select Top 100 * from dbo.DimDates
Select Top 100 * from dbo.DimTitles
Select Top 100 * from dbo.FactSales
Select Top 100 * from dbo.FactTitlesAuthors
```

4. Once you have verified that the data is loaded, create a backup of the database in its filled state using the code in Listing 8-11.

Listing 8-11. Backing Up the Filled DWPubsSales Database

```
Backup Database DWPubsSales
To Disk='C:\_BISolutions\PublicationsIndustries\DWPubsSales\DWPubsSales_AfterETL.bak'
With Init
```

In this exercise, you tested the SSIS package that filled the DWPubsSales database. SSIS can be challenging to master, but with practice you will find it is a superior tool for performing ETL processing.

░ **Note** Remember, if you had any trouble completing this exercise, you can always compare your package to the authors' completed version in the C:_BookFiles\Chapter08Files\PublicationsIndustries folder. Optionally, you can restore the filled database using the SQL database backup file: C:_BookFiles\Chapter-08Files\FilledDWPubsSalesDB.bak. We also included a SQL code file to help you restore the database from the backup file called C:_BookFiles\Chapter08Files\RestoreDWPubsSalesDB.sql.

Moving On

At this point, a BI solution developer has a choice to make: start working on reports or add OLAP cubes to the solution. The decision involves weighing a number of factors, which are best determined by what would be most logical for the particular solution you are working on. As you can see in Figure 8-26, we have chosen Create Cubes as our next step to the process.

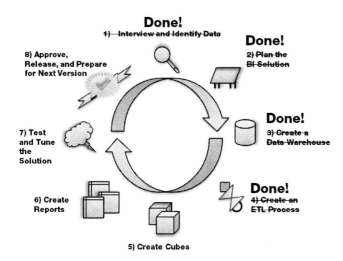

Figure 8-26. *The ETL process is complete*

In Chapter 9, we are going to cover how to create OLAP cubes using Microsoft's SQL Server Analysis Server (SSAS). Once you see what cubes do and how you create them, you will have a much better chance to make the correct decision process for your own BI solution, namely, about whether to include them in your solution or move past them and start working directly on preliminary reports.

LEARN BY DOING

In this "Learn by Doing" exercise, you will create an ETL process similar to the one defined in this chapter. This time you need to use the Northwind database. We have included an outline of the steps you performed in this chapter and an example of how the authors handled them in two Word documents. These documents are found in the folder C:_BISolutionsBookFiles_LearnByDoing\Chapter08Files. Please see the ReadMe.doc for detailed instructions.

What's Next?

We would like to show you more on SSIS logging, events, and configuration, but we will have to stop here and move on to the next process in our BI solution. Meanwhile, the data warehouse is filled with transformed data, and so the ETL process is officially complete. If you would like to learn more about using SSIS for ETL processing, we have articles, videos, and demonstrations on our website at http://www.NorthwestTech.org/ProBISolutions/ETLProcessing.

CHAPTER 9

Beginning the SSAS Project

Every new beginning comes from some other beginning's end.

—Seneca

For more than a decade now, alongside its relational database server, Microsoft has provided a multidimensional database server, also known as a SQL Server Analysis Services (SSAS) cube. In all of that time, this premier multidimensional SSAS cube server has proven itself to be both a cost-effective and powerful addition to BI solutions around the world.

The purpose of SSAS is to provide high-performance reporting data. Reports created on a multidimensional cube run faster compared to reports built upon a set of relational tables. As usual, faster performance means more complexity. Creating SSAS cubes demands that you understand dimensional data in a completely new way. It requires that you understand the difference between SSAS cubes and dimensions. It even requires you to learn new development and administrative tools, plus four programming languages (MDX, XMLA, DMX, and DAX), if you want to master SSAS.

Microsoft has combated this complexity by adding wizards to its development tools that walk you through the process of creating SSAS cubes. These wizards make it easy for new developers to start building cubes and keeps novice cube developers on the right track.

In this chapter, we begin the process of creating complex SSAS cubes, by developing the dimensions they are comprised of. We detail the difference between cubes and dimensions and examine how to use the development and management tools. To start, let's look at the differences between SQL Server and SSAS databases.

SQL Server vs. Analysis Server Databases

The most noticeable difference between a traditional SQL Server database and an SSAS database is SSAS's increase in reporting performance. SSAS is designed to create and store aggregate values, and it has a very different design structure. Let's take a moment to examine these differences.

In a SQL Server relational database, the tables and databases are logical constructs representing data stored in physical files on the hard drive (Figure 9-1). Each SQL Server database has at least two files, but the SQL server engine makes it appear as if it is one object. The master data file (.mdf) holds table data in a binary format. The log file (.ldf) holds a running account of changes made to the .mdf file. Whenever you add a table to the database or add a row of data to a table, the log file records the change and the .mdf file receives the data. This design strongly protects transactional statements that add, modify, or remove data in the database from any inconsistencies.

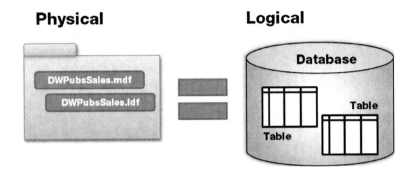

Figure 9-1. *Anatomy of a SQL Server database*

Analysis Server's structure is much different. In SSAS, each database is stored as a hard drive folder, not a master data file. This database folder contains a collection of files and subfolders that are the physical representation of the cubes and dimensions of that particular database (Figure 9-2). Each SSAS database can have many cubes and dimensions. And one Analysis Server can host many databases.

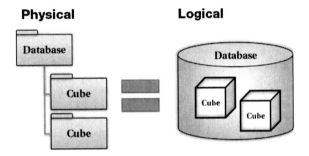

Figure 9-2. *Anatomy of a Analysis Server database*

In both cases, the main purpose is to store data in logical collections called *tables* or *cubes*, respectively. Note, however, that SSAS databases lack transactional logging. SSAS databases are focused on storing and retrieving data, with little to no transaction processing. Most of the data you find in a cube is copied from relational tables, and once copied, they never change.

The act of copying the data from the relational database to the cube database is called *processing*. Processing collects data from one or more tables, places a copy of that data into the cube, and optionally creates stored aggregate values as well. This means a cube can be thought of as a set of tables combined into a single reporting object, as you can see demonstrated in Figure 9-3.

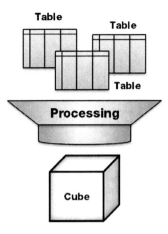

Figure 9-3. *Processing data from tables to a cube*

This concept is similar to a SQL view with one major difference. The SQL views are just named SQL select statements that combine results from one or more tables but never stores any data. In contrast, cubes hold a copy of the actual data from one or more tables. Reports created against the cube cannot access the original tables and will not be slowed by ongoing transactional activity.

WHY ARE THEY CALLED CUBES?

The term *cube* can be misleading, as it implies a cubes structure has three dimensions. SSAS cubes are more accurately described as multidimensional data structures. The word *cube* is just easier to say.

Here are some key points to help understand the structure of an SSAS cube:

- A column is a list of values and can be thought of as a single-dimensional array on a single attribute. Example: an author's last name.

- A table is a set of columns and can be thought of as a two-dimensional array on a set of attributes pertaining to a single subject. Example: an author's basic information, including address and phone number.

- A cube can be thought of as a multidimensional array on a set of subjects such as a list of authors, titles, publishers, and stores multiplied by its measurable values. Each subject represents a dimension of the cube, and each measurable value is cross multiplied to provide a distinct aggregate value for each combination of attribute and measure. The product of which is a multidimensional cube. Figure 9-4 illustrates this concept.

Figure 9-4. *An illustration of the concept of a cube. (If you have four dimensions and two measures, you will have eight categories of measured values, one for each combination.)*

The organization of data within an SSAS multidimensional cube is not the same as a three-dimensional cube known in geometry.

OLAP Cubes vs. Reporting Tables

SSAS cubes can also be referred to as online analytical processing (OLAP) cubes. Strictly speaking, the purpose of an OLAP cube is reporting, but then again, so are the tables in a data warehouse. Many companies find that reporting against data warehouse tables is sufficient for their needs. So, why use a cube? And what are the differences between reporting tables and cubes?

To answer these questions, let's take a look at Table 9-1 to see the comparison.

Table 9-1. *Cubes vs. Tables*

Reporting Tables (Data Warehouse Tables)	Cubes (SSAS/OLAP)
Simpler to create than cubes.	Complex setup.
Standard performance retrieval.	Higher performance retrieval.
Requires SQL joins and subqueries.	SQL joins and subqueries are removed.
Aggregate values are discouraged and are not easily stored.	Aggregate values are encouraged, and cubes are tuned to store them.
Accessed via standard SQL programming statements.	Can be accessed via MDX or XMLA programming languages, which are not commonly known in the industry.
Cannot always replace a poorly designed existing data warehouse database.	Is able to replace a poorly designed existing data warehouse database.

Most companies believe that tables are sufficient for their needs, at least at first. Nevertheless, as reporting activity grows over time, they move to a cube-based reporting system to gain its benefits.

Microsoft added two new SSAS server models in SQL 2012: a tabular model and a PowerPivot model. They act similar to reporting tables, but with the performance of a cube. And in many cases, the performance is even better. Both of these are somewhat easier to configure than the original multidimensional cube model, but they do not include as rich a feature set. Table 9-2 identifies some of these differences.

Table 9-2. *Features Associated with SSAS Server Models*

Feature	Multidimensional	Tabular	PowerPivot
Actions	Yes	No	No
Custom assemblies	Yes	No	No
Custom rollups	Yes	No	No
Drillthrough	Yes	No	Yes
Hierarchies	Yes	Yes	Yes
KPIs	Yes	Yes	Yes
Linked objects	Yes	No	Yes
Many-to-many	Yes	No	No
Parent-child hierarchies	Yes	Yes	Yes
Partitions	Yes	Yes	No
Perspectives	Yes	Yes	Yes
Translations	Yes	No	No
User-defined hierarchies	Yes	Yes	Yes
Writeback	Yes	No	No

■ **Tip** We do not discuss tabular or PowerPivot models in this book, but you can find out more on our website, www.NorthwestTech.org.

SQL Server vs. Analysis Server Applications

To mitigate some of the differences between reporting with tables versus cubes, Microsoft has designed many of its applications to work with both SSAS cubes and SQL Server tables in a similar manner. SQL Server Management Studio is just one example of an application that works with both.

In Figure 9-5, Management Studio is connected to the relational database engine and has returned a result set from a SQL query.

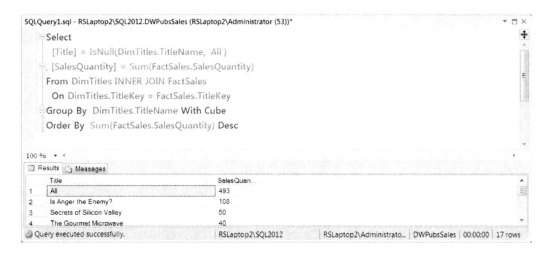

Figure 9-5. *Accessing databases and tables from Management Studio*

In Figure 9-6, the SSAS cube database engine and the result set is from an MDX statement. (For more information about MDX statements, see Chapter 14.) The results are the same, but the user interface is a bit different.

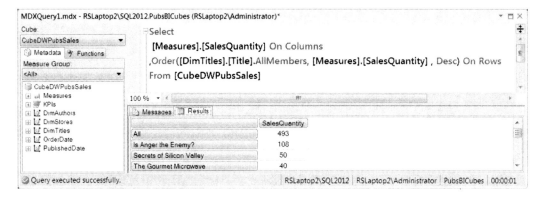

Figure 9-6. *Accessing databases and cubes from Management Studio*

SQL Server and Analysis Server are also similar when reports are made against them. In Figure 9-7, note how these Excel reports look alike even though one report is based upon a SQL view and the other, an SSAS cube. Other Microsoft applications, such as Reporting Server (SSRS), smooth out the differences as well.

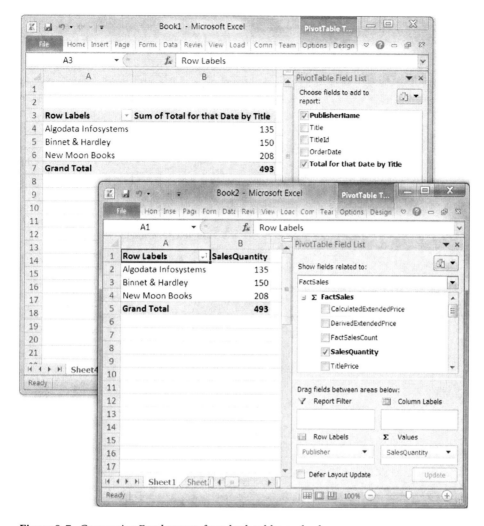

Figure 9-7. *Comparing Excel reports from both tables and cubes*

SSAS Projects

Now that you know what a cube is, the question is how to make them. You can create them programmatically using either MDX or XMLA; however, most of the time you want to use Visual Studio instead. The process of creating a cube is easier in Visual Studio because of the graphical presentation and wizards that walk you through the process.

To start creating your cubes with Visual Studio, begin by creating an SSAS project. In Figure 9-8 a new SSAS project is added to the PublicationsIndustries solution.

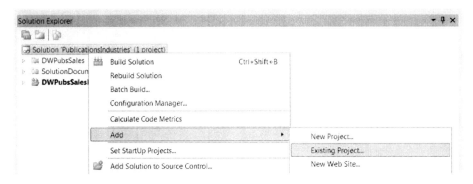

Figure 9-8. *Adding a new SSAS project to your solution*

When the Add New Project windows displays, select the Analysis Services Multidimensional option to create a traditional multidimensional cube SSAS project (Figure 9-9).

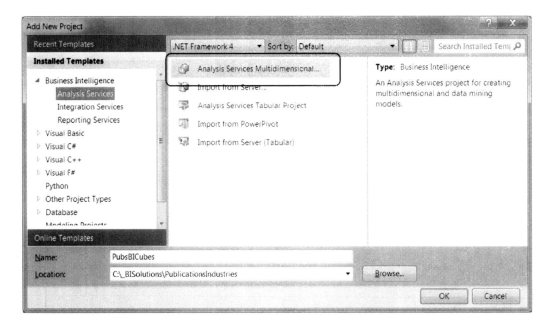

Figure 9-9. *Choosing between multidimension and tabular SSAS projects*

Once you select the template and click the OK button, the project is created, and you see a number of new project folders in Solution Explorer (Figure 9-10).

Figure 9-10. *Orginazation of an SSAS multidimensional project*

We provide a description of each folder in Table 9-3.

Table 9-3. *The SSAS Project Folders*

Folder	Description
Data Sources	A collection of XML files that hold connection strings and configurations.
Data Source Views	A collection of XML files that describe relational tables and their columns and relationships between the tables. This XML data is presented and configured in a graphical format within Visual Studio.
Cubes	A collection of XML files that describe the cubes you want to create on SSAS. This XML data is presented and configured in a graphical format within Visual Studio.
Dimensions	A collection of XML files that describe the dimensions you want to create on SSAS. This XML data is presented and configured in a graphical format within Visual Studio.
Mining Structures	A collection of XML files that describe the mining structures you want to create on SSAS. This XML data is presented and configured in a graphical format within Visual Studio.
Roles	A collection of XML files that describe the security permissions you want to create on SSAS. This XML data is presented and configured in a graphical format within Visual Studio.
Assemblies	A collection of .NET .dll files that contain custom logic to be used by SSAS.
Miscellaneous	A collection of files that you have added to the project but are not a recognized project extension. Typically these can be developer notes or a collection of MDX scripts.

Data Sources

Cube data comes from relational databases. To access these databases, you need to identify their location and supply security credentials. In SSAS this information is stored in a data source. Before you create cubes in Visual Studio, you need to have at least one data source.

To create a data source, launch the Data Source wizard by right-clicking the data source project folder in Solution Explorer, and select New Data Source from the context menu. A wizard appears to walk you through the creation process (Figure 9-11).

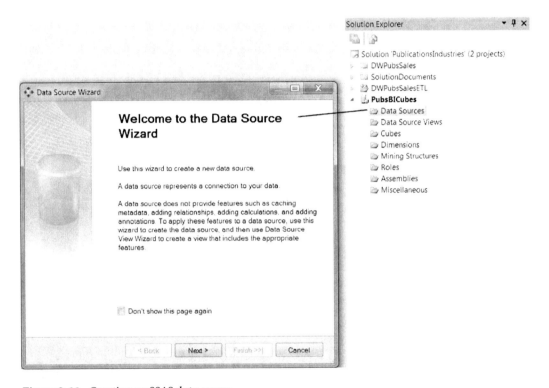

***Figure 9-11.** Creating an SSAS data source*

Define a Connection

The SSAS wizard is remarkably similar to the Data Source Wizard you used in SSIS, and it is configured the same way. If you have previously created a connection within a solution, Visual Studio already has the data connection listed on the second screen of the Data Source Wizard. If one is not listed, click the New button to create a new connection. This takes you to the Connection Manager screen.

In the Connection Manager screen, choose the provider you want to use in the top dropdown box. "OLE DB/ SQL Server native client 11.0" is the typical provider used to connect to Microsoft SQL Server 2012.

After you select the provider, select the name of the server and the name of the database you want to connect to. Clicking the Test Connection button will indicate whether a connection is successful.

If you have performed the exercises in the book so far, you already have the connection configured in your SSIS project, so you select that one and click the Next button to advance the wizard (Figure 9-12).

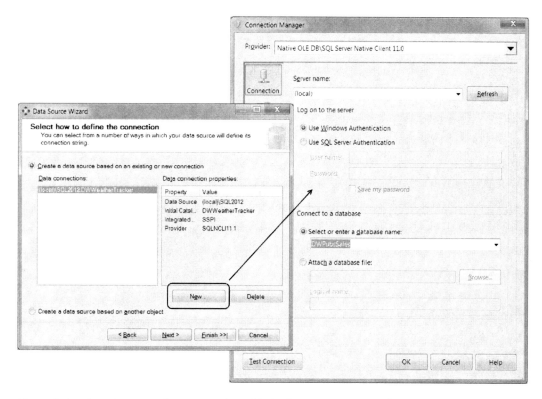

Figure 9-12. *Creating a new data connection with the Conection Manager dialog window*

Impersonation Information

The Impersonation Information window identifies the Windows account used for SSAS processing (Figure 9-13). This Windows account must have access to both the SQL Server relational database engine and the Analysis Server cube engine, because it interacts with both servers during SSAS processing.

Let's look at the breakdown of each of the four options.

■ **Note** We are using a Windows administrator account that has been given access to both SQL Server and SSAS. While this is not the best practice in a production environment, it simplifies a number of permission issues for readers. Because this book is less about troubleshooting security issues and more about understanding the nature of a BI solution, we recommend that you also use a Windows administrator account too; just make sure you are working behind a secure firewall and do not run BI servers that you are currently using. Setup instructions are in the downloadable book content under the C:_BookFiles_SetupFiles folder.

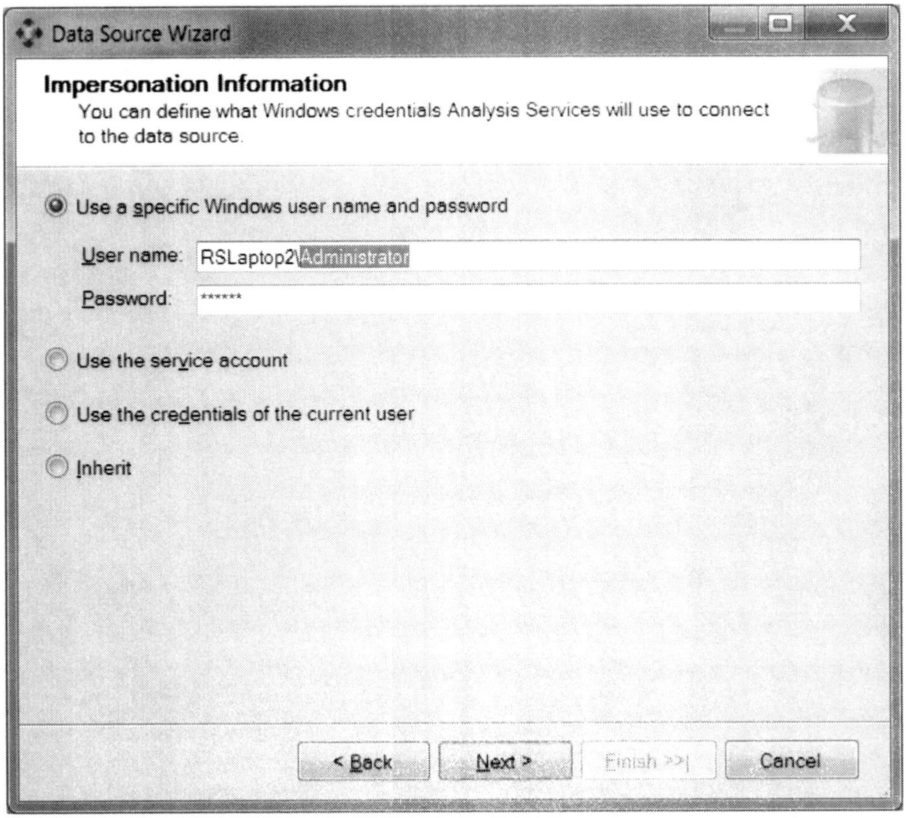

Figure 9-13. *Setting the impersonation information*

Service Account Option

The "Use the service account" option uses a Windows Security ID (SID) associated with the account that starts up the SSAS service. This account is chosen during the installation of SSAS, but it can be configured after installation as well. To verify the account used (or to change it), use Microsoft's configuration manager application. This application is opened using Start ➤ Microsoft SQL Server 2012 ➤ Configuration Tools ➤ SQL Server Configuration Manager.

The configuration manager application in Figure 9-14 lists all of the SQL Server–related services found on the local computer. In this example, there are many versions of the BI servers installed, but the one we are interested in is SQL Server Analysis Services 2012. When SQL Server 2012 is installed as a default instance, it will be labeled (MSSQLSERVER), but in our example it represents a named instance and is labeled (SQL2012).

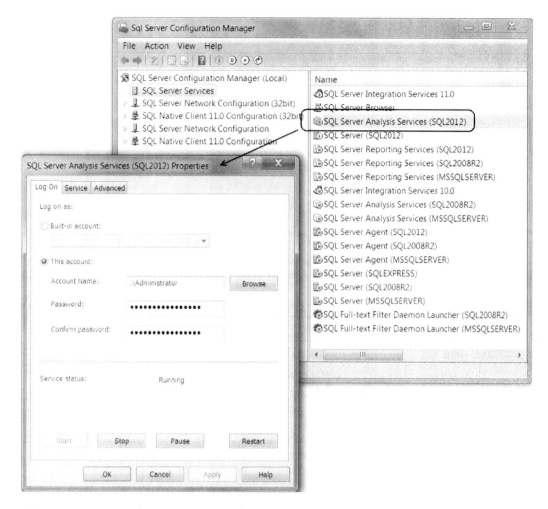

Figure 9-14. *Locating the service account information*

Double-clicking an entry brings up a Properties dialog window (Figure 9-14). The LogOn tab within this window allows you configure which account to use when starting the service.

If this service account is used for the impersonation information, it must have access to the data warehouse and the SSAS objects. If this account has permissions, then using it for impersonation works. If not, you need to give permissions to the account, select a different account, or use a different option.

Using the service account is not your best option because the permissions given to the account may be above and beyond what is actually needed. It is best to use this option for times where only simple security delineations are needed, such as on a development server.

Current User Account Option

The "Use the credentials of the current user" option (Figure 9-13) allows the connection to access the SQL Server and SSAS with the SID of the person using Visual Studio. Later, when a cube deploys from Visual Studio to the actual SSAS server, whatever account used to automate cube processing will become the current user. This is confusing and prone to configuration errors.

In general, current user account option is not a recommended choice. This option is useful only as a temporary option while you are developing cubes in Visual Studio.

Inherit Option

One option that looks attractively simple is the Inherit option (Figure 9-13). One might consider this to be the "automatically choose for me" option since that is what it does. It automatically selects the service account for some operations and the current user's credentials for others. This option may work in Visual Studio, but this typically will not work once the cube deploys to the SSAS server. Therefore, the Inherit option should not be used except within Visual Studio during development.

Specific Windows Account Option

The first option, "Use a specific Windows user name and password," is the best choice (Figure 9-13). This option allows you to use a specific Windows account by entering the name and password in the provided textboxes.

The account should have the computer name or the domain name prefaced before the individual account name. Some examples are My_Domain\My_Windows_Account and My_Computer_Name\My_Windows_Account.

■ **Important**   The permissions required are dependent upon various configuration and connections for individual cubes, and there are many possible combinations. If you are unsure about how to set up proper security, we recommend using your SQL Server and an SSAS administrator account for the exercises in this book. Learning how to use four BI servers is complex enough without adding permission issues to the mix!

Once you provide an account and its password to the wizard, click Next to proceed to the final dialog window where you can change the name of the data source, as shown in Figure 9-15.

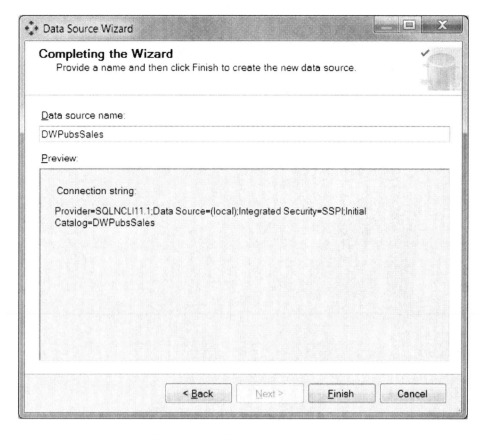

Figure 9-15. *Naming your data source and completing the wizard*

We typically do not include spaces in the name because it causes problems in some software. The choice to use spaces, however, is up to you. Whatever your decision, be consistent with your naming conventions throughout the project. Each wizard assigns spaces to the object names during the creation of cubes and dimensions. If you prefer to remove the spaces in one object's name in your project, do so for all the objects in a project.

VERIFYING THAT YOU HAVE ADMINISTRATOR ACCESS

One common mistake is believing that you have administrator access to SSAS when you do not. The easiest way to verify this is to open SQL Server Management Studio and try to connect to the SSAS server. Select the Analysis Services option in Object Explorer's Connect menu, and the Connect to Server dialog box appears. Enter the name of your SSAS server and then click the Connect button (Figure 9-16).

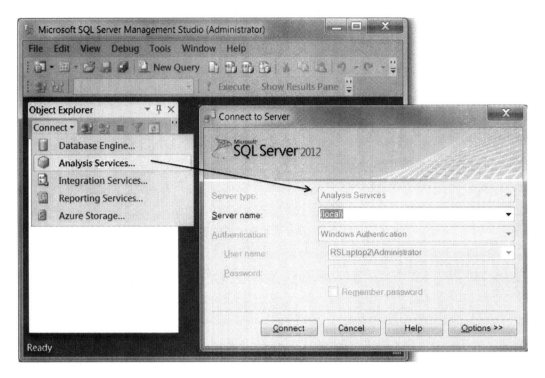

Figure 9-16. *Connecting to the SSAS server from SQL Server Management Studio*

If you are unable to make a connection to SSAS with your current Windows account, you cannot use that account for the impersonation information and must select another.

Even if you are able to connect, you must still verify that you have permissions to create databases on the SSAS server. You can do so by right-clicking the server icon in Object Explorer and selecting Properties from the context menu.

When the Analysis Server Properties dialog window appears, look to the left side for the Security page and click its icon. The security page displays a list of Windows accounts that have been granted SSAS administrator access. Your account should be listed. If it is, then you have all the permissions you need!

You can also grant different Windows accounts administrator privilege by clicking the Add button and selecting a Windows account. Figure 9-17 shows the Randal account on his local computer in the process of being added as an SSAS administrator.

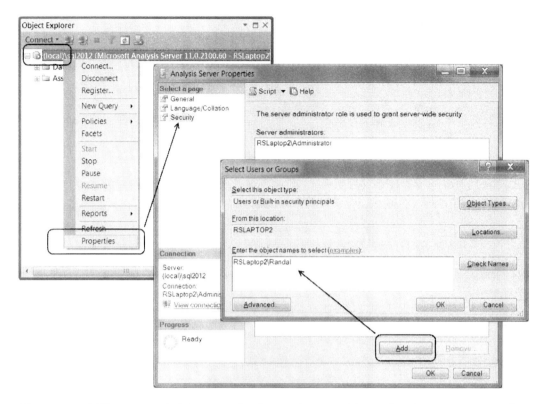

Figure 9-17. *Adding a new administrator using the security page of the Analysis Server property window*

If you cannot connect to the Analysis Server at all, it is possible that you forgot to add your user account to the list of SSAS administrators at the time you installed SQL Server. If this happens, you may want to uninstall and reinstall SSAS again. This is often the easiest way to remedy this problem. Other options exist, but research them carefully, as they are complex.

Data Source Views

Microsoft does not allow cubes to be built directly on the data warehouse tables. Instead, Microsoft decided to use an abstraction layer, which is the essence of a data source view. Therefore, before you can make your cubes, you need a data source view.

A data source view consists of a set of one or more logical tables with each table representing a saved SQL select statement. This feature has a number of advantages. It allows you to add or modify the underlying database design to fit your current needs without breaking the cubes. If the underlying database changes, modify the data source view to present the database as it was before the change. Additionally, if the data warehouse is not designed the way you would like it to be, you can modify the data source view to mimic your ideal data warehouse design.

Here is a common example: a data warehouse starts out containing a single fact table, but over time it has grown very large. So the decision is made to split the table into sections based on years (Figure 9-18). These tables are then placed on their own hard drives, thus increasing I/O performance in the data warehouse. By changing the SQL select statement in the data source view, the three tables are still presented to SSAS as a single table. And all of the cubes based on the earlier version of the FactSales table function as they always have.

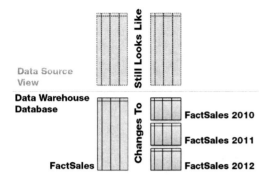

Figure 9-18. Data source views represent a logical collection of one or more tables.

The Data Source View Wizard

A data source view is created by right-clicking the Data Source Views folder in Solution Explorer and choosing New Data Source View from the context menu (Figure 9-19). This choice starts a wizard that walks you through the creation process.

Figure 9-19. Adding a new data source view to the project

Select a Data Source

The data source wizard asks you to select an SSAS data source. You can choose only one data source during the wizard, but you are able to add additional data sources to the data source view after the wizard completes. This unusual feature allows you to connect to tables from multiple databases, which can occasionally useful on occasion, but a single connection is sufficient for most needs. In our example, we chose the DWPubsSales data source we created earlier (Figure 9-20).

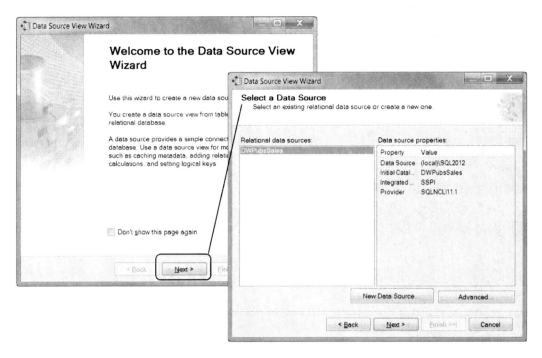

Figure 9-20. *Selecting a data source for your data source view*

Name Matching

Clicking Next button takes you to one of two dialog windows, depending on whether you have foreign key constraints in your data warehouse. If you have foreign key constants, the wizard automatically maps logical relationships between tables based on these constraints. And you will not see the dialog window shown in Figure 9-21.

If you do not have foreign keys constraints, the wizard tries to create logical relationships between the selected tables based upon these three options (as listed in Figure 9-21):

- *Same name as primary key:* A column in one table has the same name as a primary key in another table. For example, Order.CustomerID relates to the primary key column Customer.CustomerID.

- *Same name as destination table name:* A column in one table has the same name as the name of another table. For example, Order.Customer relates to the primary key column Customer.CustomerID.

- *Destination table name + primary key name:* A column name matches a composite of the table + primary key column names. For example, Order.CustomerID relates to the primary key column Customer.ID.

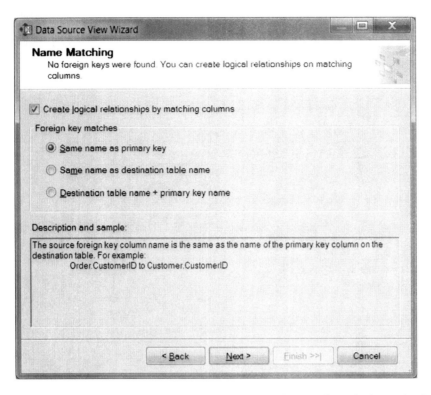

Figure 9-21. *Configuring the Data Source View Wizard to map logical relationship between tables*

■ **Tip** Because our data warehouse has foreign key constraints, do not expect to see this particular dialog window while working through the exercises in this book.

Select Tables and Views

The next wizard dialog window allows you to select which tables should be included in the data source view. You can select all the tables or just choose some among them. In our example, we select all the tables in the data warehouse and leave the view behind (Figure 9-22).

Figure 9-22. *Selecting which tables are used in your data source view*

SQL views also appear as available objects, which means you can create views in OLTP or data warehouse databases that can be used as the source of your cubes and dimensions. Using views, you can combine one or more tables to create what appears to be a single table. This is identical to the scenario shown in Figure 9-18 with one important distinction; the code in a SQL Server view is stored in the SQL database, while the code in the data source view is stored in an SSAS database. Both have equivalent performance since they are saved SQL statements after all. The choice is usually based on permissions.

As an SSAS cube developer, you probably cannot create whatever views you want in the data warehouse, but you can create what you want on SSAS. Therefore, the choice becomes pretty simple. Once you have selected the tables, click the Next button.

■ **Important**   The view in Figure 9-22 has not been created as part of the chapter exercises and should not be in your database. We added it as an example of what *may* exist in a typical data warehouse database.

Completing the Wizard

In the final dialog window (Figure 9-23), you can name the data source view. A name is suggested for you, but it does not really matter what you call it as long as it is consistent with other items in the project. You may find that using a name that describes the purpose and has no spaces between each word is the most convenient and easiest to work with.

Figure 9-23. *Naming your data source wiew*

When the wizard completes, Visual Studio opens to a graphical display of your new data source view file. Although it may not like it, the underlying file is an XML file. You can work with directly with the XML code if you choose to, but it is rare that you need to do so. Instead, you most often work with a graphical user interface (Figure 9-24).

■ **Note**   We have adjusted the lines in our data source view for display purposes, but you can expect the wizard to create a chaotic mess. You can move the table or lines to clean up the visual display, but adjusting them does not change the final outcome in the cube and dimension creation process.

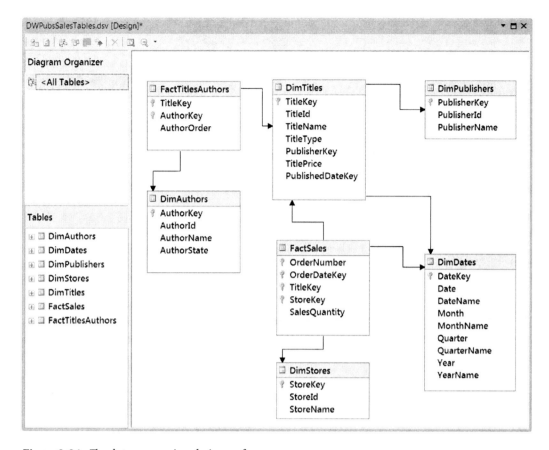

Figure 9-24. *The data source view design surface*

The Data Source View Designer

In the Data Source View Designer, both SQL Server tables and views are presented as tables. And each table on the designer surface represents an underlying SQL select statement. For example, the DimStores table represents the SQL statement `Select StoreKey, StoreId, StoreName From DimStores`. Analysis Server uses these select statements to copy data into the cube.

The Explore Data Option

You can also run the select statement by using the Explore Data option from the context menu that pops up when you right-click a table (Figure 9-25).

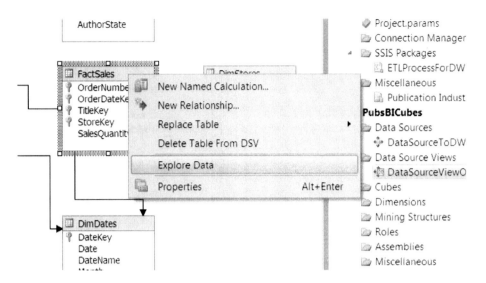

Figure 9-25. *Exploring the underlying data through the data source view*

The query results are presented, but by default only some of the rows may be displayed. You can change the number of rows displayed by clicking the button in the upper right of the Explore window and setting the sample count (Figure 9-26).

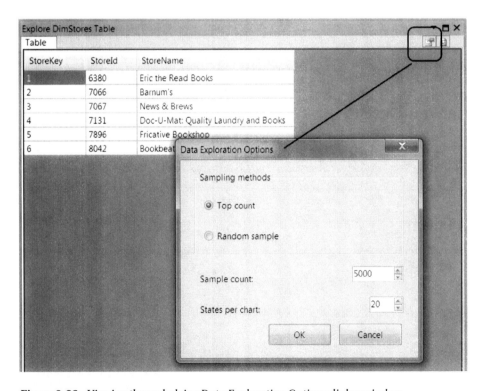

Figure 9-26. *Viewing the underlying Data Exploration Options dialog window*

■ **Tip** In previous versions, viewing this underlying data included additional features such as charting options and pivot table reports. But the developer could not create final reports from these additional views. Perhaps this is why they have now been removed in SQL 2012. Whatever the reason, this is still a convenient feature for viewing the data in a data warehouse without opening another application. It is also a quick way to verify that you can connect to the table! This is a convenience when you open a Visual Studio project created by someone else.

Friendly Names

Another interesting aspect of the data source view is your ability to change the names of both columns and tables. For example, if you decide you would like to rename the Date column in the DimDates table to FullDate, you can easily do so. We can either provide a column alias or configure a friendly name for it. In Figure 9-27, we have selected the date column in the DimDates table and applied a friendly name to the column using the property sheet. When SSAS creates dimensions using this data source view table, the friendly name is used instead of the original name.

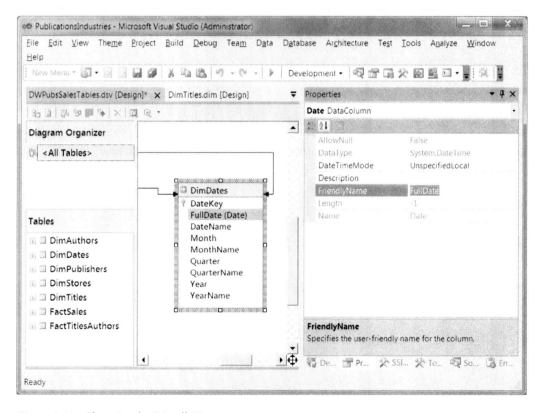

Figure 9-27. Changing the FriendlyName property

Named Queries

Each data source view table is a select statement. You can modify these select statements and join data to them from more than one SQL Server table. In our example, the Titles dimension includes both the DimTitles and DimPublishers tables in a snowflake pattern (Figure 9-28). We can combine these two tables into one data source view table using SQL code. To do this, right-click the design surface and choose New Named Query from the context menu (Figure 9-28). This launches a query designer.

Figure 9-28. *Adding a named query*

The query designers in SQL Server, SSIS, SSRS, and SSAS are all same. They allow you to create SQL code by adding one or more tables to the user interface and provide checkboxes to select the columns you intend to use. To add tables to the UI, right-click in the upper part of the Query Definition window and select Add Table.

Figure 9-29 displays a SQL join that includes DimTitles and DimPublishers. To include all the publishers (even those without any associated titles), click the connecting relationship line and select the Select All Rows from DimPublishers option from the context menu (Figure 9-29).

After you create your query and close the Create Named Query dialog window, you will have a new table in the data source view. If you created a new data source table that includes data duplicated in other tables, you must remove those other tables from the data source view. Once that is complete, these multiple tables function as a single table—at least as far as SSAS is concerned.

Tables from multiple databases can be combined as well. For example, because we did not include the city and state in the DimStores table, we could add those columns now by joining the DimStores table to the Stores table in the original Pubs database, based on the Store ID. This is not a commonly seen strategy, but it is an example of the flexibility of the data source view.

■ **Note** For our purposes, within our exercises, we are leaving the DimTitles, DimPublishers, and DimStores as they are. We continue to point out how the current design affects the BI solution throughout the book.

Figure 9-29. *The Create Named Query dialog window*

Named Calculation

Named calculations allow you to modify a portion of the underlying SQL statement that is used by a data source view. A named calculation is similar to a named query, except that you add only a SQL expression to an existing statement, rather than using a whole SQL statement.

To create a new named calculation, right-click the data source table and select New Name Calculation from the context menu. This launches the Create Named Calculation dialog window. Provide a name and description for the calculation, followed by the SQL expression that defines the results to be returned when the underlying SQL statement runs.

For example, let's say we want to have a fancy version of a month's name available for reporting purposes. By creating a named calculation (Figure 9-30), you are able to add a new column to the data source view table using this SQL code: `'The Month of'+MonthName+'in the year'+YearName`.

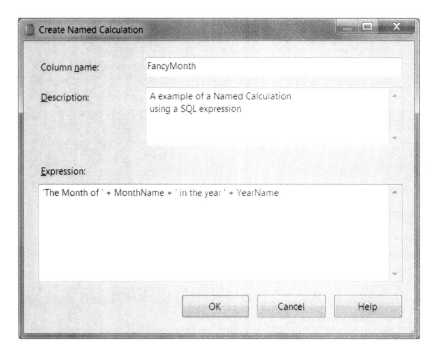

Figure 9-30. The Create Named Calculation dialog window

Once the new named calculation is added, you can see the outcome of this addition using the Explore Data option (Figure 9-31).

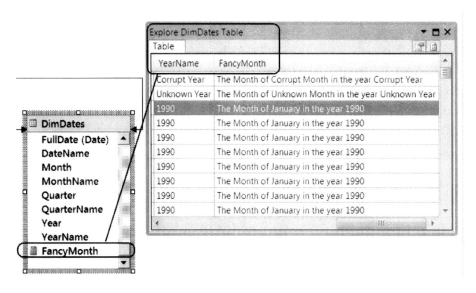

Figure 9-31. Verifying the results from a new named calculation

Relationships

Tables within a data source view must have relationships between them before they can be used with your cubes and dimensions. Normally, the relationship defined by a foreign key constraint within the data warehouse is what is used in the data source view. But, as we have seen previously, the wizard tries to create a relationship even if there was no foreign key constraint.

From the data source view's perspective, the foreign key constraint is not very important. What is important is a logical relationship between the different tables in the data source view.

Occasionally the wizard cannot map all the relationships for you. If this happens, you can easily create a relationship line by clicking an attribute column in one table and dragging it to the attribute column another.

■ **Note**　Named query relationships must be mapped manually. To do so, right-click a column (or columns) and select Set Logical Primary Key from the context menu. Drag and drop the relationship lines between the named query table and the table columns to connect them.

After a relationship line is created, verify that the correct columns were connected. If not, they can be connected by right-clicking the relationship line in the data source view and selecting Edit Relationship from the context menu (Figure 9-32).

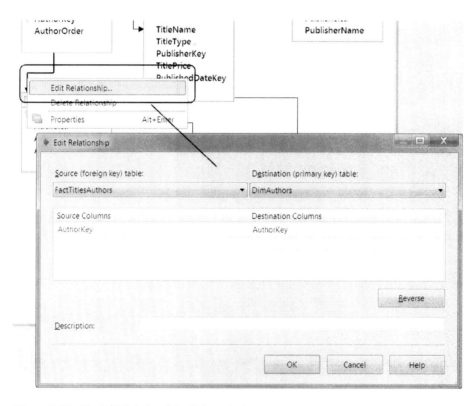

Figure 9-32. *The Edit Relationship dialog window*

In this dialog window, you can adjust or reselect the columns, reverse relationship direction, or determine that the relationship consists of multiple columns.

Now that you know how to create a data source and data source view, let's create both on your machine using the steps outlined in Exercise 9-1.

EXERCISE 9-1 CREATING A DATA SOURCE AND DATA SOURCE VIEW

In this exercise, you add a new SSAS project to your current Visual Studio Solution and create a new data source and data source view within the project. Once completed, your SSAS project should look similar to Figure 9-33.

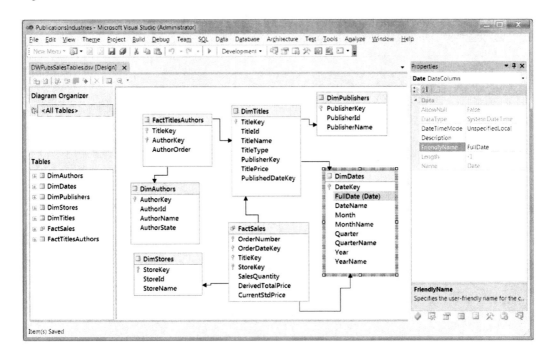

Figure 9-33. *The Edit Relationship dialog window*

Add an SSAS Project to an Existing Solution

1. Open Business Intelligence Development Studio, and select Start ➤ All Programs ➤ Visual Studio 2010.

Important: Remember to right-click the menu item and select Run as Administrator and answer Yes to close the UAC.

2. Open the PublicationsIndustries.sln file you have been working on in the previous chapter by selecting File ➤ Open ➤ Project/Solution from the main menu at the top of Visual Studio. When the Open Project dialog window opens, navigate to the

`C:\_BISolutions\PublicationsIndustries\PublicationsIndustries.sln` file
to open the solution.

3. Add a new SSAS project to the solution by selecting File ▸ Add ▸ New Project from
   the main menu at the top of Visual Studio or by right-clicking the Solution icon in
   Solution Explorer (Figure 9-8).

4. When the Add New Project dialog window appears, select the Analysis Services
   Multidimensional Cube project template and name the project **PubsBICubes**
   (Figure 9-9).

Create a Data Source

1. Create a new data source that connects to the DWPubsSales data warehouse by
   right-clicking the Data Source project folder and selecting New Data Source from the
   context menu, as shown in Figure 9-11.

2. When the data source wizard appears, click Next to access the wizard's connection
   dialog window. Create or select a connection to the DWPubsSales data warehouse
   (Figure 9-12).

3. If a connection is already available in the Data Connections pane, select the
   connection. If it is not available, click New to create a new connection in the
   Connection Manager dialog window (Figure 9-12). Enter your SQL server's name,
   select the DWPubsSales data warehouse database, and click OK to close the dialog
   window.

4. Click Next to advance the wizard to the Impersonation Information dialog window
   (Figure 9-12).

5. Configure the Impersonation Information dialog window to use a specific Windows
   account (Figure 9-13).

6. Type in a user name in this format: **your_computer_name\a_SQL_and_SSAS_
   administrator** (for example, **MyPC\BobSmith**). Type in the Windows password
   into the password textbox and click Next to advance the wizard to the final dialog
   window.

7. In the final dialog window, enter the name of the data source as **DWPubsSales**
   (Figure 9-15).

Create a Data Source View

1. Create a new data source view by right-clicking the data source view project folder
   and selecting New Data Source from the context menu (Figure 9-18).

2. When the Data Source View Wizard appears, click Next to continue to the second
   dialog window and select the data source you just created from the Relational Data
   Sources list box (Figure 9-20). Click Next to continue.

Note: Because we have foreign key relationships in our data warehouse, the next dialog window should be
the Select Table and Views dialog window rather than the Name Matching dialog window (Figure 9-22).

3. In the Select Table and Views dialog window, highlight all the dimension and fact tables from the Available Objects list box and click the arrow button to insert them into the Included Objects list box, as shown in Figure 9-22.

4. Click Next to proceed to the wizard's final dialog window and name your data source view **DWPubsSalesTables**.

5. Review all the tables on the Data Source View Designer surface. Verify that each of the relationships between tables have been correctly indicated. (In the unlikely event that a relationship is missing, drag and drop columns between tables to create any missing relationships.)

Add a Friendly Name

1. Right-click the Date column in the DimDates table, and select Properties from the context menu.

2. When the Properties window appears, change the FriendlyName property to FullDate.

At this point, your SSAS project should look similar to the one shown in Figure 9-33.

3. Save your work by selecting File ➤ Save All from the Visual Studio main menu.

4. Leave the Visual Studio solution open for the next exercise.

In this exercise, you added a new SSAS project to your current BI solution. You then created a data source and a data source view within the project. In the next exercise, you will create dimensions that use both of these items.

Dimensions

As discussed in earlier chapters, dimensions describe attributes of measured values. For example, a customer's dimension might describe the customer's name associated with a given sales quantity—as in "Bob Smith bought 15 items." Of course, this leads to questions like "What items?" and "When did he buy them?" From these questions, you might decide to create both a date dimension and a products dimension, which in turn may suggest new dimensions.

In the PubsBICubes project, we currently have a single measure we need to describe: SalesQuantity. The SalesQuantity attributes are an order number, order date, title, and store. Accordingly, we create a dimension for each of these descriptors.

At a minimum, each dimension should include a unique identifier and name. This typically is a dimension key and a column that contains a human-friendly name. SSAS only forces you to include the dimensional key column, but configuring a name column is recommended.

In addition to the key and name columns, you can include other columns that provide a way of aggregating measured values or clarifying their meaning. For example, in the PubsBICube project, we use title types for both purposes—clarifying a book's type and aggregating the sales quantity for a given type of book.

The Dimension Wizard

To create a dimension, launch the Dimension Wizard by right-clicking the Dimensions folder in Solution Explorer and selecting New Dimension from the context menu (Figure 9-34).

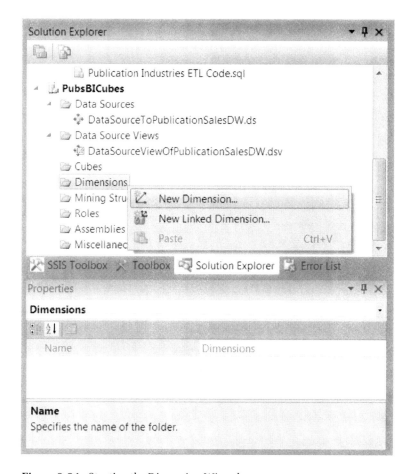

Figure 9-34. *Starting the Dimension Wizard*

Select Creation Method

When the wizard starts, you are presented with a welcome screen. Click Next to navigate to the Select Creation Method page. On this page, the default choice is to use an existing table in your data source view (Figure 9-35).

The second option is to generate a timetable in the data source. Selecting this option allows SSAS to generate and fill a new date table in your SQL Server data warehouse, much as we did during the ETL process. This option has some distinct advantages, the foremost of which is that you do not have to write any SQL code yourself. The downside is that the format the table is created in may not be to your liking.

Considering that this is an easy way to create a date-time table within the data warehouse, some developers create a disposable SSAS project solely for this purpose. For example, create a new SSAS project and add a New Date Time dimension using the "Generate a time table in the data source" option. Once the table is created in the data warehouse, delete the project.

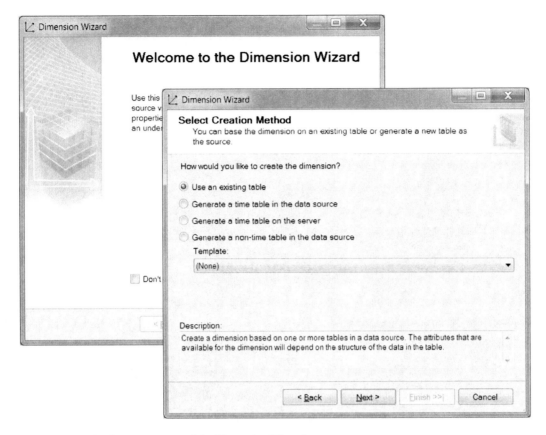

Figure 9-35. *The first two screens of the Dimension Wizard*

If you choose this option, the final dialog window of the wizard offers to immediately generate and run the SQL code in your data warehouse, or you can wait and perform this action at a later time.

The third option on the Dimension Wizard page is similar to the second. It generates a virtual time dimension table within the SSAS database, but not in the data warehouse. Although the table is not in the data warehouse, you still create a time dimension as if it was.

While this option is interesting, it is better to have a table in the data warehouse so that reports that work directly with the data warehouse can use it.

The fourth selection on this wizard page allows you to create one or more tables using templates. Microsoft supplies a number of templates, but you are free to create your own as well.

To create a template, make a data source view with your preferred dimension design. Then create a dimension based on the data source view and save the resulting XML file to the proper folder, usually `C:\Program Files (x86)\Microsoft SQL Server\110\Tools\Templates\olap\1033\Dimension Templates`.

As with the time table options, each template you create can be reversed engineered by Visual Studio to generate SQL code. This code can then be used to create tables within the data warehouse. With this feature, you could technically create the dimensions in Visual Studio before you created a data warehouse. But like the previous option, it is interesting but not commonly used.

The most common option used is the initial default option. Once you have selected your option, click Next to proceed to the third page of the wizard.

Specify Source Information

On the third page of the wizard, select which data source view contains the table or tables you are going to use for your dimension. In the PubsBICubes project, we have only one data source view to choose from (Figure 9-36).

Figure 9-36. *The Specify Source Information page in the Dimension Wizard*

After having selected the data source view, select a table from the "Main table" dropdown box. Each dimension may have only one table, which of course will be its main table. In cases where the dimension is designed in a snowflake format, such as the Titles dimension in the PubsBICubes project, the main table is the one directly connected to the fact table. This means you do not have to select both the DimTitles table and the DimPublishers tables. Only select the main DimTitles table, as shown in Figure 9-37.

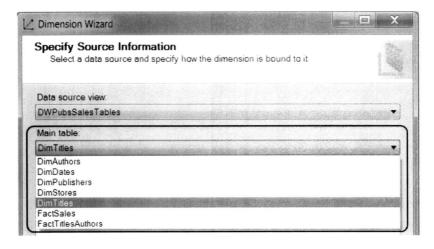

Figure 9-37. Selecting the main table for the dimension

Once you select a table, select one or more columns that define the table's dimensional keys. Ordinarily, they are singular key columns such as the TitleKey in Figure 9-38. If you have a composite key, you can select a second column by clicking the "Add key column" option (covered by the dropdown box in Figure 9-36) after you choose the initial key column.

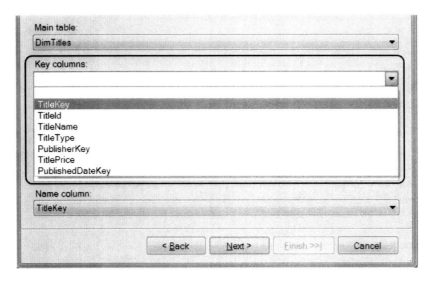

Figure 9-38. Selecting the key columns for the dimension

■ **Tip**   If you want to undo a selection, highlight the errant column, open the dropdown box and click the blank area just above the column listing. In Figure 9-38, the blank area is just above the TitleKey.

The last dropdown box on this page allows you to select a name column. The name column is the one that holds a label for the key column's value. Only one column can be selected for this assignment, even if you are using a composite key.

To select a name column, access the dropdown box and select a name column (in this case it is TitleName) from the list of columns in the table (Figure 9-39). It is possible to leave them all unselected, but when a user creates a report using that dimension, they will see only ID or key values instead of human-friendly names.

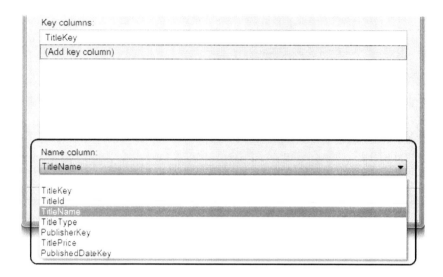

Figure 9-39. *Selecting the name column for the dimension*

Select Related Tables

Clicking the Next button advances the wizard to one of two pages: the Dimension Attributes page or the Select Related Tables page. If the only relationship line you have in the data source view is to the fact table, you will advance to the Select Dimension Attributes page. If, however, you have tables in a snowflake pattern with relationship lines connecting to other dimensional tables, you will advance to the Select Related Tables page (Figure 9-40).

In our example, The DimTitles table is connected to both the DimPublishers and DimDates tables in the data source view. Therefore, the Select Related Tables page displays both of these tables with a checked checkbox. It is optional to include a related table; therefore, the wizard allows you to uncheck the checkbox if you feel a related table is not appropriate for the dimension you are currently building. In the case of the DimTitles table, we will leave both tables checked because they include additional information that we want to include as part of the Titles dimension.

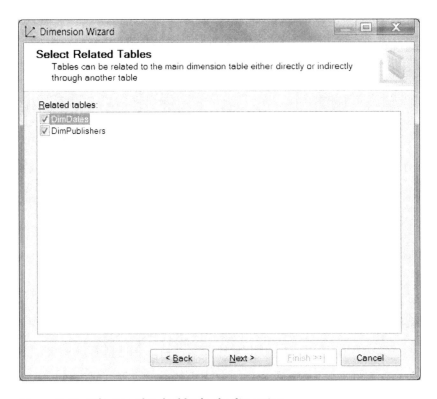

Figure 9-40. *Selecting related tables for the dimension*

Select Dimension Attributes

Dimensional tables commonly have many columns in addition to their key and name columns. In the example, we are using additional columns, such as title type, title price, and publisher name.

Each of these columns of data is copied to SSAS to form additional dimensional attributes. Remember that a dimensional attribute is a copy of data from one or more columns in a data warehouse table. For example, you may note that the TitleName does not show up in the list displayed in Figure 9-41, even though the Title Key attribute does. This is because the Title Key attribute represents both the title name and the title key columns in the data warehouse.

Later when you query the data from the SSAS dimension, you are able to specify whether the TitleKey attribute will display either the key value or the name value. From this example, you can see that a dimensional attribute is not the same as a tables column but rather a logical representation of one or more columns of data. Keep in mind that many dimensional attributes represent only one column within the data warehouse table (that is, the key and name columns are the same). The dimension key is the most common exception, but date columns are another. For example, our DimDates table includes columns for Month and MonthName, which will form a single attribute called Month (Figure 9-41).

On the Select Dimension Attributes page, you can choose to select all or some of the available attributes. By default, all attributes are included, but you may want to uncheck the checkbox to exclude them when the occasion calls for it. You can also change the selections, exclusions, and other options by using the dimension designer after completing the wizard.

We will configure DateName, MonthName, QuarterName, and YearName as the name columns of DateKey, Month, Quarter, and Year. Therefore, we will not select these as available attributes (Figure 9-41).

Figure 9-41. *Selecting dimensional attributes in the Dimension Wizard*

In our current example, the DimTitles table relates to both the DimPublishers table and the DimDates table to form a snowflake design within the data source view. Because of this, we see many columns available in the Select Dimensional Attributes page (Figure 9-41).

It makes sense to include some of these columns from each dimension table, but not necessarily all of them. Including publishers names, for example, will allow for creating reports that can group titles by publishers. The same is true of the title types. But the artificial publisher key or the artificial date key (without an associated name column) is not as useful, because it is unlikely a report will benefit from either one of these columns. They have no significance outside the context of the data warehouse design.

■ **Note**   Your attributes may be in a different order than those in Figure 9-41. The order depends upon which table the wizard decided to select first, either the DimPublishers or the DimDates table. It may be annoying, but as long as you pay attention to which attribute you are configuring, it makes no difference.

Setting Attribute Types

The wizard allows you to identify each attribute type. On the Select Dimension Attributes page, use the dropdown boxes under the Attribute Type column. All dimensional attributes are set to Regular by default. This can be left as it is.

On the occasion that you need to change the attribute type, select the row of the attribute that you want to configure. When the dropdown box appears, navigate through the dropdown treeview display until you find the appropriate type (Figure 9-42).

Figure 9-42. *Setting the attribute type on a natural key column*

Not every possible type is represented here. For example, if we try to set the TitlePrice attribute type, we find that there is no exact match within the predefined options. Most software ignores the Attribute Type setting; therefore, it has little to no impact on creating reports. Microsoft has included the setting for application programmed to use it, but leaving the selection at Regular is most often the appropriate choice.

Occasionally an attribute's type must be more definitive. For example, the TitleId attribute in the DimTitles dimension contains data from the title_id column in the original Pubs database. Including the TitleId attribute is useful for reports that display this original Identifier. It can also be useful if you are tracking changes to dimensional attribute values over time using a Slow Changing Dimension strategy (discussed in Chapter 4).

■ **Note**   We discussed Slow Changing Dimensions (SCD) in Chapter 4, but we did not implement them in our tables. We give an example of how to set one up later in Chapter 9, but for now we simply indicate that the title ID is an original ID, as shown in Figure 9-42.

When the dimensional attribute includes date data, the attribute type should always be configured to reflect the content. The date data determines how SSAS performs aggregations and how MDX functions are processed. In our current example date, month, quarter, and year attributes must be configured accordingly.

■ **Tip**   Oddly enough, the treeview has redundant type settings. In Figure 9-43 the Date dimension attribute is available in two places. Although this is confusing, it does not matter which is used to set the attribute.

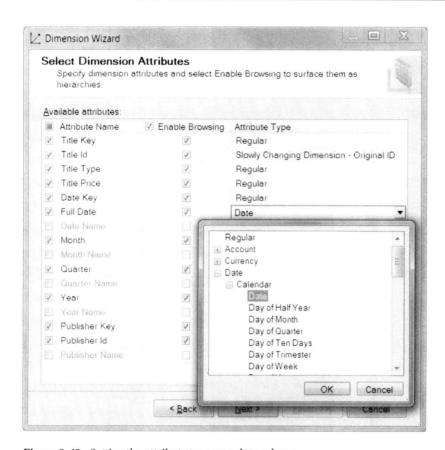

Figure 9-43. *Setting the attribute type on a date column*

After you have completed setting the attribute types, the Select Dimensional Attributes wizard page will look similar to Figure 9-44. Click Next at the bottom of this page to advance to the final page of the wizard.

Figure 9-44. *The Select Dimension Attributes page after it is configured*

Completing the New Dimension Wizard

On the final page of the wizard, a treeview display shows the dimension composition (Figure 9-45). There is not much to do on this page other than perhaps change the name of the dimension. And since we have been removing the spaces that Visual Studio inserts between Dim and Titles, it is appropriate to do so once again on this page by changing the name to DimTitles. (We remove the spaces in the attribute names in Chapter 10, but for now, we can just leave them as is.)

Figure 9-45. *Completing the Dimension Wizard*

Click Finish to close the wizard, and you will be taken to the Dimension Designer tab (Figure 9-46). In this Dimension Designer tab, you can make numerous configurations and changes to the dimension. We discuss some of these options next.

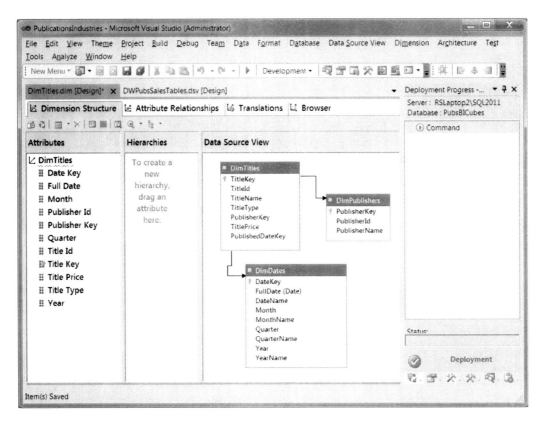

Figure 9-46. *The Dimension Designer tab*

Building the SSAS Project

Creating SSAS dimensions, data source views, and data sources all generate XML files. The format of these XML files are not compatible with SSAS. Therefore, they must be converted before uploading them to the server. Building an SSAS project in Visual Studio combines each of the individual XML files into one master file and converts it to a compatible XMLA format. Once that is complete, it can then be uploaded to the SSAS server during deployment.

■ **Note**   We discuss building and deploying projects in Chapter 10. For now, we are using this feature to check for errors within the dimensions.

You can build a Visual Studio project using the build menu item (Figure 9-47). While building your project, errors in the dimension will be reported, and if errors exist, the build process will fail. You must resolve these errors and build the project once more before you can upload any of your project files to the server.

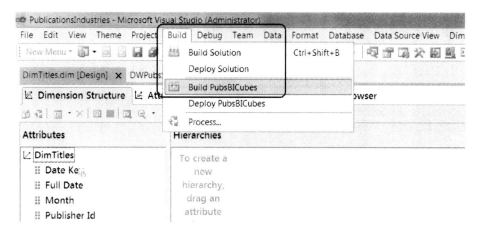

Figure 9-47. *The build menu item*

As an example, let's do something crazy like removing the TitleKey column from a dimension. When we build the PubsBICubes project, an error message displays, as shown in Figure 9-48.

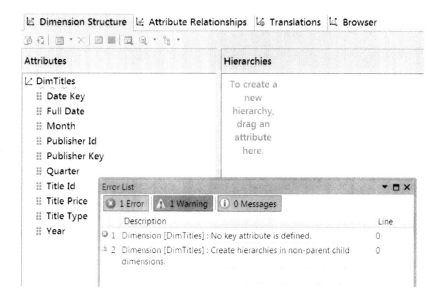

Figure 9-48. *Build errors and warnings*

Notice that in Figure 9-48, building the project displays both error and warning messages. Warning messages give you Microsoft's recommendations on how to properly configure your dimensions. Sometimes these recommendations can be ignored, but often they are a good idea to implement.

■ **Tip** Both errors and warnings can be hidden by clicking the error and warning buttons at the top of the error list window shown in Figure 9-48. If you do not see the errors or warnings, you may have to click the buttons to do so.

In Chapter 12 we discuss configuring, building, deploying, and processing dimensions, for now let's create our dimensions in the following exercise.

EXERCISE 9-2: CREATING THE DIMENSIONS

In this exercise, you create the dimensions for the SSAS project. In our current design we have five dimensions to create: titles, authors, stores, dates, and orders. Let's start with the titles dimension.

Note: If the solution from the previous exercise is closed, reopen it and remember to right-click the menu item, select Run as Administrator, and answer Yes to close the UAC.

Create the Titles Dimension

1. Start the Dimension Wizard by right-clicking the Dimensions folder in Solution Explorer, as shown in Figure 9-34.

2. When the wizard opens, click Next to advance to the Select Creation Method dialog window (Figure 9-35). Choose the Use an Existing Table radio button, and click Next to advance to the Specify Source Information dialog window.

3. Choose DWPubsSalesTables in the "Data source view" dropdown box of the Specify Source Information dialog window, as shown in Figure 9-36.

4. In the "Main table" dropdown box, choose the DimTitles table, as shown in Figure 9-37.

5. In the "Key columns" list box, choose the TitleKey column, as shown in Figure 9-38.

6. In the "Name column" dropdown, choose the TitleName column, as shown in Figure 9-39. Click Next to advance to the Select Related Tables dialog window.

7. Check both the DimDates and DimPublishers checkboxes in the Select Related Tables dialog window, as shown in Figure 9-40. Click Next to advance to the Select Dimension Attributes dialog window.

8. Check all the attributes except for the Date Name, Month Name, Quarter Name, Year Name, and Publisher Name in the Select Dimension Attributes dialog window, as shown in Figure 9-41.

9. Change the attribute types for each of the attributes listed here (Figure 9-43), and then click Next to advance to the Completing the Wizard dialog window:

    - Title Id as Slowly Changing Dimension - Original Id

    - Date Key as Calendar ➤ Date

    - Full Date as Calendar ➤ Date

    - Month as Calendar ➤ Month

- Quarter as Calendar ➤ Quarter

- Year as Calendar ➤ Year

- Publisher Id as Slowly Changing Dimension - Original Id

10. Rename the dimension from Dim Titles to **DimTitles** (Figure 9-44), and then click Finish to complete the wizard.

11. Use the Build menu to verify that the DimTitles dimension builds successfully. If it does not, troubleshoot the problem or delete the dimension from Solution Explorer and create the dimension again, being careful to include any steps you may have omitted.

Create the Authors Dimension

The Authors dimension contains only four columns. Next we will combine two columns (ID and Name) into a single attribute called Author. The other two columns will each be configured as a separate attribute of the Authors dimension.

1. Start the Dimension Wizard by right-clicking the Dimensions folder in Solution Explorer.

2. Click Next to advance to the Select Creation Method dialog window, and choose the Use an Existing Table radio button.

3. Click Next to advance to the Specify Source Information dialog window, and choose DWPubsSalesTables in the "Data source view" dropdown box, as shown in Figure 9-49.

4. In the "Main table" dropdown box, choose the DimAuthors table.

5. In the "Key columns" list box, choose the AuthorKey column.

6. In the "Name column" dropdown box, choose the AuthorName column, and click Next to advance to the Select Dimension Attributes dialog window.

Note: When we created the Titles dimension, the Select Related Tables dialog window was the next to be displayed. The wizard displays that dialog window only if the relationship lines in the data source view indicate that the main table is a parent table to a child table; therefore, you will not see this window when creating the DimAuthors table.

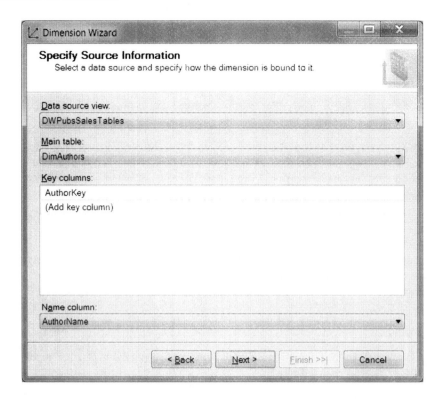

Figure 9-49. Specifying the source information for the Authors dimension

7. Check the Author Key, Author Id, and Author State checkboxes in the Select
    Dimension Attributes dialog window (Figure 9-50).

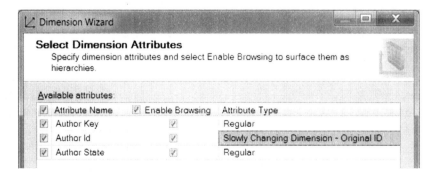

Figure 9-50. Selecting the attributes for the Authors dimension

8. Change the attribute type for the attribute listed here (Figure 9-50):

    •   Author Id as Slowly Changing Dimension - Original Id

9. When you have completed this, the dialog window should look like Figure 9-50. Click Next to advance the wizard.

10. Change the name of your new dimension to the **DimAuthors** in the Completing the Wizard dialog window by removing the space between the word Dim and Authors.

11. Click Finish to complete the wizard.

12. Use the Build menu to verify that the DimTitles dimension builds successfully. If it does not, troubleshoot the problem or delete the dimension from Solution Explorer and go through the steps again, being careful to include any steps you may have omitted.

Create the Stores Dimension

We now have two dimensions created and built. Let's work on the Stores dimension next. This process will be almost identical to steps required to create the Authors dimension.

1. Start the Dimension Wizard by right-clicking the Dimensions folder in Solution Explorer.

2. Click the Next button to advance to the Select Creation Method dialog window, and choose the Use an Existing Table radio button.

3. Choose DWPubsSalesTables in the "Data source view" dropdown box within the Specify Source Information dialog window.

4. In the "Main table" dropdown box, choose the DimStores table.

5. In the "Key columns" list box, choose the StoreKey column.

6. In the "Name column" dropdown box, choose the StoreName column, and click Next to advance to the Select Dimension Attributes dialog window.

7. Check the Store key and Store ID checkboxes in the Select Dimension Attributes dialog window.

8. Change the attribute type for the following attribute:

   • Store Id as Slowly Changing Dimension - Original Id

9. When you have completed this, the dialog window should look like Figure 9-51.

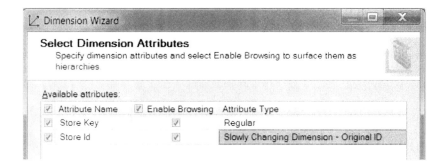

Figure 9-51. *Selecting the attributes for the Stores dimension*

10. Click Next to advance the wizard.

11. Change the name of your new dimension to the **DimStores** in the Completing the Wizard dialog window by removing the space between the word Dim and Stores.

12. Click Finish to complete the wizard.

13. Use the build menu to verify that the DimStores dimension builds successfully. If it does not, troubleshoot the problem or delete the dimension from Solution Explorer and go through the steps again, being careful to include any steps you may have omitted.

Create the Date Dimension

The Date dimension has nine columns. Most of these columns will be combined into name and key pairs so that we will end up with only five attributes. An example of t\his is combining the YearName column and year column into one dimensional attribute called year. Unfortunately, the wizard is unable to completely configure the attributes. Therefore, we select only a few columns during the wizard's creation process and configure the attributes after the wizard is complete.

1. Start the Dimension Wizard by right-clicking the Dimensions folder in Solution Explorer.

2. Click Next to advance to the Select Creation Method dialog window, and choose the Use an Existing Table radio button. Click Next to advance to the Specify Source Information dialog window.

3. Choose DWPubsSalesTables in the "Data source view" dropdown box within the Specify Source Information dialog window.

4. In the "Main table" dropdown box, choose the DimDates table.

5. In the "Key columns" list box, choose the DateKey column.

6. In the "Name column" dropdown box, choose the DateName column and click Next to advance to the Select Dimension Attributes dialog window.

7. Check the Date Key, Full Date, Month, Quarter, and Year checkboxes in the Select Dimension Attributes dialog window (Figure 9-52).

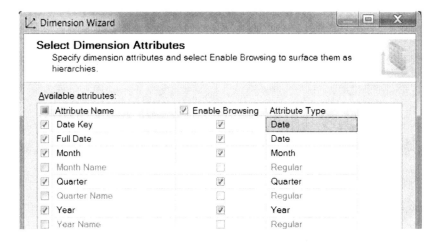

Figure 9-52. *Selecting the attributes for the Date dimension*

8. Change the attribute types for the attributes listed here (Figure 9-52):

   - Date Key as Calendar ➤ Date

   - Full Date as Calendar ➤ Date

   - Month as Calendar ➤ Month

   - Quarter as Calendar ➤ Quarter

   - Year as Calendar ➤ Year

9. When you have completed these steps, the dialog window should look like Figure 9-52. Click Next to advance the wizard.

10. Change the name of your new dimension to **DimDates** In the completing the Wizard dialog window by removing the space between the word Dim and Dates.

11. Click Finish to complete the wizard.

12. Use the build menu to verify that the DimDates dimension builds successfully. If it does not, troubleshoot the problem or delete the dimension from Solution Explorer and go through the steps again, being careful to include any steps you may have omitted.

Create the Orders Dimension

The last dimension we need to make is the Orders dimension. If you recall, this dimension is a fact dimension, meaning that all the columns we need to create the dimension are in a fact table. This dimension is also created using the wizard.

1. Start the Dimension Wizard by right-clicking the Dimensions folder in Solution Explorer.

2. Click Next to advance to the Select Creation Method dialog window, and choose the Use an Existing Table radio button. Click Next to advance to the Specify Source Information dialog window.

3. Choose DWPubsSalesTables in the "Data source view" dropdown within the Specify Source Information dialog window box, as shown in Figure 9-33.

4. In the "Main table" dropdown box, choose the FactSales table.

5. In the "Key columns" list box, choose the OrderNumber column.

Important: You need to deselect all other columns besides the OrderNumber column. To do so, highlight each errant column, open the dropdown box, and click the blank area just above the column listing. For example, in Figure 9-53, the blank area is just above the StoreKey.

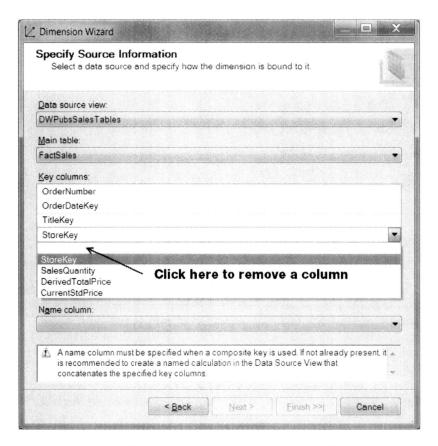

Figure 9-53. *The Specify Source Information window before configuration*

6. In the "Name column" dropdown box, choose the OrderNumber column, and click Next to advance to the Select Dimension Attributes dialog window (Figure 9-54).

Figure 9-54. The Specify Source Information window after configuration

7. In the Select Related Tables dialog window, uncheck the DimStores, DimDates, DimTitles, and DimPublishers checkboxes, as shown in Figure 9-55. Click Next to advance to the Select Dimension Attributes dialog window.

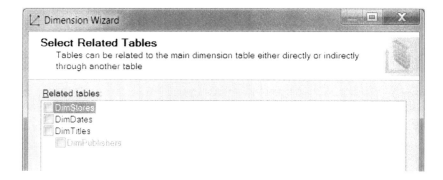

Figure 9-55. Unchecking the unrelated tables

8. On the Select Dimension Attributes dialog window, deselect the Sales Quantity, Store Key, Order Date Key, and Title Key checkboxes (Figure 9-56). Leave the Order Number checked and its attribute type at Regular. Click Next to advance to the final Dimension Wizard dialog window.

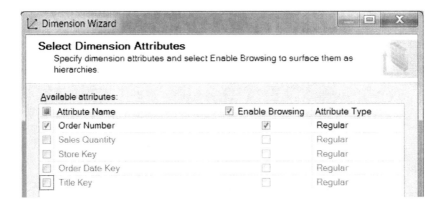

Figure 9-56. *Unchecking the unrelated attributes*

9. In the Completing the Wizard dialog window, change the name of your new dimension to **DimOrders**.

10. Click Finish to complete the wizard.

11. Use the Build menu to verify that the DimOrders dimension builds successfully. If it does not, troubleshoot the problem or delete the dimension from Solution Explorer and go through the steps again, being careful to include any steps you may have omitted.

In this exercise, we created five dimensions using the Dimension Wizard. At this point, all the dimensions are created but still need additional configurations.

Moving On

In this chapter, you saw how to create an SSAS project, data source, data source view, and five dimensions. We have covered a lot, but we are still just beginning. In Chapter 10, we configure each of the dimensions and then create a cube that uses them.

LEARN BY DOING

In this "Learn by Doing" exercise, you create a project and dimensions very similar to those defined in this chapter. This exercise uses the Northwind database. We have included an outline of the steps you performed in this chapter as well as an example of how the authors handled this exercise in two Word documents. These documents are found in the folder C:_BookFiles_LearnByDoing\Chapter09Files. Please see the ReadMe.doc for detailed instructions.

What's Next?

Getting started with SSAS can be a tricky, but it gets easier the more you are exposed to it! For further study, we have created videos that demonstrate the processes discussed in this chapter. You can find these videos at http://NorthwestTech.org/ProBISolutions/Videos.

CHAPTER 10

■ ■ ■

Configuring Dimensions with SSAS

O Marvelous! What new configuration will come next? I am bewildered with multiplicity.

— William Carlos Williams

Over the years, we have developed a proven method of creating and configuring SSAS cubes that provides the least amount of pain and the greatest chance of success. We have already begun our process within the previous chapter by preparing for the creation of our SSAS cubes by making our data source, data source view, and dimensions.

In this chapter, we continue this process by configuring SSAS dimensions using the dimension designer. We also discuss deploying and processing dimension and cubes. Along the way, we verify that the cubes and dimensions are working as expected and what to do when they are not.

Let's get started with configuring the dimension created in Chapter 9, using Visual Studio's Dimension Designer.

The Dimension Designer

In Visual Studio, dimensions must be created using the Dimension Wizard, but after each dimension is created, the Dimension Designer is used to complete its configuration. The designer is divided into four tabs (Figure 10-1):

- Dimension Structure
- Attribute Relationships
- Translations
- Brower

Let's take a look at each of these and see how they are used in the configuration process.

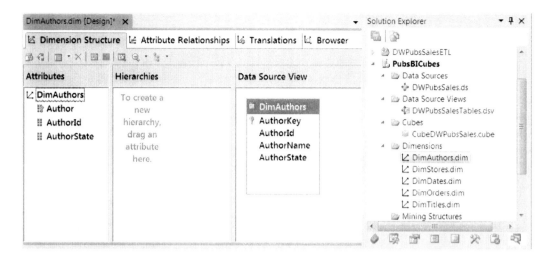

Figure 10-1. *The Dimension Designer tabs*

Dimension Structure Tab

The Dimension Structure tab allows you to add, remove, and group attributes within your dimension. It is divided into three panes:

- Attributes

- Hierarchies

- Data Source View

Let's take a closer look at each of these.

The Attributes Pane

The Attributes pane allows you to select an attribute and configure it using the Properties window of Visual Studio (Figure 10-2). To edit a property, right-click an attribute and select Properties from the context menu. This will open the Visual Studio Properties window if it is not already open or bring it to the forefront if it is already open.

As mentioned in previous chapters, dimensions are made of a collection of one or more attributes. Attributes are similar to columns in a database table, with one important exception: each attribute can be associated with one or more columns in a table. For instance, in Figure 10-2, the Properties window shows that the Author attribute is associated with the DimAuthors.AuthorsKey column (used as a key column) and the DimAuthors. AuthorName column (used as a name column).

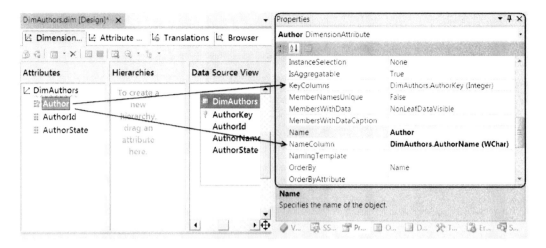

Figure 10-2. *Accessing the Properties window*

Table 10-1 describes the most commonly used properties.

Table 10-1. *Common Attribute Properties*

Property	Description
KeyColumns	A collection of one or more columns uniquely identifying one row from another in the dimension table (for example, 1, 2, or 3)
Name	The logical name of an attribute
NameColumn	The values that are displayed in the client application when a particular attribute is selected (for example, Red, Green, or Blue)
Type	The type of data the attribute represents (for example, day, month, or year)

Each attribute must have one or more KeyColumns. Most of the time you will have just one column defined in the KeyColumns property. But if a single column cannot uniquely identify one row from another, you can configure this property to use multiple columns, as we will show you later in this chapter.

Each attribute must have a name. We recommend using a name without spaces to avoid rare problems with some client applications. The name should be simple and descriptive. As an example, in Figure 10-3 we elected to change the attribute name Date in the DimTitles dimension to PublishedDate, because it is more explicit.

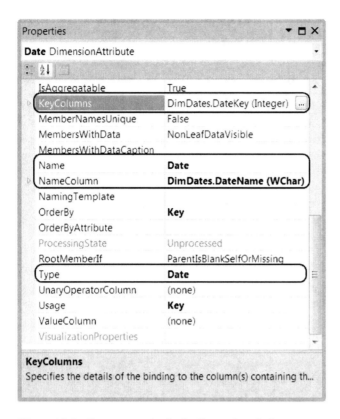

Figure 10-3. *Common setting in the Properties window*

Each attribute uses the NameColumn property to determine which values are displayed to a client application. If you do not select a column, SSAS will use the KeyColumn by default. If this happens, the results will display integer values instead of names. For example, if someone creates a report showing the quantity of sales based on published dates, that report will display the date key (such as 1) instead of the actual date (such as 01/01/2012). As you might imagine, this renders the report useless. Therefore, be sure to set a NameColumn property where appropriate. For example, in Figure 10-3 we have configured it to use the DateName column.

In Chapter 9 we discussed how most attributes types are configured as regular. A special few, such as date attributes, are set to other predefined types, such as Date, Month, or Year. If you did not set the attributes type in the Dimension Wizard, you can change it in the Properties window as shown in Figure 10-3.

The Hierarchies Pane

When you first create a dimension, each attribute is independent of each other, but user hierarchies allow you to group attributes into one or more parent-child relationships. Each relationship forms a new level of the hierarchy and provides reporting software with a way to browse data as a collection of levels-based attributes.

Creating Hierarchies

The act of creating a hierarchy is quite simple. All you do is drag and drop an attribute into the Hierarchies section, and a hierarchy will be created automatically.

In Figure 10-4 you can see that the DimAuthor's dimension contains the Author Key, Author Id, and Author State attributes, renamed to Author, AuthorId, and AuthorState.

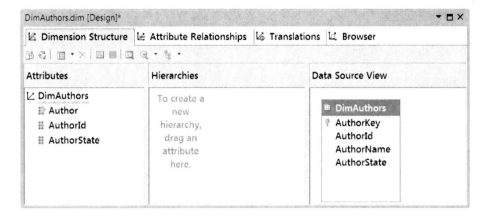

Figure 10-4. *The DimAuthors attributes*

SSAS always provides a default, invisible hierarchy for each attribute consisting of two levels. In this default design, each individual attribute of the dimension is considered a child attribute of an All attribute, as illustrated in Figure 10-5. The All level contains only one member called, well, All.

Figure 10-5. *The default attribute hierarchy*

The All member forms the root-level value of the hierarchy. Once this is set, the All level represents the total of all the members of a group (Figure 10-6). It is simply a way of organizing information. Microsoft recommends that you define user hierarchies when it makes sense to do so.

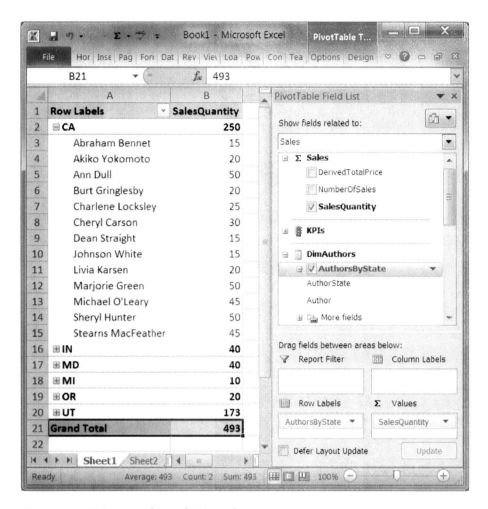

Figure 10-6. *Using a user hierarchy in Excel*

In our example, we grouped authors by states by dragging and dropping the AuthorState attribute into the Hierarchies pane and then dragging and dropping the child-level attribute, Author, beneath it (Figure 10-7). Once this is set, reporting software like Excel can easily browse the authors grouped by state.

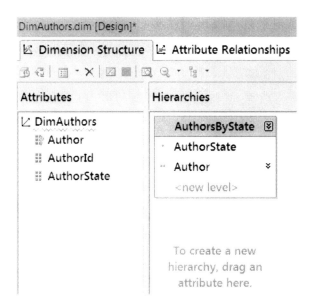

Figure 10-7. A new user-defined attribute hierarchy

Each dimension can have many hierarchies, and the implied individual hierarchy exists even after you create your own user-defined attribute hierarchies. In our example, we have three attributes and have added a user-defined hierarchy, so we now have four hierarchies total. Figure 10-8 is a conceptual example of this concept.

Figure 10-8. The DimAuthors dimension including the new user-defined attribute hierarchy

The attributes AuthorID, Author, and AuthorState still form their own hierarchical parent-child relationship with the All attribute. Microsoft calls these *attribute hierarchies* in contrast to the term *user hierarchies*.

Microsoft recommends hiding attribute hierarchies if you will not use them independently in your reporting applications. They can be hidden by setting the AttributeHierarchyVisible property to False, as shown in Figure 10-9.

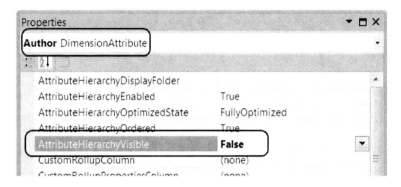

Figure 10-9. *Hiding attribute hierarchy*

Now, in Microsoft Excel, the AuthorState and Authors attributes will no longer be displayed independently of the AuthorState.

Not all client applications hide the individual attributes when this setting is applied. This may seem strange, but because the attribute hierarchy is marked only as hidden, it is not deleted (Figure 10-10). Since some software does not check to see whether the property is set to True or False, the individual attribute may appear in some reporting applications.

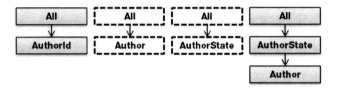

Figure 10-10. *Hidden attribute hierarchies are invisible to only some client applications*

■ **Note**   Although Microsoft recommends hiding the attribute hierarchies after you have created a user-defined hierarchy, it is not always in your best interest to do so. Often, report builders want more options instead of less. We configure the Authors dimension to hide the Authors attribute in the next exercise, and in Chapter 15 we demonstrate the impact that this choice has in Microsoft Excel.

Now that the hierarchies are defined, let's configure them.

Configuring Hierarchies

Table 10-2 describes the most common properties for user-defined hierarchies.

Table 10-2. *Common Attribute Properties*

Property	Description
AllowDuplicateNames	Allows an attribute member name to appear more than once under the same parent attribute. When set to True, a parent can have two children named Bob. This is a bad idea most of the time, but if it could conceivably happen in your data, you can leave this set to True, which is the default.
MemberKeysUnique	When set to Unique, the key values, like 1, 2, or 3, must not repeat in the user-defined hierarchy. For example, this setting would cause an error if the Author attribute has a key value of 1 and the AuthorState attribute has a key value of 1. The default setting of NotUnique is the most common selection.
MemberNamesUnique	When True, the member name values must be unique. For example, setting Mexico at the city level with Mexico at the country level causes an error. The default setting of False is the most common selection.
Name	The logical name of the user-defined hierarchy and is displayed in most client software.

No matter what the settings, attribute values must have some way of determining which child value belongs to which parent value, or the user-defined hierarchy will not work. This causes problems is when dealing with dates, because the child value of January is duplicated under the first quarter of one year as well as the first quarter of another, thus giving January two parents.

When situations like this occur, the error displayed while processing the dimension states, "A duplicate key has been found," as shown in Figure 10-11.

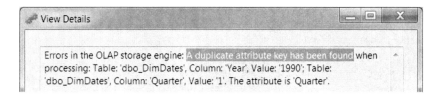

Figure 10-11. *A typical error message when duplicate keys are found*

This issue can be resolved by uniquely identifying the parent of January. Let's discuss two ways of making this distinction.

The first method of distinguishing one value of January from another is by concatenating additional text to the word *January*, such as the year. This will give each a unique value, eliminating duplication. With this method the values will no longer all be "January"; they will be "January 2000," "January 2001," and so on.

If concatenating the year with the month is inconvenient for whatever reason, the second method is to uniquely identify January's parent using a composite key. With this method, January will be identified by both its quarter and its year. Each value has a unique parent value that defines its individuality.

To set a composite key for an attribute, access the Properties window and modify the KeyColumns property by clicking the ellipsis (…) button that magically appears when you click the Properties window (Figure 10-12).

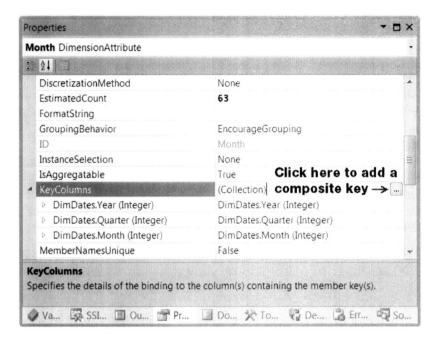

Figure 10-12. *Modifying the KeyColumns property to use a collection*

Clicking the ellipsis button launches the Key Columns dialog window (Figure 10-13); columns can be added and removed using the right and left arrow buttons. Additionally, columns can be moved up or down by using the arrow buttons on the left side of the dialog window. Be sure to move the parent column to the top followed by each level of the hierarchy in its proper order. In our example, the proper order for the Month attribute is the year, followed by the quarter, followed by the month, as shown in Figure 10-13.

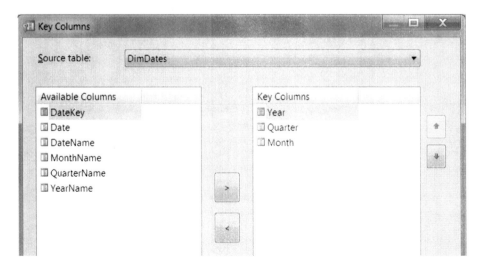

Figure 10-13. *Selecting multiple KeyColumns*

You now have a lot of new information to think about. Let's put your knowledge to work by configuring the dimensions created in the previous chapter.

EXERCISE 10-1: CONFIGURING KEYS, NAMES, AND HIERARCHIES

In this exercise you will start configuring your dimensions by verifying or changing the names, keys, and hierarchy properties.

Note: If the solution from the previous exercise is closed, open Visual Studio by clicking the Start button and navigating to All Programs ➤ Visual Studio 2010. Right-click this menu item, select Run as Administrator, and then answer Yes to close the UAC. If did not complete the project in the previous chapter, you can find our completed version in the
`C:\_BookFiles\Chapter09Files\Chapter09_Complete\PublicationsIndustries` folder.

Configure the Authors Dimension

We begin by configuring the Authors dimension.

1. Open the DimAuthors dimension editor by right-clicking the dimension in Solution Explorer and selecting View Designer from the context menu.

2. When the dimension editor opens, select each attribute and use the Properties window to change or verify the settings shown in Table 10-3.

Table 10-3. *The Initial Property Settings for the DimAuthors Dimension*

Attribute	Property Name	Value
Author Key	Name	Author
	KeyColumn	DimAuthors.AuthorKey
	NameColumn	DimAuthors.AuthorName
	AttributeHierarchyVisible	False
Author Id	Name	AuthorId
	KeyColumn	DimAuthors.AuthorId
	NameColumn	DimAuthors.AuthorId
	AttributeHierarchyVisible	True
Author State	Name	AuthorState
	KeyColumn	DimAuthors.AuthorState
	NameColumn	DimAuthors.AuthorState
	AttributeHierarchyVisible	True

3. Create a user-defined hierarchy for authors grouped by state by dragging and dropping the AuthorState attribute to the Hierarchies section of the dimension editor. This creates a new hierarchy.

4. Drag and drop the Author attribute just beneath the AuthorState attribute in the new hierarchy (Figure 10-14).

5. Rename the new hierarchy to AuthorsByState using the Properties window. When you are finished, your dimension editor should look like Figure 10-14.

6. Disregard the warning icons or wavy blue lines beneath the dimension name for now. We will address this later in the chapter.

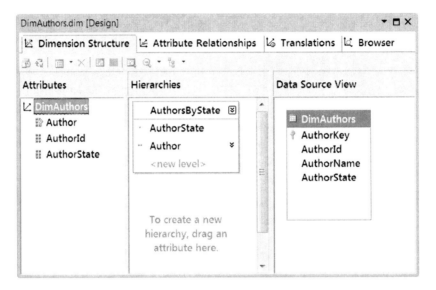

Figure 10-14. *The DimAuthors dimension after configuration*

Configure the Stores Dimension

Next we configure the Stores dimension.

1. Open the DimStores dimension editor by right-clicking the dimension in Solution Explorer and selecting View Designer from the context menu.

2. When the dimension editor opens, select each attribute and use the Properties window to change or verify the settings in Table 10-4.

Table 10-4. *The Initial Property Settings for the DimStores Dimension*

Attribute	Property Name	Value
Store Key	Name	Store
	KeyColumn	DimStores.StoreKey
	NameColumn	DimStores.StoreName
	AttributeHierarchyVisible	True
StoreId	Name	StoreId
	KeyColumn	DimStores.StoreKey
	NameColumn	DimStores.StoreId
	AttributeHierarchyVisible	True

3. Disregard any warning icons or wavy blue lines beneath the dimension name for now. When you are finished, your dimension editor should look like Figure 10-15.

Figure 10-15. *The DimStores dimension after configuration*

Configure the Orders Dimension

Now we configure the Orders dimension.

1. Open the DimOrders dimension editor by right-clicking the dimension in Solution Explorer and selecting View Designer from the context menu.

2. When the dimension editor opens, select each attribute and use the Properties window to change or verify the settings in Table 10-5.

Table 10-5. *The Initial Property Settings for the DimOrders Dimension*

Attribute	Property Name	Value
Order Number	Name	OrderNumber
	KeyColumn	FactSales.OrderNumber
	NameColumn	FactSales.OrderNumber
	AttributeHierarchyVisible	True

3. When you are finished, your dimension editor should look similar to Figure 10-16.

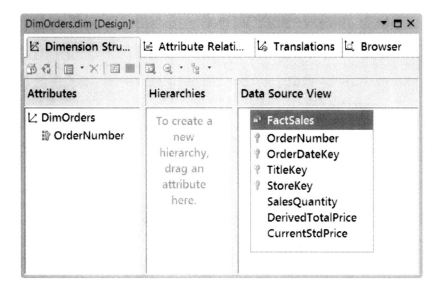

Figure 10-16. *The DimOrders dimension after configuration*

Configure the Dates Dimension

Now, it's time to configure the Dates dimension.

1. Open the DimDates dimension editor by right-clicking the dimension in Solution Explorer and selecting View Designer from the context menu.

2. When the dimension editor opens, select each attribute and use the Properties window to change or verify the settings in Table 10-6.

Table 10-6. *The Initial Property Settings for the DimDates Dimension*

Attribute	Property Name	Value
Date Key	Name	Date
	KeyColumn	DimDates.DateKey
	NameColumn	DimDates.DateName
	AttributeHierarchyVisible	True
Full Date	Name	FullDate
	KeyColumn	DimDates.DateKey
	NameColumn	DimDates.FullDate
	AttributeHierarchyVisible	True
Month	Name	Month
	KeyColumn	DimDates.Year
		DimDates.Quarter,
		DimDates.Month
	NameColumn	DimDates.MonthName
	AttributeHierarchyVisible	True
Quarter	Name	Quarter
	KeyColumn	DimDates.Year
		DimDates.Quarter
	NameColumn	DimDates.QuarterName
	AttributeHierarchyVisible	True
Year	Name	Year
	KeyColumn	DimDates.Year
	NameColumn	DimDates.YearName
	AttributeHierarchyVisible	True

3. Create a user-defined hierarchy for dates grouped by month, quarters, and years. You can do this by dragging and dropping the Year attribute to the Hierarchies section of the dimension editor. This creates a new hierarchy.

4. Drag and drop the Quarter attribute just beneath the Year attribute in the new hierarchy (Figure 10-17).

5. Drag and drop the Month attribute just beneath the Quarter attribute in the new hierarchy (Figure 10-17).

6. Drag and drop the Date attribute just beneath the Month attribute in the new hierarchy (Figure 10-17).

7. Rename the new hierarchy to Year-Qtr-Month-Day using the Properties window. When you are finished, your dimension editor should look like Figure 10-17.

8. Disregard any warning icons or wavy blue lines beneath the dimension or hierarchy names for now.

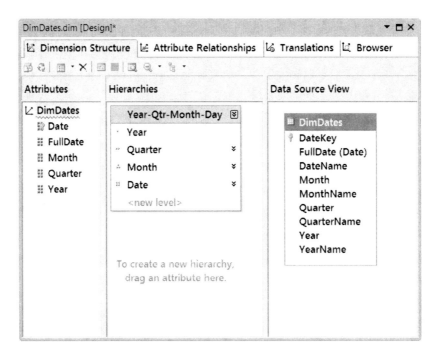

Figure 10-17. *The DimDates dimension after configuration*

Configure the Titles Dimension

It's time for the Titles dimension! This one has a lot of attributes. Be careful not to miss any.

1. Open the DimTitles dimension editor by right-clicking the dimension in Solution Explorer and selecting View Designer from the context menu.

2. When the dimension editor opens, select each attribute and use the Properties window to change or verify the settings in Table 10-7.

Table 10-7. *The Initial Property Settings for the DimTitles Dimension*

Attribute	Property Name	Value
Title Key	Name	Title
	KeyColumn	DimTitles.TitleKey
	NameColumn	DimTitles.TitleName
	AttributeHierarchyVisible	True
Title Id	Name	TitleId
	KeyColumn	DimTitles.TitleId
	NameColumn	DimTitles.TitleId
	AttributeHierarchyVisible	True
Title Price	Name	TitlePrice
	KeyColumn	DimTitles.TitleKey
	NameColumn	DimTitles.TitlePrice
	AttributeHierarchyVisible	True
Title Type	Name	TitleType
	KeyColumn	DimTitles.TitleType
	NameColumn	DimTitles.TitleType
	AttributeHierarchyVisible	True
Publisher Key	Name	Publisher
	KeyColumn	DimPublishers.PublisherKey
	NameColumn	DimPublishers.PublisherName
	AttributeHierarchyVisible	True
Publisher Id	Name	PublisherId
	KeyColumn	DimPublishers.PublisherId
	NameColumn	DimPublishers.PublisherName
	AttributeHierarchyVisible	True
Full Date	Name	PublishedFullDate
	KeyColumn	DimDates.DateKey
	NameColumn	DimDates.FullDate
	AttributeHierarchyVisible	True

(continued)

Table 10-7. *(continued)*

Attribute	Property Name	Value
Date Key	Name	PublishedDate
	KeyColumn	DimDates.DateKey
	NameColumn	DimDates.DateName
	AttributeHierarchyVisible	True
Month	Name	PublishedMonth
	KeyColumn	DimDates.Year,
		DimDates.Quarter,
		DimDates.Month
	NameColumn	DimDates.MonthName
	AttributeHierarchyVisible	True
Quarter	Name	PublishedQuarter
	KeyColumn	DimDates.Year,
		DimDates.Quarter
	NameColumn	DimDates.QuarterName
	AttributeHierarchyVisible	True
Year	Name	PublishedYear
	KeyColumn	DimDates.Year
	NameColumn	DimDates.YearName
	AttributeHierarchyVisible	True

3. Create a user-defined hierarchy for titles grouped by title types by dragging and dropping the TitleType attribute to the Hierarchies section of the dimension editor. This creates a new hierarchy.

4. Drag and drop the Title attribute just beneath the TitleType attribute in the new hierarchy, as shown in Figure 10-18.

5. Rename the new hierarchy to TitlesByType using the Properties window. When you are finished, your dimension editor should look like Figure 10-18.

6. Create a second user-defined Hierarchy for Titles grouped by Publishers. You can do this by dragging and dropping the Publisher attribute to the Hierarchies section of the dimension editor. This creates a new hierarchy.

7. Afterward, drag and drop the Title attribute just beneath the Publishers attribute in the new hierarchy, as shown in Figure 10-18.

8. Rename the new hierarchy to TitlesByPublisher using the Properties window. When you are finished, your dimension editor should look like Figure 10-18.

9. Create a third user-defined hierarchy for titles grouped by PublishedMonth, PublishedQuarters, and PublishedYears. You can do this by dragging and dropping the PublishedYear attribute to the Hierarchies section of the dimension editor. This creates a new hierarchy.

10. Afterward, drag and drop the PublishedQuarter attribute just beneath the PublishedYear attribute in the new hierarchy (Figure 10-18).

11. Now, drag and drop the PublishedMonth attribute just beneath the PublishedQuarter attribute in the new hierarchy (Figure 10-18).

12. Finally, drag and drop the Title attribute just beneath the PublishedMonth attribute in the new hierarchy (Figure 10-18).

13. Rename the new hierarchy to TitlesByPublishedDate using the Properties window. When you are finished, your dimension editor should look like Figure 10-18.

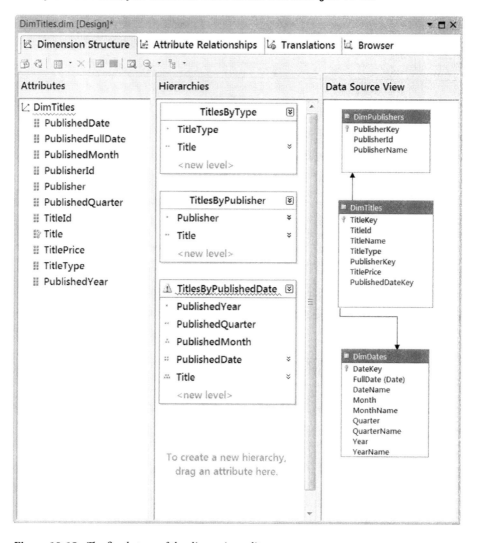

Figure 10-18. The final stage of the dimension editor

14. Disregard any warning icons and wavy blue lines beneath the dimension or hierarchy names for now. We will address this later in the chapter.

15. Use the Build menu to verify that the dimension builds successfully. If it does not, troubleshoot the problem or delete the dimension from Solution Explorer and go through the steps again, being careful to include any steps you may have omitted.

In this exercise, you began configuring your dimension. Next we will look at some additional configurations that need to be performed and how you can deploy and review your dimensions.

Attribute Relationships Tab

All attributes are related to the key attributes within their specific dimension. For example, in the Authors dimension we have two attributes in addition to the Authors key attribute. The first is AuthorId, and the second is the AuthorState attribute. The Attribute Relationships tab allows you to specify how these three separate attributes are associated.

Attributes are associated with each other in two ways. Either they are in a parent-child relationship or they are siblings. Another way to think of this is that they are either in a one-to-many relationship or in a one-to-one relationship.

■ **Note** We are somewhat hesitant to use the one-to-one or one-to-many description to describe attribute relationships, because this term can be mistaken with the same terms used to describe the data warehouse structure. Keep in mind that these relationships are between attributes within the dimension. They have nothing to do with the relationships between tables in the data warehouse.

To give an example, an author ID will only ever be associated with one author key, but one state may be associated with many authors. This is indicated on the Attribute Relationships tab in a graphical format. Sibling attributes, like AuthorId, are placed within the same rectangle as the dimensional key attribute. On the other hand, parents of the dimensional key, such as AuthorState, are placed in their own rectangle with a connecting arrow pointing toward the parent attribute (Figure 10-19).

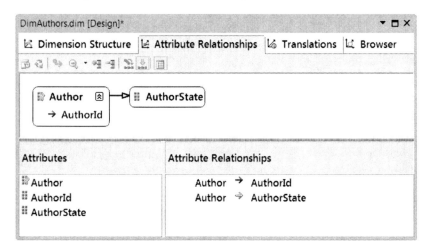

Figure 10-19. *The DimAuthors Attribute Relationships tab*

Configuring Relationships

When you create a dimension with the wizard, it tries to implicitly map the relationships for you. But more often than not, you will have to adjust them yourself.

To adjust the relationships, begin by removing the incorrect ones on the Attribute Relationships tab. Locate the Attribute Relationships pane at the bottom portion of Visual Studio (Figure 10-19), highlight the existing relationships, right-click, and select Delete from the context menu.

Once the relationships are removed, the Attribute Relationships tab will look like Figure 10-20, where the attributes are shown independent of each other.

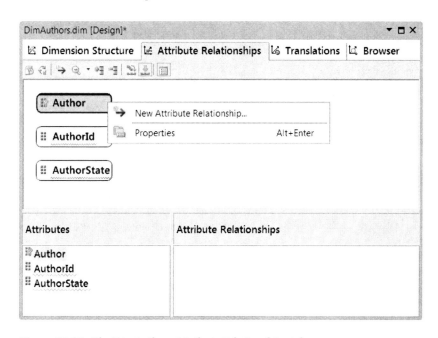

Figure 10-20. *The DimAuthors Attribute Relationships tab*

Now your job is to map the relationships correctly. A simple way to accomplish this is to right-click a child attribute and then click the New Attribute Relationship option from the context menu (Figure 10-20). A new dialog window will appear allowing you to map the relationship child to a parent (Figure 10-21). To map the relationship, select the related attribute in the Name dropdown box and then determine whether the relationship is expected to change over time or whether it should be considered an inflexible (aka rigid) relationship. Clicking OK will configure the relationship.

▪ **Important** The Edit Attribute Relationship dialog window is the same for both sibling and parent-child relationships. Visual Studio will automatically determine the type of relationship based on the attribute's appearance in a user-defined hierarchy. You may wonder how it was wrong in the first place, but occasionally it is! As you will see in the upcoming exercise, this automapping feature works well, even though it may not give you the level of control you would like. Simply delete any incorrect relations, remake them, and move on.

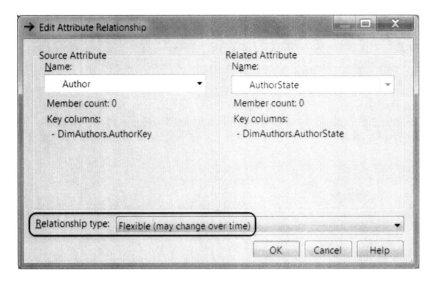

Figure 10-21. *The Author-AuthorState relationship*

Configuring the relationship to be flexible or rigid helps SSAS improve processing and query performance. SSAS has a processing algorithm that runs more efficiently if it doesn't have to look for relationships that may have varied over time. By setting the relationship to rigid, SSAS can skip over this section of the algorithm. Query performance is affected by formulating an execution plan that is more efficient with a rigid relationship as well.

In our experience, 80 percent of all the relationships we define are rigid. An example of a rigid relationship is data that does not change, such as the cities of a particular state. It is unlikely that Seattle, which is in the state of Washington this year, will somehow be in the state of Oregon next year.

A flexible relationship example is sales territories. A specific sales territory may be assigned to one employee and have a specific geographic area this year, but next year it may be assigned to a different employee.

Keep in mind that regardless of whether you choose flexible or inflexible, the processing and query results in your reports will still work. This discussion is simply about efficiency and the benefits of selecting the appropriate relationship type where it is known. When in doubt, however, you can leave it flexible. Failure to do so will have a negligible impact on your dimension unless it is very large.

To map the AuthorID attribute to the Author attribute, repeat the process (Figure 10-22).

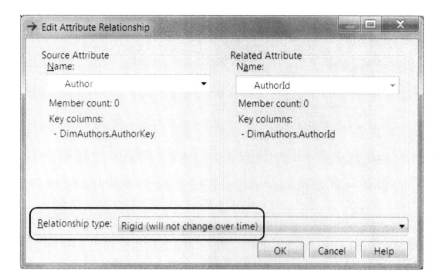

Figure 10-22. The Author-AuthorId relationship

■ **TIP** In our example, it looks quite easy to set up the attribute relationship, and in fact it is. But if you are new to working with this interface, you may have troubles at first. This is perfectly normal because the interface is somewhat unintuitive. For instance, if you drag and drop the Author attribute to the AuthorState attribute, it will form the relationship between them without having to configure it using the dialog window (Figure 10-22). But, doing so may set AuthorState as a sibling attribute instead of a parent-child one. This inadvertent misunderstanding between you and Visual Studio has the potential to cause frustration. If this happens, just delete the relationship in the lower-right side of the attribute relationship window and try again. With practice, you will find that this is easier than it appears.

Testing Your Progress

In Chapter 9, we showed you how to build a Visual Studio project using the Build menu item. While building your project, any errors in the dimension will be reported, so you will know immediately when one of your configurations has caused an error, and you can take steps to correct the problem.

Build the project by clicking Visual Studio's Build menu item and selecting Build PubsBICubes (Figure 10-23).

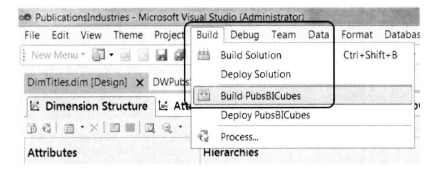

Figure 10-23. *Building the project*

After you build the project, you will want to browse the dimensions to verify that the design is correct, but we still have a few steps to accomplish before we can do that. For now let's concentrate on getting the attribute relationships configured in the next exercise.

EXERCISE 10-2: CONFIGURING ATTRIBUTE RELATIONSHIPS

In this exercise, you will configure the attribute relationships for each of the dimensions for the SSAS project. Let's start with the simplest ones and move onto the more complex as we go.

Note: If the solution from the previous exercise is closed, open Visual Studio by clicking the Start button and navigating to All Programs ➤ Visual Studio 2010. Right-click the menu item, select Run as Administrator, and then answer Yes to close the UAC.

Configure the DimAuthors Dimension

Since we just talked about it, let's start with the Authors dimension.

1. Open the DimAuthors dimension editor by right-clicking the dimension in Solution Explorer and selecting View Designer from the context menu.

2. When the dimension editor opens, access the Attribute Relationships tab (Figure 10-19).

3. Verify that Author to AuthorId is in a sibling relationship and that Author to AuthorState is in a parent-child relationship, as indicated in Figure 10-19. If not, delete the relationships from the Attribute Relationships pane and re-create them, as shown in Figure 10-20.

4. Set the Author to AuthorId relationship as Rigid and the Author to AuthorState relationship as Flexible by right-clicking each relationship and selecting the relationship types from the context menu (Figure 10-24).

Figure 10-24. *The DimStores attribute relationships*

5. Use the Build menu to verify that the dimension builds successfully (Figure 10-23). If it does not, troubleshoot the problem or delete the dimension from Solution Explorer and go through the steps again, being careful to include any steps you may have omitted.

Configure the Stores Dimension

Now for the Stores dimension. It is a very simple dimension and should already be in the configuration we want. Still, it is a good idea to verify the settings, so let's do that now.

1. Open the DimStores dimension editor by right-clicking the dimension in Solution Explorer and selecting View Designer from the context menu.

2. When the dimension editor opens, access the Attribute Relationship tab and verify that there is a sibling relationship between Store and StoreId as shown in Figure 10-25.

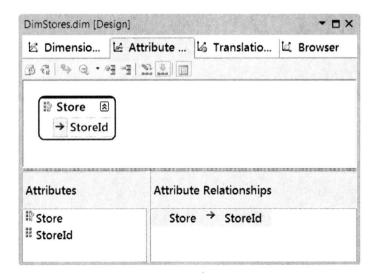

Figure 10-25. *The DimStores attribute relationships*

3. Set the Store to StoreId relationship as Rigid by right-clicking the relationship and selecting that relationship type from the context menu.

4. Use the Build menu to verify that the dimension builds successfully (Figure 10-23). If it does not, troubleshoot the problem or delete the dimension from Solution Explorer and go through the steps again, being careful to include any steps you may have omitted.

Configure the Orders Dimension

Now we turn to the Orders dimension. It is even simpler than the DimStore dimension and should not require any configuration, because there is only one attribute in this dimension. Therefore, there cannot be any attribute relationships. Your goal is to review the current design.

1. Open the DimOrders dimension editor by right-clicking the dimension in Solution Explorer and selecting View Designer from the context menu.

2. When the dimension editor opens, access the Attribute Relationship tab and verify that it looks as shown in Figure 10-26.

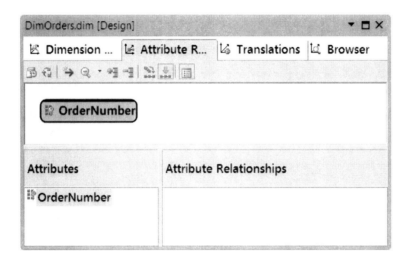

Figure 10-26. *DimOrders Attribute Relationships box is empty*

Configure the Dates Dimension

The Dates dimension has a number of attributes that need to be related correctly. Be sure to review it closely and make any changes necessary.

1. Open the DimDates dimension editor by right-clicking the dimension in Solution Explorer and selecting View Designer from the context menu.

2. When the dimension editor opens, access the Attribute Relationship tab.

3. Verify that Date to FullDate is in a sibling relationship (Figure 10-27).

4. Verify that Date to Month is in a parent-child relationship. If not, delete the relationship from the Attribute Relationships pane and re-create it, as shown in Figure 10-27.

5. Verify that Month to Quarter is in a parent-child relationship. If not, delete the relationship from the Attribute Relationships pane and re-create it, as shown in Figure 10-27.

6. Verify that Quarter to Year is in a parent-child relationship. If not, delete the relationships from the Attribute Relationships pane and re-create it, as shown in Figure 10-27.

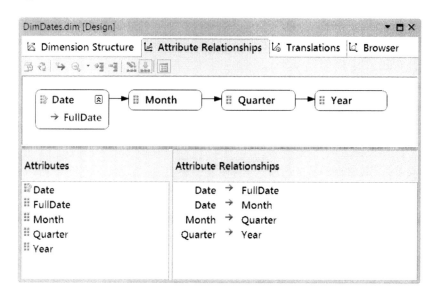

Figure 10-27. *The DimDates attribute relationships*

7. Set all of the relationships as Rigid by right-clicking each relationship and selecting that relationship type from the context menu.

8. Verify that the Attribute Relationships tab looks like Figure 10-27.

9. Use the Build menu to verify that the dimension builds successfully (Figure 10-23). If it does not, troubleshoot the problem or delete the dimension from Solution Explorer and go through the steps again, being careful to include any steps you may have omitted.

Configure the Titles Dimension

Next we configure the Titles dimension. As in the previous exercise, this is the one that is most complex. Carefully review each relationship. Remember that the relationships will automatically be correctly mapped if you remove and reconfigure them. So, all you need to do is replace them if they are currently incorrect.

1. Open the DimTitles dimension editor by right-clicking the dimension in Solution Explorer and selecting View Designer from the context menu.

2. When the dimension editor opens, access the Attribute Relationship tab.

3. Verify that Title to TitleId is in a sibling relationship (Figure 10-28).

4. Verify that Title to TitlePrice is in a sibling relationship (Figure 10-28).

Now let's work on the published dates attributes.

5. Verify that Title to PublishedDate is in a parent-child relationship. If not, delete the relationship from the Attribute Relationships pane and re-create it, as shown in Figure 10-28.

6. Verify that PublishedDate to PublishedFullDate is in a sibling relationship. If not, delete the relationship from the Attribute Relationships pane and re-create it, as shown in Figure 10-28. This will automatically remap the relationship correctly.

7. Verify that PublishedDate to PublishedMonth is in a parent-child relationship. If not, delete the relationship from the Attribute Relationships pane and re-create it, as shown in Figure 10-28.

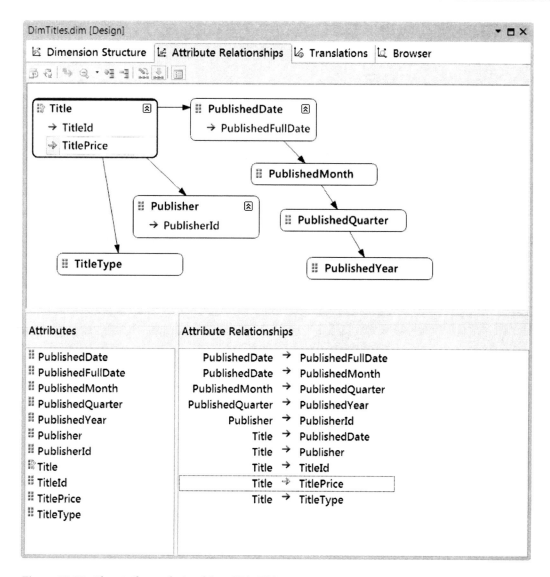

Figure 10-28. *The attribute relationships of DimTitles*

8. Verify that PublishedMonth to PublishedYear is in a parent-child relationship. If not, delete the relationship from the Attribute Relationships pane and re-create it, as shown in Figure 10-28.

With the date attributes mapped out, let's finish up with the TitleType, Publisher, and PublisherId attributes.

9. Verify that Title to Publisher is in a parent-child relationship. If not, delete the relationship from the Attribute Relationships pane and re-create it, as shown in Figure 10-28.

10. Verify that Publisher to PublisherId is in a sibling relationship. If not, delete the relationship from the Attribute Relationships pane and re-create it, as shown in Figure 10-28. This will automatically remap the relationship correctly.

11. Verify that Title to TitleType is in a parent-child relationship. If not, delete the relationship from the Attribute Relationships pane and re-create it, as shown in Figure 10-28.

12. Use the Build menu to verify that the dimension builds successfully (Figure 10-23). If it does not, troubleshoot the problem or delete the dimension from Solution Explorer and go through the steps again, being careful to include any steps you may have omitted.

In this exercise, you configured the attribute relationships for all of the dimensions. At this point, the dimensions are configured with the common settings that you will use in all of your dimensions. We still need to talk about the Translations tab and the Browser tab before we move on to cubes, so let's talk about them next.

Translations Tab

The Translations tab allows you to apply optional captions for the various attributes. The concept is that you can include a different caption for each natural language. This does not change the data in your data warehouse. Instead, it provides you with a different label for each attribute associated with a particular language.

To add a new caption under a given language, click the New Translation button, as highlighted with its associated tooltip in Figure 10-29.

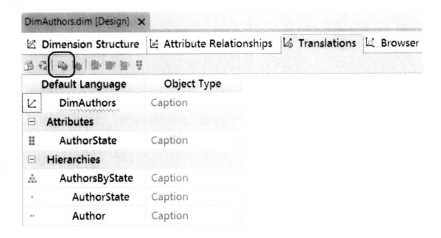

Figure 10-29. *Adding a new translation on the Translations tab*

■ **Note** Although this is a nice feature, we will skip over it to simplify an already complex chapter. If languages are important to your implementation, this is a simple option to configure, and information on the subject is easy to find on the Internet.

Browser Tab

Unlike the other tabs, the Browser tab does not provide you with much to do in its initial state. This is because—unlike the other tabs—the Browser tab does not use the XML code that is created by the Dimension Wizard. Instead, it displays information directly from the SSAS server.

When you first create a new dimension, the dimensions definition is only within the XML code and is not yet uploaded to the SSAS server. In this stage, the Browser tab will display nothing but a couple of hyperlinks, as shown in Figure 10-30.

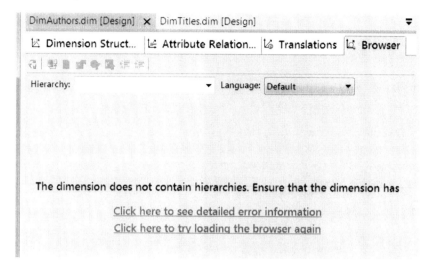

Figure 10-30. *The Browser tab in its initial state*

Build, Deploy, and Process

To utilize the Browser tab, you must first deploy the XML code to the SSAS and then process the data from the data warehouse into the SSAS folder. Afterward, you will be able to use the Browser tab to verify your dimensional design. There is actually quite a bit to understand about building, deploying, and processing, so let's take a moment to drill into all three topics.

Building

We mentioned a number of times that Visual Studio uses XML files for data sources, data source views, dimensions, and cubes, and there is a hidden master file that contains all of this code. The act of building of an SSAS project checks the validity of these files and combines them into this master file. This master file is found in the binaries folder of your SSAS project. In our example, that folder is at this location: `C:\_BISolutions\PublicationsIndustries\PubsBICubes\bin`.

If you open this folder before you build your project, it will be empty. After you build your project, you will find four files within the folder. Three of these files are configuration files for the deployment process, but one of them is the master file. You can tell the master file because of its extension `.asdatabase` with the projects name prepended. Figure 10-31 shows the path and files of this folder.

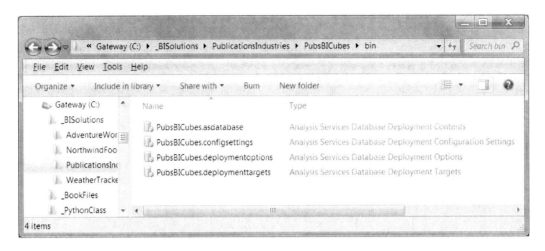

Figure 10-31. The bin folder of the PubsBICubes project

Typically, there is no need to access this folder. We have shown you the folder so that you can have a clear understanding what the build process entails. One of the few times you might use this folder is to copy the .asdatabase file and send it to a co-worker, because this one file contains all the code needed to create all the objects in an SSAS project.

Microsoft has anticipated this use and provides a wizard that walks you through utilizing an .asdatabase file to deploy an SSAS project to an SSAS server.

The wizard is found under the Start menu ➤ Microsoft SQL Server 2012 ➤ Analysis Server menu folder. The first couple screens of this wizard are shown in Figure 10-32. As you can see, the wizard allows you to specify the name of an SSAS server and database to deploy to.

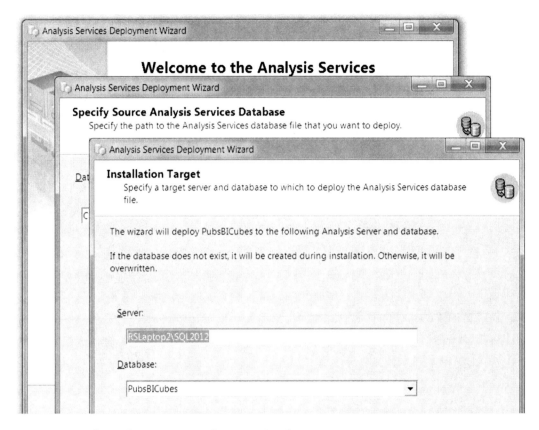

Figure 10-32. *The Analysis Services Deployment Wizard*

One advantage of the wizard is that you do not need to send your entire Visual Studio project in order for it to be deployed. This way, the SSAS database and its objects can be created on another server without having to know the ins and outs of Visual Studio. Additionally, it prevents other developers from making inconsistent changes to your project.

Deploying your project files is normally accomplished using Visual Studio without the Analysis Services Deployment Wizard. This is convenient because it saves a developer from launching an additional piece of software to deploy a project.

Deploying

The SSAS bin folder contains three configuration files (Figure 10-31). These files revolve around deployment options. They are XML files and can be configured using a text editor like Notepad, but a better way to do so is by accessing the SSAS project property pages, as shown in Figure 10-33.

To access the property pages, right-click the SSAS project icon in Solution Explorer. In our example, that is the PubsBICubes project icon. Select Properties in the context menu to display the Configuration Properties dialog window (Figure 10-33).

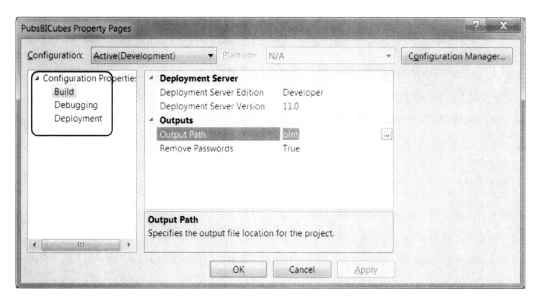

Figure 10-33. *Setting the SSAS project deployment properties*

The Configuration Properties dialog window is organized into three configuration pages: Build, Debugging, and Deployment.

The Build and Debugging Pages

On the Build page you can specify items such as which folder the build process will place the `.asdatabase` file in. On the Debugging page, you can choose which object will be validated first as the build process starts. In other words, if you try to build the entire project and you want to validate the DimTitles dimension first, you can select that object on the Debugging page.

Typically, neither one of these pages needs to be touched, but the third Deployment page bears looking into.

The Deployment Page

On the Deployment page, shown in Figure 10-34, you can define important options such as the following:

- To automatically perform processing (as soon as the build has completed)

- To incrementally apply SSAS objects

- To perform a complete refresh

We recommend setting Processing Option to Do Not Process in the dropdown box (Figure 10-33). Changing this option gives you more control over when processing will occur. This option is important because processing can take a long time when it copies large amounts of data from the data warehouse to the SSAS folders. When left at the default setting, the behavior is for Visual Studio to request to incrementally process data directly after a deployment has been successful.

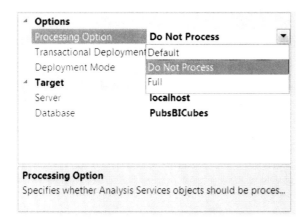

Figure 10-34. *Setting the SSAS project deployment properties*

The Deployment Mode option is configured in a similar manner. This option determines whether the XML code that is deployed will modify your existing SSAS database or whether a complete refresh of code should be performed on the database instead.

Occasionally doing an incremental refresh (which is the default) will not correctly update your SSAS database with the appropriate changes. Switching the deployment mode to perform a full refresh of the code resolves this issue. You can force the full refresh by setting the Deployment Mode dropdown box to Deploy All.

The Server and Database settings allow you to define which SSAS server and database you want to deploy to. By default Visual Studio assumes you want to connect to your local SSAS server. But this may not be the case if you are working with a developer or production server at your office or if you are working with a named instance on your personal computer.

The Database setting represents the name of the database that will be created on the SSAS target server. By default the name of the database will be the same as your project in Visual Studio, but this can be changed if you so desire.

In Figure 10-35 you can see the changes that were made on Randal's computer before he deployed the PubsBICubes project. Since he is using a named instance of SSAS, he has included the instance name.

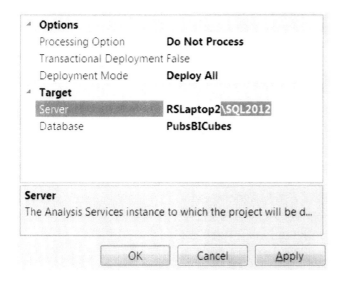

Figure 10-35. *Common settings changed in the SSAS project deployment properties*

Deployment Target Folder

When Visual Studio's XML code is uploaded to the SSAS server, all objects designed in your project will be implemented in a folder based on the location defined in the properties of your SSAS server. These properties are accessed using SQL Server Management Studio, not Visual Studio.

In a production environment you might want to change the location of the folders to a drive that would have the appropriate amount of free space available (or is configured as a RAID array of drives). Figure 10-36 shows an example of this property page and the Data Directory property's default setting.

■ **Note**   For our examples, we do not need to configure these options and will leave the default settings alone.

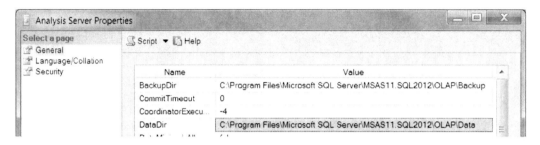

Figure 10-36. *The data directory configuration in SQL Server Management Studio*

If you navigate to the folder indicated in this property, you will see that there are a number of files already within it before you even deploy an SSAS database. After you deploy, the cube, dimensions, and other supporting files will be created as subfolders of the SSAS data folder. As an example, Figure 10-37 shows the contents the SSAS data folder after the PubsBICubes project is deployed.

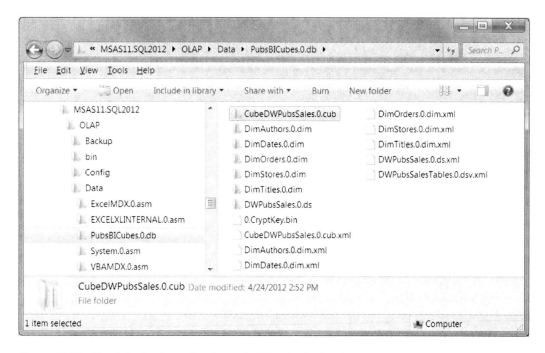

Figure 10-37. *The data directory after the project files are deployed*

Deploying from Visual Studio

To deploy your project, use the Visual Studio Build menu and access the Deploy menu option beneath it (Figure 10-38). The Deploy menu option can also be accessed by right-clicking the project icon under Solution Explorer.

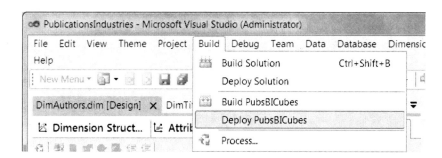

Figure 10-38. *Starting the deployment from the Build menu*

The Build menu has four common options, but it is context-sensitive, so you may have more or less options depending on your settings in Visual Studio. The four common settings are Build Solution, Deploy Solution, Build Project, and Deploy Project (Figure 10-38).

The Build Solution option allows you to build one or more SSAS projects within a single Visual Studio solution. While this may be occasionally useful, the majority of the time you will have only a single SSAS project within a Visual Studio solution.

The Deploy Solution option is much the same in that it allows you to deploy one or more SSAS projects in a single Visual Studio solution.

The Build PubsBICubes option, shown in Figure 10-38, allows you to build all the XML files within that particular SSAS project into the master .asdatabase file. And, the Deploy PubsBICubes option allows you to deploy this master file to your analysis server computer

Selecting the Deploy PubsBICubes option will automatically save any open XML files, build the master file, and attempt to deploy that project. For this reason, you do not have to build (or even save any files) before you deploy.

In Figure 10-38, you also see a fifth option called Process. This will automatically perform all actions necessary before processing can begin, including building and deploying both cubes and dimensions. However, we prefer building and deploying manually most of the time. Doing so forces error messages be displayed in an Error List window, where we can troubleshoot problems with our cube and dimension before we actually process them.

The Deployment Process Window

Once all the dimensions build successfully, the deployment phase should provide you with a successful message in the Deployment Progress window. If for some reason it does not display, you can force it to do so by accessing Visual Studio's View menu and selecting the Show Deployment Progress option (Figure 10-39).

Once deployment succeeds, a message displays at the bottom of the Deployment Progress window indicating success (Figure 10-39).

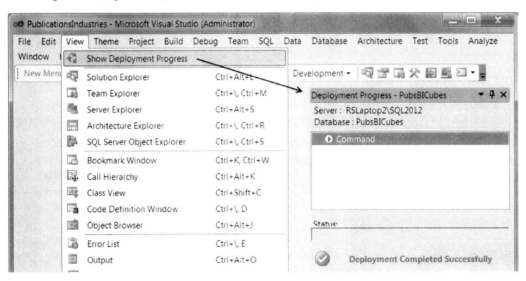

Figure 10-39. *Displaying the Deployment Progress window*

■ **Note** During the deployment phase you may be asked to retype your password. This is dependent upon whether you have made changes to the data source recently, reopened the project from a previous development cycle, or moved to project files to another computer. If the Account Password window displays, type in your password and click the OK button.

At this point we have a folder for the SSAS database and subfolders for each dimension, but there is no data within the folder. Next, we will process the project to fill the folders with data, after which we are able to browse the dimensions to verify that they are designed correctly.

WHEN DEPLOYMENT FAILS

If you work with SSAS long enough, you will at some point have a deployment that fails. Usually this can be resolved fairly quickly by checking a few common settings to resolving any issues. Figure 10-40 is an example of a typical error message during the deployment phase.

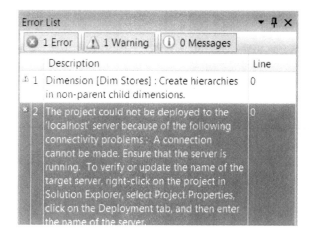

Figure 10-40. *Localhost server was not the correct server name!*

This could mean that you cannot connect because the server is not running or that your network connection has failed, but more likely, it is because the server name has been incorrectly typed into the project's deployment pages (Figure 10-35). Verify the spelling of the server. Be sure to check that you have indicated a named instance (if needed) or that you have the slash going the correct way (\). Once you find an error and correct it, try deploying again.

If you verify that the server name is typed correctly and you still have problems, the second most likely cause is an issue with the impersonation information. Unfortunately, the error message that you receive from this will not be clear, so it's a good idea to check these setting if you continue to have problems.

To access the impersonation information, navigate to Solution Explorer, and select the data source file (in our example the file name is DWPubsSales.ds). Once selected, access the designer dialog window by right-clicking the file and selecting View Designer from the context menu, and it will display the Data Source Designer dialog window. This window has two tabs, but the one you want is the Impersonation Information tab (Figure 10-41).

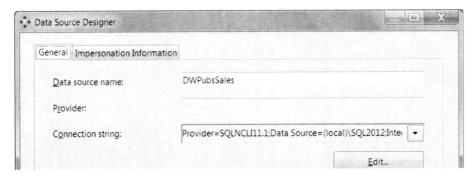

Figure 10-41. *The Data Source Designer dialog window*

Once you have access this tab, check the spelling of your username and password (Figure 10-42). One easy way to check password is to type it within the username dialog box textbox so that it is visible and then copy and paste it into the password text box.

Figure 10-42. *The Impersonation Information tab*

If you are using a domain account, make sure you typed in the domain name before the backslash symbol. For example, if you use a company laptop and the daily login that you use is part of the MyCompany domain, then type in something like **MyCompany\MyName**. (Once again, be careful that the slash is going the correct way since that is the most common mistake we see in this textbox.)

Clearly, there can be more arcane reasons why you cannot deploy, but these are by far the most common issues we encounter on a regular basis.

Processing

Processing the dimensions takes place by selecting the project and clicking Process in the context menu (Figure 10-43). This will process all of the dimensions in the project.

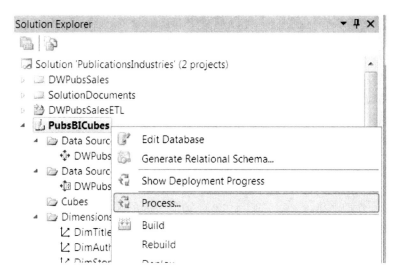

Figure 10-43. *Processing the project's cubes and dimensions*

Occasionally it is helpful to process a dimension individually, such as when you have changed settings in only a single dimension and you do not need to process all dimensions within the project. Processing an individual dimension is also useful when you are getting an error message during the processing phase and you want to isolate which dimensions are causing the problem. Sometimes the error messages do not clearly indicate the source of the issue. But, you may be able to uncover the problem by processing the dimensions individually.

To process a single dimension, right-click the dimension file (such as DimStores.dim) and select Process from the context menu.

■ **Note**   You can exclude a dimension file from your project by right-clicking the project and selecting "Exclude from project." This does not delete the file. Instead, it will no longer be included in the building, deploying, or processing of a project. If you want to add it back to you project later, right-click the Visual Studio SSAS project icon and select Add ➤ Existing Item from the context menu. This can be a great way to troubleshoot issues that may involve a particular dimension.

Clicking Run at the bottom of this window will start the processing and launch the Process Database dialog window. A list of the objects to be processed will be visible at the top of the window. In Figure 10-44, we are processing the PubsBICubes database object. When we do so, all the dimensions will be processed, because all the dimensions are inside of this database.

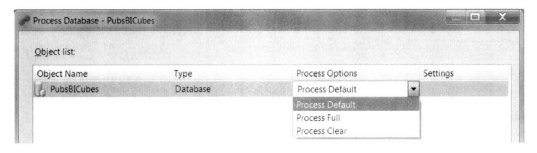

Figure 10-44. *Setting the processing options*

The Processing Options dropdown box is an important configuration within this window. In the following list, we discuss each of its settings:

- *Process Default* : Processes only objects that require processing. For example, if you have successfully processed an individual dimension and then try to process the project as a whole, the successfully processed dimension will be skipped.

- *Process Full* : Allows you to override all the existing data with fresh data. This is useful when there may be corruption within your SSAS database that can be resolved with a complete data reload. An example is when one dimension is out of sync with the others and the error message is not self-explanatory. If processing fails at the project level and the reasons are not obvious, first process the dimensions individually and then process the entire project using the Process Full option. This often clears up the issue.

- *Process Clear:* Cleans out the data within the SSAS database. This can be used to create a backup of the SSAS database without any data in it.

Process Default is likely all you need, especially in a production environment where processing can take hours. You would not use the Process Full option unless troubleshooting was necessary.

Once you have selected an option and clicked Run at the bottom of the Process Dialog window, an additional dialog window called Process Progress will appear, and the processing will truly begin (Figure 10-45).

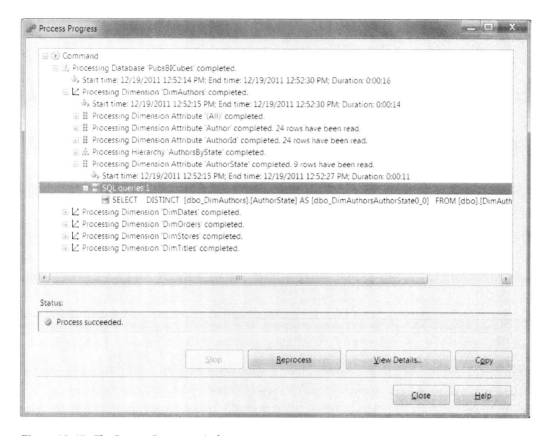

Figure 10-45. *The Process Progress window*

After processing has successfully completed a "Process succeeded" status message will display toward the bottom of the dialog window. You will also see processing report information similar to that shown in Figure 10-45. Expanding the treeview displays individual details about the processing, such as which select statements were used to copy the data from the data warehouse. This is good to understand, because it can be used to troubleshoot issues where the data in your dimensions and cubes are out of sync with the data in your data warehouse.

Browsing the Dimension

Once you have successfully deployed and processed your dimensions, you can use the Browser tab to review what you have created. Figure 10-46 displays data in the DimAuthors dimension from the Browser tab.

Let's discuss the three buttons at the top of the Browser tab from left to right, as described by their appearance:

- *The circle of green arrows button*: This button looks like a recycle symbol. Its function is to start processing that dimension. Obviously, if you have just processed this dimension, you do not need to do so again.

Figure 10-46. *Browsing a processed dimension*

- *The computer reconnection button*: This button is important because it allows you to reconnect to your analysis server computer. Remember that when you browse the dimension, you are browsing on the server and not from the XML files within Visual Studio. This is why you have to deploy and process the dimension before it can be browsed. The Browser tab caches the results of your dimensional query and may not show you the actual contents on the SSAS server unless you either refresh or reconnect.

- *The page refresh button*: This button is your standard refresh button. Its function is similar to the reconnect button, but in our experience, the reconnect button is much more likely to give you a clean refresh than this page refresh button. Therefore, we recommend clicking the middle button whenever you deploy a processor dimension.

Note the Hierarchy dropdown box. Remember that each attribute forms its own attribute hierarchy, so unless you have hidden that attribute hierarchy, you will see it listed within the dropdown box. Because we hid the Author attribute, in Exercise 10-1, it does not display in the dropdown box of Figure 10-46. You will, however, see any user-defined hierarchies that you may have created. Selecting these hierarchies allows you to browse them in a treeview just like you would a simple attribute hierarchy. And although we cannot browse the Author attribute by itself, we can see its data using the AuthorsByState user-defined hierarchy.

As the developer, you will test each dimension using a browser to verify its construction. If you find anything that looks incorrect, you fix it by accessing the appropriate dimension designer tab, make any changes, then redeploy and process. Afterward, use the reconnect (center) button to see the changes you have made.

Now that we have discussed how to deploy and process dimensions, let's do so to the dimensions we have created so far in the next exercise.

EXERCISE 10-3: DEPLOYING, PROCESSING, AND BROWSING THE DIMENSIONS

In this exercise, you will deploy, process, and browse the dimensions in your SSAS project. Let's start by configuring the project deployment settings.

Configure Your Project Deployment Settings

Before we can deploy the project we need to configure the deployment options.

1. Right-click the PubsBICubes project icon in Solution Explorer and select Show Properties from the context menu.

2. When the Properties Pages dialog window appears, navigate to the Deployment page, select the Processing Options dropdown box (Figure 10-34), and select Do Not Process.

3. Verify the server is configured to point to your SSAS server (Figure 10-35). Use (local), localhost, or your actual computer name; if you are using a named instance, type in its proper name as well.

4. Click OK to close the Properties Pages dialog window.

Build Your Project

When the configuration is verified, or reconfigured as needed, it is time for one last build to make sure that all is ready for deployment.

1. Build the project one more time and troubleshoot any errors that may occur using the Error List window (Figure 10-38).

Tip: At this point there should not be any errors, because you have been using the Build option to test your configurations over and over again. If you do find an error and the problem is not apparent, isolate each dimension to determine which one is giving you the problem by excluding a dimension from the project and building the project again. If it builds successfully without the excluded dimension, you will know that the excluded dimension is the source of the issue.

Once you determine the problematic dimension, you can add the excluded dimension back into your project by right-clicking the SSAS project icon and selecting Add Existing Items from the context menu. Now you can focus your attention on the problematic dimension to correct the issue. Worst-case scenario: delete the dimension and remake it again. This sometimes is all it takes to get it working.

Deploy Your Project

With the build successful, we can now deploy or dimensions to the SSAS server.

1. Deploy the project using Visual Studio's Build menu (Figure 10-38). If you are asked for a password, please supply the password.

2. Troubleshoot any errors that may occur using the Error List window (Figure 10-40). Refer to the "When Deployment Fails" topic within this chapter for troubleshooting tips.

3. Verify that the deployment has succeeded using the Deployment Processes window. If the window is not showing, you can access this by using the View ➤ Show Deployment Progress menu item, as shown in Figure 10-39.

Process Your Project

The dimensions are designed, built, and deployed, but they do not have data until they are processed. We will do that now.

1. Right-click the PubsBICubes project icon in Solution Explorer and select Process from the context menu, as shown in Figure 10-43.

2. When the Process dialog window appears, verify that the Processing Options dropdown box is set to Process Full, and click Run at the bottom of the dialog window.

3. When processing succeeds, click Close to close the Process Progress window, as shown in Figure 10-44.

Browse Your Dimensions

After the dimensions are processed, you can test your design using the dimension browser.

1. Access the dimension designer Browser tab by right-clicking the DimAuthors dimension in Solution Explorer and selecting Browse from the context menu.

2. Click the Reconnection button on the Browser tab's toolbar, as shown in Figure 10-45.

3. Use the treeview and the Hierarchy dropdown box to browse your dimension attributes and hierarchies (Figure 10-45).

4. Browse all of the other dimensions and check for any errors or inconsistencies.

In this exercise, you deployed, processed, and browsed the dimensions that you created earlier. It is now time to create a cube so that these dimensions can be associated with our measures. Let's discuss how that can be done in the next chapter.

Moving On

We have now completed the dimensions, but we still need to create cubes that use them! In our next chapter, we create a cube, show how cubes and dimensions are interconnected, and discuss basic cube configurations.

LEARN BY DOING

For this "Learn by Doing" exercise, configure dimensions similar to the ones defined in this chapter using the Northwind database. We have included an outline of the steps you performed in this chapter and an example of how the authors handled them in two Word documents. These documents are found in the folder C:_BookFiles_LearnByDoing\Chapter10Files. Please see the ReadMe.doc for detailed instructions.

What's Next?

At this point, you now have a good idea of how dimensions are created and configured using Visual Studio. But there is still so much more to tell. We are going to address some of these items in Chapter 11, but for more detailed information on this subject, we have found the following book to be quite useful. And although it is an older book, most of the details still apply to all versions of SSAS from SQL 2005 and on.

Expert Cube Development with Microsoft SQL Server 2008 Analysis Services]
By Marco Russo, Alberto Ferrari, Chris Webb
Publisher: Packt Publishing
ISBN-10: 1847197221

Creating and Configuring SSAS Cubes

Once I completed the cube and demonstrated it to my students, I realized it was nearly impossible to put down.

—Erno Rubik

In the previous two chapters, we prepared dimensions for our cube (or cubes, if you are creating a BI solution that uses more than one). Now it is time to make the cube and find out whether the cube and dimensions are presenting the right report data. We then resolve any issues using common SSAS configurations.

In this chapter, you learn how to create and configure Analysis Server cubes. We show you how to configure the cubes using Business Intelligence Development Studio and how to manage these objects using SQL Server Management Studio. Let's get going and create our cube now!

■ **Tip**　As you will see in this chapter, the process of creating dimensions and cubes is similar. Neither will ever interact directly with the data warehouse; instead, they interact with an SSAS Data Source View as an abstraction layer. Microsoft refers to this design as the *unified dimensional model (UDM).* This terminology can be confusing, because Microsoft also refers to SSAS cubes as UDMs as well, and uses the terms interchangeably.

Creating Cubes

We begin creating our cubes using the dimensions that we recently completed and tested. And surprise, surprise—this is performed using a wizard. Even if you are not a wizard fan, the Cube Wizard is short and to the point and does not limit your control over your cube later, so we consider it a useful tool.

To start the wizard, right-click the Cubes folder and select New Cube from the context menu (Figure 11-1). When the wizard opens, click Next to navigate from the Welcome dialog window to the Select Creation Method dialog window.

- ▲ ⬛ **PubsBICubes**
  - ▲ 🗁 Data Sources
    - ✥ DWPubsSales.ds
  - ▲ 🗁 Data Source Views
    - ▦ DWPubsSalesTables.dsv
  - 🗁 Cubes
  - ▲ 🗁 Dimens | 🗔 New Cube...
    - ⎿ Dim | 📋 Paste   Ctrl+V
    - ⎿ Dim

Figure 11-1. *Starting the Cube Wizard*

In the Select Creation Method dialog window, typically you would leave the "Use existing tables" radio button checked and click the Next button to continue (Figure 11-2). But, other options may occasionally be useful.

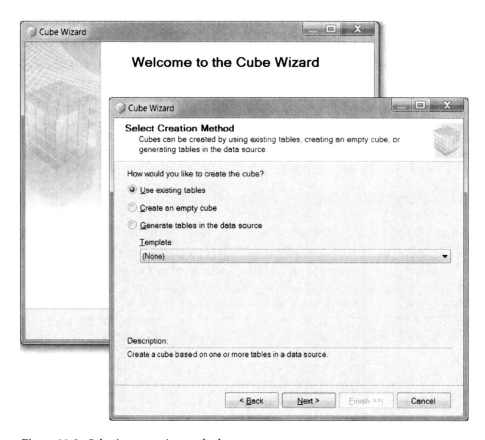

Figure 11-2. *Selecting a creation method*

The second radio button, "Create an empty cube," allows you to create an empty cube that you can add to later. This option bypasses the remainder of the wizard.

The third radio button, "Generate tables in the data source," allows you to create virtual tables in the data source view. These tables can be scripted into SQL code that can then be run on your data warehouse to create the real relational tables. This is an interesting option because it allows you to start with a blank data warehouse and creates your cube first, designing your measures and your dimensions on top of the blank data warehouse. After you have finished configuring your cube and dimensions in the SSAS project, you can have Visual Studio generate an entire SQL script that will create your data warehouse tables.

The third option, "Generate tables in the data source," includes a dropdown box that allows you to use predefined templates for your cube and dimensions. These templates are just the XML files that came from another Visual Studio SSAS project. You can make one of these templates yourself by making an SSIS project, configuring it as you like, and then copying all the files and folders for the project to a predefined folder that comes with your SQL installation. Your path may vary, but the standard installation places the folder at this location: C:\Program Files (x86)\Microsoft SQL Server\110\Tools\Templates\olap\1033\Cube Templates.

The basic concept behind this template option is that a consultant can make a predefined cube and dimension structure that the client's company recommends. The consultant would then bring it to the client's site, modify the cube and dimensions to the client's needs, and then reverse engineer the data warehouse to fit the cubes and dimensions in the SSAS project. Afterward, the developer needs to set up an ETL process to fill the new data warehouse, but this is just one more way to implement a BI project.

After you select the creation method, click Next to advance the wizard to the Select Measure Group Tables dialog window. This window allows you to choose which fact tables will be included in the cube (Figure 11-3).

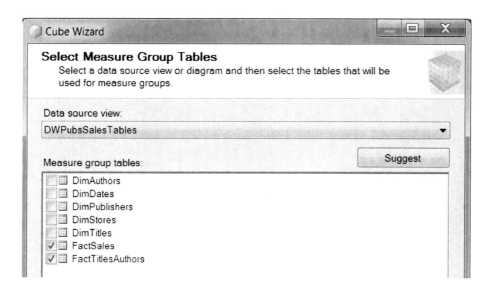

Figure 11-3. *Selecting the measure group tables*

In our example, we have two fact tables: one that contains the measures and one that is used as a bridge between the Authors many-to-many dimension. As you can see in Figure 11-3, we select both tables, because both are necessary if we want to use all of the dimensions we created.

Clicking Next moves you to the Select Measures dialog window shown in Figure 11-4. You need to decide which measures are appropriate to use. In our example, we leave them all checked.

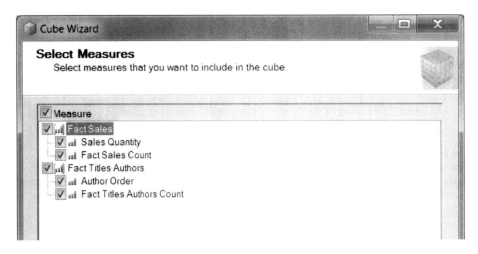

Figure 11-4. *Selecting the measures to use in the cube*

The wizard suggests columns and measures to use by comparing the columns' data types. For example, Author Order is recommended as a measure because it is a fact table and it has a numeric value. This simple criteria is how the wizard decides what should and should not be included.

As you can see in Figure 11-4, Fact Sales Count and Fact Titles Authors Count have been added. The wizard suggests these additional measures to provide row counts for Fact Sales and Fact Titles Authors for your convenience, in case you need totals for each. The wizard labels these as Count measures for you, but you can change the names if you prefer.

Count rows can always be deleted later. And, because measures can be added or subtracted after the wizard is complete, keeping the default settings as is for this dialog window will not be an issue.

Clicking Next moves you to the Select Existing Dimensions dialog window, as shown in Figure 11-5. In this dialog window, you can select one or more of the dimensions you have already created. If no dimensions have been created, the wizard skips over this dialog window to another that allows you to create simple dimensions right from the Cube Wizard. (That particular window will not appear in our example, however, because we already created the dimensions.)

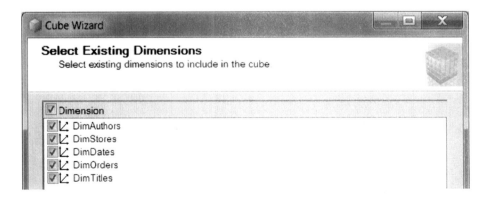

Figure 11-5. *Selecting existing dimensions*

After the dimensions have been selected, clicking Next usually brings you to the end of the wizard, as shown in Figure 11-6 (that is, unless the wizard senses that there could be additional dimensions created based on the relationship lines in the data source view). In other words, if you created some of the dimensions, let's say only four of the five dimension tables, the Cube Wizard will see that there is a missing dimension and offer to create it for you in the Select New Dimensions dialog window (Figure11-6).

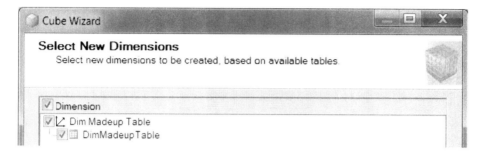

Figure 11-6. *The optional Select New Dimensions dialog window*

THE CUBE WIZARD CAN ALSO CREATE DIMENSIONS

The Cube Wizard allows you to create dimensions while you create the cube. This means that when you create a Visual Studio SSAS project, you do not necessarily have to start creating the dimensions before you create the cube. Instead, you can go right to creating the cube and have the wizard generate the basic outline of your dimensions for you using the Select New Dimensions dialog window, as shown in Figure 11-6. This option does not provide you with any additional advantage over creating your dimensions first. Therefore, we have found that using it is a somewhat redundant feature.

When you create all of the dimensions before launching the Cube Wizard, you immediately proceed to the final dialog window of the wizard from the Select Existing Dimensions dialog window. In this window, you see an overview of your design and have a chance to rename the cube to something appropriate. In our example in Figure 11-7, we named it DWPubsSalesVer1.

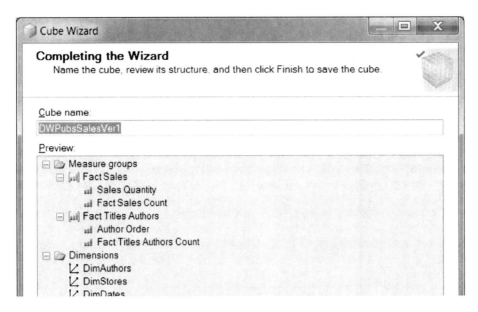

Figure 11-7. *Completing the Cube Wizard*

After you have finished the Cube Wizard, Visual Studio displays the cube designer window, as shown in Figure 11-8. From here you can configure the cube using the various tabs that are included in the designer.

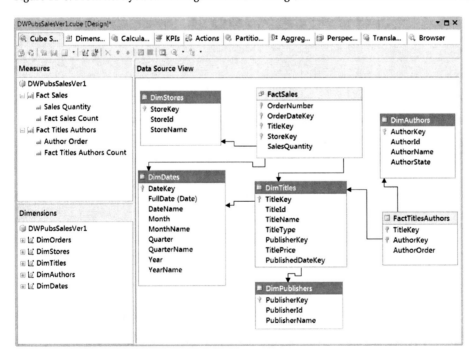

Figure 11-8. *The cube designer window*

We discuss how to use the designer in detail in just a bit. But, for now, let's create the cube for our project in the following exercise.

EXERCISE 11-1. CREATING A CUBE

In this exercise, you create a basic cube using the Cube Wizard.

Note: Open Visual Studio by clicking the Start button and navigating to All Programs ➤ Visual Studio 2010. Right-click this menu item, select Run as Administrator, and then answer Yes to close the UAC. If you did not complete the project in the previous chapter, you can find our completed version in the C:_BookFiles\Chapter10Files\Chapter10_Complete\PublicationsIndustries folder.

1. Right-click the Cubes folder in Solution Explorer, and select New Cube from the context menu. The Cube Wizard launches, as shown in Figure 11-1.

2. Click Next to advance to the Select Creation Method dialog window, and verify that the "Use existing tables" radio button is checked, as shown in Figure 11-2.

3. Click Next to advance to the Select Measure Groups Tables dialog window. Select the Fact Sales and FactTitleAuthors check boxes, as shown in Figure 11-3.

4. Click Next to advance to the Select Measures dialog window. Verify that all measures are selected, as shown in Figure 11-4.

5. Click Next to advance to the Select Existing Dimensions dialog window. Verify that all five of your existing dimensions are selected, as shown in Figure 11-5.

6. Click next to advance to the Completing the Wizard dialog window, and rename the cube to **DWPubsSalesVer1** (Figure 11-7).

7. Click Finish to exit the wizard.

In this exercise, you created a basic cube. Next, we review the cube and make any required configurations. We begin by processing the cube and examining its contents. Let's delve into this topic now!

Processing the Cube

Like dimensions, you must build, deploy, and process a cube before it can be browsed. Also like dimensions, the act of building adds the XML code you have just created using the wizard to the master .asdatabase XML file. Deploying uploads this master XML file to the SSAS server, and processing copies data from the fact tables in your data warehouse to the SSAS folders.

The easiest way to process your cube is to right-click the cube in Solution Explorer and select Process from the context menu. Once you do this, Visual Studio gives a warning indicating that the SSAS server is not synchronized with the changes to your project and offers to build and deploy it for you (Figure 11-9). If you want to process your data with the new cube, you have to click Yes to continue.

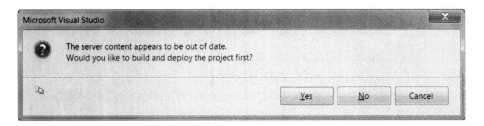

Figure 11-9. Visual Studio offers to build and deploy for you

---

■ **Note**   Visual Studio may ask for the password again. If so, enter your password and click OK to continue.

Once the deployment is successful, Visual Studio presents the Process Cube dialog window, as shown in Figure 11-10. This dialog window is almost identical to the Process Dimension and Process Database dialog windows. Leaving the options as they are and clicking Run at the bottom of the dialog window starts the processing.

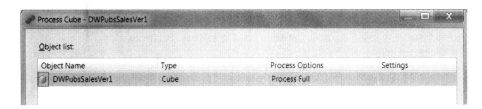

Figure 11-10. The Process Cube window

Within a few seconds a second dialog window appears displaying processing messages (Figure 11-11). The messages vary, because Visual Studio may not need to reprocess dimensions or measures that remain unchanged from previous processing.

Figure 11-11. The Process Progress window

Eventually the processing messages stop as processing completes. If there are errors, they require troubleshooting, but if all succeeds, you can then close the dialog windows.

Configuring Cubes

After the cube has been created, the cube needs to be configured and validated. Configuring a cube means modifying the cube to fit your needs, whereas validating a cube means verifying that the cube's configuration is viable.

The cube designer has various tabs that enable you to configure your cube. We describe the tabs in Table 11-1.

Table 11-1. *The Tabs of the Cube Designer*

Name	Description
Cube Structure	Sets the properties for measures and measure groups
Dimension Usage	Maps connections between each dimension and measure group
Calculations	Adds additional MDX calculations to your cube
KPIs	Adds key performance indicator measures to your cube
Actions	Adds additional actions such as drillthrough and hyperlinks
Partitions	Divides measures and dimensions for fine-tuning aggregations to be stored on the hard drive
Aggregations	Finds which aggregations should be stored on the hard drive
Perspectives	Defines SQL view–like structures for your cube
Translations	Adds additional natural languages to the captions of your cube
Browser	Browses the cube from the development environment

As you advance in your BI solution developing abilities, you will find that each of these tabs holds something of interest. The tabs we discuss in this chapter are Browser, Cube Structure, Dimension Usage, Calculations, and KPIs. These first five tabs will keep us plenty busy. We discuss the other five tabs in Chapter 12.

The Browser Tab

The Browser tab is used for validating your cube configurations. To get to the Browser tab, access the cube designer window and navigate to the last tab. Depending on your screen resolution, you may see only a few letters with an ellipsis (. . .) indicating that the rest of the title is hidden (Figure 11-12).

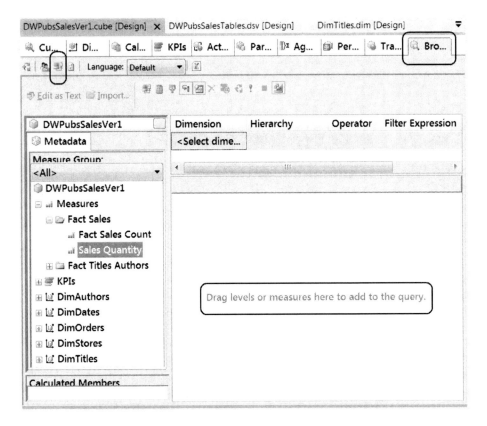

Figure 11-12. *The Browser tab*

REFRESHING THE BROWSER

To use the Browser tab, the dimensions must have already been uploaded to your Analysis Server and the cube processed. It does not work if the XML files have been generated only in Visual Studio. After processing the cube, you need to manually reconnect to SSAS to see the latest version. Do so by clicking the Reconnect button on the browser's toolbar to refresh the browser (circled in Figure 11-12). Oddly, you do not use the Refresh button as you might expect! Throughout this chapter, we make several changes and reprocess the cube. This means you click the Reconnect button a number of times.

When the Browser tab is open, its treeview is visible on the left side of the screen. Measures and dimensions can be dragged from this treeview to the center for viewing. The interface is very similar to working with a pivot table in Excel, but most developers find it to be rather unintuitive.

Begin by selecting an item from the treeview. It is best to start with a measure, such as the Sales Quantity measure shown in Figure 11-12. Drag and drop the measure to the middle of the viewing pane where it says "Drag levels or measures here to add to the query." The query referred to here is the MDX query that will be invisibly written for you by Visual Studio.

Once you have dragged and dropped your measure into the viewing pane, the grand total of the measured value will be displayed. To see a breakdown of this grand total based on a dimensional attribute, you need to drag and drop something from one of the dimensions into the viewing pane as well. The trick is to drag and drop it just to the left of where the measures are being displayed. Figure 11-13 has this area circled.

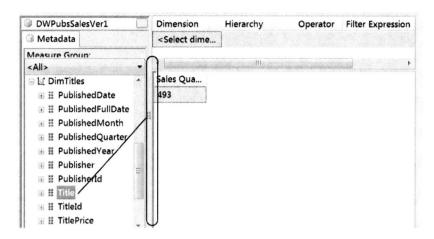

Figure 11-13. *Dragging and dropping to add items to the results*

When you grab an attribute, such as the Title attribute, and drag it into this location, you are able to see the results of your query with the sales quantities broken into subtotals based upon the individual titles, as shown in Figure 11-14.

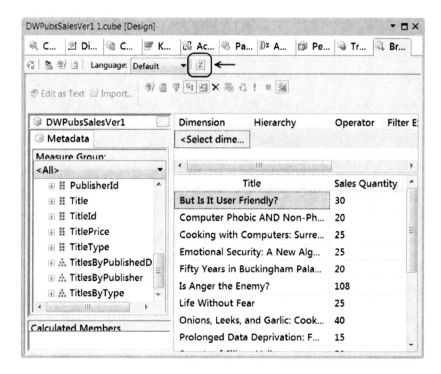

Figure 11-14. *Viewing cube measures in the Browser tab*

Keep in mind that these tools are not made for the end user. Instead, the end user will use something like an Excel spreadsheet or a Reporting Server report to view the data. This function is simply a convenience for the developer who has just finished creating a cube and wants to review and validate the cube's design. Expect to configure, process, and browse the cube repeatedly as you go through the development cycle.

Before you complete your validations, you should view the cube using a common reporting application like Microsoft Excel. In SQL 2012, Microsoft has conveniently included a way to open an Excel spreadsheet from the cube's Browser tab. The button is next to the language dropdown box and has the icon that is used for older versions of Excel. You can see it circled in Figure 11-14. Clicking this button launches Excel and presents you with a pivot table report, as shown in Figure 11-15. We discuss Excel pivot tables in Chapter 15, but for now let's look at some of the other tabs in the cube designer.

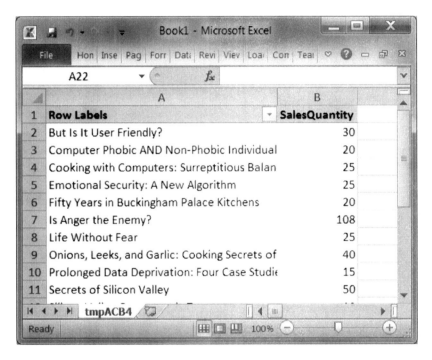

Figure 11-15. *An Excel pivot table generated with the cube Browser tab*

Validate the Measures

When checking your work in the browser, we recommend starting with the measures. One reason for this is that attempting to use the dimensions in the browser without a measure may not give you any useful results.

Drag and drop a measure to the view pane. You will see the grand total for that measure. In Figure 11-16, you can see that we have dragged all of the measures into the viewing pane and now are seeing the grand total for each one.

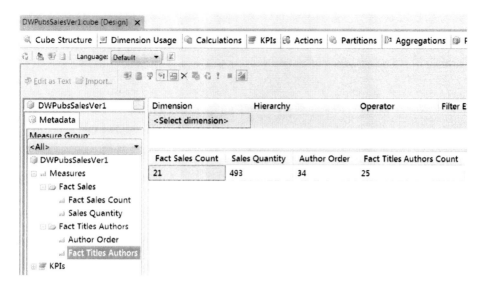

Figure 11-16. *Browsing the cube measures*

With our first look, the values are readily apparent. For example, the Fact Sales count indicates that there are 21 rows in the Fact Sales table (Figure 11-16). The Sales Quantity measure indicates that the grand quantity of sales is 493. But, the Author Order measure is showing that there are 34 "somethings"—but this is nonsensical data.

As it turns out, Author Order indicates the order in which each author appears on a title's cover—first, second, or third. Because the default behavior of a measure summarizes the values, the grand total of the sum of 1, 2, and 3 for all the rows in the Author Order is apparently 34. Now that we see this value, we know that the Author Order requires configuring before it is useful.

Similarly, the Fact Titles Authors Count indicates that there are 25 rows in the FactTitlesAuthors table. At first this may seem somewhat useless, but when combined with a specific title or a specific author, we can tell how many titles are associated with the authors using this measure. So, this measure also needs to be configured before it is useful.

Review the Dimensions

Next, let's add some dimension values to get a clearer picture of our measures. Drag the title into the viewing pane. Figure 11-17 shows an example of the results.

Dimension	Hierarchy	Operator	Filter Expression	
<Select dimension>				

Title	Fact Sales Count	Sales Quantity	Author Order	Fact Titles Authors Count
But Is It User Friendly?	1	30	1	1
Computer Phobic AND Non-P...	1	20	3	2
Cooking with Computers: Surr...	1	25	3	2

Figure 11-17. *Adding the titles attribute to the report*

From these results we can see that the "Computer Phobic . . . " title has been sold one time with a total quantity of 20 units and that there are two authors associated with this title. So far so good, but why does the Author Order report a 3? Oh, that's because it is adding the author order numbers of 1 and 2 and displaying their sum. We make a note of this and continue verifying other results.

Validating the Results

To further verify your results, you should run SQL statements on your data warehouse to validate the numbers a second time. Listing 11-1 shows some SQL code that will do that for you.

Listing 11-1. Validating the Development Report

```
Use DWPubsSales
Go

Select *
From DimTitles
Where TitleName Like 'Computer Phobic %'
-- Results in 14

Select
TitleKey
,SalesQuantity = Sum(SalesQuantity)
,[SalesCount] = Count(*)
From FactSales
Where TitleKey = '14'
Group By TitleKey
-- Results in SalesQuantity = 20 and SalesCount = 1 *Good*

Select
TitleKey
,[AuthorCount] = Count(*)
From FactTitlesAuthors
Where TitleKey = '14'
Group By TitleKey
-- Results in AuthorCount = 2 *Good*
```

■ **Tip**　In the real world, the calculations involved in your cubes can be quite complex. You may have to ask for proofs from the data analysts or accounting team members. This is an important and time-consuming step in validating your cube, and you need to plan accordingly.

In our example, the results of the SQL code confirm the results we see in the Visual Studio cube browser. We continue to examine and validate the dimensions and measures using the cube browser and SQL code until we have verified that all of the information is correct. Wherever we find a problem, we note it and change the configurations to resolve it.

The Cube Structure Tab

Modifying the properties of your measures is accomplished using the Cube Structure tab. Begin by highlighting a measure in the Measures pane, as shown in Figure 11-18. Access the Properties window by right-clicking the measure and choosing Properties from the context menu.

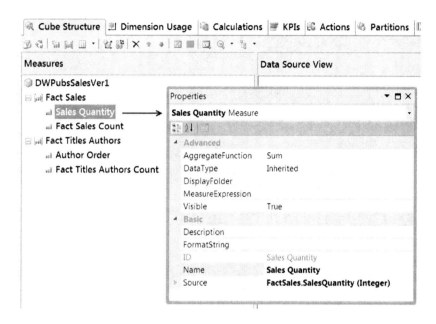

Figure 11-18. *Accessing the properties of a measure*

Common Measure Properties

Although a number of properties are available for each measure, you will use only a few of them on a regular basis. Table 11-2 describes the commonly used property settings.

Table 11-2. *Commonly Used Property Settings*

Property	Description
Aggregate Function	Allows you to select how the measure will be aggregated
Format String	Allows you to indicate how the measure will be formatted
Name	Allows you to specify the name of the measure
Visible	Allows you decide whether most clients will be able to see the measure

The first property in the Properties window is AggregateFunction. Here you can change the aggregation settings from the default to the other selections available in the dropdown box. Figure 11-19 shows the options with the dropdown box expanded.

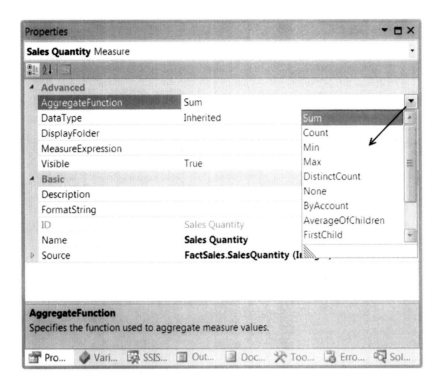

Figure 11-19. *Setting the AggregationFunction property*

While Sum is the default and most commonly used option, other options are useful on occasion. Table 11-3 briefly describes what they do (Table 11-3).

Table 11-3. *A Description of the AggregationFunction Settings*

Aggregation Function	Description
Sum	Adds values for all individual measure values into a total.
Count	Provides a count of all individual measure values.
Min	Returns the lowest value of all individual measure values.
Max	Returns the highest value of all individual measure values.
DistinctCount	Returns a count of all unique individual measure values.
None	No aggregation is performed, and null is used for places where subtotals would normally display. (It is useful for items such as the Author Order in our example.)
ByAccount	Calculates the aggregation according to an assigned custom aggregation function. (Used only for special account dimensions.)
AverageOfChildren	Returns an average of values for all nonempty individual measure values.
FirstChild	Returns the value of the first child member under a given parent. (For example, an inventory count measure on January 1 would be the first child under the parent January. So, you would see the inventory count for the first day of that January but not the total of all the other days.)
LastChild	Returns the value of the last child member under a given parent.
FirstNonEmpty	Retrieves the value of the first nonempty child member under a given parent. For example, if January 1 had no inventory values but January 2 did, you would see the value for the second day of January.
LastNonEmpty	Retrieves the value of the last nonempty child member under a given parent.

The FormatString property uses the same style of format notation as Excel spreadsheets. In this property, you can specify how you want the data to display to the client. Most client software will respect this, but some may not. In those cases, the formatting will have to be reapplied at the report level. In Figure 11-20, we show the list of predefined options, but you can type in your own as well.

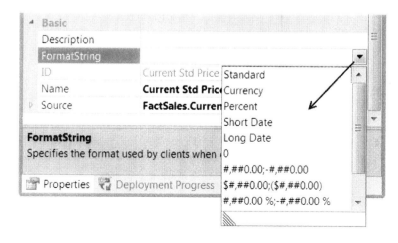

Figure 11-20. *Setting the FormatString property*

Type in a name to set the Name property. For our example, we choose to keep the names consistent with our other naming conventions by removing any spaces. We also change the name of the measure to better represent the data, such as changing Author Order to AuthorListingOrder (Figure 11-21).

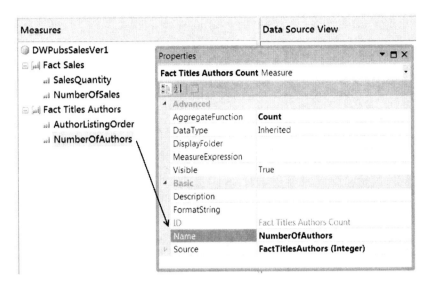

Figure 11-21. *Setting the Name property*

Visibility is controlled by the Visible property setting. The default is set to True, but it can be changed to False if you prefer your clients to ignore it. This is a convenient feature if you believe that a specific measure may be confusing to some users yet you want to have it available for power users or for additional calculations that may be performed with MDX code. In our example, we might choose to hide both measures in the Fact Titles Authors measure group, because they are useful from the context of authors and titles dimension members but have no context in regard to dates or stores.

■ **Note** Even when a measure is hidden, they are still accessible using MDX code. Some client applications do not check to see whether the visible setting is set to True; therefore, they still show up anyway.

Measure Group Properties

Measure groups are logical ways of grouping one or more measures. They make it easier for report users to find what they need. Measure groups are based on which fact table the data is coming from. By default each measure group will have the same name as the fact table. If you want to change this, you can do so by highlighting the measure group in the Measures pane and adjusting the property accordingly. For example, we have changed the measure names and the measure group names in Figure 11-22.

■ **Note** Measure groups also have an impact on processing and partitioning, which we talk about in Chapter 12.

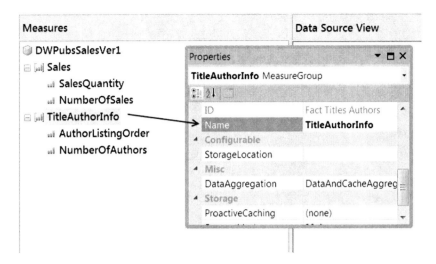

Figure 11-22. *Changing the names of the measure groups*

The Dimension Usage Tab

The Dimension Usage tab provides a grid layout that indicates how a dimension and measure group are associated with each other (Figure 11-23). The Cube Wizard is able to determine most associations, but you may still need to configure some of the associations by hand. At the very least, you will want to review this tab as part of your cube validation.

	Measure Groups	
Dimensions	Sales	TitleAuthorInfo
DimOrders	OrderNumber	
DimStores	Store	
DimTitles	Title	Title
DimAuthors		Author
DimDates	Date	

Figure 11-23. *The Dimension Usage tab*

The grid displays dimensions on the left side of the window and measure groups across the top. Each intersecting cell represents a relationship. Cells that are grayed out indicate that there is no relationship defined between that measure group and that particular dimension. In Figure 11-23, you can see that the DimStores dimension is related to the FactSales measure group based on the Store attribute. But it is not related to the other fact table, FactTitleAuthors. This is indicated by the cell beneath the FactTitleAuthors measure group and across from DimStores displaying as blank.

Configuring a Relationship

If you need to make an adjustment, click the cell you want to configure, and the ellipsis (. . .) button will appear. When you click the ellipsis, a new dialog window appears displaying the logic that defines the relationship. As you can see in Figure 11-24, it consists of a dropdown box at the top of the window that allows you to select the relationship type you want to use.

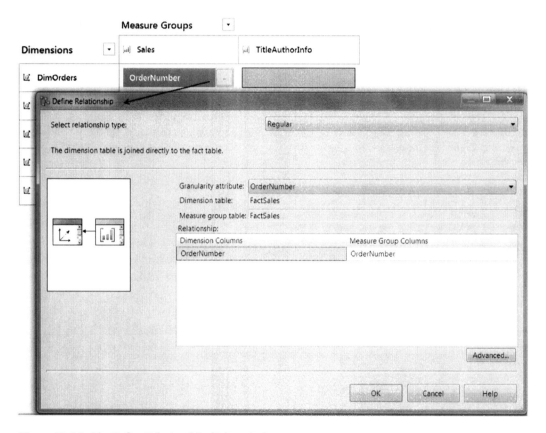

Figure 11-24. *The Define Relationship dialog window*

In Figure 11-24, you can see the different relationship choices. Regular is the wizard default, and it is also the most common relationship type. In our example, this is not the correct choice, because the Order Numbers dimension is a fact dimension. Therefore, we need to choose Fact from the dropdown list, as shown in Figure 11-25.

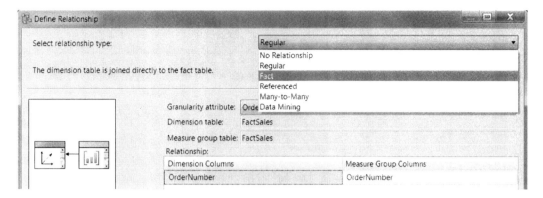

Figure 11-25. *Selecting a relationship type*

Table 11-4 describes each relationship type.

Table 11-4. *Relationship Types*

Relationship Type	Description
No Relationship	Defines a cell as not having any relationship.
Regular	Defines a cell as having a regular relationship where there is a standard star or snowflake dimension associated with that measure group.
Fact	Defines a cell as having a fact relationship where all the dimensional attributes are within the fact table.
Referenced	Defines a cell as having a reference relationship where you reference a table through another table (for example, choosing to reference the publishers table through the titles table, skipping over titles completely).
Many-to-Many	Defines a cell as having a many-to-many relationship where a bridge table is used to indirectly associate a dimension table with the central fact table. (In our example, this is how DimAuthors is constructed.)
Data Mining	Defines a cell as having a data mining relationship. If you have created a data mining object with any project and allow it to create a data mining dimension for you, it can be associated in the cube by the setting.

The reference and data mining relationship types are less common than the others. We will see an example of how the more common ones are used in the following exercise.

■ **Tip** Occasionally the Dimension Usage tab, as well as some of the other tabs, may not refresh correctly. Unfortunately, there is no refresh button on these tabs; therefore, the way to refresh them is to close the cube designer window and reopen it. You can reopen the cube designer window by right-clicking the cube in Solution Explorer and choosing View Designer from the context menu.

EXERCISE 11-2. CONFIGURING MEASURES PROPERTIES AND RELATIONSHIPS

In this exercise you begin configuring the cube you created in Exercise 11-1 using the first two tabs of the cube designer.

Note: If you have closed the cube designer window since the previous exercise, please open it now.

Configure Cube Structure Tab

With the cube designer open, select the Cube Structure tab. Highlight the various measures and adjust their properties. There are four main properties to choose from for each dimension. Please verify that the current selection is correct, and if it is not, adjust it.

1. Change or validate the properties of the measure in the Fact Sales measure group. You can do so by highlighting each object in the Measures pane and, if necessary, changing the properties in accordance with those listed in Table 11-5.

Table 11-5. *Property Settings for the Fact Sales Measure Group*

Object	Property	Value
Fact Sales measure group	Name	Sales
Sales Quantity measure	Name	SalesQuantity
	AggregationFunction	Sum
	FormatString	Standard
	Visible	True
Fact Sales Count measure	Name	NumberOfSales
	AggregationFunction	Count
	FormatString	Standard
	Visible	True

2. Change or validate the properties in the Fact Titles Authors measure group. You can do so by highlighting each object in the measures pane and, if necessary, changing the properties in accordance with those listed in Table 11-6.

Table 11-6. *Property Settings for the Fact Titles Authors Measure Group*

Object	Property	Value
Fact Titles Authors measure group	Name	TitleAuthorInfo
Author Order measure	Name	AuthorListingOrder
	AggregationFunction	None
	FormatString	Standard
	Visible	False
Fact Titles Authors Count measure	Name	NumberOfAuthors
	AggregationFunction	Count
	FormatString	Standard
	Visible	False

Configure Dimension Usage Tab

Let's go to the Dimension Usage tab and change some of the *relationship type* settings.

1. Click the Dimension Usage tab and verify that it currently looks like Figure 11-23.

2. Click the cell row named OrderNumber. It is beneath the Sales column and alongside DimOrders.

3. Click the ellipsis button when it appears (Figure 11-24), and the Define Relationship dialog window will open.

4. In the Define Relationship dialog window, change the relationship type to Fact (Figure 11-25).

5. Note how the diagram on the left side of this dialog window describes the relationships between the dimensions and the measure groups using symbols (Figure 11-26).

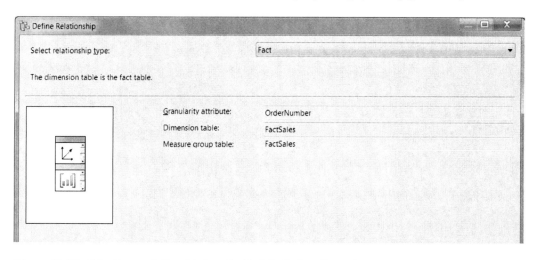

Figure 11-26. *Selecting a relationship type for the DimOrders dimension*

6. Click OK to close the Define Relationship dialog window.

7. Click the cell beneath Sales column and alongside of DimAuthors row. This cell should currently be blank (Figure 11-27).

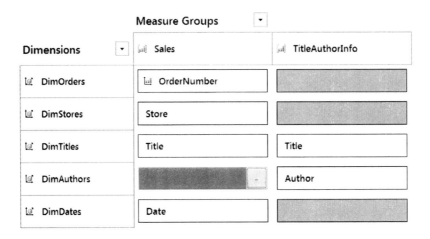

Figure 11-27. *Configuring the DimAuthors dimension*

8. Click the ellipsis button when it appears, and the Define Relationship dialog window will open.

9. Change the relationship type in the Define Relationship dialog window to Many-to-Many using the "Select relationship type" dropdown box (Figure 11-28).

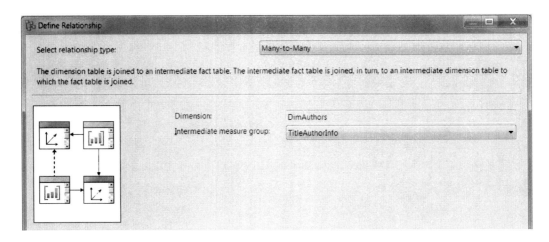

Figure 11-28. *Selecting a relationship type for the DimAuthors dimension*

10. Using the "Intermediate measure group" dropdown box, select TitleAuthorInfo, as shown in Figure 11-28.

11. Note how the diagram on the left side of this dialog window describes the relationships between the dimensions and the measure groups using symbols.

12. Click OK to close the Define Relationship dialog window.

13. Verify that the Dimension Usage tab displays, as shown in Figure 11-29.

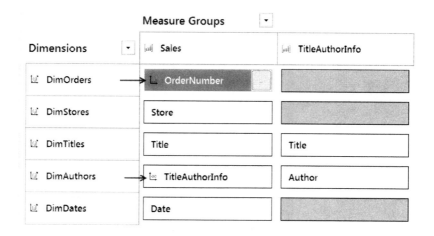

Figure 11-29. *The Dimension Usage tab after configuration*

14. Note how the icons now indicate nonregular relationship types.

Verify Your Design with the Browser Tab

Now that we have made our changes, let's process the cube and then use the Browser tab to verify that our design is correct.

1. Right-click the cube file in Solution Explorer called DWPubsSalesVer1.cube and choose Process from the context menu.

2. If a message box appears indicating that changes need to be uploaded to this deployment, click Yes to accept deployment and then Yes to accept your changes overwriting the existing cube. The Process Cube dialog window will appear.

3. Click Run to start the processing. When processing completes, click Close to close the processing windows.

4. Click the Browser tab in the cube designer window.

5. Drag and drop measures and dimensions from the treeview to create a result that looks like the one displayed in Figure 11-30.

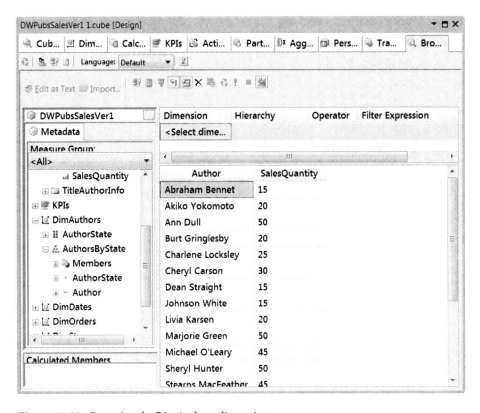

Figure 11-30. *Browsing the DimAuthors dimension*

As we conclude this exercise, review the values in the report to make sure they are accurate. We provided some SQL code for this in Listing 11-1, but for now we will assume that all are correct and move on to our next topic.

In this exercise, you made changes to the cube using the Cube Structure tab and the Dimension Usage tab. You then processed the cube and verified that the changes were successful using the Browser tab. Next we focus on the other tabs that come with the cube designer.

The Calculations Tab

The Calculations tab enables you to create named MDX expressions, also known as *calculated members*. Typically these calculated members are used to create additional measures, but they can also be used to create additional dimensional attributes.

An example of an additional measure is acquiring the total price of an individual sale by multiplying the quantity by the price of the product. An additional dimensional member example is combining countries into groups such as combining Mexico, Canada, and the United States into one member called North America.

The Calculations tab consists of three basic sections: the Script Organizer and Calculation Tools panes on the left side of your screen and the script-editing area on the right (Figure 11-31).

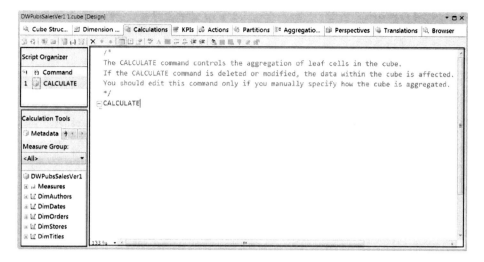

Figure 11-31. The Calculations tab

■ **Note**   The Metadata table in the lower-right side of the Calculations tab will not display the treeview until you have processed the cube at least once. After processing, you may have to use the Reconnect button on the Calculations tab's toolbar to force a refresh.

Adding a Calculated Member

To create a new member, click the New Calculated Member button, which conveniently looks like a calculator (circled in Figure 11-32). When clicked, the script editing area changes to display in a *form view*. You can switch between the Form View and the Script View in the script editing area by clicking their respective buttons (circled in Figure 11-31).

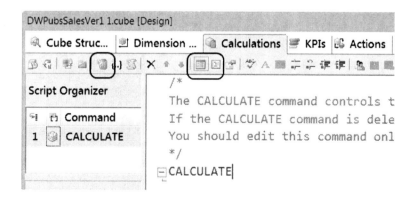

Figure 11-32. The Calculations tab toolbar

> ■ **Important**   Make sure you do not delete the CALCULATE command from the Script Organizer or from the script-editing area, or your cube will not be able to process correctly.

When the Calculations tab is displaying the Form View, it will look as shown in Figure 11-33. Here you can type an MDX expression into the Expression textbox, define its name in the Name textbox, and determine whether it is going to be part of the measures or a particular dimension in the Parent Hierarchy dropdown box.

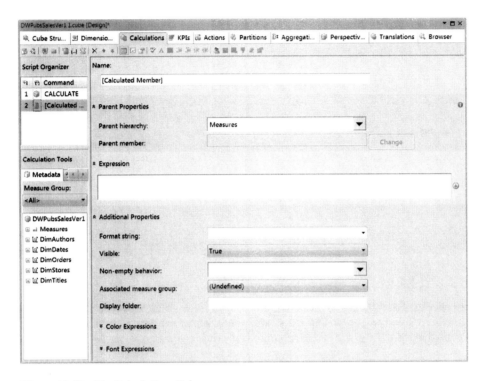

Figure 11-33. *The Calculations Tab*

By default, all new calculated members are created on the *Measures dimension*. Calculated members in the Measures dimension go by a special name, known as *calculated measures*. You will note that we are using the term *Measures dimension* to distinguish it from all of the other dimensions. This may seem confusing at first, but as you will see in Chapter 14, measures are nonhierarchical dimensions. In other words, there is no All level that acts as a parent to all the members of the Measures dimension. Therefore, the parent member dropdown box is grayed out.

Configuring a Calculated Member

To configure your calculated member, choose an appropriate name, determine the parent hierarchy, and define an MDX expression. For example, we are going to create a calculated member that gives us the extended price by multiplying the title price in the titles dimension by the sales quantity from the Measures dimension. So, we named the new calculated member Calculated Total Price, we left the Parent hierarchy set at Measures, and in the Expression textbox we defined an MDX statement that multiplies the TitlePrice by SaleQuantity (Figure 11-34).

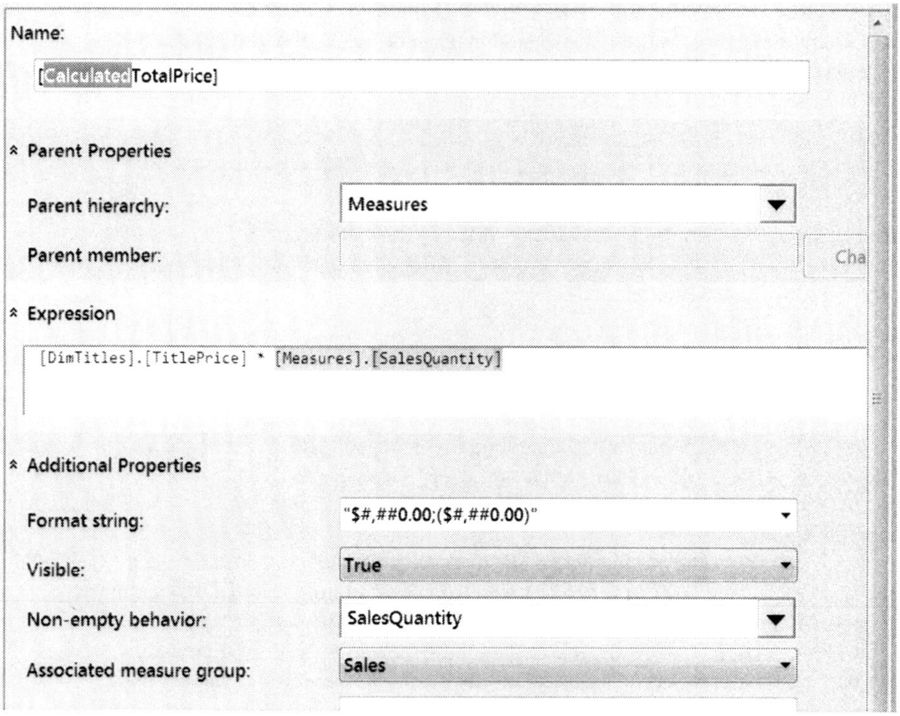

Figure 11-34. *Configuring the new calculated measure*

■ **Important** The expression we have used in Figure 11-34 will not work as expected! We explain the reason in just a bit.

We set four other commonly used properties in Figure 11-34. The "Format string" property allows you to define how the values will be displayed in client applications like Excel. By default it is blank, but as you can see in Figure 11-34, it can be adjusted using the dropdown box or typing in a pattern.

The Visible dropdown box determines whether the member will be displayed by client applications. Setting it to True can be useful when the calculated member is to be treated as an intermediate value for additional calculations; it is not designed to be used independently.

The "Non-empty behavior" setting allows you to define the value you would like to display if the calculation returns an empty value. This setting determines what will be displayed. For example, multiplying a null Title Price by 5 Sales Quantity returns a null value. Setting the nonempty behavior to SalesQuantity overrides the null and displays the Sale Quantity value of 5.

■ **Note** Our example begs the question: Why was there a null price for an item, yet the company sold five of them? It also reminds us that the Sales table in Pubs should have had a Sales Price column to store the price a title sold for, in a given sales event. Sadly, this is not the case, so we will have to work with what we have.

The Associated Measure Group dropdown box allows you to determine where the new calculated measure will be displayed within a client application. Although it may seem strange, even after you change this setting and process the cube, you will not see your new calculated member displayed in the Cube Structure tab as you would expect. However, you will be able to see it in the Browser tab, and most Microsoft clients will see it as well.

Calculated Members vs. Derived Members

In our example, we want a measure that multiplies the current price by the sales quantity. You would think that the simple MDX statement in Listing 11-2 would do this for us. After all, it multiplies the TitlePrice for the DimTitles table by the SaleQuantity in the FactSales table.

Listing 11-2. This MDX Expression Produces a Null Value

```
[DimTitles].[TitlePrice] * [Measures].[SalesQuantity] -- Will be Null due to the context!
```

Nevertheless, if you use this expression to create a calculated measure, you will find that it does not return the extended sales price because of the context of the expression. Unfortunately, SSAS cannot automatically map the TitlePrice from the DimTitles table to the SalesQuantity in the FactSales table, so you have to help it by creating a derived member in the FactSales table.

Derived members are created in SSAS by modifying the SQL code behind each table in a data source view. This can be somewhat confusing, so let's review the difference between calculated members and derived members.

Calculated members use MDX expressions to create new members to the measures or other dimensions. Derived members use SQL expressions to create new members on the measures or other dimensions.

Both are similar, but they each use a different language for the expression, and they evaluate their expressions during different events. In the case of a calculated member, the event that causes the expression to evaluate is when a client application queries the cube or dimension. In the case of a derived member, the event that causes the expression to evaluate is when a cube or dimension is processed.

Listing 11-3 is a SQL statement that can be used to create two additional columns in a data source view. One is a copy of the TitlePrice member of the Titles dimension, and one is a new member that represents the product of the price of a given title by the sales quantity of a sales event.

Listing 11-3. SQL Code for a Derived Measure

```
SELECT
  FactSales.OrderNumber
, FactSales.OrderDateKey
, FactSales.TitleKey
, FactSales.StoreKey
, FactSales.SalesQuantity
-- Adding derived measures
, DimTitles.TitlePrice as [CurrentStdPrice]
, (DimTitles.TitlePrice * FactSales.SalesQuantity) as DerivedTotalPrice
FROM FactSales
INNER JOIN DimTitles
  ON FactSales.TitleKey=DimTitles.TitleKey
```

■ **Note** We could have created both of these during the ETL process instead of here, but we chose not to do so in order to show you this feature.

Columns can be added to a table in the data source view either by adding a new Named Calculation or by replacing a table completely using a named query. For our example, it is necessary to use a named query, because we must join data from two tables, and named calculations cannot do this.

You can add members from other tables, and even databases, by replacing the table with a new named query, as shown in Figure 11-35.

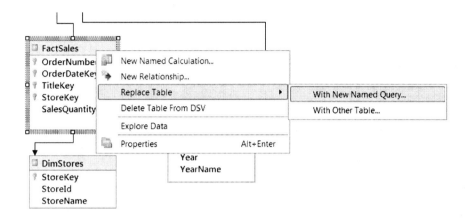

Figure 11-35. *Adding a derived measure using a named query*

▨ **Tip** Of course, you must reopen the data source view editor in order to do this. It can be reopened any time by right-clicking the data source view icon in Solution Explorer and selecting View Designer from the context menu.

Once you select this option from the context menu, you will be presented with a query designer. You can either type in the code or let the designer create it for you. Another option is to copy and paste code that has been created and tested in SQL Server Management Studio, as we have done in Figure 11-36.

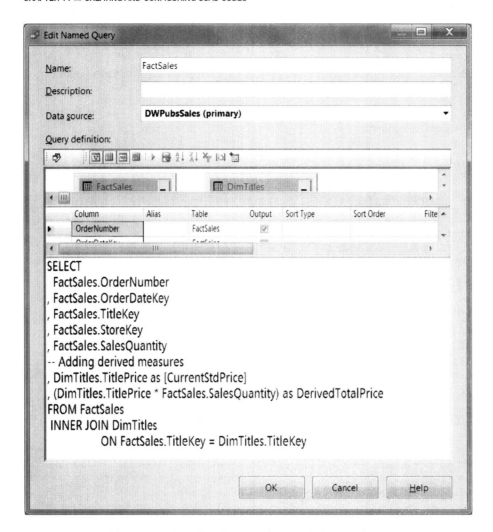

Figure 11-36. *Adding SQL code to the Edit Named Query dialog window*

Clicking OK adds the derived members to the fact table. Now the data source view will display these additional members, as shown in Figure 11-37. When the cube is processed, the SQL statement that represents the FactSales table will run and copy the new DerivedTotalPrice values to the SSAS cube folder.

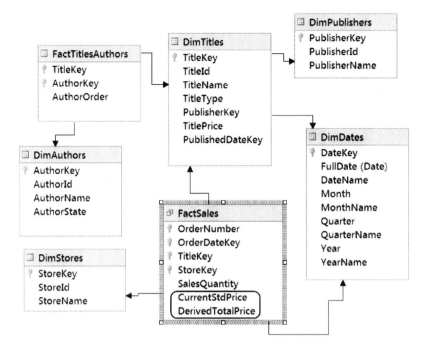

Figure 11-37. *The FactSales table now has a derived measure*

Note that this differentiates calculated from derived members. Derived members are stored in the folders managed by SSAS, but calculated members are never stored. Instead, each time a client executes a query against the cube or dimension, the calculated member expression is evaluated. This can have a performance impact since the aggregations have to be evaluated on each query request.

This might lead you to believe that the derived member is always the best choice since you likely want the best query performance for your end users. But this is not always the case, because sometimes the calculations will be incorrect when you use a derived member instead of a calculated member.

This has to do with the order of operations used to evaluate the expression. In other words, when we multiply a price by the quantity, does the multiplication happen after the summation of the individual quantity values, or does it happen before? As you know from learning mathematics in school, the order of operations directly impacts the results.

In the end, this means you must always test them to verify that the values are correct, whether you are using derived or calculated members. If the results are the same, then you can choose to use a derived member for its increased query performance. But, remember that it will be at the cost of taking longer to process (since the values have to be aggregated and stored during the processing). If the values are different, you need to figure out which is the correct value and use the member that is correct, regardless of performance.

In the upcoming exercise, we create both derived and calculated members for comparison and then validate which one is correct. Before we do that, let's talk about how you can create a copy of a working cube for testing purposes.

Making a Test Copy of a Cube

Microsoft has made it easy to create a copy of your cube. You can simply right-click the cube in Solution Explorer and choose Copy from the context menu. After that, highlight the Cubes folder, right-click the folder, and select Paste from the context menu to create your copy, and a dialog window appears that allows you to name your new cube. In Figure 11-38 we have defined the cube name by adding ForTesting to the existing name.

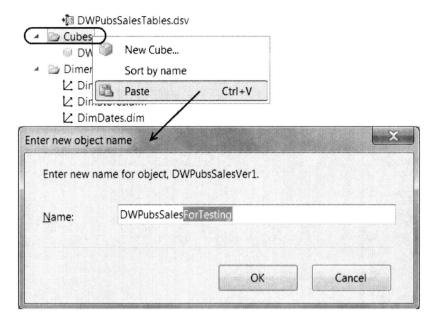

Figure 11-38. *Pasting the cube file brings up the "Enter new object name" dialog window*

At this point you have a copy of the cube, but it does not include the derived member if it has been added to the data source view after the cube's creation. To add the derived member to an existing cube, navigate to the Cube Structure tab and select the folder that represents the fact table where the derived member was created. In our example, we have added the derived member to the Fact Sales table. Therefore, it appears under the Sales measure group. By right-clicking the Sales measure group, we can add the new derived member, as shown in Figure 11-39.

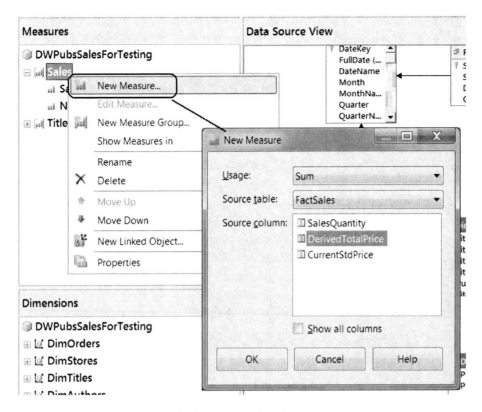

Figure 11-39. *Adding a new derived member to the cube*

Once you have added the derived member to the cube, the queue must be processed for it to appear. This is logical, because it is the act of processing that creates the derived member and its values on the SSAS server.

Let's put our knowledge to work in the next exercise by creating both a calculated measure and a derived measure and testing to see whether one or both of them are correct.

EXERCISE 11-3. ADDING CALCULATED AND DERIVED MEASURES

In this exercise, you create a copy of the cube for testing your derived and calculated measures. You also modify the copied version of the cube to include both a calculated and derived measure. You then browse the test cube to determine which value is correct and then decide which new measure to keep.

Note: If the solution from the previous exercise is closed, open Visual Studio. Remember to select Run as Administrator and answer Yes to close the UAC.

Add a Derived Measure

We need to create the derived measure in the data source view, so we navigate to the data source view and modify the fact table to include the derived member.

1. In the existing data source view, right-click the FactSales table and choose the Replace ➤ With New Named Query option from the context menu, as shown in Figure 11-35. This will launch the Create Named Query dialog window.

469

2. When the Create Named Query dialog window appears, type in the code defined in Listing 11-3. Afterward, click OK to close the dialog window, as shown in Figure 11-36.

Tip: This code can be found in `C:\_BookFiles\Chapter11Files\ListingCode\Listing 11-3`.

3. Verify that the FactsSales table now looks like Figure 11-37.

Copy the Cube

Now make a copy of the cube for testing purposes and add the derived member and calculated members to the test version of the cube.

1. Create a copy of the existing cube for test purposes and name it **DWPubsSalesForTesting**. To do so, right-click the DWPubsSalesVer1cube and select Copy from the context menu. Right-click the Cubes folder and select Paste from the context menu (Figure 11-38). Finally, type in the new name in the dialog box as it appears in Figure 11-38.

Add the Derived Member to the DWPubsSalesForTesting Cube

The two derived members now exist in the data source view. Next, add the two derived members to the cube.

1. Navigate to the DWPubsSalesForTesting's Cube Structure tab of the cube-editing window. You can do this by right-clicking the `DWPubsSalesForTesting.cube` file and selecting View Designer.

2. Highlight the Sales measure group, as shown in Figure 11-39.

3. In the Sales measure group, right-click and choose New Measure from the context menu. The New Measure dialog window will appear, as shown in Figure 11-39.

4. Select the DerivedTotalPrice, and click OK to close the New Measure dialog window.

5. Right-click Sales Measure Group again, and choose New Measure from the context menu.

6. When the New Measure dialog window appears, select the CurrentStdPrice, and click OK to close the New Measure dialog window.

7. Rename both members by removing the spaces that the designer has added to their names.

8. Verify that both new measures have been added to the Sales measure group, which should now look like Figure 11-40.

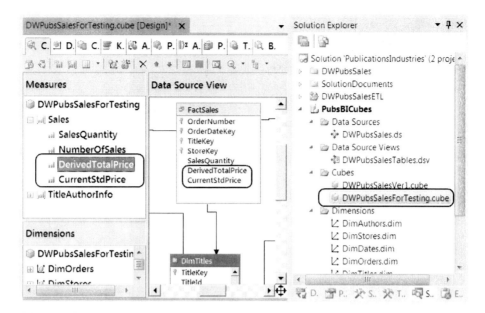

Figure 11-40. *The test cube with the new derived members added*

Add a Calculated Measure to the DWPubsSalesForTesting Cube

Now we can add a calculated member to the cube.

1. Navigate to the Calculations tab, as shown in Figure 11-32.

2. On the Calculations tab, locate the New Named Calculation button on the toolbar, as shown in Figure 11-32. Click this button to create a new calculated member. The Calculated Form Editor will appear, as shown in Figure 11-33.

3. In the Name textbox of the Form Editor window, change the name to **CalculatedTotalPrice**, as shown in Figure 11-41.

4. Verify that the parent hierarchy dropdown box displays the measures, as shown in Figure 11-41.

5. Type in the MDX expression [Measures].[CurrentStdPrice] * [Measures].[SalesQuantity], as shown in Figure 11-41.

6. Use the format string dialog window to select the U.S. dollar format string, as shown in Figure 11-41.

7. Use the "Non-empty behavior" dropdown box to select SalesQuantity, as shown in Figure 11-41.

8. Use the "Associated measure group" dropdown box to select the Sales measure group, as shown in Figure 11-41.

Figure 11-41. *The settings for the CalculatedTotalPrice*

Process the DWPubsSalesForTesting Cube

We now have both a calculated and derived member in the Measures dimension (also known as calculated and derived measures). We need to process the cube before it can be viewed in the browser. Let's do that now.

1. Right-click the cube in Solution Explorer, and select Process from the context menu. This will start the building and deployment of your changes to SSAS before processing will begin.

2. After the build and deployment are successful, a Processing dialog window will appear. Click Run and wait for it to successfully process.

3. In a few seconds, you should be able to close the Processing dialog window, and processing will be complete.

Validate the CurrentStdPrice Measure

You have created both the calculated and derived members and added them to the cube. It's time to validate whether they are giving you correct answers.

1. Navigate to the Browser tab in the cube-editing window.

2. Create a result that looks similar to the one shown in Figure 11-42.

Title	CurrentStdPrice	NumberOfSales	SalesQuantity	CalculatedTotalPrice	DerivedTotalPrice
But Is It User Friendly?	22.95	1	30	688.5	688.5
Computer Phobic AN...	21.59	1	20	431.8	431.8
Cooking with Compu...	11.95	1	25	298.75	298.75
Emotional Security: A...	7.99	1	25	199.75	199.75
Fifty Years in Buckin...	11.95	1	20	239	239
Is Anger the Enemy?	43.8	4	108	4730.4	1182.6
Life Without Fear	7	1	25	175	175
Onions, Looks, and G...	20.95	1	40	838	838

Figure 11-42. *Testing the results using the Browser tab*

Using this report, we can now evaluate the effectiveness of our derived and calculated members.

3. Locate the row with the title Is Anger the Enemy? Notice that the derived member CurrentStandardPrice indicates a value of $43.8.

4. Open the data source view, right-click the DimTitles table, and then select Explore Data from the context menu (Figure 11-43).

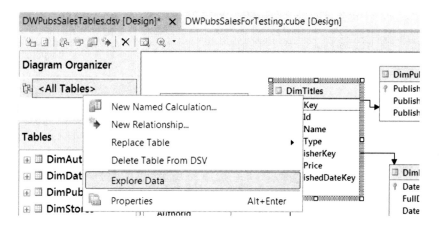

Figure 11-43. *Testing the results by exploring the data warehouse data*

5. When the table data appears, note that the price for Is Anger the Enemy? in the DimTitles table is $10.95 (Figure 11-44).

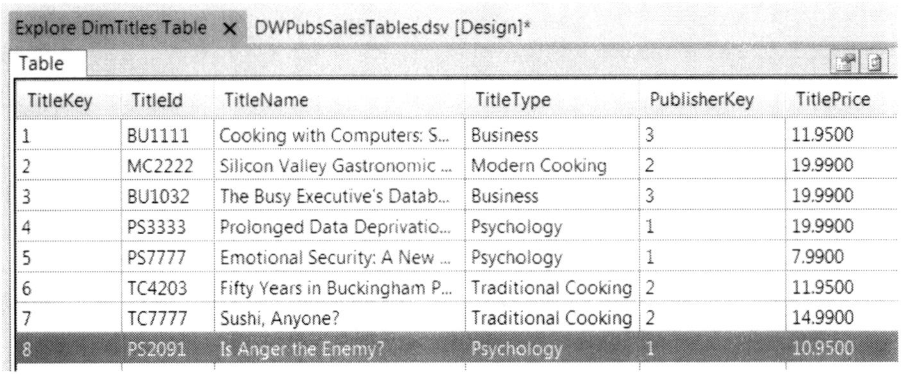

TitleKey	TitleId	TitleName	TitleType	PublisherKey	TitlePrice
1	BU1111	Cooking with Computers: S...	Business	3	11.9500
2	MC2222	Silicon Valley Gastronomic ...	Modern Cooking	2	19.9900
3	BU1032	The Busy Executive's Datab...	Business	3	19.9900
4	PS3333	Prolonged Data Deprivatio...	Psychology	1	19.9900
5	PS7777	Emotional Security: A New ...	Psychology	1	7.9900
6	TC4203	Fifty Years in Buckingham P...	Traditional Cooking	2	11.9500
7	TC7777	Sushi, Anyone?	Traditional Cooking	2	14.9900
8	PS2091	Is Anger the Enemy?	Psychology	1	10.9500

Figure 11-44. *Checking the TitlePrice*

On first inspection, you may think this value is incorrect. But looking at the NumberOfSales measure in Figure 11-43, you can see that this title sold four times. And, if we add the order number to the results, we will see that the value comes up to the correct amount of $10.95. This is because the results are reduced to the granularity of an individual order number and not the collective total of a particular title. Therefore, this means that our CurrentStandardPrice derived member is working correctly, but you might want to hide it from casual view later if you think it will confuse report developers.

Validate the CalculatedTotalPrice Measure

Next, let's look at the CalculatedTotalPrice measure (Figure 11-42).

1. Locate the row with the title Is Anger the Enemy? Notice that the derived member CalculatedTotalPrice indicates a value of 4730.4, but the DerivedTotalPrice has a value of 1182.6.

Here you can see that that CurrentStandardPrice aggregation value has been multiplied by the sales quantity, giving us an incorrect value of 4730.4. The total price of all four sales should be $1182.60. This indicates that for this particular calculation, using a calculated member will not work, but the DerivedTotalPrice is correct.

Configuring the DWPubsSalesVer1 Cube

Now that we have an understanding of what works and what does not, we must to apply the working features to version 1 of the cube.

1. Navigate to the DWPubsSalesVer1's Cube Structure tab of the cube-editing window. You can do this by right-clicking the DWPubsSalesVer1.cube file and selecting View Designer.

2. Highlight the Sales measure group, as shown in Figure 11-39.

3. In the Sales measure group, right-click and choose New Measure from the context menu. The New Measure dialog window will appear, as shown in Figure 11-39.

4. Select the DerivedTotalPrice, and click OK to close the New Measure dialog window.

5. Right-click Sales measure group again, and choose New Measure from the context menu.

6. When the New Measure dialog window appears, select the CurrentStdPrice, and click OK to close the New Measure dialog window.

7. Rename both members by removing the spaces that the designer has added to their names.

8. Verify that both new measures have been added to the Sales measure group, which should now look similar to Figure 11-45. These are now in the DWPubsSalesVer1 cube and not the DWPubsSalesForTesting cube.

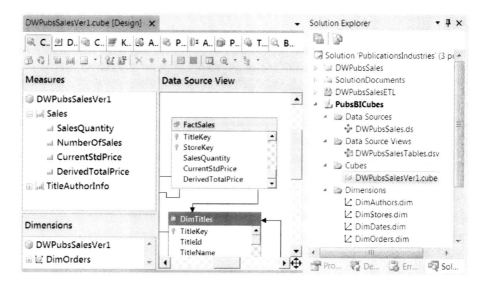

Figure 11-45. *The derived measures are now in the DWPubsSalesVer1 cube*

We have created the derived measures in our nontest cube. This is the cube we use throughout the rest of the book because its members have been tested and proven correct.

Delete the DWPubsSalesForTesting Cube

You have seen how easy it is to create a copy of a cube when you need one, but the test cube has served its purpose. So, let's delete it now.

Note: When building BI solutions outside of our examples, you could choose to either remove the testing cube or keep it for further testing. One advantage of keeping it in its current state is that it can be used to validate your findings if you find that you have forgotten why you chose a derived member over a calculated measure six months down the road, when the thrill of creating both of them has long since passed.

1. Right-click the DWPubsSalesForTesting cube, and select Delete from the context menu. This will delete the XML file from the Visual Studio project.

2. Process the project so that the cube is deleted on the server as well. You can do this by right-clicking the project icon in Solution Explorer and selecting Process from the context menu. Once the processing completes, this cube will be deleted from the server.

At this point, you have created a new derived measure for your cube that you can use in your reports. Although the calculated member did not work in this particular case, it will often work just fine and is a useful tool in many cubes.

In this exercise, you created a testing copy of our cube. You then created both derived and calculated measures and verified which one of these new measures was correct. Finally, we move the derived member to the cube we are currently working on, DWPubsSalesVer1. Now we need to continue configuring our cube by moving to the next tab, which allows us to include KPIs.

KPIs

A key performance indicator (KPI) is a way of grouping measures together into five basic categories. The five basic categories start at -1 and proceed to a +1 using an increment of (-1, -0.50, 0, +0.50, and +1). The numbering system may seem odd to some, but it has to do with the science of statistics. Because of this, SSAS uses only these five categories, and they cannot be redefined.

The idea behind a KPI is for you to reduce the number of individual values in a tabular report to the essence of those values. This is convenient when you have a large report and what you really want to see is whether something has achieved a predefined target value, exceeded it, or did not make it to that value.

KPIs can be created in programming code such as SQL, C#, or MDX. In Listing 11-4 , you can see an example of an MDX statement that defines a range of values and categorizes each value within that range as either -1, 0, or 1. To keep things simple, we excluded the .05 and -05 categories and will work with just these three categories for now.

Listing 11-4. MDX Statement That Groups Values into Three KPI Categories

```
WITH MEMBER [MyKPI]
AS
case
  when [SalesQuantity]<25 or null then -1
  when [SalesQuantity]>= 25 and [SalesQuantity]<= 50 then 0
  when [SalesQuantity]>50 then 1
end
SELECT
{ [Measures].[SalesQuantity], [Measures].[MyKPI] } on 0,
{ [DimTitles].[Title].members } on 1
From[CubeDWPubsSales]
```

The categorization comes from the MDX Case expression where the values of sales quantity are divided into the three categories of less than 25, 25 to 50, or more than 50. We chose this range at random for our example, but the range you choose for your BI solution should be based upon a range that has some significance to the business you are building it for.

In the cube designer window, you can include an MDX expression using the KPI tab, as shown in Figure 11-46. On the KPI tab, click the new KPI button (circled in Figure 11-46), and a new KPI will appear under the KPI organizer pane. Rename the KPI to something descriptive that has no spaces by typing the new name in the Name textbox. In our example, we have named it SalesQty25To50KPI.

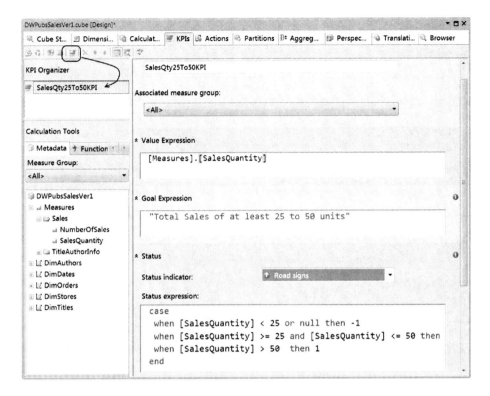

Figure 11-46. *Adding a KPI to your cube*

Beneath the Name textbox is the "Associated measure group" dropdown box. This identifies which measure group the KPI belongs to. In our example, we picked the Sales measure group.

Beneath the dropdown box is the Value Expression textbox. Indicate which measure or MDX expression you want to evaluate in this box. In our example, we evaluate the measure SalesQuantity.

The Goal Expression textbox allows you to type a description that will appear in some client applications such as Excel. This is to give your end users a better understanding of what the KPI describes.

The status indicator dropdown box allows you to choose between several icons that come with SSAS. These icons represent suggestions as to what should be displayed to client applications in a report using the KPI. Until recently, most clients ignored these settings. But newer versions of Excel can read them from SSAS and display an icon that looks very similar to the one suggested.

If a client application (such as Excel) does not have an icon that looks exactly like the SSAS icon, it will use another one. These can be adjusted on the client application. When an application is unable to read the suggested icon from SSAS, the report builder will have to determine which icons to use on their own.

The Status Expression textbox is where the MDX expression is placed to create the KPI value categories. Notice that this is not a full MDX statement. Instead, it shows the expression that evaluates as the KPI categories of -1 to 1.

More settings can be configured, but these are the core settings necessary to create a KPI. Once you have defined these settings, you can review the output of the KPI by switching to the browser view. To switch between the browser view and the form view of the KPI tab, toggle between the two toolbar buttons circled in Figure 11-47.

Once you switch to the KPI browser, Visual Studio will display the KPI based on the grand total of the value expression. Our value expression is the sales quantity, and the grand total is 493. It displays this value in the Value column in the results section.

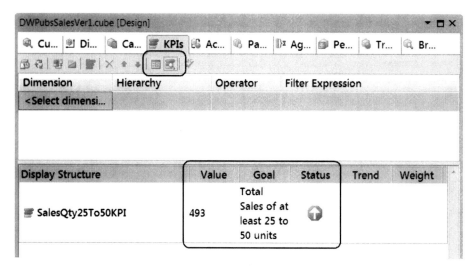

Figure 11-47. *Reviewing the KPI results*

■ **Important** A cube must be processed before Visual Studio can display the results. The reason for this is that the data is not derived from the XML files in Visual Studio but rather from the database on your SSAS server. Notice that the Reconnect button is available on the KPI tab just as it is on the cube Browser tab. You can use this to reconnect to your server after the cube has been processed.

The number 493 is greater than 25 to 50; therefore, the status indicator is an upward-pointing arrow. This particular result is somewhat meaningless, because what we would really like to see is how a particular title fared within that sales quantity range. To adjust this, we need to filter out the value that we want to test from the cube. We can do this by defining the dimension hierarchy operator and filter expression just above the results pane, as shown in Figure 11-48. Each of these columns holds a dropdown box allowing you to choose which item to use. By selecting a dimension hierarchy operator and a particular title, we can see what category of sales quantity that particular title falls under.

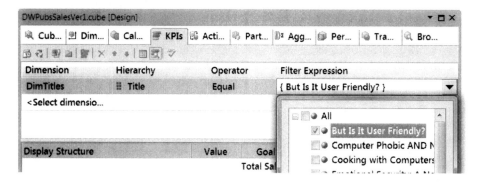

Figure 11-48. *Filtering the KPI results*

Once you have chosen a particular title, clicking back in the results pane refreshes the screen and displays your KPI value (Figure 11-49). Sadly, this tool shows you only a single report value at a time, but you can check the first three titles by changing the filter expression one title at a time.

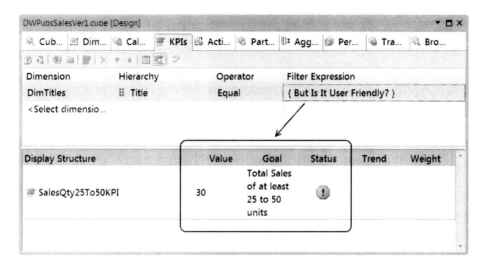

Figure 11-49. *Viewing the filtered KPI results*

The end user does not have this restriction. Applications such as Reporting Server and Microsoft Excel can display the KPI values for all titles at once, as shown in Figure 11-50. (Notice that the icons in this figure are not an exact match for the icons used in SSAS. This is because Excel has its own set of icons and reads the suggestions that you have configured on SSAS only to determine which icons should best be used.)

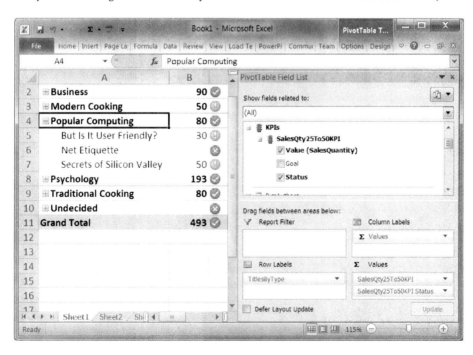

Figure 11-50. *Using Excel to view the KPI results*

479

We talk more about making Excel reports in Chapter 15, but for now let's create a KPI for our cube in the following exercise.

EXERCISE 11-4. ADDING A KPI TO YOUR CUBE

In this exercise, you add a KPI to the DWPubsSalesVer1 cube. Once you have configured the KPI, we use the browser feature that comes with Visual Studio to test a few titles.

Note: If the solution from the previous exercise is closed, open Visual Studio. Remember to select Run as Administrator and answer Yes to close the UAC.

Creating the KPI

1. Navigate to the KPI tab of the cube designer window, as shown in Figure 11-46.

2. Click the New KPI button on the toolbar (circled in Figure 11-46). This creates a new KPI.

3. Change the name in the Name textbox to **SalesQty25to50KPI**, as shown in Figure 11-46.

4. In the "Associated measure group" dropdown box, select the Sales measure group (Figure 11-46).

5. In the Value Expression textbox, type in **[Measures].[SalesQuantity]**, as shown in Figure 11-46.

6. In the Goal Expression textbox, type in **"Total Sales of at least 25 to 50 units"** (including the quotation marks), as shown Figure 11-46.

7. Under the "Status indicator," select the road signs icon using the dropdown box (Figure 11-46).

8. In the "Status expression" textbox shown in Figure 11-46, type in the MDX expression indicated in Listing 11-5.

    Listing 11-5. Grouping KPI Values into Three Categories

    ```
    case
      when [SalesQuantity]<25 or null then -1
      when [SalesQuantity]>= 25 and [SalesQuantity]<= 50 then 0
      when [SalesQuantity]>50 then 1
    end
    ```

Process the Cube

1. Right-click the cube in Solution Explorer, and select Process from the context menu. This builds, deploys, and processes your cube and sends the changes for your new KPI to the server.

2. Complete the processing step and close all the dialog boxes when it has completed.

Testing the KPI

1. Click the Browser View button on the KPI toolbar of the KPI tab (as circled in Figure 9-47).

2. Change the filter so that only the quantity for a single title is shown. You can do this by setting Dimension to DimTitles, Hierarchy to Title, Operator to Equal, and the filter expression to a single title, as shown in Figure 9-48.

3. Click the results pane to evaluate the new filter expression. The results should now show the sales quantity of a single title and indicate whether it met the goal of 25 to 50 units.

In this exercise, you created and tested a KPI. KPIs are useful additions to reports, and many managers prefer them to tabular pivot table reports.

Moving On

We have now created a cube and examined the first five tabs of the cube designer window, but we still have another five tabs to go! In the next chapter, you learn how these other tabs are used to configure a cube; we also show you some additional dimension configurations that are not needed for our current DWPubSales project but are commonly used in the industry.

LEARN BY DOING

In this "Learn by Doing" exercise, you configure dimensions similar to the ones defined in this chapter using the Northwind database. We have included an outline of the steps required as well as an example of how we handled them in two Word documents. These documents are found in the folder C:_BookFiles_LearnByDoing\Chapter11Files. Please see the ReadMe.doc file for detailed instructions.

What's Next?

You now know how dimensions and cubes are interconnected and have learned several cube configurations. In the previous chapter, we recommended the book *Expert Cube Development with Microsoft SQL Server 2008 Analysis Services* by Marco Russo, Alberto Ferrari, and Chris Webb (Packt PublishingISBN-10: 1847197221). We recommend that book once again for the subject matter covered in this chapter.

CHAPTER 12

Additional Cube and Dimension Configurations

There are two mistakes one can make along the road to truth . . . not going all the way, and not starting.

—Gautama Siddharta

In the previous chapter, we handled the first half of the cube configuration tabs. This chapter is a continuation of Chapter 11, and we discuss the second half of the configuration tabs. We also cover some additional dimensional designs that, although not required in the Publication Industries cubes, are common to BI solutions.

Additional Cube Configurations

Of the ten cube configuration tabs, we still need to talk about the Actions, Partitions, Aggregations, Perspectives, and Translations tabs. Let's examine them now.

Note Only one of these configurations is needed for our Publication Industries BI solution, as you will see in the exercise near the end of this chapter. We did, however, make sure to show you the steps necessary to perform these configurations on your own if you feel like experimenting. Just remember to use a copy of your solution if you do! You can also use the copy of the solution we provided in the `C:\_BookFiles\Chapter12Files\Chapter11_Complete` folder.

Actions

Actions give SSAS cubes the ability to drill down into the values of a dimension or utilize other applications to provide more information beyond what is contained within your cube.

This feature has been in SSAS for more than a decade. And yet, after all this time, only a few clients are designed to use actions. But there is hope! Excel 2007 and Excel 2010 support actions, and we anticipate other clients becoming more savvy in the future as well. Let's take a look at how actions are configured.

To add an action to your cube, activate the Action tab and click one of the three buttons (New Action, New Drillthrough Action, and New Report Action) circled in Figure 12-1.

Figure 12-1. *The Actions tab and toolbar*

Tip   You can see each button title in a tooltip when you hover over them.

URL Actions

The first circled button, New Action, allows data to be sent from the cube to another application for processing. One common example of a URL action is to add dimension values to a web application using an HTTP query string.

To create a new action of this type, click the first of the three buttons to open the configuration options. The tooltip calls this button New Action. Type a logical name for your action into the Name textbox. We titled ours **URL Action Demo** in Figure 12-2. Setting the "Target type" dropdown box to Cell allows you to click a report cell to activate the action.

Filter out which cells allow actions to occur by using the "Target object" dropdown box. In Figure 12-2, we kept the default "All cells."

Define the action's type using the Type dropdown box. This determines the manner in which data is passed to an application. The URL type is one of the most useful. It allows you to build a string that will be passed to the client's default web browser.

Note   All textboxes can be expanded and collapsed using the double arrow button next to each textbox. We have collapsed the Condition textbox in Figure 12-2 to more easily display the configuration values in the other textboxes.

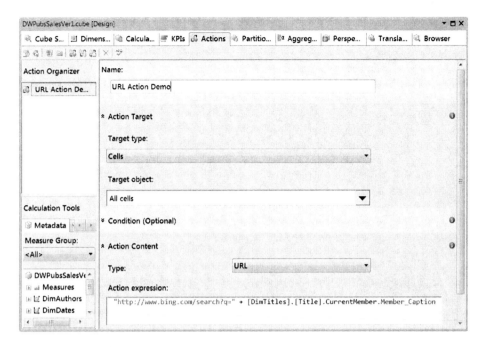

Figure 12-2. *Configuring a URL action*

To build the URL string, enter the code from Listing 12-1 into the Action Expression textbox.

Listing 12-1. An Action Expression Example

```
"http://www.bing.com/search?q=" + [DimTitles].[Title].CurrentMember.Member_Caption
```

You can test the action in Excel. This example action functions as follows: when a value is clicked in an Excel report, a web browser launches and passes the action expression as a URL address string. The string in our example uses the Member_Caption property to extract a title name and passes it onto the website www.bing.com.

This works well as long as the string you send matches the format required by the receiving web server. Our current example works for Bing; it will not work for Google, because Google requires a different query string format in the URL. If you adjust it to match Google's requirement, however, it will work for that site as well.

Another, and perhaps more powerful option, is to pass queried data to a custom ASP.NET application. Using this option, you can create your own custom ASP.NET application that uses the string data to launch custom reports or perform additional actions.

To launch a URL action in Excel, begin by creating a pivot table report that accesses your cube. Then create a result that includes measures and the dimensional attribute your action is associated with. In our case, that is the Titles dimension and the Title attribute. Next, right-click the measure cell and select the Additional Actions option from the context menu. Finally, choose which action you want to launch (Figure 12-3).

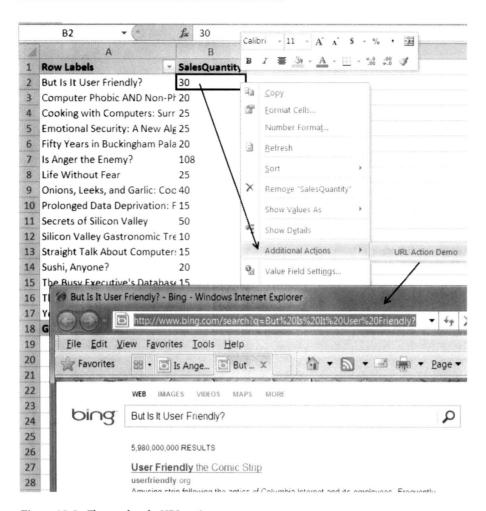

Figure 12-3. *The results of a URL action*

Drillthrough Actions

The second action button in the toolbar (circled in Figure 12-2) creates a drillthrough action. These actions allow you to query the cube for additional data. The data that is returned is defined in the Drillthrough Columns grid, which consists of several dropdown boxes that allow you to choose which dimensions, and members of those dimensions, to query (Figure 12-4).

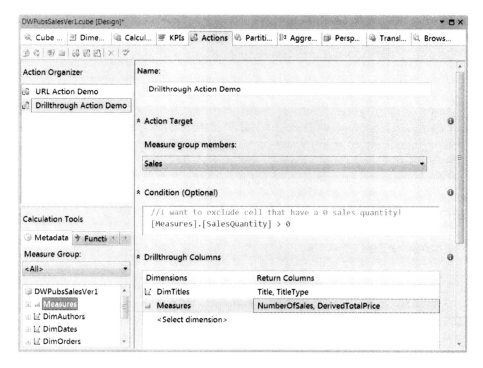

Figure 12-4. *Configuring a drillthrough action*

When a client application activates a drilldown action, the results of the dimensions and members are returned to the client applications. This is like submitting an MDX query to the SSAS server. The displayed results are dependent upon the client application. For example, in Excel 2010, the results will display on a new worksheet.

To configure the drillthrough action, enter an appropriate name in the Name textbox. Then, identify the measure group that you want the drillthrough to be activated on. As you can see in Figure 12-4, we have identified the Sales measure group and named it **Drillthrough Action Demo**.

We have also added the condition that sales quantities must have a value greater than zero before the drillthrough action can be activated. This is an optional step, but it is helpful. For example, when drilling down to an item with zero sales, no results will be returned.

To specify what the results are going to be, identify which dimensions and attributes you want to drill into. Ask yourself, what type of report do you expect to get when the drillthrough action is activated?

You can choose one or all dimensions, including the Measures dimension, and one or all of the attributes from each dimension. Some restraint is recommended, because the report that is generated must make sense to the end user. In our example, we have chosen two columns from the DimTitles dimension and two columns from the Measures dimension to display in our report.

Launch this action by clicking a cell, and select Additional Actions from the context menu (similar to Figure 12-3, but this time choose Drillthrough Action Demo).

■ **Note**   To make the new action appear in Excel, you must process the cube and refresh the data in the Excel report using the Refresh button on the Data tab (Figure 12-5).

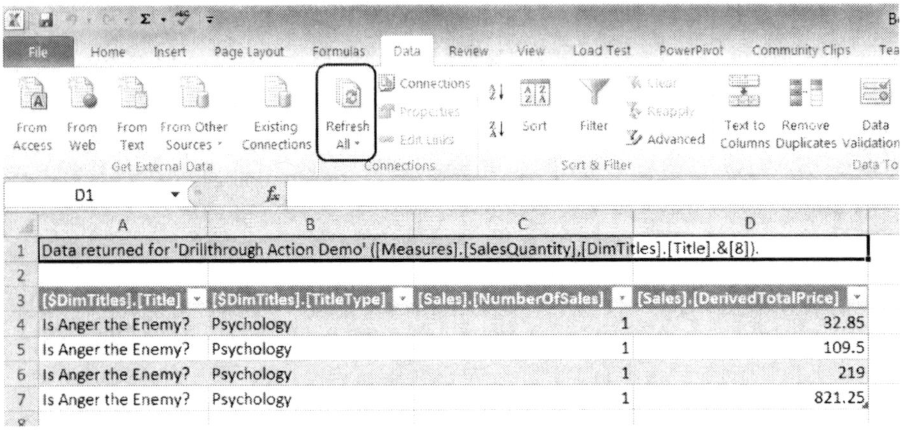

Figure 12-5. *A drillthrough action report in Excel 2010*

Report Actions

Report actions are similar to URL actions with the exception that they are designed to pass query string data to a Reporting Server report. For that reason, you must have Reporting Server (SSRS) installed and the SSRS web service running (which we discuss further in Chapter 16).

To create a report action, click the third button, New Report Action (circled in Figure 12-1). This creates a new report action and displays an editing window.

Add a name in the Name textbox. In our example, we called it *Report Action Demo*. Next, choose the target type. We used cells in Figure 12-6, which means that a user clicks a report cell to launch the report. We also configured "Target object" to "All cells," but they can be filtered. You can also provide a conditional statement, but we will not be doing so here.

The *Server name* and *Report path* settings must be exact for it to find the report page and its folder location. If you are unfamiliar with SSRS, this may not have a lot of significance to you, but we will be discussing this in Chapter 13.

In the configuration shown in Figure 12-6, we have identified the SSRS installation on Randal's computer, which uses a named instance of SQL. The standard setting when using an SSRS installation on your computer would be something like `Localhost/ReportServer/Pages/ReportViewer.aspx?`. And, the name localhost would be configured to your reporting server in a production environment.

The path to the report, in the "Report path" textbox, is dependent upon the name of your report and the SSRS folder it was deployed to. In our example, CubeOptionsActionReport is the name of the SSRS folder, and TitlesReport is the name of the report we created and deployed for this demonstration.

Beneath that is the optional Parameters section, which can be expanded using the double arrow symbol. Here you identify the parameter data to pass onto your SSRS report. In our example, we are passing the TitleName, so the report generates different output based upon an individual title from the cube.

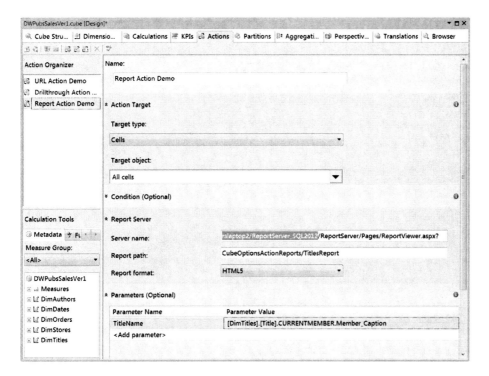

Figure 12-6. *Configuring a Report Action*

Once the action has been configured and the cube has been processed, you can create reports from Excel. By right-clicking an item and choosing Additional Actions from the context menu, you can launch the Report Action Demo, which in turn launches your default web browser and navigate to the Reporting Server report, as shown in Figure 12-7.

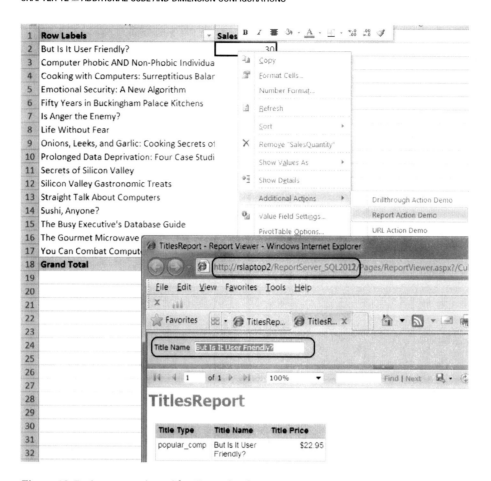

Figure 12-7. *A report action with a Reporting Server report*

■ **Note**   Prior to SQL 2012, you could also test all of these actions using a cube's Browser tab, but SQL 2012 uses a different browser than previous versions. The new browser looks very similar at first, but over time you will notice subtle differences, such as the one just noted.

Partitions

The Partitions tab allows you to create and configure cube partitions. Partitions are a way of dividing a cube into one or more folders. These folders are usually placed on one or more hard drives and independently configured for increased performance.

Classic examples include configuring one partition to store a large number of aggregations, while other partitions store little to no aggregations. The idea is that the first partition holds data, and aggregations, that are queried on a regular basis, and that having stored aggregations increases reporting performance. Other partitions, which are not queried on a regular basis, would not need to be stored, because normal report for performance should be sufficient.

Every time you create a cube, you will have one default partition for each fact table. If you want to create more partitions, do so by configuring the default partition to exclude some of the data. An example of this would be to create a partition that held only the most recent year's data by configuring the *Partition Source* dialog window to filter out all other data.

Partition Sources

The Partitions Source dialog window is used to define which source data will be included in a particular partition. You can access this dialog window by navigating to the Partitions tab and clicking on the ellipsis button under the Source column as shown in Figure 12-8. When you first open this tab the ellipsis button is not displayed, but clicking on a cell under the Source column forces the ellipsis button to appear. Once it does you can click it to launch the Partitions Source dialog window.

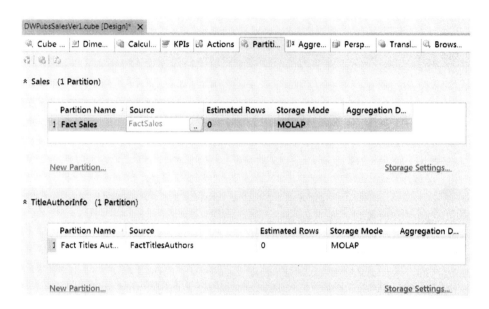

Figure 12-8. *The Partitions tab of the cube designer*

To use the Partition Source dialog window, start by configuring the partition's *Binding type*. The default binding type is *Table Binding*. With this binding type, select the table that holds your partition data and do not define any other filters. For example in Figure 12-9 we configure a partition to use all of the FactSales table for its data.

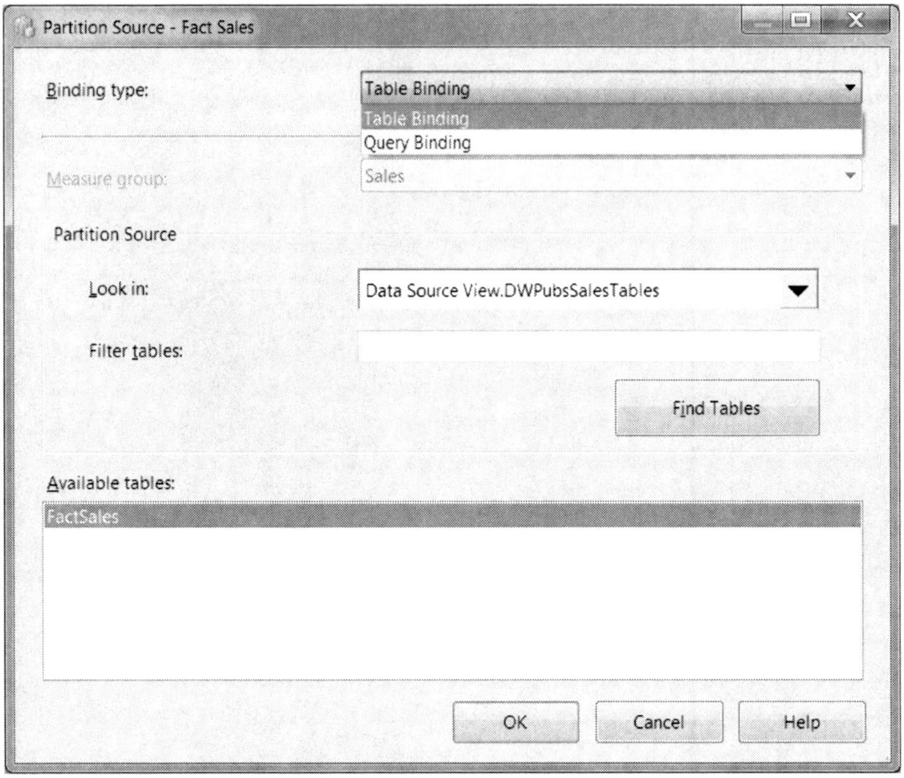

Figure 12-9. *The Partition Source dialog window*

If you want to change the binding type, use the dropdown box and select Query Binding. With query binding, enter a SQL statement that represents the results of a query against your data warehouse tables. In Listing 12-2, we created a SQL statement that filters out any data from the fact table that is not equal to the year 1994.(The last year in the Pubs database.)

Listing 12-2. A Statement that Filters Data from a Fact Table

```
SELECT
  FactSales.OrderNumber
, FactSales.OrderDateKey
, FactSales.TitleKey, FactSales.StoreKey
, FactSales.SalesQuantity, DimTitles.TitlePrice AS CurrentStdPrice
, DimTitles.TitlePrice * FactSales.SalesQuantity AS DerivedTotalPrice
FROM FactSales
INNER JOIN DimTitles
  ON FactSales.TitleKey=DimTitles.TitleKey

INNER JOIN DimDates
  ON FactSales.OrderDateKey=DimDates.DateKey
WHERE DimDates.[Year]=1994
```

After creating and testing the query, we can change the Partition Source to Create Binding and add the SQL statement to the query textbox, as shown in Figure 12-10.

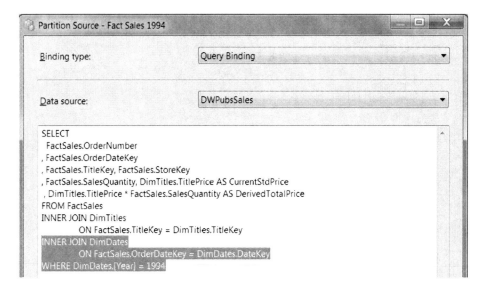

Figure 12-10. *Using query binding to filter a partition source*

Now the partition will only hold data that is associated with the year 1994. No other data in that fact table will be included in the cube with this current configuration. For data to be included for years other than 1994, you will have to create another partition.

The Partition Wizard

To create another partition, click the New Partition hyperlink, as shown in Figure 12-11. This launches the Partition Wizard.

As you navigate through the Partition Wizard, the third screen allows you to restrict the rows by entering a SQL query. The SQL statement you will use will be quite similar to the one shown in Listing 12-2, but this time we include the argument `Where Not DimDates.Year=1994`, as shown in Figure 12-11.

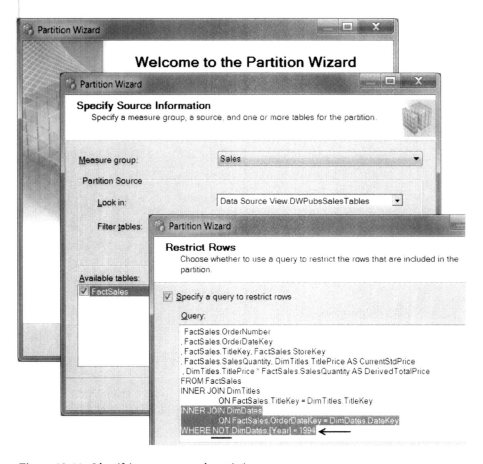

Figure 12-11. *Identifying a source and restricting rows*

Now you can click Finish and accept all the default options for the next two screens, or you can select Next to move to the Processing and Storage Locations screen, as shown in Figure 12-12. From this screen, you can see that "Storage location" defines which hard drive and folder path the partition data is stored on. Notice the drive letter in Figure 12-12 is currently C:\, which is also the drive used by the Windows operating system. Placing partition folders on the same drive as the OS decreases performance, because Windows and SSAS must then compete for access to hard drive resources. Placing a partition folder on its own hard drive and/or dividing partition folders across mutiple hard drives are two common techniques for increasing performance.

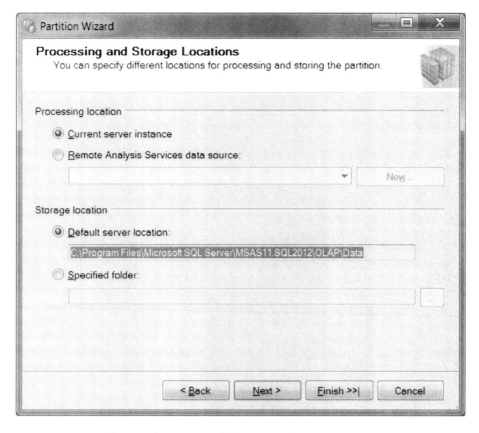

Figure 12-12. Configuring the partitioned data storage location

---

▪ **Tip** Another technique used with partition folders is to spread them across multiple hard drives using a RAID array. A RAID array increases performance implicitly without the administrative overhead of defining partition locations explicitly.

After you have identified the location of your partition's folder, you can click Next, which takes you to the Completing the Wizard window. On this screen, you are given the option to name your partition. In Figure 12-13, we have named ours **Fact Sales Older than 1994**.

In this window, you can also design your aggregations immediately or delay them. Designing aggregations involves determining which totals and subtotals to include within the cube partition.

The radio button default setting is "Design aggregations for the partition now." Keeping the default setting and clicking Finish starts the Aggregation Design Wizard.

If you check the "Design aggregations later" radio button (Figure 12-13), you can launch the Aggregation Wizard from the cube designer's Aggregations tab at a later point in time. This option is often a good choice, because it allows you to return later to create all your cube partitions at one time.

Figure 12-13. *Configuring the location and aggregation settings*

The last radio button, "Copy the aggregation design from an existing partition," allows you to do just what it says, copy a previously created partition. If this radio button is checked, then the "Copy from" dropdown box will activate, and you can select the aggregation to copy. You can then click Finish to complete the wizard.

After the wizard completes, the Partitions tab shows the name of your new partition along with either the table or the select statement that identifies its source, the current estimated number of rows in the partition, and the storage design (Figure 12-14).

DWPubsSalesVer1.cube [Design]* ✕

Cube ... | Dimen... | Calcul... | KPIs | Actions | Partiti... | Aggre... | Persp... | Transl... | Browser

⌃ Sales   (2 Partitions)

	Partition Name	Source	Estimated Rows	Storage Mode	Aggregation Design
1	Fact Sales 1994	SELECT FactSales.Ord...	0	MOLAP	
2	Fact Sales Older than 1994	SELECT FactSales.Ord...	0	MOLAP	

New Partition...   Storage Settings...

Figure 12-14. *The Partitions tab after two partitions have been created on the Fact Sales table*

One thing to remember is that typically, a single partition is all you really need. If at some point you realize that you have defined more partitions than you need, you can delete a partition by right-clicking the partition and selecting Delete from the context menu (Figure 12-15).

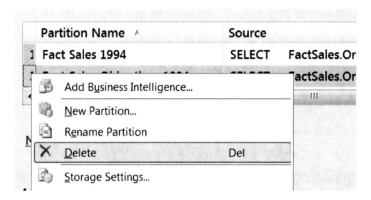

Figure 12-15. *Deleting an existing partition*

■ **Caution**   Deleting the partition does not mean the data the partition used to contain will be magically added to the other partitions. You must change the source description or SQL code on any existing partitions to make sure you are obtaining all the data. For example, we could delete both partitions defined in Figure 12-14, but we would then have to make a new partition that covers all years by selecting the entire Fact Sales table as the source.

Partition Storage Designs

Earlier, in Figure 12-14, you may have noticed two hyperlinks on the Partitions tab. The first hyperlink allows you to launch the Partition Wizard, but the second hyperlink opens the Partition Storage Setting dialog window.

In this dialog window you can choose how data is stored for a particular partition. Partition data falls into three categories of data: leaf level (individual values), aggregated total and subtotal values, and metadata that describes the partition.

Data storage designs also fall into three categories: MOLAP, HOLAP, and ROLAP (Figure 12-16). Storage locations, however, fall into two categories: an SSAS database folder or tables in a relational database.

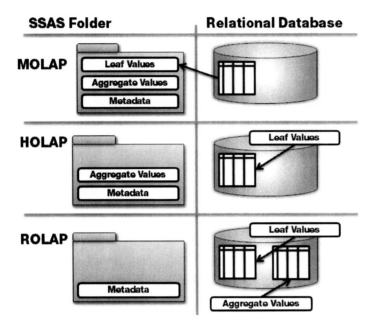

Figure 12-16. *Partition storage design categories*

The partition data leaf-level values are the row values within the data warehouse tables. They are normally copied from the data warehouse into the SSAS database folder. Aggregated total and subtotal values are created by SSAS, and are normally stored within the SSAS database folder as well, but are created when the cube is processed. Partition metadata is always stored by SSAS in the database folder and is created when the cube is deployed and then updated when the cube is processed.

By default all three data categories are stored in the SSAS database folder. This particular storage design is referred to as MOLAP for short and stands for Multidimensional Online Analytical Processing. That is quite a mouthful for a name that is somewhat meaningless. It simply indicates that you are storing all the data into folders managed by a multidimensional OLAP cube server such as SSAS.

It is also possible to store some partition data elsewhere. For example, you can choose to keep the leaf-level values within the tables of the data warehouse and store only the aggregate values in the SSAS database folder. When this is done, the configuration is considered a hybrid approach, or HOLAP for short (Figure 12-16).

A third possibility is to store both the leaf-level values and the aggregate values within the relational database. This is known as a ROLAP configuration (Figure 12-16).

As we mentioned, the SSAS database metadata will always be stored within the database folder regardless of your configuration choices.

We recommend using a MOLAP design for most situations (which is the default). This design gives you the best report performance at the cost of storing a copy of all the leaf-level values within the SSAS database folder. In other words, if you have 6 million rows of data in your data warehouse tables, you are going to have to store the equivalent of 6 million rows of data in the SSAS database folder.

ROLAP and HOLAP

SSAS is very good about compressing leaf-level data, so for most situations, this is still considered a good option. In those rare cases where having two copies of the data is just too much of a strain on your environment, you can choose to use either a hybrid OLAP or a relational OLAP design (HOLAP and ROLAP).

Another occasion where a ROLAP design may be convenient is when report data needs to be as real time as possible. For example, consider a situation where a cube is built on top of an OLTP database that tracks patient medication data for a hospital. If a doctor checks a report to find out what medication a patient has been given, the time between entering the medication dispensing data into the database and the time that it shows on the report must be as short as possible.

When the cube partition is designed as ROLAP, report queries sent to the cube are passed through the SSAS server to the data warehouse in real time. In our example, the data warehouse is an OLTP database that records patient medications; therefore, reports through SSAS will show medication data as soon as it is entered into the OLTP database.

With a ROLAP design, you may ask yourself, "Couldn't I get the same reports directly from the OLTP database?" The answer is, "yes!" And, in fact, if you only occasionally need immediate reports, creating a ROLAP cube can be considered redundant.

Still, there some advantages to ROLAP. It is true that you bypass all the performance gains normally associated with the SSAS server, but you still gain the capacity to rename columns, create user-defined hierarchies, and associate dimensions with cube measures.

In the BI world, ROLAP designs are rare, and the same is true of the hybrid approach, HOLAP. HOLAP cubes are useful on occasions when reports deal with aggregate values and not individual leaf values. Examples would be reports designed as dashboards with KPIs. These reports require only the higher level, pre-aggregated values, and not the leaf level data (which is stored back in the OLTP database). Therefore, report performance is improved without having to store a copy of the leaf-level data on the SSAS computer.

Partition Storage Settings

The Partition Storage Settings dialog window allows you to configure a partition's storage design (as shown in Figure 12-17). MOLAP is the most common choice, and it is divided into four categories. Microsoft has included

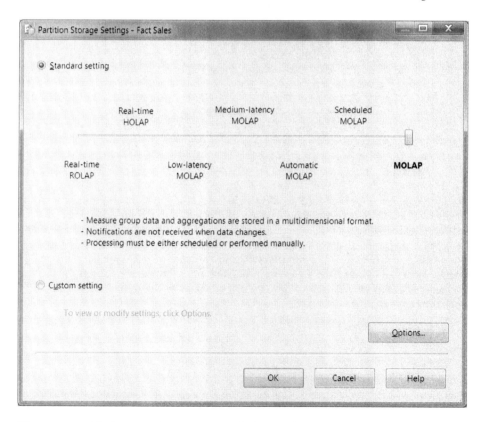

Figure 12-17. The Partition Storage Settings dialog window

a sliding bar user interface to choose between the different storage settings. Each of these categories provides a set of predefined settings that can be customized even further by selecting the Custom Settings radio button and clicking the Options button at the bottom of the dialog window.

Once you have clicked the Options button, a Storage Options dialog window appears, as shown in Figure 12-18. In this dialog window, you can choose between using MOLAP, HOLAP, or ROLAP using the "Storage mode" dropdown box. This has the same effect as using the slider bar in the previous dialog box.

In the default configuration, each partition is set to the storage mode of OLAP without the proactive cache option enabled.

Selecting "Enable proactive cache" checkbox (circled in Figure 12-18) provides a way of updating data between the data warehouse and the partition in an automated manner. For example, if you look at the settings in Figure 12-18, we have checked the box that allows SSAS to update the cache whenever data changes within the data warehouse.

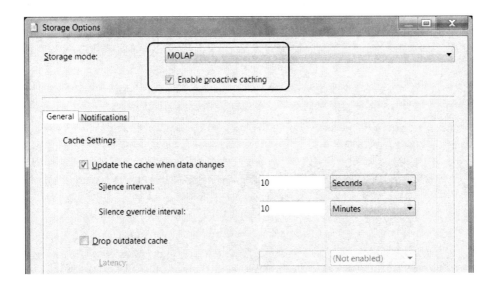

Figure 12-18. *The Storage Options dialog window*

The "Silence interval" option indicates the time SSAS should wait before it attempts to process the partition after a transaction has been sensed within the data warehouse tables. The default is ten seconds. This means that SSAS waits for ten seconds of no transaction activity before it attempts to process new data into your SSAS partition.

If transactions are happening so rapidly that a ten-second pause is not occurring, the "Silence override interval" setting will force processing to begin after ten minutes (while in the default configuration).

Automated updates may seem like a pretty good idea. And in fact they are, if they are set up correctly. But, there are a few things that need to be considered. Notification that the data has changed within the data warehouse is done in one of three ways:

- Setting up a job to poll for changes

- Creating an application that polls for changes

- Creating database triggers

For many developers, none of these options seem all that attractive. Because of this, many companies do not enable the proactive caching with automatic updates, but instead rely on scheduled processing that occurs each night or several times a day. In these cases, employees are notified that the report data is "up-to-date" based on the last time the cube was processed and is not real-time data. This is by far the most common approach, which is probably why it is also the default setting whenever you create a new partition!

Aggregations

One of the reasons for using an SSAS cube is that it provides increased reporting performance. Cube based reports are faster, in part due to SSAS's ability to create and store preprocessed totals and subtotals (collectively referred to as aggregations). However, by default, an SSAS cube does not create and store aggregations. Instead, you have to configure them yourself and then deploy and process them.

To do this, navigate to the cube designer's Aggregations tab and use one of the buttons on the toolbar or on the context menu (Figure 12-19). Both launch the Aggregation Design Wizard.

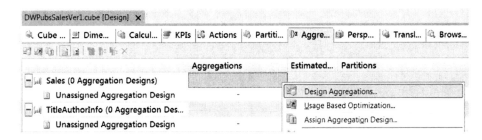

Figure 12-19. *The Aggregations tab*

The Aggregation Design Wizard

The Aggregation Design Wizard opens with a typical welcome dialog window. After clicking the Next button, you are presented with the Review Aggregation Usage dialog window pictured in Figure 12-20. Indicate which, if any, dimensions you want to use. If you choose to include aggregations for a dimension, the aggregate totals and subtotal will be created and stored for each of the cube's measures associated with that dimension.

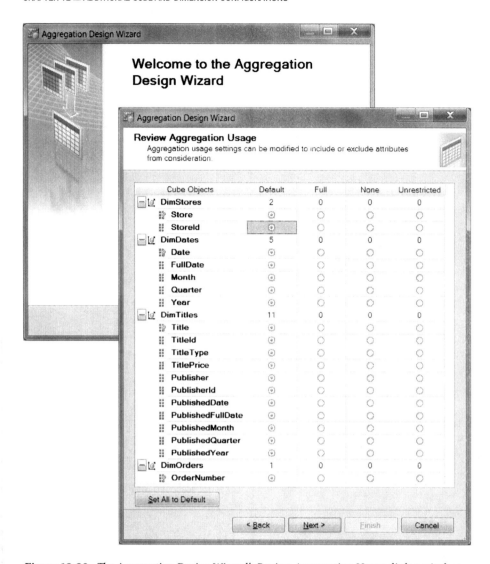

Figure 12-20. *The Aggregation Design Wizard's Review Aggregation Usage dialog window*

As shown in Figure 12-20, there are four radio button options for each dimensional attribute. Table 12-1 gives an overview of each option.

Table 12-1. *The Review Aggregation Usage Options*

Option	Description
Default	The attribute may qualify for aggregation. The designer chooses aggregations for you using a Microsoft algorithm, which is based upon the type setting of attribute and dimension. The chosen aggregation will be considered as the wizard continues.
Full	The attribute will qualify for aggregation, and all aggregations will be considered as the wizard continues. This setting can cause excessive processing time if there are a lot of rows in the dimension tables.
None	The attribute will not qualify for aggregation, and no aggregation will be considered as the wizard continues.
Unrestricted	This uses a modified version of Microsoft's default algorithm, where no restrictions are put on the aggregation designer; however, the attribute must still be evaluated to determine whether it is a valuable aggregation candidate.

Clicking *Set All to Default* at the bottom of this window resets all of the settings for each attribute to the default radio button. In most cases, this is an appropriate choice, because Microsoft's algorithms have proven to be competent at choosing which aggregations are most useful.

Clicking Next brings you to the Specify Object Counts dialog window (Figure 12-21) that contains Count and Stop buttons. For the wizard to continue, you must let SSAS count how many rows are in each dimension table. Clicking the Count button starts this procedure.

Usually it takes only a few minutes or less for it to count the rows, but it will take a very long time if there are many rows in each table. As such, Microsoft has provided the Stop button in case this takes too long. But be aware that you cannot continue through the other screens of the wizard without having created an estimated count for each table, so stopping the count is only useful if you plan to try once again, when there is less activity on the network. Therefore, once you click Stop, you must exit the wizard and create the aggregations at another point in time.

■ **Note** This may seem quite tragic until you realize that cube based reports work just fine without stored aggregations. The difference is that they are just not as fast as they could be.

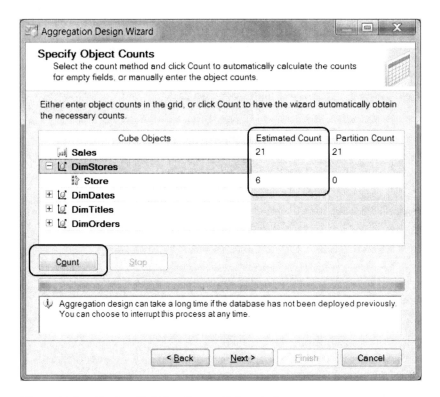

Figure 12-21. The Specify Object Counts dialog window

The Set Aggregation Options Dialog Window

Clicking Next takes you to the Set Aggregation Options dialog window where you are able to choose how many aggregations you want in the partition. Microsoft has created an algorithm that prioritizes which aggregations are most beneficial, so the point of this dialog window is to determine how much space you intend to devote to the aggregations that Microsoft's algorithm chooses for you.

By default, this dialog window is set to the maximum devoted hard drive space in megabytes, but you can also specify gigabytes. You can see this setting pictured in Figure 12-22. When you first access the dialog window, the estimated storage size will be set at 100 MB. Once you click Start, the same dialog window will replace the image on the right side of the window with a graph indicating how much space is being used on the hard drive compared to how much performance you may have gained.

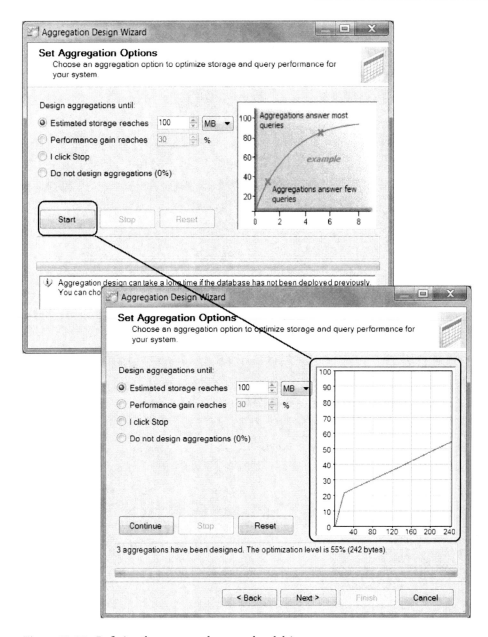

Figure 12-22. *Defining the space used on your hard drive*

On the left side or y-axis of this graph, the numbers are from 0 to 100 percent performance gain. The bottom of the graph, or x-axis, indicates the amount of estimated space necessary for this particular amount of performance gain. Just beneath the graph is a text version describing how much space is being used.

You may have noticed in Figure 12-22 that an estimated 55 percent optimization will be reached in the first 242 bytes of storage space on the hard drive. The amount in today's world is pretty miniscule; however, the wizard was created before the year 2000, and over the last decade it has never been updated.

Table 12-2 outlines the additional options in this dialog window.

Table 12-2. *The Review Aggregation Usage Options*

Option	Description
Estimated storage reaches	Limits the aggregations based on the maximum hard drive space devoted to them, in either megabytes (MB) or gigabytes (GB).
Performance gain reaches	Limits the aggregations based on the estimated performance gain percentage.
Clicking Stop	Limits the aggregations by clicking Stop during the selection process. (This option was more useful a decade ago when the selection process took longer to run.)
Do not design aggregations (0%)	Specifies that there are no aggregations. This option allows you to clear an existing aggregation.

The buttons at the bottom of this dialog window were not necessarily obvious when you first encountered this dialog window, but they make sense in the context of how the aggregation selection process works. The Start button, of course, begins the aggregation selection process, and the Stop button ends this process. However, note that the Start button changes to the Continue button while the selection process is active.

After the selection process has run its course, some aggregations will have been chosen. Clicking Next moves you to the Completing the Wizard dialog window. If you would like to make a change, such as increasing the space available for aggregations and running the selection process one more time, the Reset button allows you to clear all the aggregation choices and begins running Microsoft's selection algorithm once more.

In the last wizard dialog window, you can specify the name of this particular aggregation design. In Figure 12-23 we have chosen to call it **SimpleAggregationDesign** to indicate that it is a preliminary choice that we may revisit at a later point in time. In the lifecycle of the cube's partitions, you may need to adjust the aggregation rules to fine-tune the report versus processing performance. Remember that increasing stored aggregations may increase report performance, but it will be at the cost of the time it takes to process the partition, because the aggregations are created and stored during the processing.

■ **Important**   It may take several attempts to get the balance right in a production environment. Remember that your reports will always be able to pull aggregations from the SSAS server. No matter what you do, your reports will still look the same. What is being configured is how many aggregations will be stored on the hard drive, not how many aggregations will be available for your reports.

Figure 12-23. *Completing the Aggregation Design Wizard*

As the wizard completes, you can also choose whether to deploy and process the cube when you hit the Finish button or to delay starting the processing until a later point in time. The two radio buttons on the dialog window allow you to choose between these options, as shown in Figure 12-23.

Processing can take several hours; therefore, most companies automate processing to run at night. Most developers will be designing the aggregations during the day; therefore, choosing to save the aggregations without processing them immediately will often be a good choice.

When the aggregation wizard completes, the named aggregation design is listed on the cube designer's Aggregations tab (Figure 12-24). You may notice that only three aggregations were chosen for this partition. This number is so small because there are very few attributes and measures in our cube's partitions. And even if we had set the allowable hard drive space to 1000 GB, Microsoft's algorithm would not have been able to find enough aggregates to fill up that space.

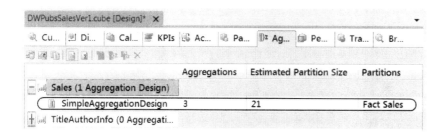

Figure 12-24. *The named aggregation design on the cube designer's Aggregations tab*

In general, the number of aggregates that the algorithm considers to be useful is rather small. But your report users won't notice, because non-stored aggregates are calculated very quickly when a report query runs.

Note Microsoft's website has some surprisingly good articles about this subject that are well worth viewing for further reading on this topic. Search the web using the keywords, "SSAS Aggregation Design."

Perspectives

Perspectives are a named selection of cubes, dimensions, and measures. These are similar to a SQL view in that they allow you to filter what can and cannot be seen by the perspective. By default, whenever you create a cube, users can see every dimension and measure within that cube. If the cube has a lot of measures and dimensions, this can be confusing for your cube report builders. You could make several cubes in the same database, but perspectives allow you to select a subset of cubes, dimensions, and measures.

To create a perspective, navigate to the Perspective tab, and click the *New Perspective* button on the toolbar (circled in Figure 12-25). This allows a new column to appear on the Perspective tab. In this new column, called Perspective Name, provide a logical name for the collection of dimensions and measures to be displayed. Then select your measures and dimensions by selecting the checkboxes provided below this new column. For example, we unchecked the measures in the TitleAuthorInfo measure group, shown in Figure 12-25. Now, no one using this perspective in a client application like Excel will see those measures.

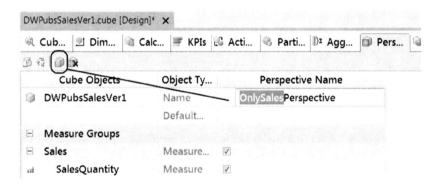

Figure 12-25. *Unchecking the measures in the TitleAuthorInfo group*

Unlike a view in SQL Server, an SSAS cube perspective cannot be used as a security mechanism. In other words, you cannot determine that only some users have access to the perspective and others do not. If a user has access to the cube, they have access to all of the perspectives in that cube. If you want to restrict users from one or more measures and dimensions, you need to create an additional cube instead of a perspective.

Translations

The cube designer allows you to define captions used in the cube for different natural languages. For example, in Figure 12-26, we added Spanish to the cube. Of course, just adding a language to the cube designer does you no good if you do not then add the captions for that language. So, we need to add the Spanish word for each caption, such as *Ventas* for Sales.

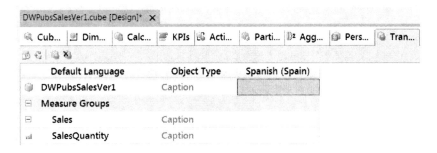

Figure 12-26. *Completing your configurations*

Browser

After configuring your cube using the various tabs, you need to deploy and process the cube before most changes will take effect. Afterwards, use the Browser tab to verify that all your configurations have been successful.

Additional Dimension Configurations

We are almost done with our look at how to create and configure SSAS cubes and dimensions, but there are still a few additional dimension configurations that we think you should know about. These configurations include the following:

- Parent-child dimensions

- Role-playing dimensions

- Reference dimensions

In Figure 12-27, we display a database diagram of the tables we use as examples for each dimension type. The DimEmployees table is used to create a parent-child dimension, the DimCategories table is used to create a reference dimension, and the DimDates dimension table is used to create role-playing dimensions.

■ **Note**   We will be using examples from a data warehouse we created for the Northwind database. This has the advantage of helping you with the "Learn by Doing" exercise at the end of this chapter.

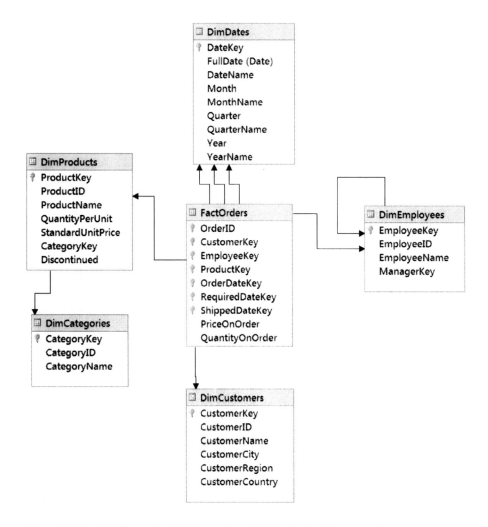

Figure 12-27. *The "Learn By Doing" version of the Northwind data warehouse*

Parent-Child Dimensions

Parent-child dimensions are created using a table that is self-referencing. The DimEmployees table has a manager key that references the employee key (Figure 12-28). This relationship indicates which employee manages other employees.

For example, the record for Nancy Davolio and several others have a ManagerKey of 11. And the record for Andrew Fuller has an EmployeeKey value of 11. Therefore, number 11 (Andrew Fuller) is listed as the manager for Nancy, Janet, Margaret, Steven, and Laura. One employee, Stephen Buchan—who is managed by Andrew Fuller—is the manager of three other employees: Michael, Robert, and Anne. This means there are three levels of organization between the employees: a manager, an assistant manager, and individual employees. And apparently Andrew Fuller has no one else overseeing him. He is his own boss.

EmployeeKey	EmployeeID	EmployeeName	ManagerKey
10	1	Nancy Davolio	11
11	2	Andrew Fuller	11
12	3	Janet Leverling	11
13	4	Margaret Peacock	11
14	5	Steven Buchanan	11
15	6	Michael Suyama	14
16	7	Robert King	14
17	8	Laura Callahan	11
18	9	Anne Dodsworth	14

Figure 12-28. The data in the DimEmployees table

When you have a table with self-referencing columns like this, you can create a parent-child dimension from it. An employee table is a classic example, but other examples exist such as tables that hold accounting data, such as the DimAccount table in Microsoft's AdventureWorks data warehouse.

Whenever you create a new dimension table that is self-referenced in the data source view, the Dimension Wizard automatically designs the dimension as a parent-child dimension. The most important property setting, and the one that distinguishes the parent-child dimension from a normal dimension, is the Usage property of the Parent attribute (Figure 12-29).

In our example, the Parent attribute is the ManagerKey, whose Usage setting has been defined as Parent. The most visual aspect of this setting is that the icon under the attribute list will change immediately to reflect that it is now considered part of a hierarchy. This configuration defines whether a dimension is a parent-child dimension or just a regular dimension.

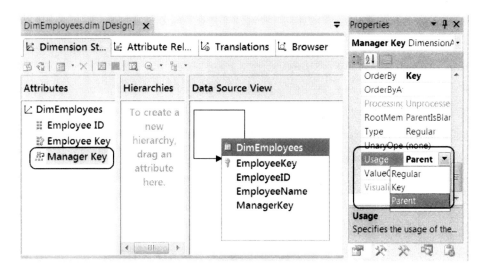

Figure 12-29. The DimEmployees dimension

You do not have to create a user-defined hierarchy in a parent-child dimension. The hierarchy is implied and will display in the dimension browser as expected once you have deployed and processed the dimension. Figure 12-30 shows an example of what you can expect to see in the browser. Notice that the hierarchy is named after the attribute, but you can still expand and see the various levels just as you would in a user-defined hierarchy.

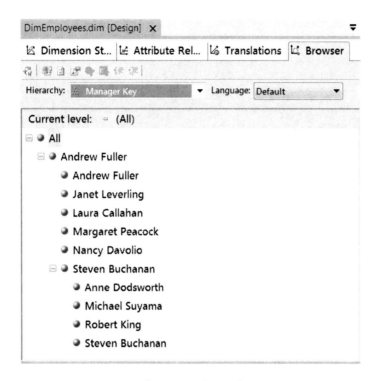

Figure 12-30. *Browsing the DimEmployees dimension*

When you first create the parent-child dimension, there will likely be a few adjustments to make. Perhaps the first adjustment is renaming the attributes to something more self-explanatory. In our example, we have renamed the ManagersKey to Employees, as shown in Figure 12-31. Now when the dimension is browsed, the word *employees* will represent a hierarchy of all employees.

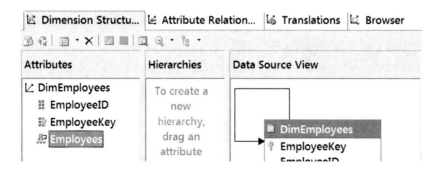

Figure 12-31. *Changing the attribute names*

Here are two examples you can change in the Parent attribute to make your parent-child dimensions more readable:

- *Set the MemberWithDataCaption property on the Employees attribute to * (Mgr)*: This adds an additional description indicating the value comes from the parent of a child member. So, in Figure 12-32, the manager has been appended to both Steve Buchanan and Andrew Fuller's values.

- *Set the Naming Template property to* Mgr;Asst Mgr;Emp: This changes the level names from the defaults level 02, level 03, level 04, and more, to a more descriptive name (Figure 12-32).

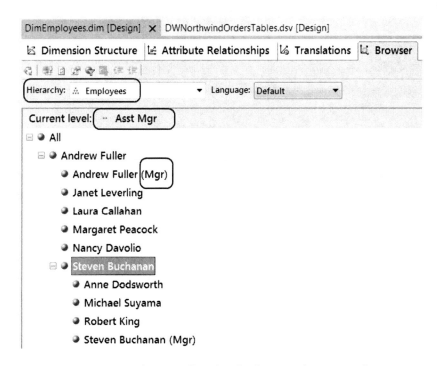

Figure 12-32. *Property changes reflected in the dimension's Browser tab*

Changes like these also show up in client applications such as Microsoft Excel and can make your reports more readable.

Adding a parent-child dimension to a cube is just like adding a regular dimension. There are no additional configurations necessary to make a parent-child dimension work with the cube, because the relationship is defined in the dimension and not within the cube. This is not the case for all dimensions. In dimensions such as role-playing and reference dimensions, the configurations are performed in the cube designer and not the dimension designer.

Role-Playing Dimensions

Role-playing dimensions allow you to reuse a single dimension multiple times. A role-playing dimension starts off as a normal dimension within the data warehouse and is created using the Dimension Wizard like any other.

However, when you use the Visual Studio's Cube Wizard, it determines if a single dimension, in our case the DimDates dimension, is represented multiple times based on the relationships defined in the data source view.

Because the data source view we are demonstrating has three relationships defined between the DimDates table and the FactOrders table, the wizard recognizes these and adds the DimDates dimension three times to the cube. And thus the DimDates dimension becomes a role-playing dimension. It is that easy; the only configuration you need to perform with role-playing dimensions is to make sure you rename them appropriately.

As you can see in Figure 12-33, the date dimension represents the Shipping Date, Order Date, and Required Date. From the perspective of the cube, they are considered three different dimensions. But, from the perspective of the SSAS database as a whole, it is only one dimension, DimDates.

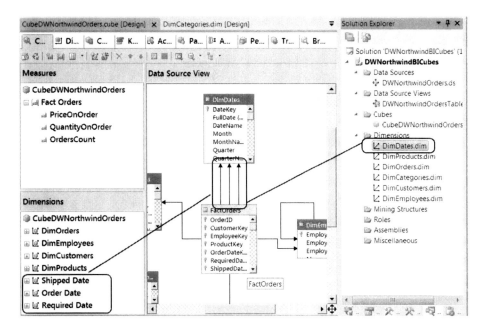

Figure 12-33. *The DimDates dimension is being used three times within the cube*

If you access the cube-editing window's Dimension Usage tab, you can see that all three dimensions are mapped from DimDates to the FactOrders table as regular relationships based upon the Date attribute, as shown in Figure 12-34.

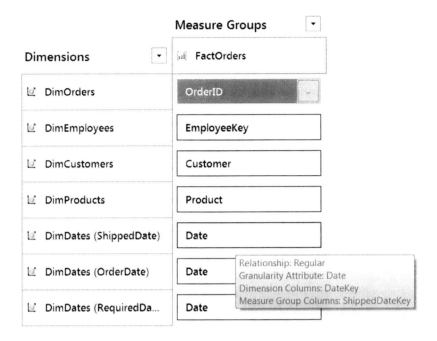

Figure 12-34. *The Dimension Usage tab indicates how the DimDates dimension table and fact table are related*

Reference Dimensions

Reference dimensions allow you to create a dimension that is indirectly linked to a fact table. To accomplish this, you must identify an intermediate dimension table. As an example, we use the DimCategories table, which is indirectly connected to the FactOrders table through the DimProducts table.

To create a reference dimension, start by creating the dimension as you would a regular dimension (Figure 12-35). Like role-playing dimensions, the configurations that turn a regular dimension into a reference dimension do not happen in the dimension editor. They happen within the cube. Therefore, there are no additional configurations to worry about in the dimension editor.

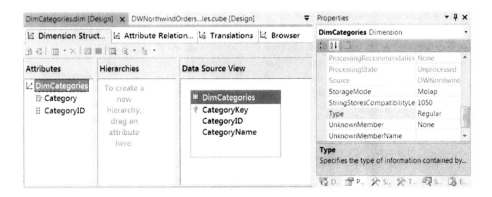

Figure 12-35. *Reference dimensions start out as regular dimensions*

In Figure 12-36 you can see that after our new cube was created, the Categories dimension was not automatically added to the cube. The Cube Wizard does not automatically create a reference dimension for you, as it does with role-playing and parent-child dimensions. The Cube Wizard identifies which dimension will be part of a cube based on direct connections from a dimension table to a fact table in a data source view. Because the DimCategories table and the FactOrders table do not have a direct connection, the wizard assumes that they are not related (Figure 12-27). This means that you must manually add a reference dimension to an existing cube, without the help of the Cube Wizard.

You can add it yourself by right-clicking in the Dimensions pane and choosing the Add Cube Dimension option from the context menu.

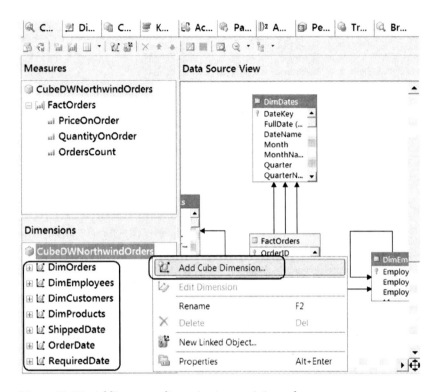

Figure 12-36. *Adding a new dimension to an existing cube*

Once the dimension has been added to the cube, access the Dimension Usage tab to complete its configuration. Unlike the parent-child and role-playing dimensions, a reference dimension is defined as a reference relationship. To configure this, click the ellipsis (. . .) as circled in Figure 12 -37 to access the Defined Relationship dialog window. Then use the Select Relationship type dropdown box and configure it to use the *Referenced* option, as indicated in Figure 12-37.

You also have to identify which table is used as the intermediate dimension. In our example, the intermediate dimension table is DimProducts.

The image shown on the left of the dialog window is useful to understand how the relationship exists from the perspective of the cube. Notice how the dotted line represents the logical relationship between the fact table and the dimension table and how the solid line represents the relationships as defined in the data source view.

The cube designer also needs to know exactly which dimensional attributes link the tables together. In this instance, that would be the Categories attribute, whose underlying column is the Category key in the data warehouse. Figure 12-37 shows how the "Reference dimension attribute" and the "Intermediate dimension attribute" settings are both configured to the Category attribute.

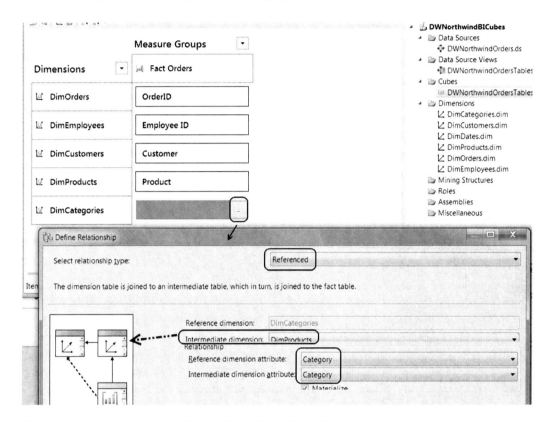

Figure 12-37. *Configuring the reference dimension's relationship*

Once we have added DimCategories to the cube, we can access category information within our DimProducts dimension or through the DimCategories dimension (Figure 12-38).

And in fact, if all we needed in a cube was the category data, we could exclude the products data from it altogether. You could simply hide the intermediate dimension by highlighting "Intermediate dimension" under the Cube Structure tab of the cube designer and then changing its visible property to False. Once that is set, most client applications will not see the intermediate DimProducts dimension at all.

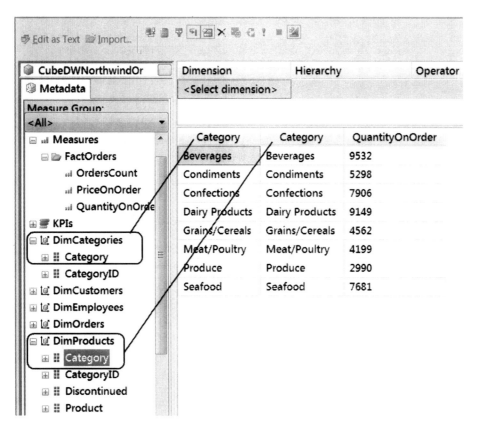

Figure 12-38. *Testing the reference dimension*

EVEN MORE DIMENSION CONFIGURATIONS

Many other dimensional configurations are possible, such as configuring slow changing dimension (SCD) attributes. These dimensions require additional columns in the data warehouse tables for tracking changes over time. For example, we could modify the DimTitles table to look like Figure 12-39.

DimTitles

	Column Name	Data Type	Allow Nulls
🔑	TitleKey	int	☐
	TitleId	int	☐
	TitleName	nvarchar(50)	☐
	CategoryName	nvarchar(50)	☐
	RecordStartDate	datetime	☐
	RecordEndDate	datetime	☑
	IsCurrent	int	☐
			☐

Figure 12-39. The DimTitles table modified to include SCD columns

Once these additional columns were added, you would change the dimension's configuration to include the new columns and set the attribute type of each column, as shown in Figure 12-40.

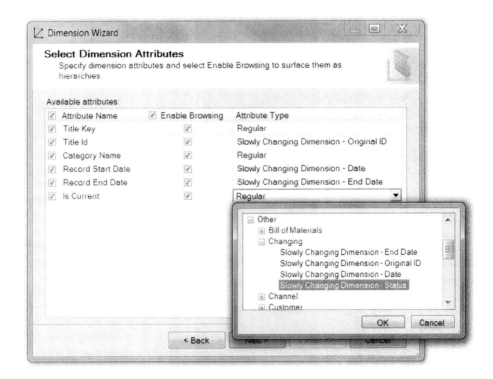

Figure 12-40. The DimTitles dimensions modified to include SCD attributes

You can find additional instruction on how to configure SCD and other dimensions on the author's website: www.NorthwestTech.org/ProBISolutions/SSASDemos

Managing Your Cubes and Dimensions

Our SSAS chapters thus far have focused on your role as a developer. Once the development of an SSAS database is completed, the administrator role needs to be addressed. SSAS database administrators can manage cubes and dimensions using SQL Server Management Studio or Microsoft's Visual Studio.

SQL Server Management Studio

Most administrators use SQL Server Management Studio for managing SQL Servers, and the same is true for an SSAS server. In fact, to connect SQL Server Management Studio to Analysis Server, you perform the same actions you would as if you were connecting to the SQL Servers database engine.

Make a connection by opening SQL Server Management Studio and clicking the Connection option in SQL Server Management Studio's object Explorer window. You will see a dialog window that allows you to choose which server you want to connect to, as shown in Figure 12-41. The server type must be set to Analysis Server instead of the database engine option that you normally see.

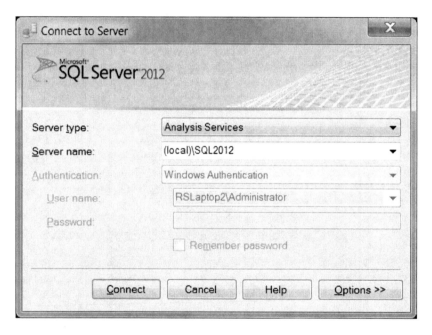

Figure 12-41. *Connecting to SSAS with SQL Server Management Studio (note the server type)*

After typing in the correct server name, click Connect. Once connected, Object Explorer displays the connection as shown in Figure 12-42. From here, you can expand the list to see the various databases that have been deployed to the server. And within each database, you expand the various folders to see your data sources, data source views, cubes, and dimensions.

Right-clicking an object in the Object Explorer treeview brings up the context menu that allows you to interact with that object. For example, in Figure 12-42, we have selected the Process option in the context menu of the CubeDWPubsSales object. This feature begins processing that cube.

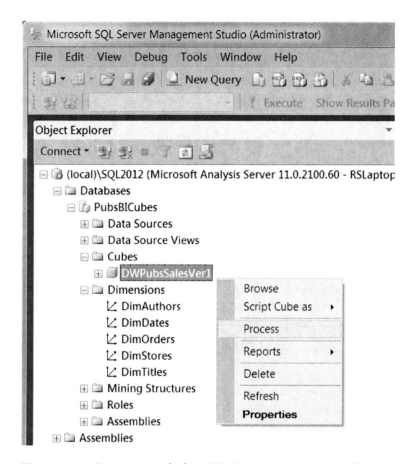

Figure 12-42. *Processing a cube from SQL Server Management Studio*

In the same menu, you can see many other options available such as the ability to browse the cube using a browser window that is similar to the one we have seen in Visual Studio. You also see the Script Cube option, which allows you to create XMLA code that re-creates the cube if necessary.

Processing a cube is considered an administrative task. When you choose the Process option from the context menu, a dialog window opens that offers processing options. One of the most important options allows you to create an XMLA script that processes the cube when it is executed (Figure 12-43). This is important because you use XMLA code like this to create automated SQL Server jobs. Afterwards, these jobs can be scheduled to perform the processing tasks each night using Microsoft's *SQL Server Agent* software.

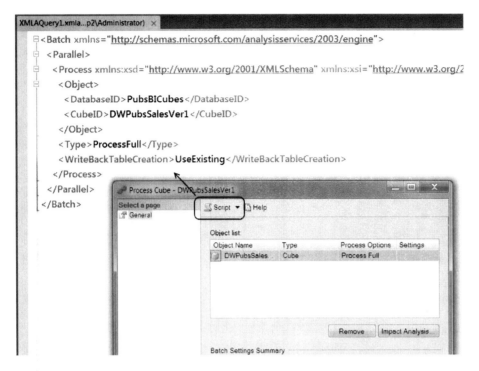

Figure 12-43. *Generating XMLA Code used for Processing*

Visual Studio (Live)

Most management tasks can be performed with SQL Server Management Studio. However, oddly enough, there are some activities that can only be performed in Visual Studio or with XMLA programming.

In Visual Studio you work with SSAS in one of two ways: live or disconnected. In the normal, disconnect approach, you create a Visual Studio solution, add an SSAS project to it, and then manipulate the XML code files that are part of that project. Afterwards you build and deploy your changes, then finally process the cubes and dimensions to complete your changes. And this is the way we have used Visual Studio so far.

Another way is to use Visual Studio to create a live connection to an existing SSAS database and then make modifications to it using Visual Studio. This works much like the disconnect option we have been using, but the difference is that it is a live connection. Therefore, there is no need to deploy the underlying XML code from Visual Studio to the SSAS server.

In live mode, you are manipulating the dimensions and cubes in real time, any changes are immediately applied to the cubes and dimensions on the SSAS server.

■ **Note**   Obviously a live connection is a very powerful option but can be quite dangerous to a BI solution in the wrong hands. Luckily, only SSAS administrators can create live connections. But even so, this option should always be used with caution.

To connect to the live SSAS server from Visual Studio, use the File ➤ Open ➤ Analysis Server databases menu option, as shown in Figure 12-44.

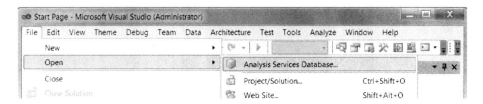

Figure 12-44. *Connecting to the live SSAS server from Visual Studio*

After clicking the Analysis Services Database option, you can see the *Connect To Database* dialog window, as shown in Figure 12-45.

In the Connect To Database dialog window, enter the name of your SSAS server and identify which database to use (Figure 12-45). Once you identify the server and database, click OK to connect to the SSAS server.

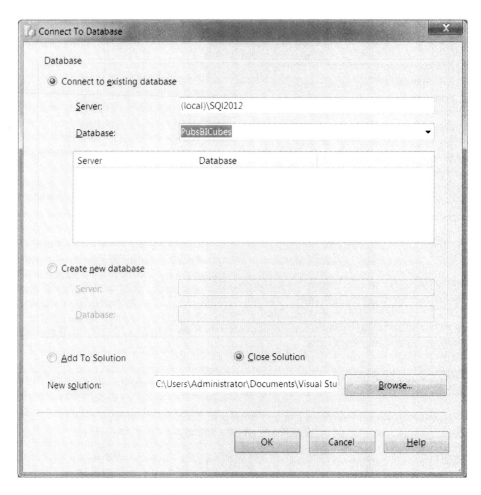

Figure 12-45. *Configuring the live connection*

If you navigate to the folder where the Visual Studio solution files are located, you will find that there are no project files associated with them. Instead, the project that you see in Visual Studio's Solution Explorer treeview is a virtual project.

This is also true of the data source, data source view, cubes, and dimensions. If you look closely at Figure 12-46, you will notice that there are no extensions shown in Solution Explorer for each of these objects (as would be with a standard Visual Studio project). Another difference is that you cannot right-click these object's icons and use the View Code option from the context menu to see the underlying XML code, because it simply does not exist for these virtual objects.

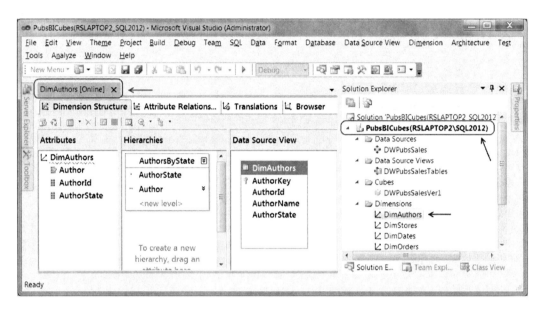

Figure 12-46. *Some visual clues indicating a live connection*

Just as with any other project, after you have made your changes, save your work. But remember, it is live. Therefore, there is no need to build and deploy it, because there are no code files to build or deploy to SSAS.

Most options and property settings in the online Visual Studio project are the same for normal, offline, SSAS projects. Because of this, the techniques that we discussed in Chapters 9 through 12 allow you to perform any changes necessary, when working in Visual Studio Live.

EXERCISE 12-1. CHANGING THE CUBE'S NAME

In this exercise, you use a live connection to SSAS to change the name of the DWPubsSalesVer1 cube to CubePubsSales. Afterward, the dimensions and cube will have prefixes of Dim and Cube, respectively.

Note: This exercise is important to perform, because later in this book we refer to the new name!

Open a Live Connection to SSAS

You cannot change the name of a cube from the context menu in SQL Server Management Studio, but you can in Visual Studio. Let's use Visual Studio to make a live connection to the SSAS server.

1. Open Visual Studio by clicking the Start button and navigating to All Programs ➤ Visual Studio 2010. Right-click this menu item, select Run as Administrator, and then answer Yes to close the UAC.

2. Open a live connection to SSAS by selecting Visual Studio's File ➤ Open ➤ Analysis Services Database menu (Figure 12-44).

3. When the Connect To Database dialog window opens, type in your server's name and the name of the database in the appropriate textboxes, as shown in Figure 12-45. Remember, unlike the picture, you use only **(local)** if you are connecting to your default instance of SSAS.

4. Click OK to create the live connection to the SSAS server.

Rename the Cube

Now that you have a live connection established, you can change the cube's name.

1. Locate the cube in Solution Explorer (Figure 12-47). Right-click the cube's icon and select Rename from the context menu.

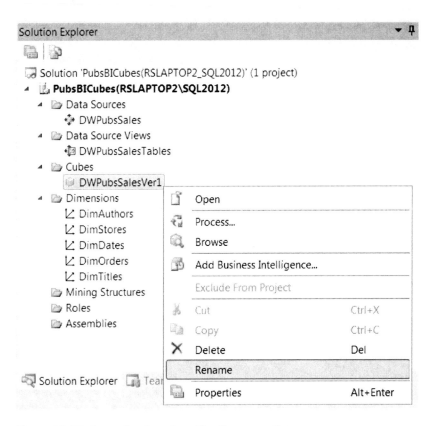

Figure 12-47. *Renaming the cube with a live connection*

2. Change the cube's name to CubePubsSales. A dialog window will briefly appear telling you that your change is being updated to the SSAS server (Figure 12-48).

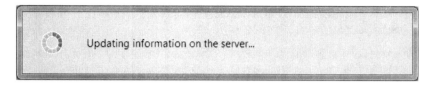

Figure 12-48. Renaming the cube with a live connection

Verify the Change in SQL Server Management Studio

1. Open SQL Server Management Studio (SSMS) and connect to the SSAS server, as shown in Figure 12-41.

2. When SSMS opens, locate the PubsBICubes database folder icon in the Object Explorer treeview.

3. Expand the PubsBICubes folder icon and locate the Cubes folder icon (Figure 12-49).

4. Expand the Cubes folder icon and verify that the cube is now named CubesPubsSales, as shown in Figure 12-49.

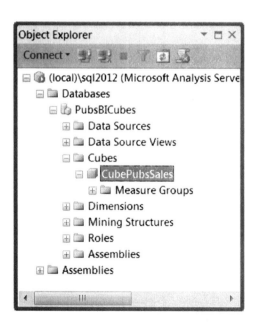

Figure 12-49. Verifying the cube has been renamed

In this exercise, we changed the name of the DWPubsSalesVer1 cube to CubePubsSales using a live connection to SSAS.

Moving On

The techniques you learned in this chapter and the previous three provide you with all you need to get started in creating your own SSAS projects. But as always, there is still more to discover. The "Learn by Doing" exercise at the end of this chapter further enhances your understanding. We hope you will give it a try.

At this point, we are more than halfway through the lifecycle of a BI solution! (See Figure 12-50.) Next we begin looking at the reporting aspects of a BI solution.

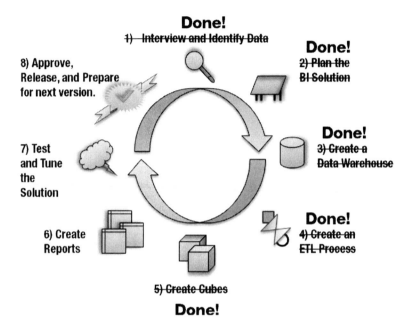

Figure 12-50. *Our progress through the BI solution lifecycle*

LEARN BY DOING

In this "Learn by Doing" exercise, you configure the cube you made in Chapter 8 similar to the way we configured our example in this chapter. Use the Northwind database to perform this exercise. We have included an outline of the steps you performed in this chapter and an example of how the authors handled them in two Word documents. These documents are found in the folder C:_BookFiles_LearnByDoing\ Chapter12Files. Please see the ReadMe.doc file for detailed instructions.

What's Next?

SSAS can be a powerful tool for creating and managing cube data. If you need more information than what we have covered so far, we recommend the following titles:

Expert Cube Development with Microsoft SQL Server 2008 Analysis Services
By Marco Russo, Alberto Ferrari, Chris Webb
Publisher: Packt Publishing
ISBN-10: 1847197221

Microsoft SQL Server 2008 Analysis Services Unleashed
Irina Gorbach, Alexander Berger, Edward Melomed
Publisher: Sams
ISBN-10: 0672330016

CHAPTER 13

Creating Reports with SQL Queries

I'm convinced you can combine this with reporting integrity and accuracy.

—Jack Brickhouse

With the data warehouse completed and filled with data, it is time to realize some of the fruits of your efforts by making your first preliminary reports.

In this chapter, we take a look at the process of creating report code using the SQL programming language and some of the decisions that you have to make regarding this code. We walk you through the process of creating report queries from start to finish.

A data warehouse can have thousands of reports against it, and all reports have aspects in common:

- Reports are viewed with user applications such as Excel, PowerPivot, or Reporting Server Reports.

- The underlying code is most often written in SQL or MDX (query languages used for relational and OLAP databases).

- All report code should be consistent and well formed.

- Abstraction layers are used to keep code maintenance costs low.

Reporting queries can become complex very quickly, and complex queries can be overwhelming for developers who do not do a lot SQL programming; therefore, we do what we can to keep them simple. And, to make things more understandable, we break the SQL code into bite-size chunks and break down each of the features that typically make up a standard SQL report query. Using the methods in this chapter, you will soon be writing polished, professional, and accurate queries.

Note Examples in this chapter are SQL-based. MDX is the other common reporting language in use. We cover MDX next, in Chapter 14.

Identifying the Data

The first step is to determine the type of report your client needs. This falls into the category of "identifying the data." After having been through the interview process with your client, you should have at least some idea of what they are looking for.

At this point, it is common for the client to be somewhat unsure of what they need. But after they have seen an initial report, it can help them be more specific. This process becomes more exact after you have presented your preliminary versions and examples to them. After your first reports, you will likely be refining them in versions 2, 3, or even 4. This is part of the process of learning what questions to ask before diving into creating the first version of the report.

We recommend getting started by creating a commented header section at the beginning of each SQL script. The comment might look something like the one shown in Listing 13-1.

Listing 13-1. A Script Header

```
/****************************************
Title:SalesByTitlesByDates
Description: Which titles were sold on which dates
Developer:RRoot
Date: 6/1/2012

Change Log: Who, When, What
CMason,6/2/2013,fixed numerous grammatical errors
*****************************************/
```

The script header is similar to what we have used in the past, but most companies have their own standards of what information they require to be inserted into this header. If the company does not already have a standard for this, now would be a good time to establish one. Once you have a header outlining what you want to accomplish, you need to locate the data within the data warehouse.

Listing 13-2 is a very simple SELECT statement against the fact table from which most reports will originate. Notice that we have formatted the SELECT statement to be more legible by stacking the column listings in a vertical fashion. Over the years Microsoft seems to have settled on this being a best practice, and we agree that it does make things easier to read when you have to go through a large amount of code.

Listing 13-2. A Basic Starter Query

```
SELECT
    OrderNumber
, OrderDateKey
, TitleKey
, StoreKey
, SalesQuantity
FROM DWPubsSales.dbo.FactSales
```

One simple addition that makes reporting easier long-term is using fully qualified names for objects. This example includes not only the name of the table but its database name of DWPubsSales and schema name of DBO as well. Maintenance on reports includes tracking which reports are connected to which databases and database objects. Using fully qualified names in your SQL queries can help with this process and is a simple addition that takes little time to implement. Besides, you also get a small gain in performance, because the database engine does not have to resolve the object name implicitly. As shown in Figure 13-1, the results you get back are not particularly pleasing to the eye.

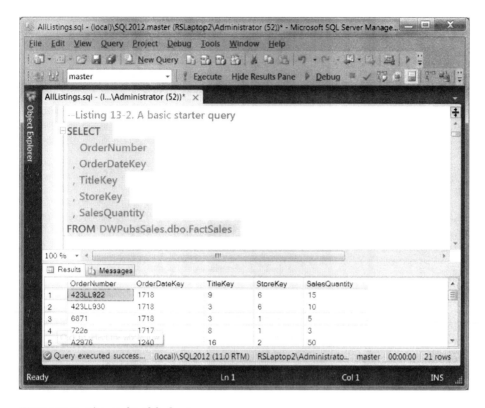

Figure 13-1. *The results of the basic query*

Additionally, it is not easy to understand. Notice that the order dates, titles, and stores show only numeric key values. Clearly, these reports are not quite user friendly yet, but they will be as soon as you enhance your query with more information. Let's look at some ways of doing so in the following sections.

Joining Table Data

In Listing 13-3 you can see an example of the basic query having been modified to include the title ID and title name from the DimTitles table. Note that you need to fully identify any columns that appear in both tables, like the TitleKey, by prefixing them with the table name. For clarity and code maintenance, we also prefix the TitleID and TitleName columns.

Listing 13-3. Adding a Table to the Query

```
SELECT
    DWPubsSales.dbo.DimTitles.TitleId
,   DWPubsSales.dbo.DimTitles.TitleName
,   OrderNumber
,   OrderDateKey
,   DWPubsSales.dbo.FactSales.TitleKey
```

```
, StoreKey
, SalesQuantity
```

```
FROM DWPubsSales.dbo.FactSales
JOIN DWPubsSales.dbo.DimTitles
  ON DWPubsSales.dbo.FactSales.TitleKey = DWPubsSales.dbo.DimTitles.TitleKey
```

When this query is run, you receive data not only from the FactSales table, but also the DimTitles table, as shown in Figure 13-2. The database engine knows that you want data from both tables because of the JOIN operator in the FROM clause, as well as how the tables are linked because of the ON operator that compares the title key in both tables.

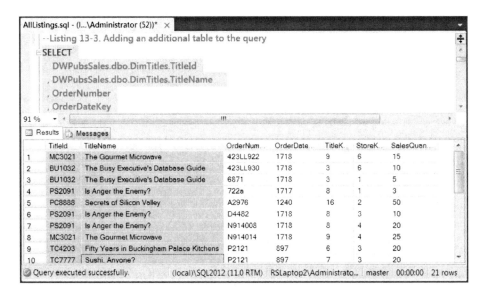

Figure 13-2. *Adding the TitleId and TitleName columns to the results*

■ **Tip** Notice in Figure 13-2 that we "stack" queries in the Management Studio window so to retain the previous working version of our query as well as the new version. It is a good idea to keep multiple versions in the window, highlighting one execute at a time. Doing so allows you to easily revert to an earlier working iteration in the event of trouble.

It is recommended to use fully qualified names to identify your tables, but a very long query can get quite tedious to both read and write. One solution is to use table aliases to represent fully qualified names. In Listing 13-4, you can see the use of the AS keyword to create a table alias. Using the AS keyword makes code much clearer than creating table aliases without it. (That is in contrast to simply putting a space after the table name and then typing in more letters to create the alias.)

Listing 13-4. Creating Table Aliases with the AS Keyword

```
SELECT
    FS.TitleKey
, DT.TitleName
, OrderNumber
, OrderDateKey
, StoreKey
, SalesQuantity
FROM DWPubsSales.dbo.FactSales AS FS
INNER JOIN DWPubsSales.dbo.DimTitles AS DT
  ON FS.TitleKey = DT.TitleKey
```

Ordering Results

Another improvement for your report may be reordering the column listings so that they make more sense to the end user. Most modern reporting software can also modify the display of the columns in the user interface, but here we are dealing with just the raw SQL code. Therefore, modifying this display is a simple matter of moving the columns in the select list to their more appropriate positions.

Since this report is about titles, we have moved the title information to the top of the select. We have also placed the TitleName in the first column in the select list, because this provides the most user-friendly column results.

At the same time, we order the results by both title and dates (see Listing 13-5).

Listing 13-5. Reordering the Columns and Rows for Better Results

```
SELECT

    DT.TitleName
, DT.TitleId

, FS.TitleKey
, OrderNumber
, OrderDateKey
, StoreKey
, SalesQuantity
FROM DWPubsSales.dbo.FactSales AS FS
INNER JOIN DWPubsSales.dbo.DimTitles AS DT
 ON FS.TitleKey = DT.TitleKey
ORDER BY [TitleName], [OrderDateKey]
```

To improve the report, some developers like to include column aliases to further simplify or elaborate on the column definitions. Listing 13-6 further improves the outcome of our query by adding two column aliases. Notice the ORDER BY clause works with either the column alias such as [Title] or the column name such as [OrderDateKey].

Listing 13-6. Adding Column Aliases for Better Results

```
SELECT
    [Title] = DT.TitleName
, DT.TitleId
, [Internal Data Warehouse Id] = FS.TitleKey
```

```
, OrderNumber
, OrderDateKey
, StoreKey
, SalesQuantity
FROM DWPubsSales.dbo.FactSales AS FS
INNER JOIN DWPubsSales.dbo.DimTitles AS DT
  ON FS.TitleKey = DT.TitleKey
ORDER BY [Title], [OrderDateKey]
```

Figure 13-3 shows the results of this query. It includes the reordered columns and rows, as well as the aliased column names.

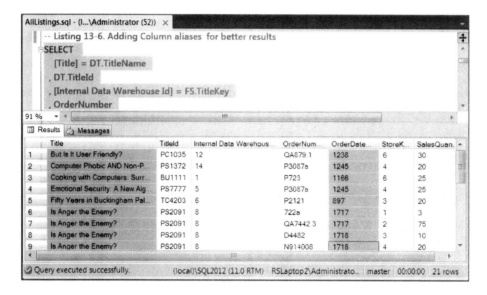

Figure 13-3. *Using column alisas and ORDER BY for our results*

Currently our results include the OrderDateKey, but we do not have an understandable date format. This is because the date data is in the DimDates table, and we have not yet included that in our query. We do so in Listing 13-7.

Listing 13-7. Adding Data from the DimDates Table

```
SELECT
  [Title] = DT.TitleName
, DT.TitleId
, [Internal Data Warehouse Id] = FS.TitleKey
, OrderNumber
, OrderDateKey
, [OrderDate] = DD.[Date]
, StoreKey
, SalesQuantity
```

```
FROM DWPubsSales.dbo.FactSales AS FS
INNER JOIN DWPubsSales.dbo.DimTitles AS DT
  ON FS.TitleKey = DT.TitleKey

INNER JOIN DWPubsSales.dbo.DimDates AS DD
  ON FS.OrderDateKey = DD.DateKey

ORDER BY [Title], [OrderDate]
```

Figure 13-4 is an example of the results of adding data from DimDates.

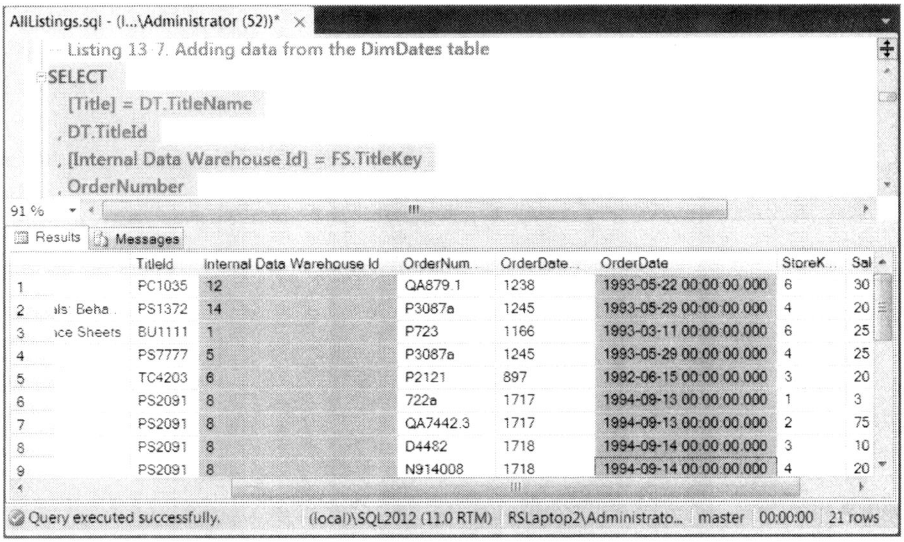

Figure 13-4. *Adding data from DimDates*

The original request did not specifically ask for reports that group titles by their individual publishers, but it is likely to benefit the client. Adding publisher names to the queries would link a fourth table to the results, as shown in Listing 13-8.

Listing 13-8. Adding Publisher Names to the Results

```
SELECT
    DP.PublisherName
, [Title] = DT.TitleName
, DT.TitleId
, [Internal Data Warehouse Id] = FS.TitleKey
, OrderNumber
, OrderDateKey
, [OrderDate] = DD.[Date]
, StoreKey
, SalesQuantity
FROM DWPubsSales.dbo.FactSales AS FS
INNER JOIN DWPubsSales.dbo.DimTitles AS DT
```

```
  ON FS.TitleKey = DT.TitleKey
INNER JOIN DWPubsSales.dbo.DimDates AS DD
  ON FS.OrderDateKey = DD.DateKey INNER JOIN DWPubsSales.dbo.DimPublishers AS DP
  ON DT.PublisherKey = DP.PublisherKey
ORDER BY DP.PublisherName, [Title], [OrderDate]
```

The results now include data from all four tables, as shown in Figure 13-5.

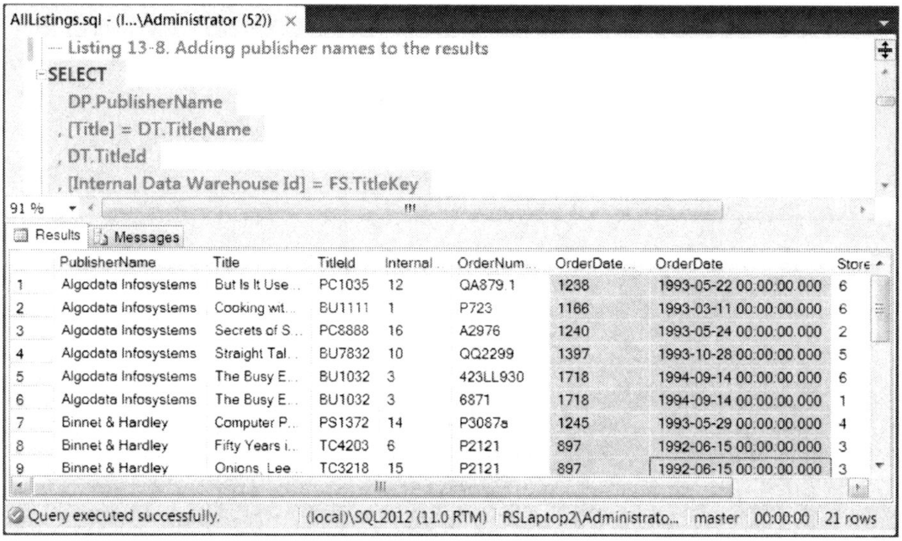

Figure 13-5. Adding data from DimPublishers

If our current query contains columns that are unimportant to our report, they can easily be excluded by either deleting them or commenting them out. For this report we will not need the TitleKey, OrderNumber, OrderDateKey, or StoreKey. We removed these four columns from the select list in Listing 13-9 by commenting them out.

Listing 13-9. Removing Columns Not Needed for Our Query

```
SELECT
    DP.PublisherName
, [Title] = DT.TitleName
, DT.TitleId
--, [Internal Data Warehouse Id] = FS.TitleKey
--, OrderNumber
--, OrderDateKey
, [OrderDate] = DD.[Date]
--, StoreKey
, SalesQuantity
FROM DWPubsSales.dbo.FactSales AS FS
```

```
INNER JOIN DWPubsSales.dbo.DimTitles AS DT
  ON FS.TitleKey = DT.TitleKey
INNER JOIN DWPubsSales.dbo.DimDates AS DD
  ON FS.OrderDateKey = DD.DateKey
INNER JOIN DWPubsSales.dbo.DimPublishers AS DP
  ON DT.PublisherKey = DP.PublisherKey
ORDER BY DP.PublisherName, [Title], [OrderDate]
```

■ **Tip** Leaving columns in your original code and simply commenting them out, rather than completely removing them, is often helpful for troubleshooting and validating your code. After the code is stabilized, they can then be removed permanently. For instructional purposes and clarity, we remove them from the future listings in this chapter.

Formatting Results Using SQL Functions

To make the data more user-friendly, we use the SQL function CONVERT to change the data into a string of characters with the typical United States presentation. The CONVERT function was designed by Microsoft to do just this. Microsoft has also incorporated a set of numbers to determine which U.S. format to use. For example, format number 110 gives a date with the *day – month – year* format, while format number 101 gives you a date with the *day/month/year* format (Listing 13-10).

Listing 13-10. Using the CONVERT Function

```
SELECT
    DP.PublisherName
  , [Title] = DT.TitleName
  , [TitleId] = DT.TitleId
  , [OrderDate] = CONVERT(varchar(50), [Date], 101)
  , SalesQuantity
FROM DWPubsSales.dbo.FactSales AS FS
INNER JOIN DWPubsSales.dbo.DimTitles AS DT
  ON FS.TitleKey = DT.TitleKey
INNER JOIN DWPubsSales.dbo.DimDates AS DD
  ON FS.OrderDateKey = DD.DateKey
INNER JOIN DWPubsSales.dbo.DimPublishers AS DP
  ON DT.PublisherKey = DP.PublisherKey
ORDER BY DP.PublisherName, [Title], [OrderDate]
```

The results now include formatted dates, as shown in Figure 13-6.

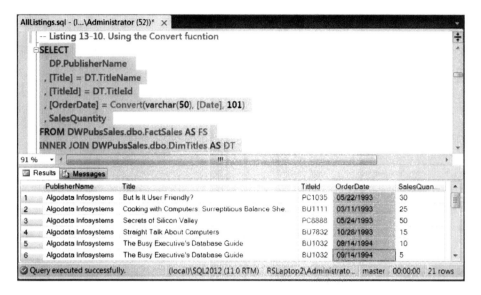

Figure 13-6. *The current results with formatted dates*

Filtering Results

Many reports filter data so that instead of showing all of the data in a table, you can select specific data to be made visible. This is done using the SQL WHERE clause, as shown in Listing 13-11.

Listing 13-11. Using a WHERE Clause to Filter the Results

```
SELECT
    [Title] = DT.TitleName
, [TitleId] = DT.TitleId
, [OrderDate] = Convert(varchar(50), [Date], 101)
, SalesQuantity
FROM DWPubsSales.dbo.FactSales AS FS
JOIN DWPubsSales.dbo.DimTitles AS DT
  ON FS.TitleKey = DT.TitleKey
JOIN DWPubsSales.dbo.DimDates AS DD
  ON FS.OrderDateKey = DD.DateKey
WHERE [TitleId] = 'PS2091'
ORDER BY [Title], [Date]
```

The results now include only filtered data, as shown in Figure 13-7.

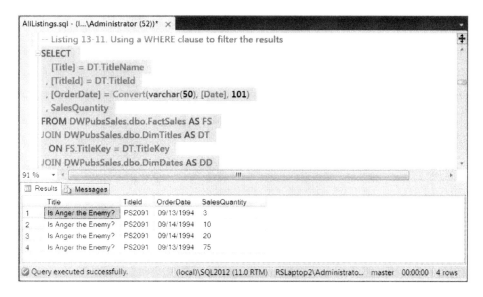

Figure 13-7. *Results filtered by a given title ID*

You may want to filter not only the title but also the date. To do this, you would change the WHERE clause to search for a particular date, as shown in Listing 13-12.

Listing 13-12. Using a WHERE Clause to Filter Based on a Given Date

```
SELECT
    DP.PublisherName
, [Title] = DT.TitleName
, [TitleId] = DT.TitleId
, [OrderDate] = Convert(varchar(50), [Date], 101)
, SalesQuantity
FROM DWPubsSales.dbo.FactSales AS FS
INNER JOIN DWPubsSales.dbo.DimTitles AS DT
  ON FS.TitleKey = DT.TitleKey
INNER JOIN DWPubsSales.dbo.DimDates AS DD
  ON FS.OrderDateKey = DD.DateKey
INNER JOIN DWPubsSales.dbo.DimPublishers AS DP
  ON DT.PublisherKey = DP.PublisherKey

WHERE [Date] = '09/13/1994'

ORDER BY DP.PublisherName, [Title], [OrderDate]
```

The results will now only show data where the exact date is found (Figure 13-8).

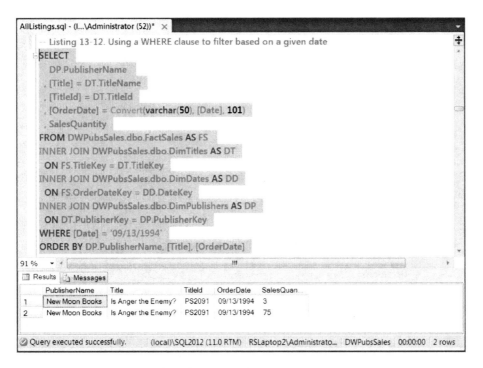

Figure 13-8. Filtering results based on a given date

When you are working with filters, it is often necessary to add more flexibility in your queries by using tools like the SQL wildcard searches. One classic example of a wildcard search is where you use the percentage sign wildcard symbol (%) in conjunction with the SQL LIKE operator to indicate that zero or more missing characters should be ignored for the purpose of a pattern match. In Listing 13-13 you can see an example of this where the query asks SQL Server to find all the data where the letters *PS* are followed by zero or more number of characters.

Listing 13-13. Filtering Results Based on Title ID with a Wildcard Symbol

```
SELECT
    DP.PublisherName
  , [Title] = DT.TitleName
  , [TitleId] = DT.TitleId
  , [OrderDate] = CONVERT(varchar(50), [Date], 101)
  , SalesQuantity
FROM DWPubsSales.dbo.FactSales AS FS
INNER JOIN DWPubsSales.dbo.DimTitles AS DT
    ON FS.TitleKey = DT.TitleKey
INNER JOIN DWPubsSales.dbo.DimDates AS DD
    ON FS.OrderDateKey = DD.DateKey
INNER JOIN DWPubsSales.dbo.DimPublishers AS DP
    ON DT.PublisherKey = DP.PublisherKey

WHERE [TitleId] LIKE 'PS%' -- % means zero or more characters

ORDER BY DP.PublisherName, [Title], [OrderDate]
```

Now the results include only data with a prefix of *PS*, as shown in Figure 13-9.

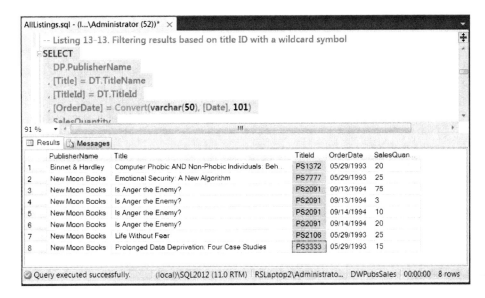

Figure 13-9. *Results filtered with a wildcard search*

The LIKE operator is certainly useful, and you will find plenty of opportunities to use it. Here are two other operators we have found useful over the years. The first is the IN operator. It filters results based on a list of possible choices. When a row has data that matches one of the listed items, the row is returned as part of the result set. Listing 13-14 shows an example of the IN operator being used.

Listing 13-14. Using the IN Operator

```
SELECT
    DP.PublisherName
, [Title] = DT.TitleName
, [TitleId] = DT.TitleId
, [OrderDate] = CONVERT(varchar(50), [Date], 101)
, SalesQuantity
FROM DWPubsSales.dbo.FactSales AS FS
INNER JOIN DWPubsSales.dbo.DimTitles AS DT
  ON FS.TitleKey = DT.TitleKey
INNER JOIN DWPubsSales.dbo.DimDates AS DD
  ON FS.OrderDateKey = DD.DateKey
INNER JOIN DWPubsSales.dbo.DimPublishers AS DP
  ON DT.PublisherKey = DP.PublisherKey
WHERE [Date] IN ( '09/13/1994' , '05/29/1993' )
ORDER BY DP.PublisherName, [Title], [OrderDate]
```

Figure 13-10 shows the results of the query.

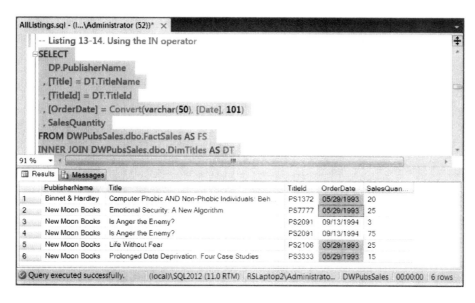

Figure 13-10. *Results filtered with the IN operator*

The second operator that is worthy of mentioning is the BETWEEN operator. It finds data that is within a range, such as numeric or alphabetical. Listing 13-15 demonstrates this operator using dates.

Listing 13-15. Using the BETWEEN Operator

```
SELECT
    DP.PublisherName
, [Title] = DT.TitleName
, [TitleId] = DT.TitleId
, [OrderDate] = CONVERT(varchar(50), [Date], 101)
, SalesQuantity
FROM DWPubsSales.dbo.FactSales AS FS
INNER JOIN DWPubsSales.dbo.DimTitles AS DT
  ON FS.TitleKey = DT.TitleKey
INNER JOIN DWPubsSales.dbo.DimDates AS DD
  ON FS.OrderDateKey = DD.DateKey
INNER JOIN DWPubsSales.dbo.DimPublishers AS DP
  ON DT.PublisherKey = DP.PublisherKey
WHERE [Date] BETWEEN '09/13/1994' AND '09/14/1994'

ORDER BY DP.PublisherName, [Title], [OrderDate]
```

As you can see in Figure 13-11, all rows of data are returned that are within the range of dates specified by the BETWEEN operator.

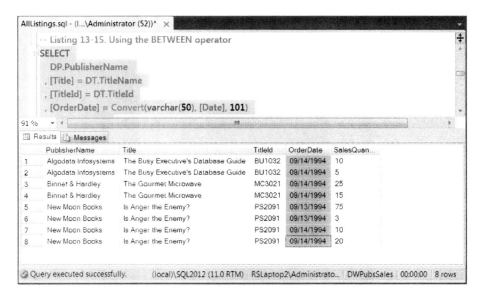

Figure 13-11. *Results filtered by with the BETWEEN operator*

■ **Tip**   Note that with the BETWEEN operator the results are inclusive. This is not always true of all SQL operators, so whenever you use an operator that defines results based on two points, you should always test to check whether the results are inclusive or exclusive.

You can also combine operators in the WHERE clause to fine-tune your result set. Listing 13-16 shows an example of using both the BETWEEN and LIKE operators in the same WHERE statement.

Listing 13-16. Combining Multiple Operators in a WHERE Statement

```
SELECT
    DP.PublisherName
, [Title] = DT.TitleName
, [TitleId] = DT.TitleId
, [OrderDate] = Convert(varchar(50), [Date], 101)
, SalesQuantity
FROM DWPubsSales.dbo.FactSales AS FS
INNER JOIN DWPubsSales.dbo.DimTitles AS DT
 ON FS.TitleKey = DT.TitleKey
INNER JOIN DWPubsSales.dbo.DimDates AS DD
 ON FS.OrderDateKey = DD.DateKey
INNER JOIN DWPubsSales.dbo.DimPublishers AS DP
 ON DT.PublisherKey = DP.PublisherKey
WHERE
 [Date] BETWEEN '09/13/1994' AND '09/14/1994'
 AND
 [TitleId] LIKE 'PS%'
ORDER BY DP.PublisherName, [Title], [OrderDate]
```

Figure 13-12 shows the results of this query.

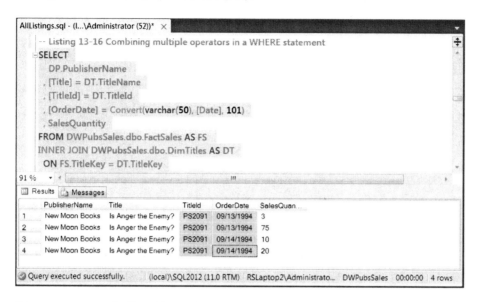

Figure 13-12. *Results filtered by multiple operators*

Adding Dynamic Filters with Parameters

In many cases, report builders prefer to set the wildcard or date values themselves. You can allow this by defining a set of parameters to further customize the report results. This can be programmed by creating SQL variables to use for report parameters. Listing 13-17 shows an example.

Listing 13-17. Adding Variables to Your Query

```
DECLARE
    @StartDate datetime = '09/13/1994'
, @EndDate datetime = '09/14/1994'
, @Prefix nVarchar(3) = 'PS%'

SELECT
    DP.PublisherName
, [Title] = DT.TitleName
, [TitleId] = DT.TitleId
, [OrderDate] = CONVERT(varchar(50), [Date], 101)
, SalesQuantity
FROM DWPubsSales.dbo.FactSales AS FS
INNER JOIN DWPubsSales.dbo.DimTitles AS DT
  ON FS.TitleKey = DT.TitleKey
INNER JOIN DWPubsSales.dbo.DimDates AS DD
  ON FS.OrderDateKey = DD.DateKey
INNER JOIN DWPubsSales.dbo.DimPublishers AS DP
  ON DT.PublisherKey = DP.PublisherKey
WHERE
 [Date] BETWEEN @StartDate AND @EndDate
```

```
AND
[TitleId] LIKE @Prefix
ORDER BY DP.PublisherName, [Title], [OrderDate]
```

The report parameters can now be used in the WHERE clause by replacing the hard-coded values with our variables. Now, if we run our report query, we will see rows of data based on whatever dates and prefix we supply to the query, as shown in Figure 13-13.

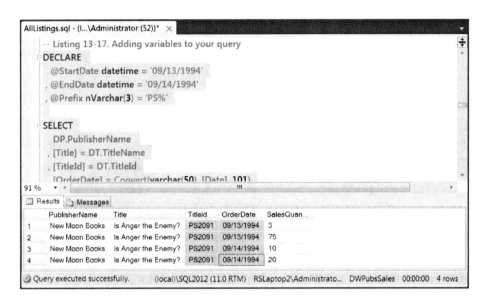

Figure 13-13. *Results filtered with parameters*

You can expect that some clients want the option of showing every row on some occasions while showing only some of the rows at other times. There are a number ways to do this, but one way is to add a parameter that serves as a flag to indicate whether you want to show all the results or just filtered results.

In Listing 13-18, you can see that we have added a parameter called @ShowAll. This parameter's job is to toggle between showing all the results or only some of the results.

Listing 13-18. Adding a Parameter Flag to Show All Data as Needed

```
DECLARE

    @ShowAll nVarchar(4) = 'True'

, @StartDate datetime = '09/13/1994'
, @EndDate datetime = '09/14/1994'
, @Prefix nVarchar(3) = 'PS%'
SELECT
    DP.PublisherName
, [Title] = DT.TitleName
, [TitleId] = DT.TitleId
, [OrderDate] = CONVERT(varchar(50), [Date], 101)
, SalesQuantity
```

```
FROM DWPubsSales.dbo.FactSales AS FS
INNER JOIN DWPubsSales.dbo.DimTitles AS DT
  ON FS.TitleKey = DT.TitleKey
INNER JOIN DWPubsSales.dbo.DimDates AS DD
  ON FS.OrderDateKey = DD.DateKey
INNER JOIN DWPubsSales.dbo.DimPublishers AS DP
  ON DT.PublisherKey = DP.PublisherKey
WHERE

@ShowAll = 'True'
OR

[Date] BETWEEN @StartDate AND @EndDate
AND
[TitleId] LIKE @Prefix
ORDER BY DP.PublisherName, [Title], [OrderDate]
```

When this query is executed, the results are similar to those shown in Figure 13-14.

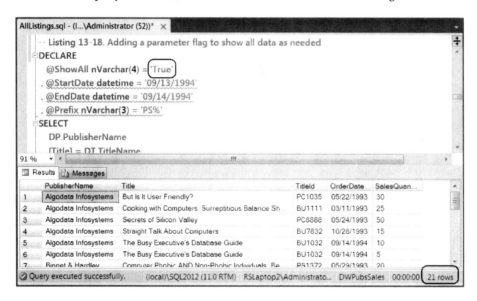

Figure 13-14. *Using a parameter flag to show all results*

Adding Aggregations

One common request is for reports to display totals and subtotals. In SQL queries, this can be accomplished by using aggregate functions. To demonstrate this, let's go ahead and summarize the sales quantities using the SUM aggregate function, as shown in Listing 13-19.

Listing 13-19. Adding Aggregate Values to Our Results

```
DECLARE
    @ShowAll nVarchar(4) = 'False'
  , @StartDate datetime = '09/13/1994'
  , @EndDate datetime = '09/14/1994'
  , @Prefix nVarchar(3) = 'PS%'
```

```
SELECT
    DP.PublisherName
  , [Title] = DT.TitleName
  , [TitleId] = DT.TitleId
  , [OrderDate] = CONVERT(varchar(50), [Date], 101)

  , [Total for that Date by Title] = SUM(SalesQuantity)

FROM DWPubsSales.dbo.FactSales AS FS
INNER JOIN DWPubsSales.dbo.DimTitles AS DT
  ON FS.TitleKey = DT.TitleKey
INNER JOIN DWPubsSales.dbo.DimDates AS DD
  ON FS.OrderDateKey = DD.DateKey
INNER JOIN DWPubsSales.dbo.DimPublishers AS DP
  ON DT.PublisherKey = DP.PublisherKey
WHERE
 @ShowAll = 'True'
 OR
 [Date] BETWEEN @StartDate AND @EndDate
 AND
 [TitleId] LIKE @Prefix

GROUP BY
    DP.PublisherName
  , DT.TitleName
  , DT.TitleId
  , [Date]
ORDER BY DP.PublisherName, [Title], [OrderDate]
```

Figure 13-15 shows the results of the aggregations.

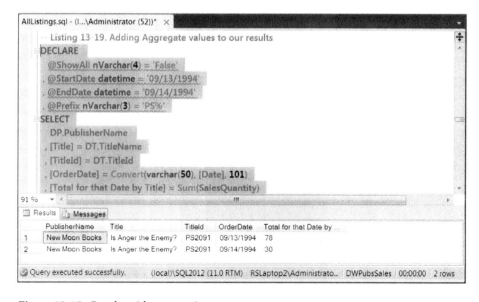

Figure 13-15. *Results with aggregations*

■ **Note**   Often, reporting software will add the aggregations for you. When this happens, do not include the aggregate values in your SQL code. This is also true for parameters; some applications filter the results using parameters only after all the results have been returned to the application. A common example for both of these can be found within Microsoft Excel. Although it is possible to submit parameter values and consume aggregate data, you are usually better off letting Excel do the aggregation and filtering for you after all the data has been returned.

EXERCISE 13-1. CREATING REPORT QUERIES

In this exercise, you create your own custom report query similar to the one we just walked though. To make this process simpler, we do not include parameters at this point. You combine data from three tables, aggregate the values, and then format your results. If you build the query a little at a time, as described in this chapter, you will find your query will be completed in no time at all!

1. Review the data in the FactSales table by executing the following SQL code in a query window:

```
SELECT * from FactSales
```

2. Review the data in the DimStores table by executing the following SQL code in a query window:

```
SELECT * from DimStores
```

3. Review the data in the DimDates table by executing the following SQL code in a query window:

```
SELECT * FROM DimDates
```

4. Combine the data from these tables into a report query that returns store names, store IDs, order dates, and the sum quantity by store and date. Your results should look like Figure 13-16.

If you get stuck, the answers for this exercise are in the downloadable example files for this book. Look in the folder for Chapter 13.

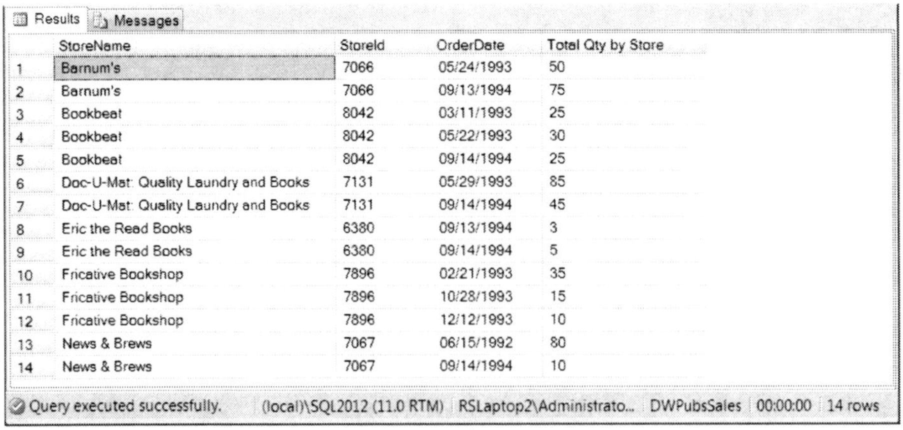

	StoreName	StoreId	OrderDate	Total Qty by Store
1	Barnum's	7066	05/24/1993	50
2	Barnum's	7066	09/13/1994	75
3	Bookbeat	8042	03/11/1993	25
4	Bookbeat	8042	05/22/1993	30
5	Bookbeat	8042	09/14/1994	25
6	Doc-U-Mat: Quality Laundry and Books	7131	05/29/1993	85
7	Doc-U-Mat: Quality Laundry and Books	7131	09/14/1994	45
8	Eric the Read Books	6380	09/13/1994	3
9	Eric the Read Books	6380	09/14/1994	5
10	Fricative Bookshop	7896	02/21/1993	35
11	Fricative Bookshop	7896	10/28/1993	15
12	Fricative Bookshop	7896	12/12/1993	10
13	News & Brews	7067	06/15/1992	80
14	News & Brews	7067	09/14/1994	10

Query executed successfully.   (local)\SQL2012 (11.0 RTM)   RSLaptop2\Administrato...   DWPubsSales   00:00:00   14 rows

Figure 13-16. *The results needed for the sales by stores report*

In this exercise, you wrote a report query that captured sales data based on sales by stores. In the next exercise, you use this code to create a reporting stores procedure. But, we have a few other items to talk about first.

Using Subqueries

Multiple queries can be used at the same time to achieve a single result. There are a couple ways to do this, and using subqueries is a popular choice. Subqueries are often used in a WHERE clause to retrieve a set of values that can be examined conditionally. Another use for subqueries is to fill up the variables with data based on the results of the subquery. Listing 13-20 shows an example of this.

Listing 13-20. Fill from Your Querying Parameters with Subqueries

```
DECLARE
    @ShowAll nVarchar(4) = 'False'
  , @StartDate datetime = '09/13/1994'
  , @EndDate datetime = '09/14/1994'
  , @Prefix nVarchar(3) = 'PS%'
  , @AverageQty int

    SELECT @AverageQty = Avg(SalesQuantity) FROM DWPubsSales.dbo.FactSales

SELECT
    DP.PublisherName
  , [Title] = DT.TitleName
  , [TitleId] = DT.TitleId
  , [OrderDate] = CONVERT(varchar(50), [Date], 101)
  , [Total for that Date by Title] = Sum(SalesQuantity)

  , [Average Qty in the FactSales Table] = @AverageQty

FROM DWPubsSales.dbo.FactSales AS FS
INNER JOIN DWPubsSales.dbo.DimTitles AS DT
  ON FS.TitleKey = DT.TitleKey
INNER JOIN DWPubsSales.dbo.DimDates AS DD
  ON FS.OrderDateKey = DD.DateKey
INNER JOIN DWPubsSales.dbo.DimPublishers AS DP
  ON DT.PublisherKey = DP.PublisherKey
WHERE
 @ShowAll = 'True'
 OR
 [Date] BETWEEN @StartDate AND @EndDate
 AND
 [TitleId] LIKE @Prefix
GROUP BY
    DP.PublisherName
  , DT.TitleName
  , DT.TitleId
  , Convert(varchar(50), [Date], 101)
ORDER BY DP.PublisherName, [Title], [OrderDate]
```

Now we have report code that uses two SELECT statements. The primary query gets the report data, but the secondary SELECT statement acts as a subquery to the primary. The purpose of this subquery is to gather the average of all the sales quantity values in the fact table and then add that result to the next SELECT statement's results. If we were to use the AVERAGE function within the second SELECT statement, we would get only the average of the filtered values, not all of the quantity values in the fact sales table. Figure 13-17 shows the results.

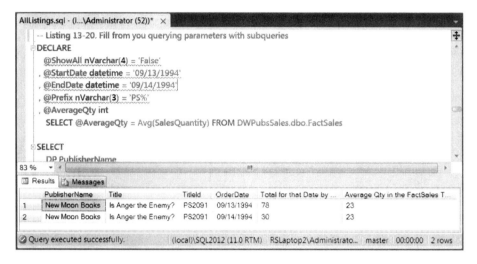

Figure 13-17. *Results with parameters filled by a subquery*

Creating KPI Queries

As reporting software has become more available, it is common for users to have to review many reports during their day-to-day work. This often results in information overload. One way to help alleviate this issue is to create some kind of indicator in your report that can be used to provide a quick determination of the value's meaning.

One way of accomplishing this is to provide key performance indicator (KPI) values within your result set. The CASE operator is a common method of including KPI's in a SQL statement. Listing 13-21 shows the use of this operator.

Listing 13-21. Adding KPIs

```
DECLARE
    @ShowAll nVarchar(4) = 'True'
, @StartDate datetime = '09/13/1994'
, @EndDate datetime = '09/14/1994'
, @Prefix nVarchar(3) = 'PS%'
, @AverageQty int
    SELECT @AverageQty = Avg(SalesQuantity) FROM DWPubsSales.dbo.FactSales

SELECT
    DP.PublisherName
, [Title] = DT.TitleName
, [TitleId] = DT.TitleId
, [OrderDate] = CONVERT(varchar(50), [Date], 101)
```

```
, [Total for that Date by Title] = Sum(SalesQuantity)
, [Average Qty in the FactSales Table] = @AverageQty

, [KPI on Avg Quantity] = CASE
WHEN Sum(SalesQuantity)
   between (@AverageQty- 5) and (@AverageQty+5) THEN 0
WHEN Sum(SalesQuantity) < (@AverageQty- 5) THEN -1
WHEN Sum(SalesQuantity) > (@AverageQty + 5) THEN 1
END
FROM DWPubsSales.dbo.FactSales AS FS
INNER JOIN DWPubsSales.dbo.DimTitles AS DT
  ON FS.TitleKey = DT.TitleKey
INNER JOIN DWPubsSales.dbo.DimDates AS DD
  ON FS.OrderDateKey = DD.DateKey
INNER JOIN DWPubsSales.dbo.DimPublishers AS DP
  ON DT.PublisherKey = DP.PublisherKey
WHERE
 @ShowAll = 'True'
 OR
 [Date] BETWEEN @StartDate AND @EndDate
 AND
 [TitleId] LIKE @Prefix
GROUP BY
    DP.PublisherName
 , DT.TitleName
 , DT.TitleId
 , CONVERT(varchar(50), [Date], 101)
ORDER BY DP.PublisherName, [Title], [OrderDate]
```

Figure 13-18 shows the results with KPIs.

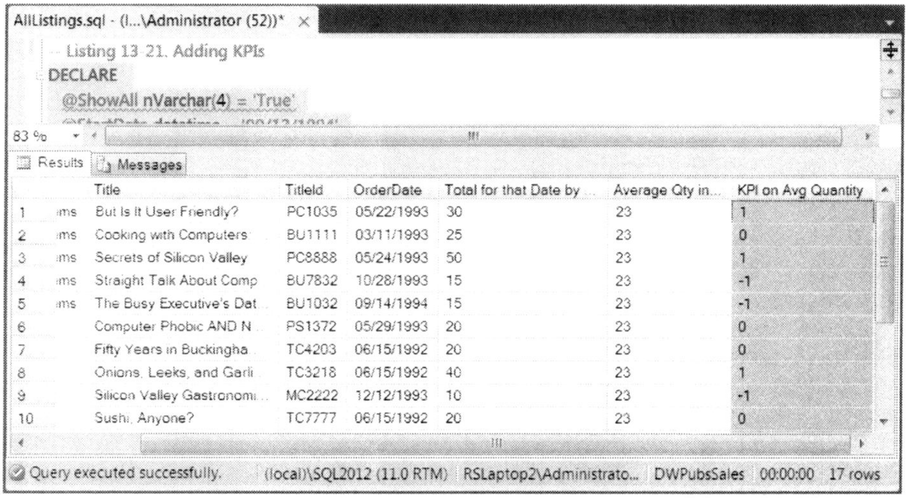

Figure 13-18. *Results with KPIs*

Adding Abstraction Layers

Microsoft and other vendors have often cautioned people against the direct use of table objects from their database systems. Instead, the preferred model is to use abstraction layers whenever possible. Abstraction layers for Microsoft SQL Server include the use of views, functions, and stored procedures. Let's take a look the two most common ones, views and stored procedures.

Using Views

The most basic abstraction tool is a *view*. It is also the most limited. Views are saved SELECT statements that are stored in Microsoft's SQL Server system tables. These hidden system tables store the text as well as binary valued conversions, though the mechanism of how the SQL code is stored is immaterial to its use. When you create a view, SQL Server stores it within the database, and you can use that SELECT statement again by just referring to it in other SQL code.

Listing 13-22 shows an example of a view being wrapped around our current SELECT statement. Some important things to note are that SQL views cannot contain a number of features we have included in our SQL code so far. For example, you cannot include variables inside a view's definition, so the variable declarations must be excluded. In addition, any use of those variables within the view will not work, and all those lines must be excluded as well. Finally, you cannot use an ORDER BY clause within a view. As such, the ORDER BY clause must also be removed.

Listing 13-22. Creating a View

```
CREATE VIEW vQuantitiesByTitleAndDate
AS

  --DECLARE
  --  @ShowAll nVarchar(4) = 'True'
  --, @StartDate nVarchar(10) = '09/13/1994'
  --, @EndDate nVarchar(10) = '09/14/1994'
  --, @Prefix nVarchar(3) = 'PS%'
  --, @AverageQty int
        --SELECT @AverageQty = Avg(SalesQuantity) FROM DWPubsSales.dbo.FactSales

SELECT
    DP.PublisherName
  , [Title] = DT.TitleName
  , [TitleId] = DT.TitleId
  , [OrderDate] = CONVERT(varchar(50), [Date], 101)
  , [Total for that Date by Title] = Sum(SalesQuantity)

  --, [Average Qty in the FactSales Table] = @AverageQty
  --, [KPI on Avg Quantity] = CASE
              --WHEN Sum(SalesQuantity)
              --  between (@AverageQty- 5) and (@AverageQty + 5) THEN 0
              --WHEN Sum(SalesQuantity) < (@AverageQty- 5) THEN -1
              --WHEN Sum(SalesQuantity) > (@AverageQty + 5) THEN 1
              --END

FROM DWPubsSales.dbo.FactSales AS FS
INNER JOIN DWPubsSales.dbo.DimTitles AS DT
  ON FS.TitleKey = DT.TitleKey
```

```
INNER JOIN DWPubsSales.dbo.DimDates AS DD
  ON FS.OrderDateKey = DD.DateKey
INNER JOIN DWPubsSales.dbo.DimPublishers AS DP
  ON DT.PublisherKey = DP.PublisherKey

--WHERE
-- @ShowAll = 'True'
-- OR
-- [Date] BETWEEN @StartDate AND @EndDate
-- AND
-- [TitleId] LIKE @Prefix
GROUP BY
    DP.PublisherName
  , DT.TitleName
  , DT.TitleId
  , CONVERT(varchar(50), [Date], 101)

--ORDER BY DP.PublisherName, [Title], [OrderDate]
```

> **Note** With all these restrictions, you might wonder why one would ever use a VIEW statement, but keep in mind that many reporting applications can provide features you cannot include in the view itself once the view has returned all of the unfiltered results to the application.

Once a view has been created, you can use the view by selecting it as if it were a table. The code to do so will look similar to Listing 13-23.

Listing 13-23. Using Your View Without Filters

```
SELECT
    PublisherName
  , [Title]
  , [TitleId]
  , [OrderDate]
  , [Total for that Date by Title]
 FROM vQuantitiesByTitleAndDate
```

When the SELECT Statement is run, you receive all the results from the view including the aggregated daily totals. Figure 13-19 shows these results.

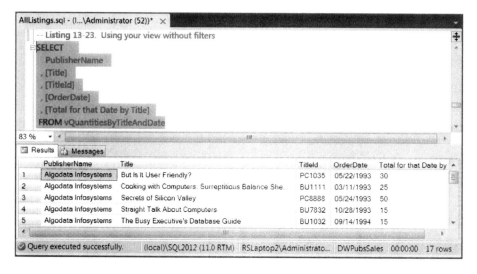

Figure 13-19. *Unfiltered Results from the view*

The ORDER BY clause can be used when you retrieve results from the view, which can be quite helpful since a view cannot store variables. Listing 13-24 shows an example of filtering the results of your view using the same options we used before the view.

Listing 13-24. Using Your View with the Previous Filters

```
DECLARE
    @ShowAll nVarchar(4) = 'False'
, @StartDate datetime = '09/13/1994'
, @EndDate datetime = '09/14/1994'
, @Prefix nVarchar(3) = 'PS%'
, @AverageQty int
    SELECT @AverageQty = AVG(SalesQuantity) FROM DWPubsSales.dbo.FactSales

SELECT
    PublisherName
, [Title]
, [TitleId]
, [OrderDate]
, [Total for that Date by Title]
, [Average Qty in the FactSales Table] = @AverageQty
, [KPI on Avg Quantity] = CASE
                WHEN [Total for that Date by Title]
                    between (@AverageQty - 5) and (@AverageQty + 5) THEN 0
                WHEN [Total for that Date by Title] < (@AverageQty - 5) THEN -1
                WHEN [Total for that Date by Title] > (@AverageQty + 5) THEN 1
                END

FROM vQuantitiesByTitleAndDate
```

```
WHERE
 @ShowAll = 'True'
 OR
 [OrderDate] BETWEEN @StartDate AND @EndDate
 AND
 [TitleId] LIKE @Prefix
ORDER BY PublisherName, [Title], [OrderDate]
```

Now we get the same results as before, but client reporting applications that expect to only use views will have them available as needed.

One change that should be noted is that since the view contains an alias column, our SQL code must be modified to include the new column names. You can see an example of this in the modified version of the WHERE clause shown in Listing 13-24. Whereas before you were using the [Date] column, you must now use the [OrderDate] column as defined by the view. The same is true in places where we reuse the Sum(SalesQuantity) in our calculations inside of the CASE operator; we must now use the [Total for that Date by Title] column of the view.

When this SELECT statement is run, all the results from the view will still be gathered in memory, but the final results will include only filtered data. Figure 13-20 shows what this looks like.

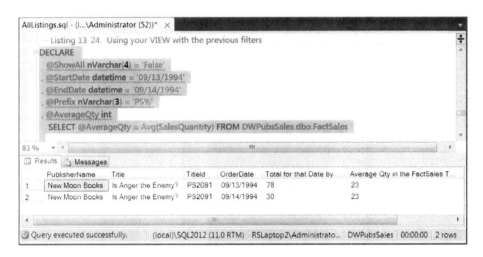

Figure 13-20. *Fitered results from the view with the KPIs added and results ordered*

Note that the result set looks remarkably similar to the one before you started using the view (see Figure 13-18). This is because we are getting the same results. The only difference is that now we are using a view.

Using Stored Procedures

Perhaps the most flexible tool you can use for abstracting your report data is SQL stored procedures. Stored procedures can include almost all SELECT clauses, built-in functions, operators, and even multiple SELECT statements. Because of this flexibility, chances are very high that you will want to create stored procedures for your reporting applications.

Listing 13-25 shows an example of a stored procedure that uses the code we have built so far.

Listing 13-25. Creating a Stored Procedure

```
-- Using Stored Procedures
CREATE PROCEDURE pSelQuantitiesByTitleAndDate
 (

  -- 1) Define the parameter list:
  -- Parameter Name, Data Type, Default Value --

    @ShowAll nVarchar(4) = 'True' -- 'True|False'
  , @StartDate datetime = '01/01/1990' -- 'Any valid date'
  , @EndDate datetime = '01/01/2100' -- 'Any valid date'
  , @Prefix nVarchar(3) = '%' -- 'Any three wildcard search characters'
  --, @AverageQty int
      --SELECT @AverageQty = Avg(SalesQuantity) FROM DWPubsSales.dbo.FactSales
 )
AS

BEGIN -- the body of the stored procedure --

 -- 2) Set the @AverageQty variable here since you cannot use subqueries in the
 -- a stored procedures parameter list.
 DECLARE @AverageQty int
   SELECT @AverageQty = Avg(SalesQuantity) FROM DWPubsSales.dbo.FactSales

 --3) Get the Report Data

SELECT
DP.PublisherName
, [Title] = DT.TitleName
, [TitleId] = DT.TitleId
, [OrderDate] = CONVERT(varchar(50), [Date], 101)
, [Total for that Date by Title] = SUM(SalesQuantity)
, [Average Qty in the FactSales Table] = @AverageQty
, [KPI on Avg Quantity] = CASE
  WHEN Sum(SalesQuantity)
    BETWEEN (@AverageQty- 5) AND (@AverageQty + 5) THEN 0
  WHEN Sum(SalesQuantity) < (@AverageQty- 5) THEN -1
  WHEN Sum(SalesQuantity) > (@AverageQty + 5) THEN 1
END
FROM DWPubsSales.dbo.FactSales AS FS
INNER JOIN DWPubsSales.dbo.DimTitles AS DT
  ON FS.TitleKey = DT.TitleKey
INNER JOIN DWPubsSales.dbo.DimDates AS DD
  ON FS.OrderDateKey = DD.DateKey
INNER JOIN DWPubsSales.dbo.DimPublishers AS DP
  ON DT.PublisherKey = DP.PublisherKey
WHERE
  @ShowAll = 'True'
  OR
  [Date] BETWEEN @StartDate AND @EndDate
```

```
AND
  [TitleId] LIKE @Prefix+'%'
GROUP BY
  DP.PublisherName
  , DT.TitleName
  , DT.TitleId
  , CONVERT(varchar(50), [Date], 101)
ORDER BY DP.PublisherName, [Title], [OrderDate]
END -- the body of the stored procedure --
```

Key points of interest are the definition of the parameter list at the beginning of a stored procedure and before the AS keyword. This serves the same purpose as declaring variables to be used to create dynamic queries.

One interesting aspect of stored procedure parameters is your ability to define default values. When a stored procedure is executed and a value is not supplied for the parameter, the default will be used. If a value is given, however, then the stored procedure will use the given value.

■ **Note** It is common for developers to include additional comments in their stored procedures. Listing 13-25 indicates the kind of data that is expected within the parameters. Other additions that are not shown here are: a header directly after the AS keyword that indicates information about who created the stored procedure, when it was first created, which objects it interacts with, and a list of changes that have occurred over time. This header information, while important, is not shown here in order to keep the code as small as possible. But we show an example of what it should look like in just a moment.

After the parameters have been defined for the stored procedure, the AS keyword is used to identify the beginning of the stored procedure body. The body of the stored procedure contains your reporting queries and any additional queries, and might be required to obtain the report data. Although not required, BEGIN and END are used to identify where the body of the stored procedure starts and finishes, and it is a recommended practice to include both.

Once the stored procedure has been created, you can execute it by using the EXECUTE keyword, which can be written out or, as shown in Listing 13-26, can be executed using just the first four letters of the keyword, EXEC.

Listing 13-26. Using Your Stored Procedure with Default Values

```
EXEC pSelQuantitiesByTitleAndDate
```

When this code runs, the stored procedure utilizes all the default values since no additional parameter values were given. In this case, the default value of True for the @ShowAll parameter is sufficient for the query to return all the results unfiltered, as shown in Figure 13-21.

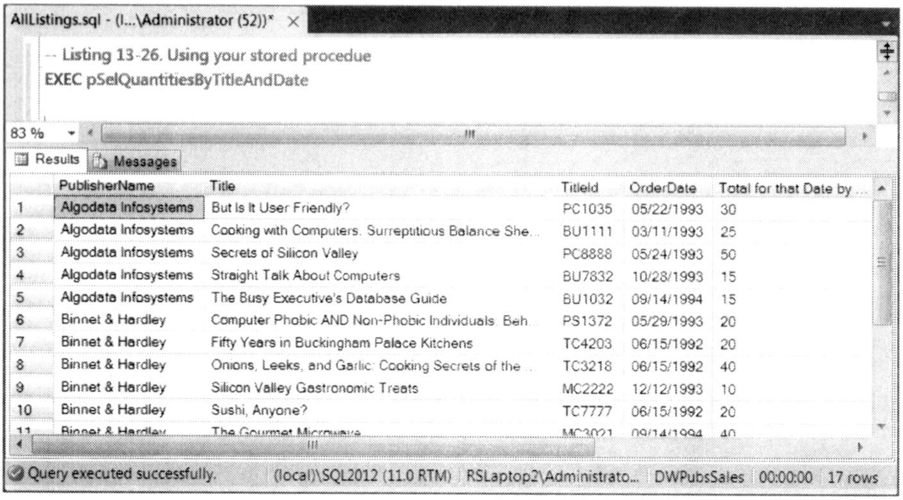

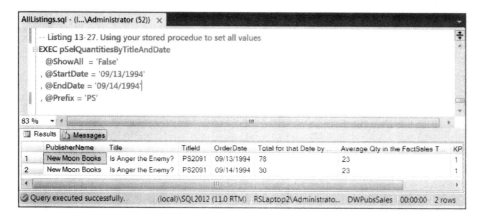

Figure 13-21. *Results for the stored procedure using the default values*

Of course, you can also supply values for each parameter, as shown in Listing 13-27.

Listing 13-27. Using Your Stored Procedue to Set All Values

```
EXEC pSelQuantitiesByTitleAndDate

    @ShowAll = 'False'
,   @StartDate = '09/13/1994'
,   @EndDate = '09/14/1994'
,   @Prefix = 'PS'
```

The results of this query display only two rows because of the filters placed on the query (Figure 13-22).

Figure 13-22. *Results for the stored procedure using provided parameter values*

Using default values for your parameters can be convenient because when executing the stored procedure, not all parameter values need to be set to be able to use any one of them. In Listing 13-28 we are setting only two of the four possible parameter values.

Listing 13-28. Accepting Defaults for Some Parameters

```
EXEC pSelQuantitiesByTitleAndDate

    @ShowAll = 'False'
, @Prefix = 'PS'
```

When this query runs, the results include anything that has the *PS%* prefix regardless of the date the order was placed. Figure 13-23 shows the results.

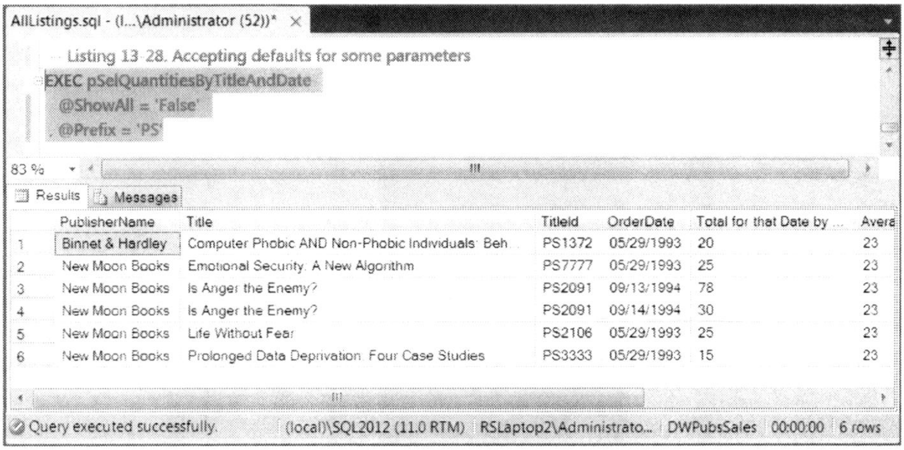

Figure 13-23. *Results for the stored procedure setting some values*

After your stored procedures are created and tested, you may want to make some changes. For example; production stored procedures commonly include a header at the beginning of the stored procedure, much like the one that was at the beginning of our script file. Listing 13-29 places a header just before the beginning of the stored procedure body.

Listing 13-29. Adding a Header to Your Stored Procedure

```
ALTER PROCEDURE pSelQuantitiesByTitleAndDate
  (
   -- 1) Define the parameter list:
   -- Parameter Name, Data Type, Default Value --
     @ShowAll nVarchar(4) = 'True' -- 'True|False'
   , @StartDate datetime = '01/01/1990' -- 'Any valid date'
   , @EndDate datetime = '01/01/2100' -- 'Any valid date'
   , @Prefix nVarchar(3) = '%' -- 'Any three wildcard search characters'
  )
AS

/****************************************
Title:pSelQuantitiesByTitleAndDate
Description: Which titles were sold on which dates
Developer:RRoot
```

```
Date: 9/1/2011
Change Log: Who, When, What
CMason,9/2/2011,fixed even more grammatical errors
*****************************************/

BEGIN -- the body of the stored procedure --
...
```

You can alter the stored procedure code to include the header using the ALTER keyword. When this code is run, the new version of the stored procedure is saved in the database. Whenever the database is backed up, the stored procedure will be backed up as well, making it easier to maintain your report queries.

EXERCISE 13-2. CREATING STORED PROCEDURES

In this exercise, you create your own custom report stored procedure similar to the one discussed in this chapter. This time, you include parameters that filter the results based on a start date and end date. Try using building the query a little at a time, as described in this chapter. You may find that creating a report procedure is easier than you think.

1. Create a stored procedure that allows report users to filter the results based on a range of dates using the code generated in Exercise 13-1.

    Execute the stored procedure to verify that it works. (Your results should look like Figure 13-24.)

If you get stuck, the answers for this exercise can be found in the example files for this book. Look in the folder for Chapter 13.

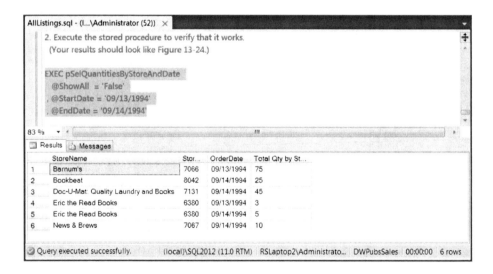

Figure 13-24. *Results showing sales by store*

In this exercise, you wrote a report query that captured sales data based on sales by stores.

Using Your Code in Reporting Applications

Once you have created your code, views, or stored procedures, you can use them in a number of reporting applications. For example, Microsoft Excel allows you to easily connect to a database and access both views and tables, as shown in Figure 13-25.

Name	Owner	Description	Modified	Created	Type
vQuantitiesByTitleAndDate	dbo			8/15/2011 7:40:10 PM	VIEW
DimAuthors	dbo			7/7/2011 4:26:09 PM	TABLE
DimDates	dbo			7/7/2011 4:26:09 PM	TABLE
DimPublishers	dbo			7/7/2011 4:26:09 PM	TABLE
DimStores	dbo			7/7/2011 4:26:09 PM	TABLE
DimTitles	dbo			7/7/2011 4:26:09 PM	TABLE
FactSales	dbo			7/7/2011 4:26:10 PM	TABLE
FactTitlesAuthors	dbo			7/7/2011 4:26:10 PM	TABLE

Figure 13-25. *Connecting to a view from Microsoft Excel*

In Excel you can display the information in either a tabular or pivot table format. The pivot table format provides options such as sorting, filtering, and subtotals, and it easily generates charts, making it a popular choice. This is significant, because as we have seen, many of these features cannot be stored directly as part of a SQL view's code. Figure 13-26 shows an example of a pivot table using the view we created.

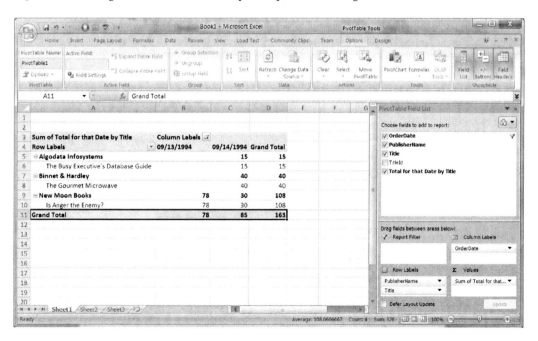

Figure 13-26. *Displaying the report with an Excel pivot table*

561

Excel can use SQL stored procedures as well, but it's not as straightforward as it could be. Other applications such as Microsoft Reporting Server (SSRS), however, have no such problems. In SSRS you can effortlessly launch a wizard and include the stored procedure name you want to use to capture your report data. Figure 13-27 shows the SSRS Report Wizard on its final dialog window, which list the details about the wizard configuration. Note that the stored procedure name was used as a query.

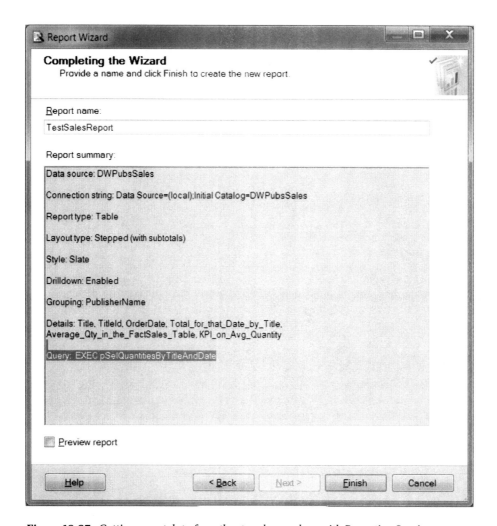

Figure 13-27. *Getting report data from the stored procedure with Reporting Services*

When the wizard completes, your SSRS project will include a new report that contains data generated from your stored procedure (Figure 13-28).

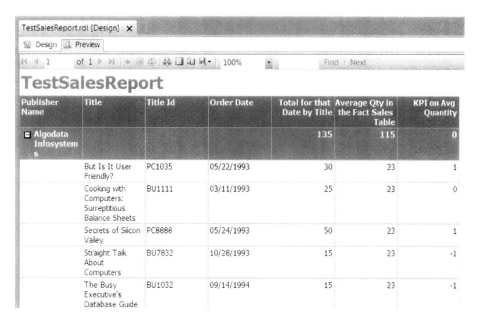

Figure 13-28. *Displaying the results of the stored procedure in Reporting Services*

We will delve deeper into both of these scenarios in the chapters on Excel reporting and reporting services.

Moving On

You will need quite a bit of practice to become an expert at writing report queries. Report stored procedures often included hundreds of lines of code. These stored procedures take many days to write and test correctly. When stored procedures become this large, it is very easy to make mistakes that can cost a lot of time and money. The methodology of incrementally building and reviewing your results, using consistent formatting and abstraction layers, make reporting queries easier to maintain and cost effective.

LEARN BY DOING

In this "Learn by Doing" exercise, you create several SQL queries using the Northwind database. We included an outline of the steps to perform and an example of how the authors handled them in two Word documents. These documents are found in the folder
`C:\_BISolutionsBookFiles\_LearnByDoing\Chapter13Files`. Please see the `ReadMe.doc` file for detailed instructions.

What's Next?

There is a lot to understand to become good at writing report queries. This includes proper syntax, performance considerations, and checking validity. While experience is your best teacher, we also recommend this book, that may help you along the way.

SQL Server 2008 Transact-SQL Recipes
A Problem-Solution Approach
By Joseph Sack
ISBN13: 978-1-59059-980-8

CHAPTER 14

Reporting with MDX Queries

But for me, it was a code I myself had invented! Yet I could not read it.

—Erno Rubik

Multidimensional Expressions (MDX) is a Microsoft-owned specification, not an open standard language. But, it has been adopted by many vendors and is considered an industry standard for querying and manipulating multidimensional data for SSAS.

If you are working with OLAP cubes, you will likely be working with MDX at some point. While you do not have to be an expert in MDX to work with SSAS cubes, acquiring a more in-depth understanding of the language is helpful in a number of ways. For example, calculated members, KPIs, and security structures within SSAS require MDX expressions. And while many client applications can work to create MDX code for you, you may find yourself troubleshooting to find out why the MDX code you are using is not returning the data you want. If you are unable to read MDX, you will not be able to manage troubleshooting problems.

Superficially, MDX is similar to the SQL language; however, MDX also has many differences. In this chapter, we explore these similarities and differences, and show you how to read and create MDX queries.

Key Concepts and Terms

You do not have to know everything about OLAP cubes to begin programming in MDX, but there are a few key concepts and terms that you need to know before you begin. Let's begin by reviewing some of them.

Consider a typical date dimension hierarchy, as shown in Figure 14-1. The dimension hierarchy is made up of levels such as year, month, and quarter. The dimension attributes are made up of members, such as 1992, 1993, and 1994.

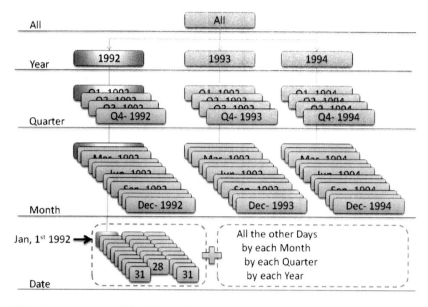

Figure 14-1. *A typical date hierarchy*

When creating a date dimension in SSAS, the attributes and hierarchies are defined as shown in Figure 14-2.

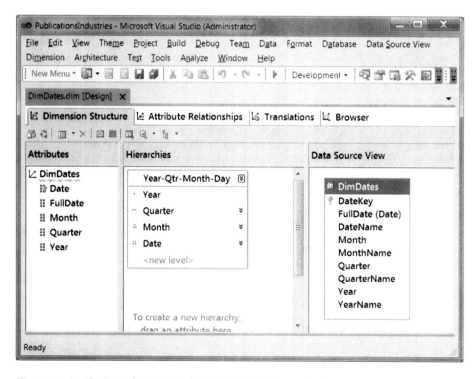

Figure 14-2. *The Date dimension of the PubsSalesCubes project*

Table 14-1 lists common MDX dimension terms and definitions. These terms are important to understand, because we use them to discuss MDX report queries.

Table 14-1. *MDX Concepts*

Concept	Description	Details/Examples
Dimensions	A way of grouping attributes.	Title names and publisher names are values. Dimensions are groupings of these values.
Measures	A column of values in the fact table described by the dimensional keys.	Each measure represents a column or calculated column from a fact table. An example is SalesQuantity from our fact table in the DWPubsSales data warehouse.
Member	Individual attributes of a dimension, hierarchy, or level. Each attribute within a hierarchy is a member.	Let's say you have a Date dimension with the levels of Years, Quarters, and Months; June would be an individual member of the Month level.
Measures dimension	The special dimension that all measures are grouped within.	The Measures dimension has only a single level and does not include an *All* level/member.
Measure groups	A way of organizing measures.	One SSAS cube can contain many measure groups.
All member	A grand total of all items within a dimension (not including the Measures dimension).	Every dimension (except the Measures dimension) has an All member. For example, in the Date dimension of Figure 14-1, there is an implied All Dates member.
Attribute	The property of a member or combination of members.	Attributes in SSAS are mapped to one or more columns in your data warehouse tables. For example, the Date key and the Date name combine to make up the Date attribute within the Date dimension. In MDX you can reference an attribute by either its key or its name.
Hierarchy	The manner in which data is grouped together and structured into levels.	For example, dates may be grouped by months, months by quarters, and quarters by years. Each dimension (except the Measures dimension) is made up of one or more hierarchies. For example, titles are grouped in one hierarchy by publishers and in a second hierarchy by types, creating more than one hierarchy.
Parent	The family relationship within a hierarchy that lower-level members are derived from.	For example, the month of January 1992 is the parent of the date January 1, 1992.

(continued)

Table 14-1. (*continued*)

Concept	Description	Details/Examples
Attribute hierarchy	The manner in which attributes are grouped together and structured into levels. This can be implied by the location of the data within a dimension.	Even when attributes are not grouped together explicitly, each attribute within an SSAS dimension is implicitly considered a hierarchy of one attribute. For example, in Figure 14-2, the Year attribute forms a Year hierarchy, but it is also part of the DatesByMonthQtrYear hierarchy.
Level	The manner in which hierarchies are structured, in a family type of parent–child relationship.	Figure 14-1 illustrates five levels of a hierarchy: All, Year, Quarter, Month, and Date. In Figure 14-2 the All level is implied.
Leaf level	The lowest level of data within a dimension's hierarchy.	For example, Date is the leaf level in Figure 14-2.
Cell	An individual member value within a cube.	Cell locations can be referenced by member and dimension coordinates within a cube. For example, each cell in the CubePubsSales cube can be located by indicating a specific value of a member within each of the five dimensions—Measures, DimAuthors, DimTitles, DimDates, and DimStores. An example of cells themselves are the SalesQuantity values for a given author, title, date, and store within each listed dimension.
A default member	A member used to calculate an expression when an attribute is left out of a query.	The default member is usually the All member of that dimension, although it can be reconfigured.

Programming with MDX

To program with MDX, it is helpful to understand the differences and similarities between MDX and SQL languages. We have several topics to discuss, so let's get right to it.

Comments

There are three ways to comment in MDX. Listing 14-1 shows an example of each.

Listing 14-1. MDX Comments

```
-- MDX and SQL Comment
// MDX Only Comment
/* MDX and SQL Block Comment */
```

Basic and Raw Syntax

The basic syntax of an MDX select statement starts like a standard SQL select statement, but soon the differences are quite apparent. Listing 14-2 outlines the basic MDX select syntax.

Listing 14-2. Basic MDX Select Syntax

```
Select
  { <a set of attributes> } On Columns,
  { <a set of attributes> } On Rows
From< cube name>
Where ( <member> );
```

The following are some key differences:

- Instead of using the table name in the FROM clause, MDX uses the cube name. This makes sense, because the cube is equivalent to a set of one or more tables in a relational database.

- Curly braces and semicolons are not commonly used in SQL programming but are used often in MDX programming.

- MDX uses the keywords ON COLUMNS and ON ROWS to indicate how the results should be displayed within an Analysis Server client application. Analysis Server applications can use 128 different axis positions to return results. Most applications, like SQL Server Management Studio, are designed only to handle the first two axis (columns and rows). In custom applications, you may see more axis used that indicate the next two positions after columns and rows, such as pages and chapters.

Listing 14-3 shows an example of a raw MDX statement that queries the CubePubsSales cube.

Listing 14-3. A Raw MDX Statement

```
Select
  { [Measures].[SalesQuantity] } On Columns, -- Axis 0
  { [DimTitles].[Title].[Is Anger the Enemy?] } On Rows -- Axis 1
From [CubePubsSales]
Where ( [DimStores].[Store].[Eric the Read Books] );
```

▦ **Note** We recommend typing each code sample, but for your convenience we have included a script file in our downloadable content called Chapter14 All MDX Listings.mdx that contains all of the MDX code used in this chapter. We have also included a processed backup of the PubsBICubes database for SSAS so that if you did not complete your cube in previous chapters, you would be able to restore the database and cube, allowing you to run the code in this chapter. The instructions on how to restore this type of database can be found in the readme file along with the backup file. You will find both items within C:_BookFiles\Chapter14Files\.

Running Your MDX Code

To run your MDX code from SQL Server Management Studio, you must either create a new MDX script or open an existing one. If you have opened an existing script with code in it, highlight the MDX statement to run and click the "! Execute" button at the top of the toolbar. To make a new MDX script, click the New MDX Query button indicated in Figure 14-3.

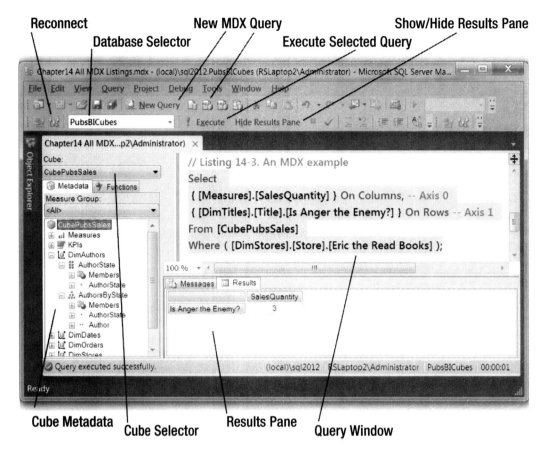

Figure 14-3. *The MDX query window*

Figure 14-3 shows a typical layout within SQL Server Management Studio while creating an MDX query. If you are familiar only with running SQL code in SQL Server Management Studio, things will look quite different here.

Table 14-2 gives a brief description of key objects used from SQL Server Management Studio while creating and executing MDX statements.

Table 14-2. *Features of the MDX Editor Within SQL Server Management Studio*

Feature	Description
Database Selector dropdown	Allows you to select which database for your query window to use. Since the SSAS server can host many databases, you must select which database to use by selecting it in this dropdown prior to executing MDX code for that database.
Cube Selector dropdown	Allows you to choose which cube in the database your Cube Metadata tab will use. One SSAS database can have many cubes; therefore, the correct one must be selected to display. This does not affect the query window.
New Query and MDX Query buttons	Creates a new MDX query window. The New Query button is context sensitive, and creates an MDX query window only if you are currently connected and focused on SSAS in Object Explorer. Otherwise a standard SQL query window is created instead. (The MDX query button always creates an MDX query window.)
Query window	Provides a location to type in one or more MDX statements, but only one statement can be executed at a time. It should be noted that this is very different from the SQL query window, which can execute several statements at a time.
Execute Selected Query button	Sends the highlighted MDX code to the SSAS database engine for processing.
Results pane	Shows the results of the executed query.
Show/Hide Results button	Allows you to toggle between showing and hiding the Results pane. This button is not included in the UI by default and must be added by customizing the toolbar. Instruction on how to do so can be found by searching for *Customizing Visual Studio 2010 toolbar*. The Ctrl + R shortcut can be used as an alternative to this button.
Cube Metadata tab	Allows you to browse the cube's members, levels, hierarchies, and dimensions. You can drag items from this tab into the query window, and its location within the cube will be typed out for you.
Reconnect button	Allows you to reconnect to the SSAS server. If you change, deploy, or process the cube or dimensions, you must reconnect using this button before those changes show up on the Cube Metadata tab. Since this tab is quite useful in creating your MDX code using its drag-and-drop functionality, it is a very good button to know about.

Optional Syntax

In MDX, parentheses, braces, brackets, and semicolons are optional syntax items. We created a comparison using MDX code in Listing 14-4.

Listing 14-4. Examples in MDX with Optional Syntax Items

```
Select
 { [Measures].[SalesQuantity] } On Columns,
 { [DimTitles].[Title].[Is Anger the Enemy?] } On Rows
From [CubePubsSales]

Where ( [DimStores].[Store].[Eric the Read Books] );
```

-- The same code with optional syntax items removed

```
Select
Measures.SalesQuantity On Columns,
```

-- Curly Braces {} are optional: Defines a Set of Results

```
  DimTitles.Title.[Is Anger the Enemy?] On Rows
```

-- Square Brackets [] are optional: Defines an Object Identifier

```
From CubePubsSales
```

-- Parentheses () are optional: Defines Function Parameters or Set of Coordinates

```
Where DimStores.Store.[Eric the Read Books];
```

-- Semi-colons; are optional: Defines the End of a Statement

Additionally, as in SQL, there is an optional WHERE clause. MDX code is not case-sensitive, and it isn't sensitive to white spaces within your code. Listing 14-5 gives three examples that highlight this point.

Listing 14-5. Three More Syntax Examples in MDX

-- The Where clause is also optional

```
Select
  [Measures].[SalesQuantity] On Columns,
  [DimTitles].[Title].[Is Anger the Enemy?] On Rows
From [CubePubsSales]
```

-- MDX is not sensitive to white spaces

```
Select [Measures].[SalesQuantity]
 On Columns     ,
   [DimTitles].   [Title].
                             [Is Anger the Enemy?]
           On Rows
  From [CubePubsSales]
```

-- MDX is not Case Sensitive

```
SelECt
[MeaSURes].[SALESQUANTITY] On CoLUMns,
[DimTITLes].[Title].[Is ANGER THE ENemy?] ON ROWs
From [CubePubsSales]
```

Of course, these examples are for demonstration purposes, and we do not recommend writing MDX code in this manner. Your MDX code should be consistently typed so that it will be easier to work with. To structure your code in a professional manner, we recommend using curly braces and square brackets even when they are not required. We also recommend using sentence, or Pascal, casing when defining objects and using white spaces consistently to organize your code.

■ **Tip** It is considered an unofficial industry standard to use curly braces and square brackets at all times in MDX code. Most printed material demonstrating code, such as magazine articles, books, and web pages, do so as well.

Default Members

The code samples so far have included members of the Titles and Measures dimensions but have not explicitly defined members from the others. Yet, the measured values returned in the results include members from each and every dimension. Dimensions that are not explicitly specified within your MDX query are implicitly defined using a default member.

The way this works is that a default member exists for each dimension, and unless you specify otherwise, it will be the grand total of all the members within the dimensions. The grand total is referred to as the All member. For example, if we do not specify which author we want information about, MDX will assume we want information about all authors. And any totals that we receive as part of the measure will be an aggregation of all authors combined.

It is possible to set a specific member as a default for a dimension, but this is probably not the best way to handle it. For example, if you define the current year as the default member of the date dimension and somebody forgot to include a specific date within their MDX query, Analysis Server returns data only for the current year. If everyone in the client's company understood that rule, then everything would be OK. Otherwise, your users will be under the assumption that they are receiving the total of every year, not just the current one, and the report will be inaccurate. For this reason, we recommend leaving the default setting for your dimensions, which is the total of all members for each dimension.

To explicitly tell Analysis Server to use the default member of a dimension in MDX, you can use the DefaultMember function, as shown in Listing 14-6. Note that we are using the name of the dimension, and the name of the leaf member in this example. But we could have used any of the attributes or user-defined hierarchies defined in each dimension, because they all have the same ultimate parent, the All member.

Listing 14-6. The DefaultMember Function

```
Select
From [CubePubsSales];

-- Same As...
Select
{ [Measures].DefaultMember } On Columns,
{( -- Parentheses are required to Define a SET of Coordinates
[DimAuthors].[Author].DefaultMember
,[DimDates].[Date].DefaultMember
,[DimOrders].[OrderNumber].DefaultMember
,[DimStores].[Store].DefaultMember
,[DimTitles].[Title].DefaultMember
)} On Rows
From [CubePubsSales];
```

> ■ **Note** In most MDX documentation, both functions and properties are referred to as *functions*. So, DefaultMember may look like a property, but in MDX it is called a function.

You can specify the All member and still receive the same results as the default (Listing 14-7). The Measures dimension is the single exception to this rule, because it does not have an All member. Instead, the first member of the Measures dimension is used implicitly whenever you do not explicitly specify which measure you want.

Listing 14-7. *The All vs. the Default Member Dimension*

```
Select
{
 [Measures].DefaultMember
 -- You cannot use the [All] member on the Measures dimension.
 -- So, we are still using the default member here.
} On Columns,
{( -- Parentheses required to Define a SET of Coordinates
 [DimAuthors].[All]
 ,[DimDates].[All]
 ,[DimOrders].[All]
 ,[DimStores].[All]
 ,[DimTitles].[All]
 )} On Rows
From [CubePubsSales];

-- In this case, since the [SalesQuantity] attribute is
-- the default member in the cube, it's the same as…
Select
{[Measures].[SalesQuantity]} On Columns,
{(
 [DimAuthors].[All]
 ,[DimDates].[All]
 ,[DimOrders].[All]
 ,[DimStores].[All]
 ,[DimTitles].[All]
 )} On Rows
From [CubePubsSales];
```

You can set the default measure in Business Intelligence Development Studio by selecting the measure within the cube-editing window and using the up and down arrows provided on the toolbar, as shown in Figure 14-4. The measure that is placed closest to the top of the Measures pane will be used as the default measure.

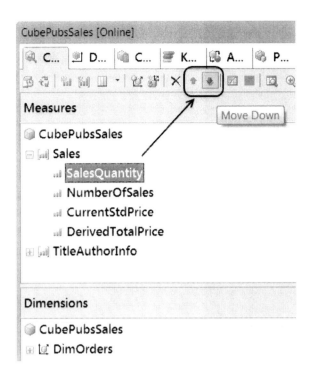

Figure 14-4. *Setting the default measure in a cube*

Using Key vs. Name Identifiers

To identify an individual member, you can use either a key or a name. To use a key, insert a numeric value that represents the member's location within that dimensional level. In other words, if we have a Titles dimension, it might include not only titles but also the publishers. If so, the publisher as well as the titles will each have a dimensional level. The key value of 1 on the publisher's level indicates the first member of that level. In the PubsSales cube, the first member of the publishers level is New Moon Books.

We can also use the publisher name instead of its key value. Listing 14-8 shows both examples. Notice that a key value is indicated by using the symbol "&" in front of the number.

Listing 14-8. Examples Using the Key vs. Name Identifiers

```
Select
{ [Measures].[SalesQuantity] } On Columns,
{ -- Dimension - Hierarchy - Level - Member by Key
```

[DimTitles].[TitlesByPublisher].[Publisher].&[1] -- Using Key

```
} On Rows
From [CubePubsSales];

-- Same result, but uses the Name of the attribute and not the key
Select
{ [Measures].[SalesQuantity] } On Columns,
{ -- Dimension - Hierarchy - Level - Member by Name
```

[DimTitles].[TitlesByPublisher].[Publisher].[New Moon Books] -- Using Name

```
} On Rows
From [CubePubsSales];
```

■ **Important** One of the biggest differences between a SQL code window and an MDX code window is that the MDX version allows you to execute only one query per batch! If you try to run more than one statement at a time, you will receive a syntax error. You can, however, separate multiple statements into their own batches using the word GO. And by the way, GO is case sensitive in the MDX editor. We show you an example in the next exercise!

We have covered quite a few of the basics. It's time to try some MDX code on your own. This next exercise will get you started.

EXERCISE 14-1. EXECUTING MDX QUERIES

In this exercise, you create a new MDX query similar to the one we just walked though. You also write another query that returns the sales quantity for the store called BookBeat. Your final results should look similar to Figure 14-7.

1. Open SQL Server Management Studio. You can do so by clicking the Start button and selecting All Programs ➤ Microsoft SQL Server 2012 ➤ SQL Server Management Studio.

Important: Remember to right-click the menu item, select Run as Administrator, and then answer Yes to close the UAC.

2. When SQL Server Management Studio opens, it will display a dialog window similar to the one in Figure 14-5.

3. Select Analysis Server from the "Server type" drop-down box.

4. Type the name of your server in the "Server name" textbox.

Tip: This will usually be (local) unless you use a named instance of SQL Server, as shown in Figure 14-5.

Figure 14-5. *The connection to Analysis Server*

5. Click Connect to open a connection to the SSAS server in Object Explorer, as shown in Figure 14-6.

6. Click the New Query button to create a new MDX coding window, as circled in Figure 14-6.

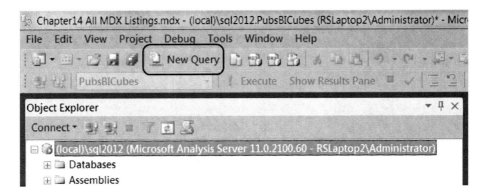

Figure 14-6. *Launching a new MDX query window*

7. When the new Query window opens, type in the MDX code shown in Listing 14-9.

Listing 14-9. Code for Exercise 14-1

```
-- Write a Query that returns the Sales Qty by Store "BookBeat"

-- Version 1: Using a member's name
Select
{ [Measures].[SalesQuantity] } on Columns,
{ [DimStores].[Store].[Store].[BookBeat] } On Rows -- Using the Name
From [CubePubsSales]
GO
-- Version 2: Using a member's key
Select
{ [Measures].[SalesQuantity] } on Columns,
{ [DimStores].[Store].&[6] } On Rows -- Using the Key
From [CubePubsSales]
GO
-- Version 3: Using a member's key and the default measure
Select [DimStores].[Store].&[6] On Columns
From [CubePubsSales]
GO
```

Highlight the first version of the select statement from the code you typed in, right-click the highlighted code, and select Execute. Alternately, you can run the query by clicking the "! Execute" button on the toolbar.

8. Verify that your results match the results shown in Figure 14-7.

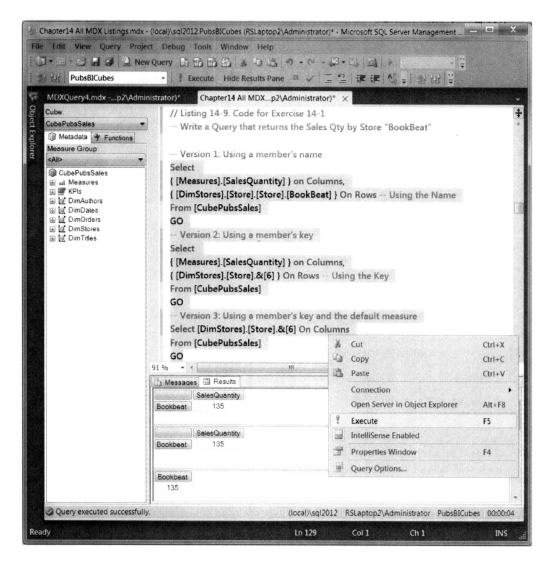

Figure 14-7. *The results of Exercise 14-1*

9. Run all three queries at once and verify that they return the same results.

Tip: Remember that the word GO is case sensitive! You will receive an error when attempting to run all three queries at once if GO is not all caps.

In this exercise, you wrote and executed three MDX queries. Now let's work on expanding your knowledge of the MDX language.

Using the Axis 0 and 1 Instead of Column and Row

When writing queries in SQL Server Management Studio, your results are returned using columns and rows. The words COLUMNS and ROWS are aliases for the position numbers. In MDX, these equate to the positions 0 and 1 in MDX queries; also written as (0) and (1). Columns are position 0, and rows are position 1. Although some developers prefer to use words, many prefer to write the MDX code using the numeric identifier instead. Listing 14-10 shows the three different styles used to identify the positions.

Listing 14-10. Identifying Axis Positions Within Queries

```
Select

 { [Measures].[SalesQuantity] } On Columns,

 { [DimTitles].[TitlesByType].[TitleType] } On Rows
From [CubePubsSales];

-- Same as...
Select

 { [Measures].[SalesQuantity] } On Axis(0), -- Using Axis Key

 { [DimTitles].[TitlesByType].[TitleType] } On Axis(1)
From [CubePubsSales];
-- Note that the Axis collection uses Parentheses (like VB)
-- and not Brackets (like C#)

-- same as before, but does not use the Axis keyword
Select

 { [Measures].[SalesQuantity] } On 0, -- Using Axis Key

 { [DimTitles].[TitlesByType].[TitleType] } On 1
From [CubePubsSales];
```

It is worth noting that you cannot skip a position in your MDX code. While some client applications allow you to display the results across columns and across rows, the MDX code must specify 0 before 1. Position 0 cannot be skipped. Listing 14-11 shows an example of this.

Listing 14-11. Positions Cannot be Skipped in Columns and Rows

```
-- This WILL NOT work, because we are not including the 0 axis
Select

 { [DimTitles].[TitlesByType].[TitleType] } On 1

From [CubePubsSales];

--ERROR="Axis numbers specified in a query must be sequentially specified, and cannot contain gaps."

-- Must be written as ...
Select

 { [DimTitles].[TitlesByType].[TitleType] } On 0
From [CubePubsSales];
```

Cells and Tuples

Microsoft's help pages define a tuple with the following statement: "A tuple uniquely identifies a slice of data from a cube." Perhaps the best way to think of a tuple is a single result from an MDX query. Understanding the difference between cells and tuples can be confusing, because cells contain a single value and tuples are a single result. Nevertheless, the two are quite different. You can write an MDX query that returns a tuple from a single cell, but you can also return a tuple that is the aggregate value of many cells.

Listing 14-12 shows three queries. The top two queries return one tuple, but the first uses data from a single cell while the second uses an aggregate from multiple cells. The last query returns two results (tuples) by aggregating data from multiple cells. Figure 14-8 displays the result of the last query.

Listing 14-12. Queries That Return One or More Tuples

-- This query returns one tuple from a single cell

```
Select
{
 [Measures].[SalesQuantity]
}
On Columns,
{(-- Parentheses are required since we specify a SET of Coordinates
 [DimAuthors].[Author].[Cheryl Carson]
,[DimStores].[Store].[Bookbeat]
,[DimTitles].[Title].[But Is It User Friendly?]
,[DimDates].[Year-Qtr-Month-Day].[Date].[Saturday, May 22 1993 12:00 AM]
)}
On Rows
From [CubePubsSales];
```

-- This query also returns one tuple, but is an aggregate of many cells

```
Select
{
 [Measures].[SalesQuantity]
}
On Columns,
{( -- Parentheses are NOT required since it no longer specifies a SET of Coordinates
[DimStores].[Store].[Bookbeat]
)}
On Rows
From [CubePubsSales];
```

-- This query returns two tuples, each is an aggregate of many cells

```
Select
{
 [Measures].[SalesQuantity]
}
On Columns,
{-- Braces are required when more than one tuple is returned
 ([DimStores].[Store].[Bookbeat]) -- Returns one tuple
-- Remember each item in parentheses implicitly includes the other dimensions
,([DimStores].[Store].[News & Brews]) -- Returns another tuple
}
On Rows
From [CubePubsSales];
```

▪ **Important**   Unless you are using a function, curly braces are required when your query returns more than one tuple.

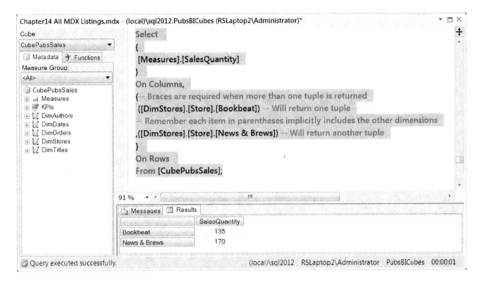

Figure 14-8. *A query that returns two tuples*

Be careful not to identify the multiple members from the same dimension as coordinates for a tuple. When you do so, you will receive an error. In Listing 14-13 you see that we are using two members, [Business] and [Psychology], from the same [DimTitles].[TitleType] hierarchy. Running this query gives the error message "The 'TitleType' hierarchy appears more than once in the tuple."

Listing 14-13. Improper Coordinates for a Single Tuple

```
-- This query will not work!

Select
{
  [Measures].[SalesQuantity]
, [Measures].[NumberOfSales]
} On Columns,

{ ( -- This parenthesis indicates you are defining multiple coordinates
  -- for a SINGLE tuple (a.k.a. a result value )

  [DimTitles].[TitleType].&[Business]

  -- A second coordinate won't work if it is from the same dimension!

, [DimTitles].[TitleType].&[Psychology]
) } On Rows
From [CubePubsSales];

//Error = The 'TitleType' hierarchy appears more than once in the tuple.

-- This works though!
Select
{
  [Measures].[SalesQuantity] -- 1st tuple
, [Measures].[NumberOfSales] -- 2nd tuples
} On Columns,
```

```
{ -- Do NOT include parentheses when
  -- you want TWO tuples (a.k.a. result values) from the SAME dimension-heirarchy
  [DimTitles].[TitleType].&[Business] -- 3rd tuple
, [DimTitles].[TitleType].&[Psychology] -- 4th tuple
 } On Rows
From [CubePubsSales];
```

Figure 14-9 shows the results of the working query.

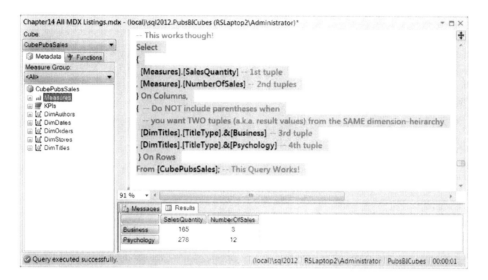

Figure 14-9. *The results from Listing 14-13*

The same is true when working with the Measures dimension. If you use parentheses to encompass two multiple measures, as demonstrated in Listing 14-14, SSAS assumes you are trying to indicate two coordinates on the Measures dimension. Doing so gives you the following error: "The 'Measures' hierarchy appears more than once in the tuple."

Listing 14-14. Using Parentheses to Encompass Two Multiple Measures

```
Select
{  ( -- This will not work!
  [Measures].[SalesQuantity]
, [Measures].[NumberOfSales]
) } On Columns
From [CubePubsSales];
```

//Error = The 'Measures' hierarchy appears more than once in the tuple.

This aspect of MDX does not come into play if you are asking for a single tuple based on the coordinates from two different dimensions. Listing 14-15 shows an example that returns a single tuple across the rows axis by identifying coordinates from both the DimTitles and DimStores dimensions. Figure 14-10 shows the results.

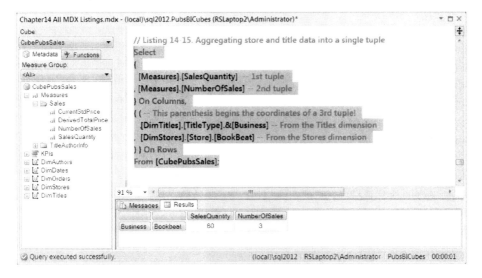

Figure 14-10. *The results from Listing 14-15*

Listing 14-15. Aggregating Store and Title Data into a Single Tuple

```
Select
{
  [Measures].[SalesQuantity] -- 1st tuple
, [Measures].[NumberOfSales] -- 2nd tuple
} On Columns,

{ ( -- This parenthesis begins the coordinates of a 3rd tuple!
  [DimTitles].[TitleType].&[Business] -- From the Titles dimension
, [DimStores].[Store].[BookBeat] -- From the Stores dimension
) } On Rows

From [CubePubsSales];
```

EXERCISE 14-2. RETURNING MULTIPLE TUPLES

In this exercise, you write a query that shows the number of SalesQuantity and NumberOfSales for the BookBeat and Eric The Read Books.

Note: If you have closed SQL Server Management Studio since the previous exercise, please open it. You can do so by clicking the Start button, navigating to All Programs ➤ Microsoft SQL Server 2012 ➤ SQL Server Management Studio, and clicking this menu item.

1. With a connection to the SSAS server established, click the New Query button to create a new MDX coding window.

2. When the new Query window opens, type in the MDX code shown in Listing 14-16.

Listing 14-16. Code for Exercise 14-2

```
Select
{
  [Measures].[SalesQuantity]
, [Measures].[NumberOfSales]
} On Columns,
{
 ( [DimStores].[Store].[BookBeat] )
,( [DimStores].[Store].[Eric The Read Books] )
} On Rows
From [CubePubsSales];
```

3. Highlight the first version of the select statement from the code you entered, right-click the code, and choose Execute. Alternately, you can run the query by clicking the "! Execute" button on the toolbar.

4. Verify that your results match the results shown in Figure 14-11.

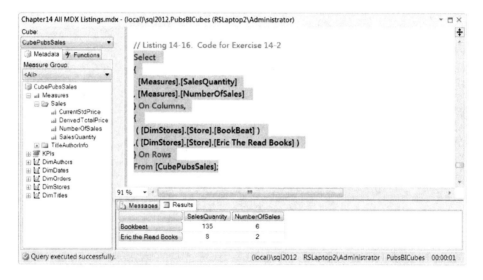

Figure 14-11. *The results of Exercise 14-2*

In this exercise, you wrote and executed an MDX query that returned four tuples. Let's continue our coverage of the MDX language by looking at various aspects of dimensional members and levels.

Calculated Members

You can add calculated members to your cube from Visual Studio, and you can also add calculated members to the results of an MDX query. The difference is that MDX's calculated member code query is not stored within the cube. It is added as an expression each time you write a query.

Listing 14-17 is an example of a query that includes a calculated member expression. Notice that the query begins with the WITH MEMBER clause and then indicates which dimension the member is to be placed within. Figure 14-12 displays the query results.

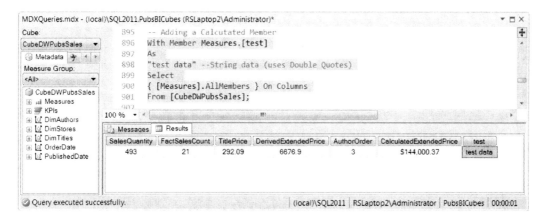

Figure 14-12. *The results of the example in Listing 14-17*

Listing 14-17. Adding a Calculated Member

```
-- Adding a Calculated Member
With Member Measures.[test]
As

"test data" --String data (uses Double Quotes)
-- Note: using Single Quotes causes an Error

Select
{ [Measures].AllMemebers } On Columns
From [CubePubsSales];
```

■ **Note**   Many developers expect the calculated member syntax to indicate the end of an expression. It is clear that the expression begins after the As keyword, but there is no clear indicator of where it ends. Because of this, many new developers try including curly braces to encompass the expression. Doing so causes an error.

Calculated members are evaluated as either string or numeric data. Use double quotes to indicate that the value is a string. This is different from SQL programming that uses a single quote to indicate a string. Use literal values for numeric data.

For mathematical expressions, specify the literal values and operator as you would any other programming language. It is good practice to include parentheses around the expressions, but they are not strictly necessary.

Although it may seem odd, you can use single quotes to surround an expression. Remember that in MDX, double quotes indicate a string of characters, not single quotes.

The three examples in Listing 14-18 highlight this calculated member syntax. The first query and third queries return the same result (shown in Figure 14-13).

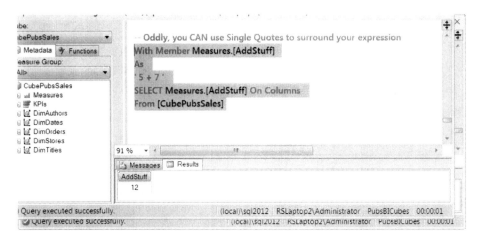

Figure 14-13. *The results of the first and last examples from Listing 14-18*

Listing 14-18. Expression Syntax

```
--1) You can use parentheses to surround expressions
With Member Measures.[AddStuff]
As

(5 + 7) -- Numeric data

SELECT Measures.[AddStuff] On Columns
From [CubePubsSales]

--2) However, you CANNOT use Braces to surround your expression
With Member Measures.[AddStuff]
As
{ 5 + 7 } -- Causes an #Error
SELECT Measures.[AddStuff] On Columns
From [CubePubsSales]

--3) Oddly, you CAN use Single Quotes to surround your expression
With Member Measures.[AddStuff]
As

'5 + 7'

SELECT Measures.[AddStuff] On Columns
From [CubePubsSales]
```

Calculated members are most commonly placed on the Measures dimension, but this is not a requirement. They can be placed on any dimension. Listing 14-19 gives some examples using the Date dimension for our calculated member instead of the Measures dimension. The result of this query is the total sales quantity for the combined years of 1992 and 1993. Figure 14-14 displays these results.

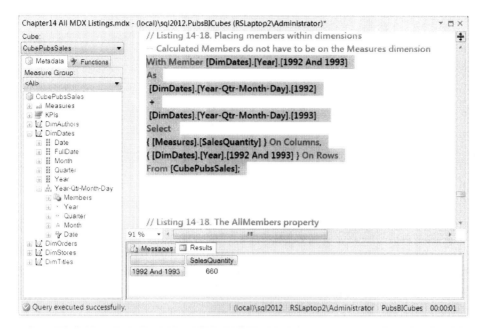

Figure 14-14. *The results of the example in Listing 14-19*

Listing 14-19. Placing Members within Dimensions

```
-- Calculated Members do not have to be on the Measures dimension

With Member [DimDates].[Year].[1992 And 1993]
As
 [DimDates].[Year-Qtr-Month-Day].[1992]
 +
 [DimDates].[Year-Qtr-Month-Day].[1993]
Select
 { [Measures].[SalesQuantity] } On Columns,
 { [DimDates].[Year].[1992 And 1993] } On Rows
From [CubePubsSales];
```

Member Properties

MDX uses a number of properties and functions to return a set of one or more tuples. Two common properties are the Members property and the AllMembers property. The Members property returns standard members. The AllMembers property returns standard members as well as any calculated members. Listing 14-20 gives an example of each. Figure 14-15 displays the query results.

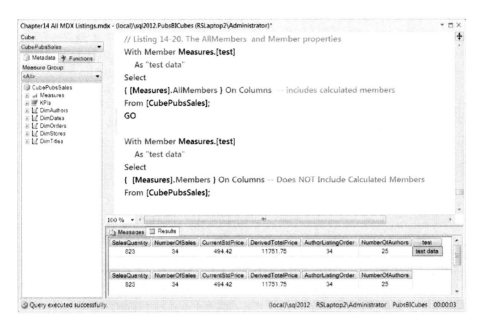

Figure 14-15. *The results of the example in Listing 14-20*

Listing 14-20. The Members and AllMembers Properties

```
With Member Measures.[test]
    As "test data"
Select
{
 [Measures].AllMembers
} On Columns -- includes calculated members
From [CubePubsSales];
GO

With Member Measures.[test]
    As "test data"
Select
{
 [Measures].Members
} On Columns -- Does NOT Include Calculated Members
From [CubePubsSales];
```

■ **Important**   When using a function or property that returns multiple tuples, your MDX code does not require braces, but most developers include them anyway. This MDX statement: `Select [Measures].Members On Columns From [CubePubsSales];` is equivalent to this MDX statement with curly braces: `Select { ( [Measures].Members ) } On Columns From [CubePubsSales];`.

Members and Levels

Each dimension consists of members. These members can be grouped into user-defined hierarchies. Each user-defined hierarchy consists of two or more levels, with the highest level being the All level. (The only exception to this rule is the Measures dimension, which is completely flat and has no hierarchical structure of this type).

To display every member of the hierarchy or level, specify its name and use the AllMembers function. To specify the members of a level, include the dimension, hierarchy, and level. The Analysis Server will implicitly use the AllMembers function for you. Listing 14-21 outlines these concepts. Figure 14-16 displays the query results.

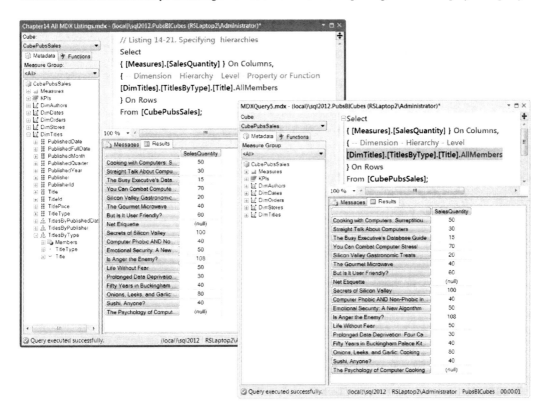

Figure 14-16. *The results of the examples from Listing 14-21*

Listing 14-21. Specifying the Hierarchy and a Level Implicitly Returns All Members

```
Select
{ [Measures].[SalesQuantity] } On Columns,
{ -- Dimension - Hierarchy - Level - Property or Function
[DimTitles].[TitlesByType].[Title].AllMembers
} On Rows
From [CubePubsSales];

Select
{ [Measures].[SalesQuantity] } On Columns,
{ -- Dimension - Hierarchy - Level
 [DimTitles].[TitlesByType].[Title]
} On Rows
From [CubePubsSales];
```

Because SSAS considers each attribute a hierarchy of its own (remember the warning in the designer about "hiding" these?), using the name of an attribute and a level also returns a SET of tuples. Listing 14-22 shows an example of this, and Figure 14-17 displays the results.

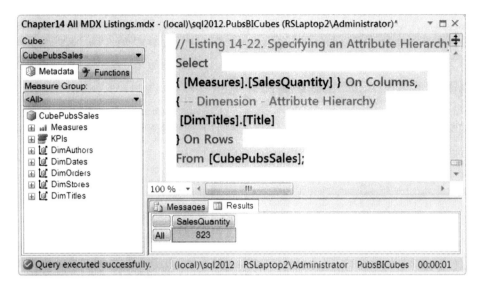

Figure 14-17. *The results of the example in Listing 14-22*

Listing 14-22. Specifying an Attribute Hierarchy Does Not Implicitly Return All Members

```
Select
{ [Measures].[SalesQuantity] } On Columns,
{ -- Dimension - Attribute Hierarchy
 [DimTitles].[Title]
} On Rows
From [CubePubsSales];
```

The NonEmpty Function

Sometimes when you execute a query, a lot of null values are included in the results. For example, in the Pubs database, many publishers have no sales associated with them. Therefore, the sales quantity for that publisher is displayed as null, which in MDX is equivalent to an empty set. The query in Listing 14-23 demonstrates this. The results are displayed in Figure 14-18.

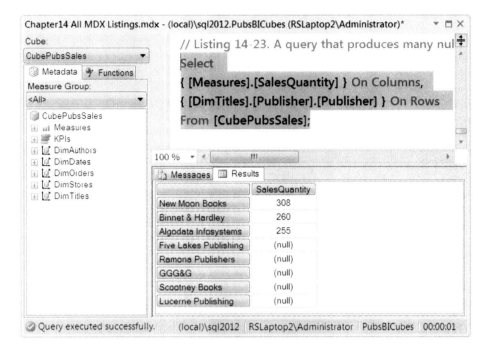

Figure 14-18. *The results of the example in Listing 14-23*

Listing 14-23. A Query That Produces Many Null Values

```
Select
{ [Measures].[SalesQuantity] } On Columns,
{ [DimTitles].[Publisher].[Publisher] } On Rows
From [CubePubsSales];
```

If you do not want null values displayed, use the `NonEmpty` function to remove nulls from the result set. Listing 14-24 shows an example of this function. See Figure 14-19 for the results.

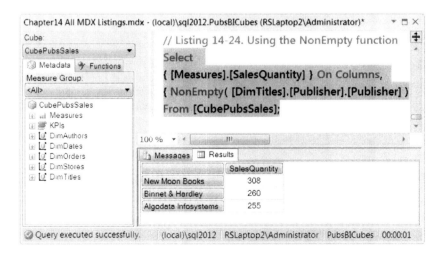

Figure 14-19. *The results of the example in Listing 14-24*

Listing 14-24. Using the NonEmpty Function

```
Select
{ [Measures].[SalesQuantity] } On Columns,
{ NonEmpty( [DimTitles].[Publisher].[Publisher] ) } On Rows
From [CubePubsSales];
```

Compare the results in Figure 14-19 to the results in Figure 14-18. Notice how the NonEmpty function has removed the null values.

The Non Empty Clause

Nulls can also be removed from a result set using the Non Empty clause in MDX statements. Both the NonEmpty function and the Non Empty clause can be combined to remove null values from your results. Listing 14-25 shows an example of this option. See Figure 14-20 for the results.

Figure 14-20. *The results from Listing 14-25*

▓ **Note**   The NonEmpty function in MDX uses parentheses similar to most programming languages. The Non Empty clause does not. Additionally, as you may have noticed, there is no space in the name of the NonEmpty function, but there is a space in the Non Empty clause!

Listing 14-25. Using the Non Empty Clause and the NonEmpty Function

```
Select -- Start with Lots of nulls in eight columns and six rows
{ [DimDates].[Year-Qtr-Month-Day].[Year].AllMembers } On Columns,
{ [DimTitles].[TitlesByType].[TitleType].AllMembers } On Rows
From [CubePubsSales]
GO

Select Non Empty -- Now has less nulls (5 columns removed)
{ [DimDates].[Year-Qtr-Month-Day].[Year].AllMembers } On Columns,
{ [DimTitles].[TitlesByType].[TitleType].AllMembers } On Rows
From [CubePubsSales]
GO

Select Non Empty -- Even Less nulls (1 more row removed)
{ NonEmpty( [DimDates].[Year-Qtr-Month-Day].[Year].AllMembers ) } On Columns,
{ NonEmpty( [DimTitles].[TitlesByType].[TitleType].AllMembers ) } On Rows
From [CubePubsSales]
```

Member and Level Paths

Developers may find it challenging that MDX code can be written with many variations. For example, to locate a particular member or level, you provide its name, like *June*, or a full path, like *DimTime.1992.June*. And the path can include various combinations of the dimension, hierarchy, level, member, and various functions.

Unfortunately, when you incorrectly indicate a path to a member, SSAS does not return an error. It returns an empty result set. This oddity makes developers scratch their head all the time, wondering what exactly went wrong.

One way to avoid path issues is to memorize typical patterns used to access a member or level. The pattern to memorize is this: **Dimension**[Optional].**Hierarchy**[Optional].**Level**[Optional]. **Member**.< ChildMember >.< ChildMember>

The code in Listing 14-26 displays examples of what the different paths may look like. Notice that each statement in this listing returns the same result, as shown in Figure 14-21.

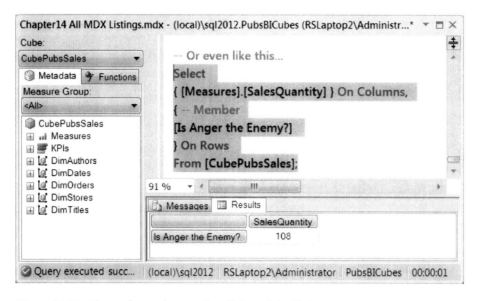

Figure 14-21. *The results are the same for all the code in Listing 14-26.*

Listing 14-26. Various Paths to Access a Member or Level

```
-- Starts like this...
Select
{ [Measures].[SalesQuantity] } On Columns,
{ -- Dim.Hierarchy.Level.Member.Member (long path)
  [DimTitles].[TitlesByPublisher].[Publisher].[New Moon Books].[Is Anger the Enemy?]
} On Rows
From [CubePubsSales];
GO

-- Can change to this...
Select
{ [Measures].[SalesQuantity] } On Columns,
{ -- Hierarchy.Level.Member.Member
  [TitlesByPublisher].[Publisher].[New Moon Books].[Is Anger the Enemy?]
} On Rows
From [CubePubsSales];
GO

-- Or this...
Select
{ [Measures].[SalesQuantity]} On Columns,
{ -- Dim.Member.Member
  [DimTitles].[New Moon Books].[Is Anger the Enemy?]
} On Rows
From [CubePubsSales];
GO

-- Or this...
Select
{ [Measures].[SalesQuantity] } On Columns,
```

```
{ -- Dim.Member
  [DimTitles].[Is Anger the Enemy?]
} On Rows
From [CubePubsSales];
GO

-- Or even like this...
Select
{ [Measures].[SalesQuantity] } On Columns,
{ -- Member
  [Is Anger the Enemy?]
} On Rows
From [CubePubsSales];
```

In Listing 14-27, we provide some examples of what *not* to do when typing MDX member paths. The following list describes what is happening in the examples in this listing:

1. A hierarchy is left out of the path and goes straight to the level and members.

2. A path to a member is broken by skipping over a level.

3. A path of members fails to be properly chained together because the level is in front of the members.

Listing 14-27. MDX Errors to Avoid

```
-- 1. Skipping the hierarchy
Select
{ [Measures].[SalesQuantity]} On Columns,

{ -- Dim.<Skipped hierarchy>.Level.Level.Member
  -- DOES NOT WORK (but does NOT give an error!)

[DimTitles].[Publisher].[Title].[Is Anger the Enemy?]
} On Rows
From [CubePubsSales];

-- 2. Skipping over a level breaks the path to the member
Select
{ [Measures].[SalesQuantity]} On Columns,

{ -- Dim.Level.<Skipped Level>.Member
  -- DOES NOT WORK (but does not give an error either!)

[DimTitles].[All].[Is Anger the Enemy?]
} On Rows
From [CubePubsSales];

-- 3. Multiple members after a level
Select
{ [Measures].[SalesQuantity]} On Columns,

{ -- Dim.Level.Member.Member
  -- DOES NOT WORK (and of course no error either!)

[DimTitles].[Publisher].[New Moon Books].[Is Anger the Enemy?]
} On Rows
From [CubePubsSales];
```

One interesting pattern that happens to work is the use of the All level followed by a path of multiple members. An example is shown in Listing 14-28.

Listing 14-28. An Exception to the Rule

```
-- This works!
Select
{ [Measures].[SalesQuantity]} On Columns,
{ -- Dim.Member.Member.Member (and not Dim.Level.Member.Member)
 [DimTitles].[All].[New Moon Books].[Is Anger the Enemy?]
```

-- Works because [All] is both a member and a level.

```
} On Rows
From [CubePubsSales];
```

What makes this last example seem so odd is that the All level is both a level and a member, so it looks like you are getting away with using an incorrect path of *Dimension. < Skipped hierarchy >.Level.Member.Member,* when you are actually using *Dimension.Member.Member.Member.* Isn't MDX fun?!

Common Functions

Like any programming language, MDX has a number of useful functions. Let's take a look at some of the ones you are likely to encounter.

■ **Tip** In Microsoft's MDX documentation, what other programming languages refer to as methods, properties, or operators are all called *functions.* Consider these terms interchangeable here.

PrevMember and NextMember Functions

These two functions return the previous or next member from the same level. For example, the three MDX examples in Listing 14-29 all return the same results, as shown in Figure 14-22.

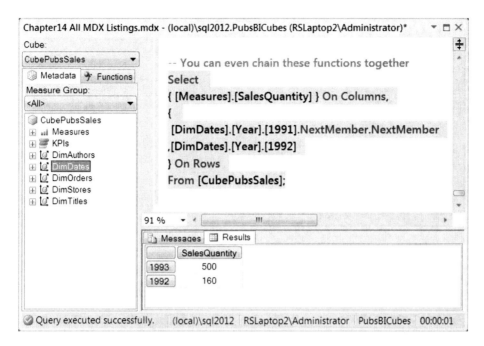

Figure 14-22. The results of the examples in Listing 14-29

Listing 14-29. PrevMember and NextMember Functions

```
Select
{ [Measures].[SalesQuantity] } On Columns,
{
  [DimDates].[Year].[1993]
, [DimDates].[Year].[1993].PrevMember
} On Rows
From [CubePubsSales];
GO

Select
{ [Measures].[SalesQuantity] } On Columns,
{
  [DimDates].[Year].[1992].NextMember
, [DimDates].[Year].[1992]
} On Rows
From [CubePubsSales];
GO

-- You can even chain these functions together
Select
{ [Measures].[SalesQuantity] } On Columns,
{
 [DimDates].[Year].[1991].NextMember.NextMember
,[DimDates].[Year].[1992]
} On Rows
From [CubePubsSales];
```

The Children Function

To use the Children function, indicate the coordinates of a member that represents the parent of the child members to be returned.

You can use this function multiple times in the same statement to access a set of child tuples from different members in the same dimension. For example, Listing 14-30 is using the Date dimension to identify the children of the years 1993 and 1992. The results are shown in Figure 14-23.

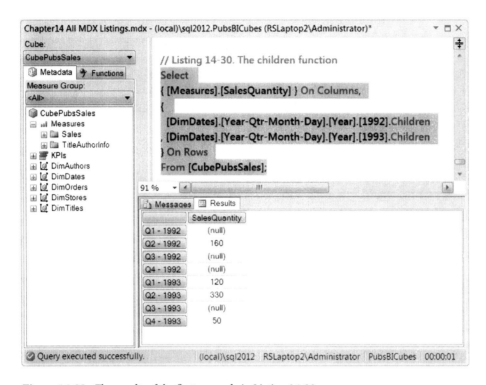

Figure 14-23. *The results of the first example in Listing 14-30*

Listing 14-30. The Children Function

```
Select
{ [Measures].[SalesQuantity] } On Columns,
{
  [DimDates].[Year-Qtr-Month-Day].[Year].[1992].Children
, [DimDates].[Year-Qtr-Month-Day].[Year].[1993].Children
} On Rows
From [CubePubsSales];
```

The Parent Function

The Parent function allows you to specify a member and receive its parent members' value. In Listing 14-31 we are specifying the first quarter of 1993 and requesting its parent. This means the value for the year 1993, which is the parent of that first quarter, will be returned. For the results, see Figure 14-24.

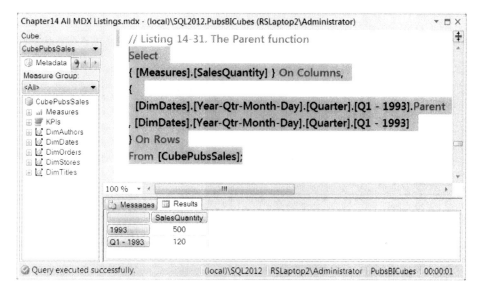

Figure 14-24. *The results of the example in Listing 14-31*

Listing 14-31. The Parent Function

```
Select
{ [Measures].[SalesQuantity] } On Columns,
{
 [OrderDate].[Year-Qtr-Month-Day].[Q1 - 1993].Parent
,[OrderDate].[Year-Qtr-Month-Day].[Q1 - 1993]
} On Rows
From [CubePubsSales];
```

The CurrentMember Function

The CurrentMember function returns values based on the member of a dimension currently under focus. To understand this, consider how SSAS must resolve a query. Each time you ask for a given member of a dimension, the SSAS query engine must check each member to see whether it is a match for your request. If it is not, it moves to the next member until it finds all the members you requested in your MDX statement. Each time the SSAS engine checks a member, that member is in focus and represents the current member of that dimension.

The CurrentMember function is rarely required for most situations in your MDX code. Listing 14-32 gives two examples; the first uses the CurrentMember function, and the second does not. Note that both queries return the exact same results, as shown in Figure 14-25.

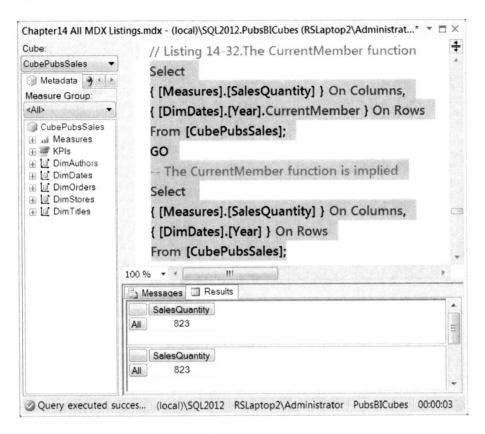

Figure 14-25. *The results of the examples in Listing 14-32*

Listing 14-32. The CurrentMember Function

```
Select
{ [Measures].[SalesQuantity] } On Columns,
{ [DimDates].[Year].CurrentMember } On Rows
From [CubePubsSales];
GO
-- The CurrentMember function is implied
Select
{ [Measures].[SalesQuantity] } On Columns,
{ [DimDates].[Year] } On Rows
From [CubePubsSales];
```

Occasionally, when you are working with MDX expressions in Visual Studio or in a reporting application such as SSRS, the CurrentMember function must be explicitly typed out for an expression to work properly. Both of these tools usually let you create the MDX expression with a designer interface that programs the MDX code for you, but not always. Therefore, keep this option in mind when creating MDX expressions in either one of these programs.

░ **Tip** Do not worry about when this will or will not occur. Just add the `CurrentMember` function to an expression whenever you get an error about an expression being ambiguous, and this will often resolve the issue. It is not very scientific, but it is easy to remember.

The Order Function

Ordering results is a common task in SQL programming. This is also true of MDX programming. Many of the client software applications that work with SSAS do the sorting for you after receiving the results of an MDX query. But, on occasion, you may want to sort the results beforehand. You can do so by using the `Order` function.

The `Order` function in MDX works slightly different from the `Order By` clause in SQL programming because of the hierarchical nature of a cube. To understand this, let's first take a look at the results of a query with no `Order` function, as shown in Listing 14-33. See Figure 14-26 for some of the results.

Messages Results	
	SalesQuantity
All	823
New Moon Books	308
Emotional Security: A New A...	50
Is Anger the Enemy?	108
Life Without Fear	50
Prolonged Data Deprivation:...	30
You Can Combat Computer ...	70
Binnet & Hardley	260
Computer Phobic AND Non-...	40
Fifty Years in Buckingham P...	40
Onions, Leeks, and Garlic: ...	80
Silicon Valley Gastronomic T...	20
Sushi, Anyone?	40

Figure 14-26. The partial results of the example in Listing 14-33

Listing 14-33. An MDX Query Without the `Order` Function

```
Select
{ [Measures].[SalesQuantity] } On Columns,
{ [DimTitles].[TitlesByPublishers].AllMembers } On Rows
From [CubePubsSales];
```

In the Pubs database, New Moon Books and Binnet & Hardley are publishers. Figure 14-26 shows that the results of the code in Listing 14-28 have been automatically grouped based on the publishers that the books fall under. Currently, the order the titles appear in is based on the position of each member's key value. The unsorted results display the title with a lower key value before the title with a higher key value. If you look up the key values, you will find that the title *Is Anger the Enemy?* has a key value of 8, while the title *Life without Fear* has a key value of 13. Therefore, the title *Is Anger the Enemy?* is displayed first.

If you want to order the results based on SalesQuantity's numeric values, you can do so with the query shown in Listing 14-34.

The first order function argument indicates the members to be sorted. The second argument indicates the values to sort by. And the third argument indicates whether you want to sort them in ascending or descending order.

Listing 14-34. The Order Function

```
-- Ordered by Sales Quantity, descending
Select
{ [Measures].[SalesQuantity] } On Columns,
{
 Order( [DimTitles].[TitlesByPublisher].AllMembers
      , [Measures].[SalesQuantity]
      , Desc
      )
} On Rows
From [CubePubsSales];
```

Figure 14-27 shows the results of the second query. As you can see, the values are now ordered sequentially and in descending order, but they are still grouped based on each publisher.

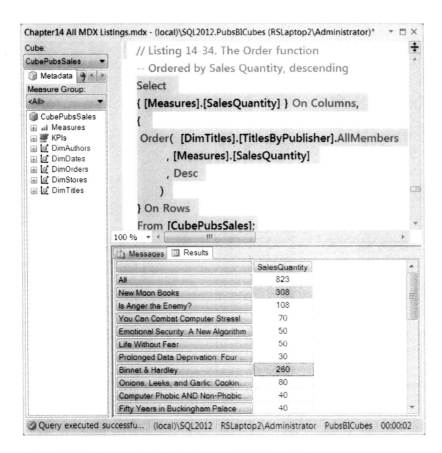

Figure 14-27. *The results of the last example in Listing 14-34*

If you do not want your values to be grouped based on hierarchy, you can use the Break Hierarchy version of the Order function. To do this, specify in the third argument that you want to use break ascending or break descending, as shown in Listing 14-35. Figure 14-28 shows the results.

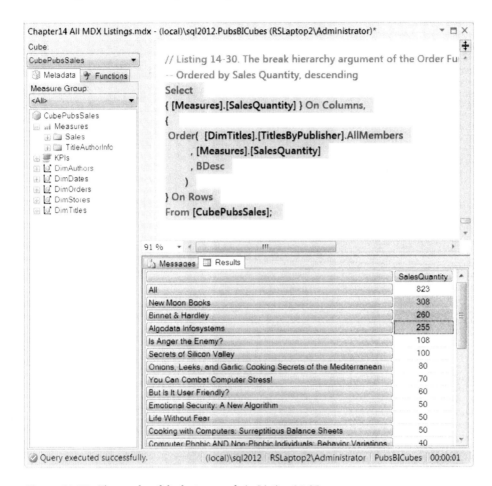

Figure 14-28. *The results of the last example in Listing 14-35*

Listing 14-35. The Break Hierarchy Argument

```
Select
{ [Measures].[SalesQuantity] } On Columns,
{
 Order( [DimTitles].[TitlesByPublisher].AllMembers
      , [Measures].[SalesQuantity]
      , BDesc
      )
} On Rows
From [CubePubsSales];
```

When the second query is run, the results will look like those shown in Figure 14-28. Notice that each of the values in this example are sorted out sequentially regardless of the hierarchy.

The CrossJoin Operator (*) and Function

As expected, there will be times when you would like your data to include subtotals based on a combination of multiple dimensions. For example, you can use the CrossJoin function to request the subtotals of sales quantity for all years and all titles. The CrossJoin function can be represented either with the * operator or with a function call. Listing 14-36 shows an example of both the cross join * operator and the function call. The results returned are the same no matter which version you use and are displayed in Figure 14-29.

Figure 14-29. *The results of the Listing 14-36*

Listing 14-36. The CrossJoin Operator and Function

```
Select
{ [Measures].[SalesQuantity] } On Columns,
{ NonEmpty(
  [DimDates].[Year].AllMembers

  * -- This symbol is the cross join operator

  [DimTitles].[TitlesByType].[Title].AllMembers
) } On Rows
```

```
From [CubePubsSales];
GO

Select
{ [Measures].[SalesQuantity] } On Columns,
{ NonEmpty(
  CrossJoin( [DimDates].[Year].AllMembers
           , [DimTitles].[TitlesByType].[Title].AllMembers
           )
) } On Rows
From [CubePubsSales];
```

■ **Tip**   Naturally, a number of null values will be returned when using this function. For instance, if a title did not sell in a particular year, the results returned will be a null value. Because of this, it is common to combine a cross join operator with the NonEmpty function as we did in Listings 14-31 and 14-32.

Joining More than Two Dimensions

It is possible to combine results from more than two dimensions at a time. In Listing 14-37 we combine results from the Years, Titles, and Stores dimensions. The code results are shown directly after the listing in Figure 14-30. Both queries return the same result.

Figure 14-30. *The results of the last example in Listing 14-37*

> ■ **Note** When using the * operator for multiple cross joins, it is important to use parentheses to indicate the order of operation. Parentheses are not necessary when using the CrossJoin function on its own.

Listing 14-37. Joining More Than Two Dimensions

```
-- Using the CrossJoin operator
Select
{ [Measures].[SalesQuantity] } On Columns,
{ NonEmpty(
    [DimDates].[Year].AllMembers

  * -- Combine one set of results

    [DimTitles].[TitlesByType].[Title].AllMembers
  )

  * -- Then cross join to those results

    [DimStores].[Store].AllMembers
} On Rows
From [CubePubsSales];

-- Using the CrossJoin function
Select
{ [Measures].[SalesQuantity] } On Columns,
{
 NonEmpty(
   CrossJoin(
    [DimDates].[Year].AllMembers
   ,[DimTitles].[TitlesByType].[Title].AllMembers
   ,[DimStores].[Store].AllMembers
   )
  )
} On Rows
From [CubePubsSales];
```

Where Clause

After receiving a large amount of data, you may want to focus on a specific slice of the result. You can use the Where clause to do so, as shown in Listing 14-38. The results of this query will exclude anything that is not part of the Store member News & Brews, as shown in Figure 14-31.

		SalesQuantity
All	Is Anger the Enemy?	10
All	Fifty Years in Buckingham Palace Kitchens	40
All	Onions, Leeks, and Garlic: Cooking Secrets of the Mediterranean	80
All	Sushi, Anyone?	40
1992	Fifty Years in Buckingham Palace Kitchens	40
1992	Onions, Leeks, and Garlic: Cooking Secrets of the Mediterranean	80
1992	Sushi, Anyone?	40
1994	Is Anger the Enemy?	10

Figure 14-31. *The results of the example in Listing 14-38*

It is worth noting that unlike the SQL programming language, the Where clause does not include conditional expressions in MDX. Instead, simply include which member or members you want to return.

Listing 14-38. The Where Clause

```
Select
{ [Measures].[SalesQuantity] } On Columns,
{ NonEmpty(
  CrossJoin(
    [DimDates].[Year].AllMembers
   ,[DimTitles].[TitlesByType].[Title].AllMembers

  -- Since the Store hierarchy cannot be used twice in the same statement,
  -- we have to comment this next line out of our code
  //,[DimStores].[Store].AllMembers

  )
 )
} On Rows
From [CubePubsSales]

Where( [DimStores].[Store].[News & Brews] );
```

It's an odd rule, but you cannot include a dimension hierarchy more than once in an MDX query. However, if you have two hierarchies that contain the same attribute, you can still use an attribute twice. For example, in Listing 14-39, we refer to the Title attribute both in the CrossJoin function and in the Where clause. This works, because we used two different hierarchies in the attributes path. The results of the query are shown in Figure 14-32.

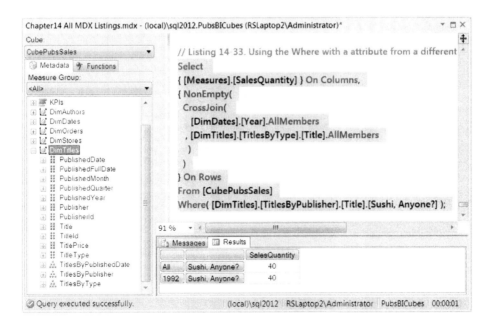

Figure 14-32. The results of the example in Listing 14-39

Listing 14-39. Using the Where Clause with an Attribute from a Different Hierarchy

```
Select
{ [Measures].[SalesQuantity] } On Columns,
{ NonEmpty(
  CrossJoin(
    [DimDates].[Year].AllMembers
  , [DimTitles].[TitlesByType].[Title].AllMembers
    )
  )
} On Rows
From [CubePubsSales]
Where( [DimTitles].[TitlesByPublisher].[Title].[Sushi, Anyone?] );
```

So far, most of our queries have included the Measures dimension across the columns, but this does not have to be the case. For example, the query in Listing 14-40 does not use the measures on the columns axis. Instead, we have used the years. Remember that even if we do not specify the SalesQuantity measure, the measure is used as the default member of the Measures dimension. If we want to show a different measure, we can add the Where clause to the query to indicate which member of the Measures dimension we want. The result of this query is shown in Figure 14-33.

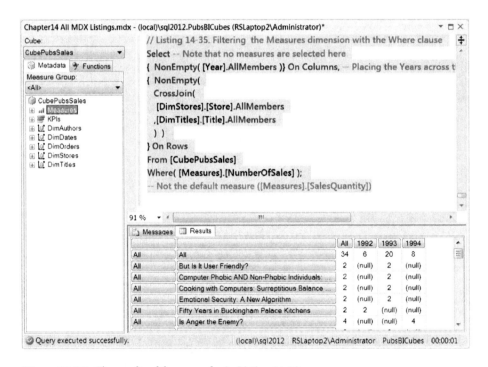

Figure 14-33. *The results of the examples in Listing 14-40*

Listing 14-40. Filtering the Measures Dimension with the Where Clause

```
Select -- Note that no measures are selected here
{ NonEmpty( [Year].AllMembers )} On Columns, -- Placing the Years across the columns
{ NonEmpty(
  CrossJoin(
```

```
    [DimStores].[Store].AllMembers
    ,[DimTitles].[Title].AllMembers
    ) )
} On Rows
From [CubePubsSales]

Where( [Measures].[NumberOfSales] ); -- Not the default measure ([Measures].[SalesQuantity])
```

Using Your Code in Reporting Applications

At this point, you probably have had enough of MDX to last for quite a while. While there are plenty of other functions and statements that we could talk about, it is not necessary to memorize them all, or even any of them. Early on, Microsoft realized that very few developers in the industry can program in MDX. As such, most of its reporting applications do not require you to do so. For example, as you will see in Chapter 15, you can use Microsoft's Excel pivot table tools to create cube-based reports without ever typing in any MDX code.

Another application that will generate MDX code for you is Microsoft's reporting server, SSRS. As we will see in Chapter 17, this can be easily accomplished using a query designer that looks remarkably similar to the cube browser found on the Browse tab of the cube designer.

You can even generate MDX code using the 2012 version of the cube browser in either Visual Studio or SQL Server Management Studio (SSMS). We have already seen the Visual Studio option in previous chapters; therefore, we use SSMS to demonstrate this feature. But both tools work the same. Let's try this in the next exercise!

EXERCISE 14-3. CREATING MDX WITH THE CUBE BROWSER

In this exercise, you generate MDX code using the cube browser that comes with SSAS 2012. You then copy the code to a new MDX code window, review the code, and remove nonuseful clauses and properties.

Note: If you have closed SQL Server Management Studio since the previous exercise, please open it. You can do so by clicking the Start button, navigating to All Programs ➤ Microsoft SQL Server 2012 ➤ SQL Server Management Studio, and clicking this menu item.

Opening the Cube Browser

1. With a connection to the SSAS server established, use the Object Explorer window to locate the CubePubsSales cube (Figure 14-34).

2. Right-click the cube's icon, and select Browse from the context menu (Figure 14-34). This will launch the cube browser.

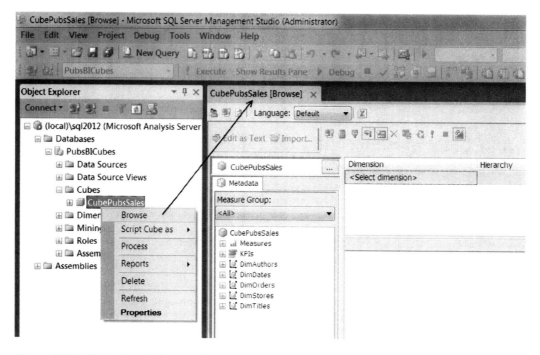

Figure 14-34. *Using the cube browser from SSMS*

Tip: We discussed using the cube browser in Chapter 11. So, if this interface seems unfamilar, you may want to review Chapter 11.

3. When the cube browser window appears, click the Design Mode button shown in Figure 14-35. This changes the cube brower's display from design mode to code mode.

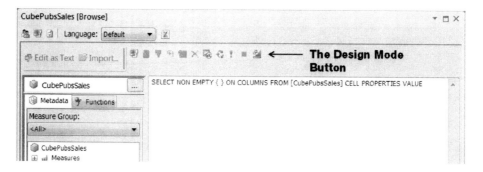

Figure 14-35. *Changing the cube browser from design mode to code mode*

Creating an MDX Query with the Design Mode

At this point, the MDX code in Figure 14-35 does not include any dimension attributes. We are going to change that by going back to the Design mode and adding in a measure and two dimensional attributes.

1. Click the Design Mode button again. This changes the cube browser's display back to design mode.

2. Drag and drop the Year, Title, and SaleQuantity members from the Metadata pane to the Query pane, as shown in Figure 14-36.

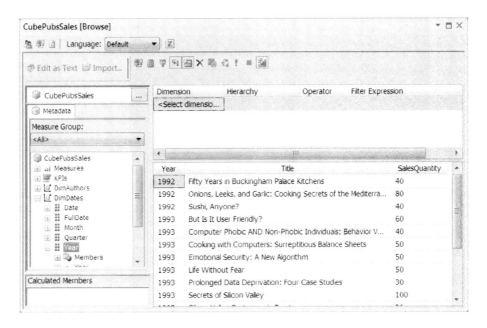

Figure 14-36. *Creating an MDX query using the cube browser's design mode*

Reviewing the New Code

As you drag and drop measures into the results pane, MDX code is being written for you. It is not necessarily good MDX code, but it is code you did not have to write yourself. Let's see what the new MDX code looks like.

1. Click the Design Mode button to reveal the MDX code the cube browser has generated (Figure 14-37). To make the code easier to see, we've displayed it in Listing 14-41 as well.

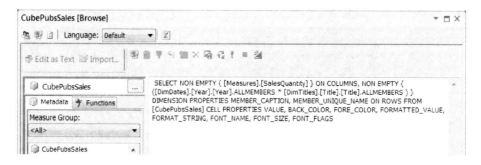

Figure 14-37. *MDX code automatically generated by the cube browser*

Listing 14-41. The Automatically Generated MDX Code

```
SELECT NON EMPTY { [Measures].[SalesQuantity] } ON COLUMNS, NON EMPTY {
([DimDates].[Year].[Year].ALLMEMBERS * [DimTitles].[Title].[Title].ALLMEMBERS ) }
DIMENSION PROPERTIES MEMBER_CAPTION, MEMBER_UNIQUE_NAME ON ROWS FROM [CubePubsSales]
CELL PROPERTIES VALUE, BACK_COLOR, FORE_COLOR, FORMATTED_VALUE, FORMAT_STRING,
FONT_NAME, FONT_SIZE, FONT_FLAGS
```

Reusing the Code

As you can see, the code needs a little work before we would want to use it as our own MDX code. Let's clean it up and test it in a regular MDX query window.

1. Highlight all of the MDX code in the designer window, right-click, and select Copy from the context window.

2. Click the New Query button to create a new MDX coding window, as circled in Figure 14-6. Paste the MDX code into the new MDX coding window that opens. The code will be presented on a single line, without any carriage returns.

3. Add carriage returns to the MDX code so that it displays like the code shown in Listing 14-42.

Listing 14-42. Reformatting the Automatically Generated MDX Code

```
SELECT
   NON EMPTY { [Measures].[SalesQuantity] } ON COLUMNS
, NON EMPTY { (
      [DimDates].[Year].[Year].ALLMEMBERS
  *
      [DimTitles].[Title].[Title].ALLMEMBERS
  ) }   DIMENSION PROPERTIES MEMBER_CAPTION, MEMBER_UNIQUE_NAME
ON ROWS
FROM [CubePubsSales]
CELL PROPERTIES VALUE, BACK_COLOR, FORE_COLOR, FORMATTED_VALUE, FORMAT_STRING,
FONT_NAME, FONT_SIZE, FONT_FLAGS
```

4. Highlight the select statement within your reformatted code, right-click, and select the Execute option. Alternately, you can run the query by clicking the "! Execute" button on the toolbar.

5. Verify that your results match the results shown in Figure 14-36.

Final Code Cleanup

As you can see, the code works as expected. But it contains a lot of extra, unnecessary code. This extra code, such as the *foreground color*, specifies settings to be used in client applications. Standard Microsoft client applications ignore these settings. Therefore, they are not necessary if you are using your MDX code to create reports in SSRS or Excel.

1. Reformat the code to exclude all client properties added by the designer. Your new code should look like Listing 14-43.

   Listing 14-43. Reformatting the Automatically Generated MDX Code

   ```
   SELECT
       NON EMPTY { [Measures].[SalesQuantity] } ON COLUMNS
   , NON EMPTY { (
       [DimDates].[Year].[Year].ALLMEMBERS
       *
   [DimTitles].[Title].[Title].ALLMEMBERS
       ) }
       ON ROWS
       FROM [CubePubsSales]
   ```

2. Highlight the select statement within your reformatted code and execute the query.

3. Verify that your results still match the results shown in Figure 14-36.

In this exercise, you generated MDX code using the cube browser, copied the code to a new MDX code window, and reformatted and tested the code. While there is still much to learn about MDX, we hope you can now read and work with the code generated by the designer!

Moving On

At this point, you should be able to write and read basic MDX queries. While there is always more to learn, the good news is that most developers can work for many years in the BI field with the basic MDX knowledge covered in this chapter.

You may ask, "Then why should I learn this at all?" The answer is simple; MDX code created by applications is rarely the most efficient code to use. And sometimes the code generated by applications is incorrect for what you need to accomplish.

In addition, your MDX skills are a powerful tool for troubleshooting. Finally, you will need to use MDX expressions in many places throughout the SSAS management and development tools. While you will seldom use whole queries in these places, understating how MDX expressions work within an MDX query will help you create and troubleshoot there as well.

LEARN BY DOING

In this "Learn by Doing" exercise, you create several MDX queries using the Northwind database. We included an outline of the steps to perform and an example of how the authors handled them in two Word documents. These documents are found in the folder C:_BISolutionsBookFiles_LearnByDoing\ Chapter14Files. Please see the ReadMe.doc file for detailed instructions.

What Next?

Becoming good at writing MDX is not an easy task. At the time of this writing, few books and web pages exist for beginners to learn from. A web search will reveal a few, however. Here is one example:

Microsoft SQL Server 2008 MDX Step by Step
MS Press
Bryan C. Smith and C. Ryan Clay
ISBN-13: 978–0735626188

CHAPTER 15

Reporting with Microsoft Excel

When you're working in front of an audience, you have incentive to excel.

—Dave Van Ronk

Excel is Microsoft's frontline reporting software. Starting originally as a commercial spreadsheet application, it has become much more.

Excel organizes data in a grid of rows and columns. These rows and columns can manipulate numeric type data by performing various calculations. Excel then is capable of displaying that data in different forms such as in charts, graphs, or other forms of reports. Excel even has a programming aspect that uses Visual Basic for Applications to perform tasks. This information can then be reported to the spreadsheet or used in other ways.

This range of capabilities allows programmers to perform advanced BI reporting, but it is not limited to professional developers. Excel reports are most commonly created by users with little to no programming skills, allowing nonprogrammers the ability to create quick and efficient ad hoc reports.

In this chapter, we discuss Microsoft Excel and how it is used to create personal and departmental BI reports. We see how easy it is to create reports from your data warehouse and your cubes. And, we look at a simple way to distribute reports among co-workers using a .pdf file. Before we begin all that, let's overview Microsoft's reporting applications and see where Excel fits within the scheme of things.

Microsoft's BI Reporting

Microsoft has spent a lot of money creating and improving BI reporting applications over the last several years. It now offers a great many reporting applications to choose from.

Table 15-1 gives an overview of some of the most current ones.

Table 15-1. *BI Reporting Categories*

Application	Description	Pros	Cons
Excel 2010	The core spreadsheet application includes more reporting features than ever. Microsoft now considers this a vital part of its BI applications stack.	Easy to learn, and earlier, similar versions have been used for more than a decade.	Must be purchased and installed on all PCs and is designed to be a single-user application.
PowerPivot	A free add-on to Excel 2010 and SharePoint 2010. It adds quite a number of advanced reporting features to both applications.	Easy to install and use. It allows for professional-looking ad hoc report building.	Must have Excel 2007 or 2010 installed on all PCs and is designed as a single-user application.
Visual Studio	A professional development tool for creating server-based reports.	Provides full development tools for creating web-based reports.	Comparably steep learning curve similar to learning SSIS and SSAS.
Reporting Builder	A free application that is a simplified development tool for creating Reporting Server reports.	Easy to install and use.	Cannot do as much as Reporting Server's full development tools.
SharePoint	Allows users to share documents and develop custom applications and components for, among other things, reporting.	Anyone with a web browser can build and view reports from SharePoint. Provides a comprehensive tool set for creating an organization's internal web portal.	Comparably high setup requirements. Comparably steep learning curve similar to learning SSIS and SSAS.
Performance Point	Discontinued as a stand-alone application and is now included in SharePoint 2010.	Adds a lot of advanced reporting features to SharePoint. Anyone with a web browser can build and view reports from SharePoint.	Comparably high setup requirements.

(continued)

616

Table 15-1. Continued

Application	Description	Pros	Cons
Excel Services and Visio Services	Add-ons to SharePoint. You can share Excel and Visio documents through a SharePoint server that can then be reviewed with a web browser.	Anyone with a web browser can see the reports from SharePoint.	Comparably high setup requirements. Excel Services are available only on SharePoint Server 2010 Enterprise. A complex way of sharing documents.
Power View	Combines features from Excel, PowerPivot, Report Builder, Report Server, and SharePoint.	Allows for professional-looking ad hoc report building. Anyone with a web browser can view reports from SharePoint.	Comparably difficult setup requirements. Microsoft SharePoint Server 2010, SQL 2012 data engine, Reporting Services 2012, Power View add-in for Reporting Services, and PowerPivot add-in for SharePoint are all required for installation.

Most of these applications' features overlap with each other, and it is likely that Microsoft will continue to combine them in the future. But, currently all of them may be considered valid options for BI reporting.

Digging through Microsoft's documentation and web pages reveals that Microsoft has classified its reporting software into three basic categories: personal, departmental, and organizational. Figure 15-1 shows a breakdown of Microsoft's most common BI reporting tools and which category they are associated with.

Figure 15-1. Mircrosoft's BI reporting software categories

You can see that many applications overlap between the different categories. To give you a better understanding of each category, take a look Table 15-2.

Table 15-2. BI Reporting Categories

Category	Description	Pros	Cons
Personal BI	Software used to create ad hoc reports for individuals or small groups of users in a self-serve environment. Often created by managers and leads for quick glimpses of current events.	Development time in hours. Immediate access to reports. Encourages data exploration and what-if scenarios. Training required to create basic reports is in days.	Everyone creating and viewing these reports needs read access to the BI views, stored procedures, or cubes. Conflicting reports are a common occurrence. No standard look or feel, and reports are often very basic. Often the person who creates the report is the only person who will use it.
Departmental BI	Software used to create reports for small to midsize groups with a mix of self-served and delivered reports.	Development time of days. Access to views, stored procedures, and cubes can be limited to only report developers and power users. Conflicting reports are less common. One report is used by many users. Departmental look and feel builds confidence. Reports tell a single story throughout the department. Reports are often advanced.	Reports are not immediately available. Requires at least peer testing and validation time. Discourages data exploration and what-if scenarios. Requires team members to dedicate some or all of their time to creating and maintaining reports. Training required to create advanced reports can take weeks.
Organizational BI	Software used to create reports used by multiple departments within an organization.	Access to views, stored procedures, and cubes can be limited to only report developers. One report is used by many users. Conflicting reports are rare. Organizational look and feel builds confidence. Reports tell a single story throughout the organization. Reports can be made available outside of organization with confidence. Reports are often interrelated and very advanced.	Development time takes weeks. Reports are not immediately available. Requires a dedicated testing and validation team. Discourages data exploration and what-if scenarios. Requires a dedicated development team. Training required to create very advanced reports takes months.

As you saw in Figure 15-1, Excel is situated as a tool for both personal and departmental BI solutions. Let's get started learning how to use it!

Excel Reports from the Data Warehouse

Excel is capable of creating reports from many different data sources including text files and Microsoft Access database and web services, to name a few. Excel becomes even more powerful when you couple it with either Microsoft SQL Server or Microsoft's Analysis Server. This coupling feature, which has been available to Excel since 1998, is one of the primary reasons for Excel's prominence as a reporting tool.

A number of additional reporting features were included with Office 2007. Office 2010 was given even more features and is required for a number of the current BI add-ons, such as PowerPivot.

Creating a connection to either data warehouses or cubes is quite simple. We walk you through both options within this chapter. We begin with Microsoft SQL Server and the data warehouse you created in Chapter 5.

Creating a Connection

To create a connection to a data warehouse on Microsoft SQL Server, the first thing to do is open an existing Excel spreadsheet or create a new one. From the open spreadsheet, go to the Data tab on Microsoft's Ribbon interface. Locate the Get External Data button group and click From Other Sources to access an additional set of options (Figure 15-2).

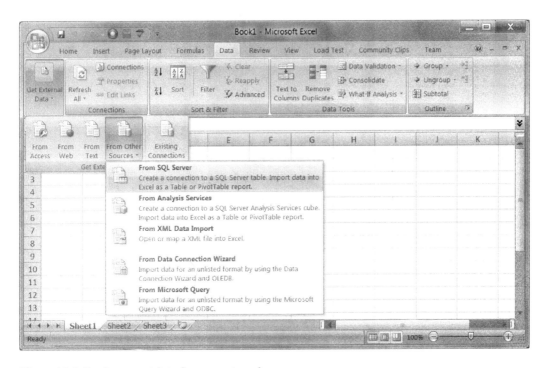

Figure 15-2. *Getting report data from an external source*

> **■ Tip** The Ribbon interface exists in Office 2007 and 2010. If you are not using one of those versions of Office, you can still connect to SQL Server or Analysis Server using a very similar process. The major difference is that your environment will appear differently than it does in this chapter.

Please note that if your screen is small, the Get External Data button group may appear as a single button (Figure 15-2). If that is the case, click the single button for the collection of other buttons to become available. Once the additional set of options is visible, select the From SQL Server option.

THE MICROSOFT OFFICE RIBBON

We will be showing several screen shots in this chapter that include the Microsoft Office Ribbon. The Ribbon dynamically changes as the window size is increased or decreased. This means that the Ribbon may look quite different between one screen shot and the next. For example, compare the two views of the Ribbon in Figure 15-3. Both are of the same window. The difference is simply that the window size has been changed.

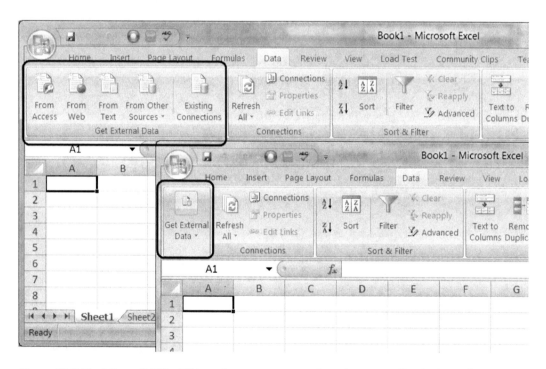

Figure 15-3. The Microsoft Office Ribbon changes appearance based on an application's window size

As you saw in Figure 15-2, a number of connection options are available. Once you select a connection option, Excel's Data Connection Wizard begins the process of creating a connection file for you. This connection file stores both a connection string and a SQL command on your hard drive so that it can be reused in future reports; this is what the Existing Connection button in Figure 15-3 is used for.

The Data Connection Wizard

Once the Data Connection Wizard launches, it asks for the database server name (Figure 15-4).

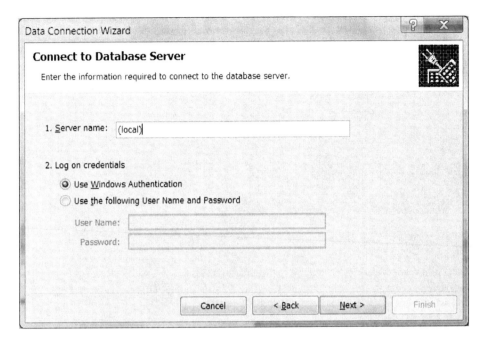

Figure 15-4. *Making a connection to your database server*

Click Next, and you are given an option to select a table or view from the data warehouse to use for your report. Unfortunately, you cannot select more than one table, so the most obvious choice is to use a view that delivers data from multiple tables. In Chapter 13, we created a number of reporting queries, including one view called vQuantitiesByTitlesDate. As Figure 15-5 shows, we can select that from the list.

■ **Important**   All the code needed for the views demonstrated in this chapter can be found in the file ViewsAndStoredProceduresForChapter15.sql. Simply open the file in SQL Server Management Studio, and execute the SQL code. Doing so allows you to follow along as we go through each example.

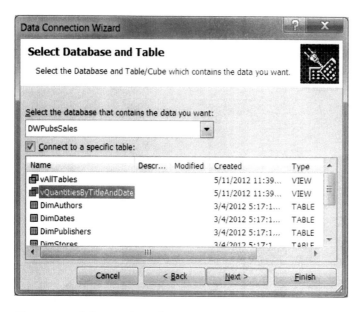

Figure 15-5. Selecting objects from your SQL Server

After you have selected the view and clicked Next, the wizard asks where you would like to save your connection file (Figure 15-6). The default name and settings is perfectly fine 99 percent of the time. But, if you would like to choose another location, you can click the Browse button and navigate to the new location.

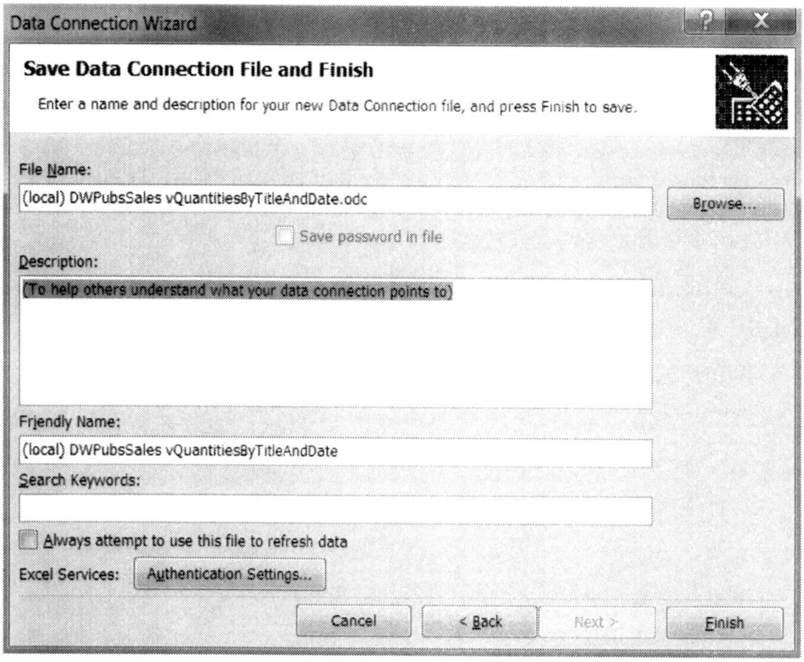

Figure 15-6. Saving your connection information and completing the wizard

Click Finish to create the file, close the Data Connection Wizard, and create a connection to the database.

WHAT IS AN .odc FILE?

In Figure 15-6, the connection file ends with an .odc extension, which indicates it is an OleDB data connection file. An OleDb data connection (.odc) file contains a connection string and query information in an XML format. Figure 15-7 shows the contents of an .odc file.

Figure 15-7. The contents of an .odc file

Since .odc files contain all the information for both a connection and a query, you can use these in other reports at a later time by clicking the Existing Connection button, as shown in Figure 15-2.

Creating a Report

After the wizard closes, the Import Data dialog window appears. The Import Data dialog window allows you to select between a tabular report, a pivot table report, and a pivot chart + pivot table report (Figure 15-8). However, some options may not be available depending on what type of connection you created.

Figure 15-8. *Selecting a PivotTable report*

This dialog window also gives you an option to create a new worksheet and select the location of cells within an existing worksheet to create the report object. Because cells can be relocated, we leave the default setting and click OK to have Excel create the report.

When you choose a pivot table report, the final outcome after the connection wizard closes will look like Figure 15-9. We have seen this in earlier chapters, but now let's look at it in more detail.

▪ **Note** Notice the PivotTable Field List area on the right side of Excel in Figure 15-9. Oddly, this disappears if you click away from the PivotTable. Each time we do a demo of this feature in class, someone shouts out, "Hey, where did it go?" Just click on the PivotTable once more to view it again.

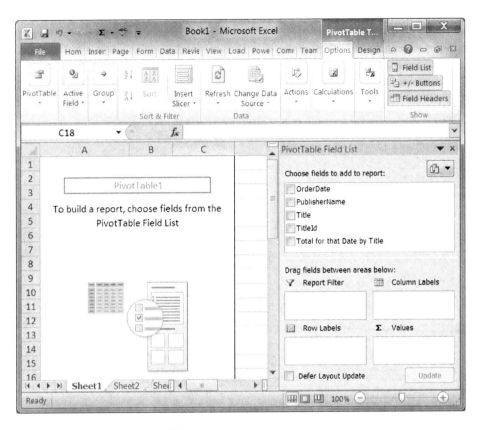

Figure 15-9. *An empty PivotTable report*

Configuring a Report

When using a pivot table or chart, you need to identify the data from the data source to be included in your report. Report data can be selected by checking the checkbox in the pivot table field list or by dragging and dropping the field list items to the pivot table or chart, or to the designer areas below the field list.

As an example, in Figure 15-10 we have created a simple report by dragging the PublisherName and Title fields into the Row Labels designer area and the Total for that Date by Title field into the Values designer area.

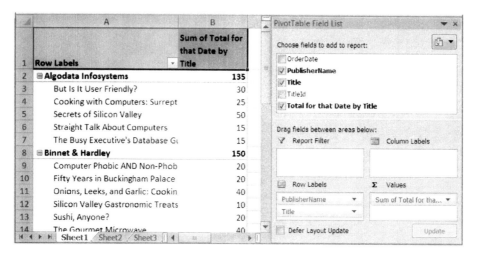

Figure 15-10. *Choosing fields for your report*

In the next exercise, you get a chance to try it yourself.

EXERCISE 15-1. CREATING A SQL VIEW REPORT

Use the steps within this chapter to connect to the data warehouse and build a report that looks like the one shown in Figure 15-11.

Creating the SQL Views and Stored Procedures

Before we make the Excel report, let's make sure that all of the views and stored procedures needed for this chapter are present in the database.

1. Open SQL Server Management Studio 2012. You can do so by clicking the Start button and navigating to All Programs ➤ Microsoft SQL Server 2012 ➤ SQL Server Management Studio. Right-click SQL Server Management Studio 2012, and click the Run as Administrator menu item. If the UAC message box appears asking, "Do you want the following program to make changes to this computer?" click Yes (or Continue depending upon your operating system) to accept this request.

2. With Management Studio open, choose the Connect to the Database Engine option in the Server Type dropdown box. Then click Connect.

3. Use the File ➤ Open ➤ File menu option to open the file that contains all the SQL code needed to create the views and stored procedures for this chapter, called `C:\_BookFiles\Chapter15Files\ViewsAndStoredProceduresForChapter15.sql`.

4. When the file opens in SQL Server Management Studio, execute the code using the "! Execute" button on the toolbar.

5. Close SQL Server Management Studio. We will no longer need it for this chapter.

Creating the Excel Report

1. Open Excel. You can do so by clicking the Start button and selecting All Programs ➤ Microsoft Office ➤ Microsoft Excel 2010.

Important: Remember to right-click the menu item, select Run as Administrator, and then answer Yes to close the UAC.

2. Go to the Data tab, as shown in Figure 15-2.

3. Create a new data source using the From SQL Server option (Figure 15-2).

4. Specify the name of your SQL Server in the Data Connection Wizard (Figure 15-4).

5. Select the DWPubsSales database from the dropdown box shown in Figure 15-5.

6. Select the view called vQuantitiesByTitlesDate (Figure 15-5).

7. Select a pivot table report (Figure 15-8). An empty pivot table will be created.

8. Using the PivotTable Field List window, drag and drop or select the checkboxes to make your report look like Figure 15-11.

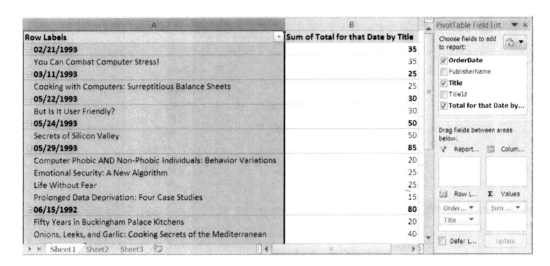

Figure 15-11. *The report for Exercise 15-1*

Save Your Report

You have created you first report for the Publications Industries BI solution. It is time to save your work in a subfolder within the Publications Industries Visual Studio solution folder.

1. Save your report using Excel's File ➤ Save As menu item. When you do, a dialog window appears similar to Figure 15-12.

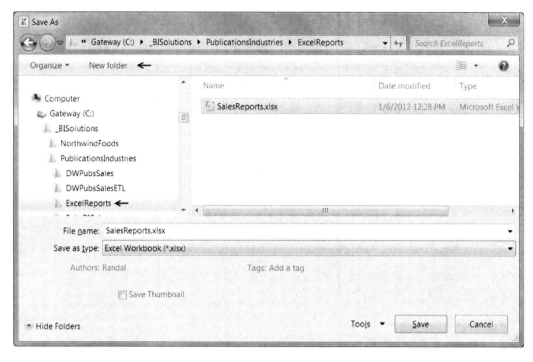

Figure 15-12. *The Excel Save As dialog window*

2. Navigate to the `C:\BISolutions\PublicationsIndustries` folder (Figure 15-12).

3. Click the "New folder" button at the top of the Save As dialog window to create a new folder named `ExcelReports` (Figure 15-12).

4. Select the `ExcelReports` folder and save the Excel report file as `SalesReport.xlsx` (Figure 15-12).

Now that you have created a report based on a SQL view, we will move on to creating SSAS cube-based reports, so please leave Excel open for now.

Changing Connection Properties

If Excel forced you to use a single SQL view per spreadsheet, it would be unreasonably restrictive, because each view may have only some of the data you need. Luckily, you can either add another report to the same Excel spreadsheet or reconfigure your existing report.

To add a different report to the spreadsheet, restart the wizard and select another view. You can then use the data in the view to make another report on the same or separate worksheet tab.

To change an existing pivot table report, reconfigure the connection property to point to a different view within the data warehouse.

Reconfiguring a Connection

To reconfigure the connection, your first step must be to make sure you have selected the pivot table in the Excel worksheet. Then navigate to the Data tab and click the Properties button in the Ribbon. Once you do this, a properties dialog window appears (Figure 15-13).

■ **Tip** Remember that the Ribbon's buttons change appearance depending on the screen resolution, so your Ribbon may look slightly different!

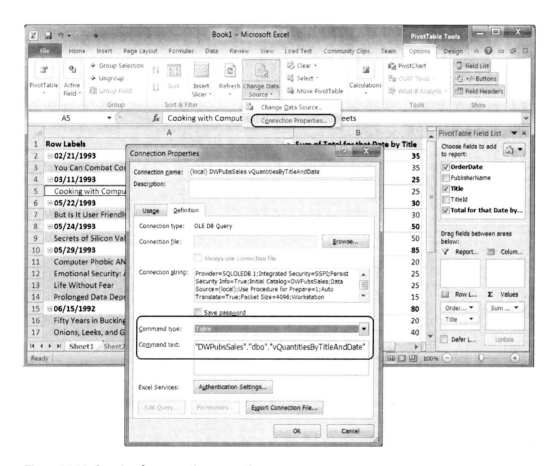

Figure 15-13. *Opening the connection properties*

You can access and change the command text by navigating to the Definition tab in the Connection Properties dialog window. In Figure 15-13, the existing command is pointing to the view called vQuantitiesByTitlesDate. If we change this to point to another view, like the vAllTables (created in this chapter by the SQL script file), we can access the information from it and create a pivot table based on its data. Figure 15-14 shows this change.

Figure 15-14. *Changing the command text to the vAllTables view*

If you look closely at Figure 15-14, you can see that the command type is set to Table. This setting includes both tables and views, so currently nothing needs to be changed to make this work. Changing this setting to use a SQL statement, however, allows you to get different data for your report by adding your own custom SQL code.

This means the SQL table or view used in the report will be entirely different than what was originally stored in the .odc file. A dialog box warning you of this appears after you change the SQL statement and click OK. The message looks pretty official, but if you want to see the new data, you will have to click Yes to continue making the change. Figure 15-15 shows an example of this warning message. (It's a very long message, so we cut some of it out.)

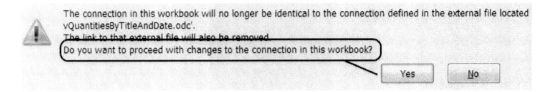

Figure 15-15. *Accept the warning to make the change*

■ **Note** This will not save your query changes in the `.odc` file, but it will allow you to change it for the moment. If you want to make a more permanent change, navigate to a new worksheet and start the Connection Wizard again. You can also save your changes by clicking the Export Connection File button, as shown in Figure 15-14, if you think you will want to reuse this connection and statement for a later report.

Next is to design the report. Any fields that were common between the two views may still be in the pivot table report, and they can easily be removed by dragging and dropping them back from the report and into the field list. But, afterward, creating a new report is just as simple as creating the first. By dragging and dropping the fields from the field list into the Pivot Table Report area or using the designer areas below the field list, you can create a report that looks something like the one shown in Figure 15-16, which now has the additional data warehouse columns, such as TitleName, TitleType, and YearName.

Row Labels	1992	1993	1994	Grand Total
business				
Cooking with Computers: Surreptitious				
Balance Sheets	50			50
Straight Talk About Computers	15			15
The Busy Executive's Database Guide		30		30
You Can Combat Computer Stress!	35			35
mod_cook				
Silicon Valley Gastronomic Treats	10			10
The Gourmet Microwave			80	80
popular_comp				
But Is It User Friendly?	30			30
Secrets of Silicon Valley	100			100
psychology				
Computer Phobic AND Non-Phobic				
Individuals: Behavior Variations	40			40

Figure 15-16. *A PivotTable report that uses the vAllTables view*

Using Stored Procedures

Views are only one example of how you can interact with SQL Server. You can also make Excel reports using a stored procedure. This may not be obvious since the user interface does not show you that as an option, but by going to the connection properties once again, you can change the settings to accomplish this. First you need to change the command type from a table to a SQL statement and then change the command text from the name of a view to the name of a stored procedure (Figure 15-17). Configuring just these two settings will allow you to use a stored procedure for your report.

Figure 15-17. *Changing the command type to SQL and command text to use a stored procedure*

■ **Note** Since stored procedures often use parameters, it seems odd that the Parameters button is not available. This button is enabled only when there are parameters defined using the Microsoft Query add-on for Excel. This add-on is, more often than not, unavailable on most computers. As such, we will not be including examples of how to use this feature. As you can see in Figure 15-16, you can still provide parameter values as part of your SQL code.

Once you have clicked OK, the data in the pivot table report will change based on the results of the select statement inside the reporting stored procedure. (You created the stored procedure by running the SQL code in Exercise 15-1.) The results of this procedure included publisher, title, dates, and a number of measured values including a KPI value, filtered by only titles that start with the letter *C* (Figure 15-18).

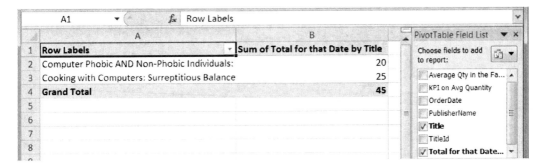

Figure 15-18. *Reporting from a stored procedure*

Working with Excel Reports from a Cube

The stored procedure and view are common ways to generate reports from a data warehouse. The stored procedures and views can be very restricted unless all the data from all the tables in the data warehouse are queried. The second view, vAllTables, did select data from all the tables in the data warehouse, but this is impractical for most occasions. If the data warehouse gets very large with lots of tables, columns, and rows, your view will be slow and cumbersome.

One of the advantages of using a cube is that all the data from the cube is easily accessible from client software such as Excel. From Excel's perspective, it is as though all the tables have been combined into one single object—the cube. As stated previously, you can think of a cube as a set of one or more relational tables combined. You can really see how this analogy fits from the perspective of reporting applications.

Connecting to Your Cube

To connect to the cube, access the Data tab and select the From Other Sources button. Next, click the From Analysis Services button to connect to SSAS (Figure 15-19).

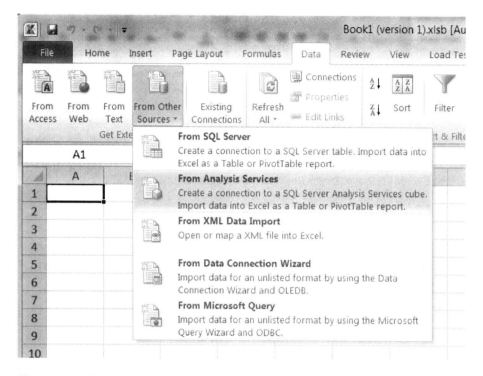

Figure 15-19. *Starting a connection to SSAS*

The Data Connection Wizard appears and walks you through the same process as when you were making a connection to the SQL Server data warehouse. The initial dialog window looks identical to the dialog window that connects you to SQL Server (Figure 15-20). Enter the name of your SSAS server, and use Windows Authentication to connect to it.

■ **Tip** SSAS can use Windows Authentication only. This is not true of the SQL Server database engine, which can use both text-based and Windows security logins. Because the Data Connection Wizard is designed to work with both types of connections, the option, mistakenly, looks like it is available.

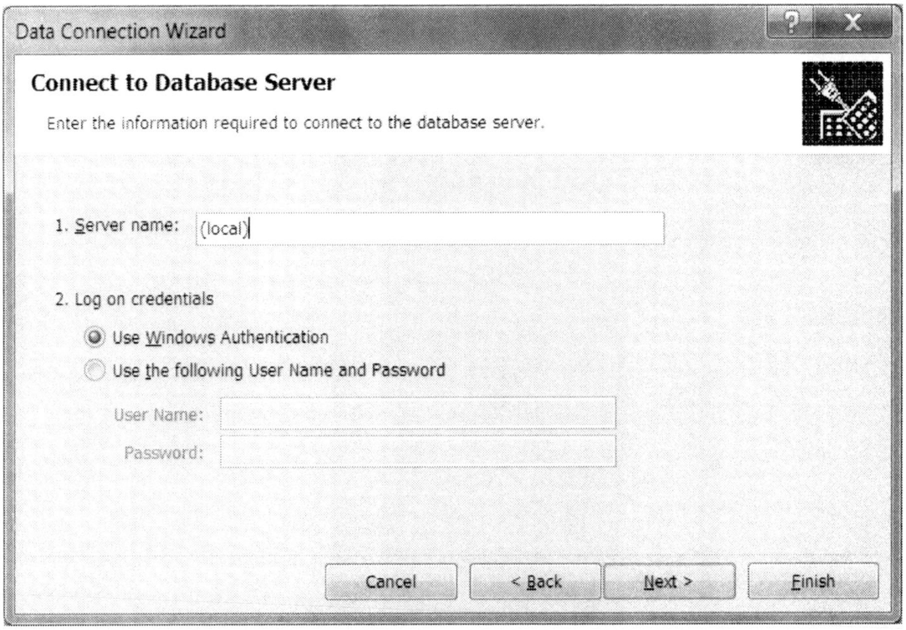

Figure 15-20. *Making a connection to SSAS*

You will have verification that you are connecting to SSAS rather than SQL Server on the following screen. In the dialog window shown in Figure 15-20, select the cube that you want to use rather than a view or table. Like tables, cubes are grouped into databases, so you must select the appropriate database before the cube can be displayed. Figure 15-21 shows that we are connecting to the PubsBICubes database and selecting the cube called CubePubsSales, which in this case is the only available option.

When you click Next, you will be asked where to save the .odc file once again. The default values will work fine for almost all occasions. Clicking Finish ends the Data Connection Wizard.

Figure 15-21. *Selecting objects from your SSAS server*

After clicking Finish, you can create a report just like you did using a view or stored procedure in the SQL Server database engine. The difference is that items in the PivotTable Field List area are now better organized by measures, dimensions, hierarchies, levels, and attributes. This makes it much easier for report users to create the reports they want. Figure 15-22 shows an example report made from the cube.

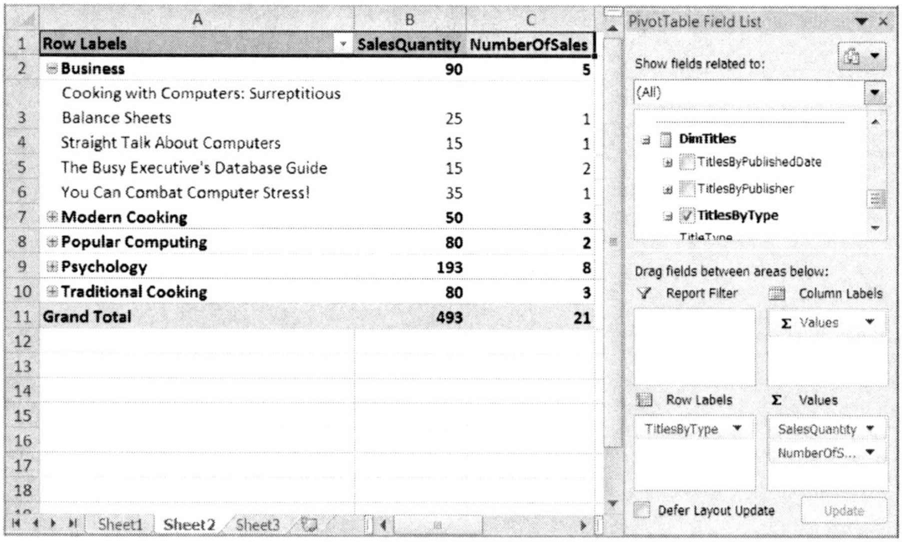

Figure 15-22. *A report using the cube*

Testing Your Reports

Do not expect your reports to be perfectly accurate. It is always important to validate your data. In Chapter 11, we cautioned you about the possibility of making incorrect aggregations when we reviewed the Calculations tab of the Cube Editor. The issue concerned members that were added to the cube as either derived or calculated members. It is very important to make sure your reports are correct; therefore, let's take a moment to review the differences between these two member types.

A derived member is added by including SQL code in an SSAS data source view. When the data is pulled from SQL Server into the cube during processing, the expression that makes up the derived member is executed. Think of it this way: a first aggregation is performed by SQL Server, and the results are stored on the hard drive as part of the cube. After that, a second aggregation is performed when a report is created using the cube. For example, if we created a derived member that multiplies the vales for sales quantity by a title's price, that expression would look something like this:

```
[FactSales].[SalesQuantity] * [DimTitles].[TitlePrice]
```

Imagine that we have a sales quantity value of 100 and a title's price value of $30. The expression would evaluate to $300. And this new value would be stored in the cube as a derived member. After the cube stored the values, it would later aggregate to get totals and subtotals when the report was created. Figure 15-23 shows an example.

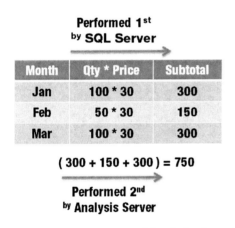

Figure 15-23. *Aggregates with a derived member*

Things are a bit different with calculated members. A calculated member never stores data in the cube. Instead, it stores only the MDX code for the expression. The evaluation of the expression happens each time a report is made. Figure 15-24 highlights this process.

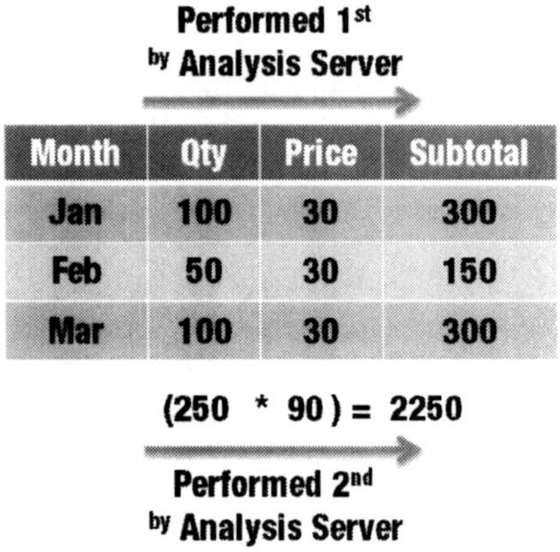

Performed 1ˢᵗ
ᵇʸ Analysis Server

Month	Qty	Price	Subtotal
Jan	100	30	300
Feb	50	30	150
Mar	100	30	300

$$(250 \ * \ 90) = 2250$$

Performed 2ⁿᵈ
ᵇʸ Analysis Server

Figure 15-24. *Aggregates with a calculated member*

Both the derived and calculated members have their places, and both are useful tools when used under the right circumstances. In general, you can use derived members to store the aggregate values within SSAS and speed up the report generation, or you can use calculated members to decrease the amount of time it takes to process aggregates to the hard drive. And, always make your final decision about which to use depending on which one gives you the correct answer.

■ **Tip** Just like in a math problem that uses both addition and multiplication, if the order of operations is not specified correctly, the answer will be incorrect. In Chapter 11, Exercise 11-3, we created both a derived measure and a calculated measure and then verified that only one was correct. You may be asking, "How will I know which one to use?" While you can take the time to examine each calculated member and manually crunch the numbers to determine which would be appropriate, the one to use is the one that, when executed, actually works. If they both give the same answer, choose derived for report performance and calculated for processing performance.

EXERCISE 15-2. CREATING A CUBE-BASED REPORT

In this exercise, you create your own report by connecting the SSAS cube and building a report that looks like the one shown in Figure 15-25.

1. With Excel open to the workbook you created in Exercise 15-1, navigate to Sheet2 by clicking the tab at the bottom of the screen.

2. Go to the Data tab of the Ribbon, as shown in Figure 15-19, and create a new data source for SSAS.

3. When the Connect to Database Server dialog window appears, specify your SSAS server's name (Figure 15-20). Then click Next to continue.

4. When the Select Database and Table dialog window appears, select the PubsBICubes database and the cube called CubeDWPubSales (Figure 15-21). Then click Next to continue.

5. When the Data Connection File and Finish dialog appears, accept the default settings and click Finish. This launches the Import Data dialog window.

6. When the Import Data dialog window appears, select the Pivot Table Report option.

7. Create a report table that looks like Figure 15-25, just as you did in the previous exercise.

8. Save your report using Excel's File ➤ Save menu item.

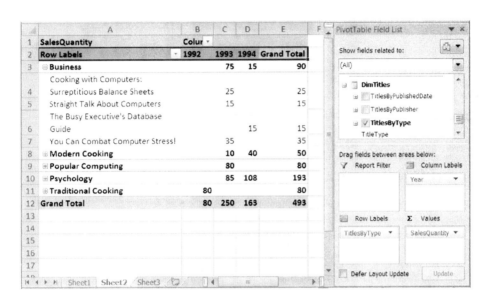

Figure 15-25. *The report for Exercise 15-2*

Now that you have created a cube-based report, we will move on to creating chart reports, so leave Excel open for now.

Creating Charts

Nothing in reporting can convey more information in less space than a chart. Excel makes creating charts easy with the inclusion of PivotCharts.

To create a PivotChart, define the data to chart using a PivotTable. One common mistake is to include all the data available. Doing so makes the chart quite messy. Instead, it is best to define charts that have only a few items.

For example, if we wanted a chart to show sales by bookstore, we would modify the pivot table to show only stores and sales quantity (Figure 15-26).

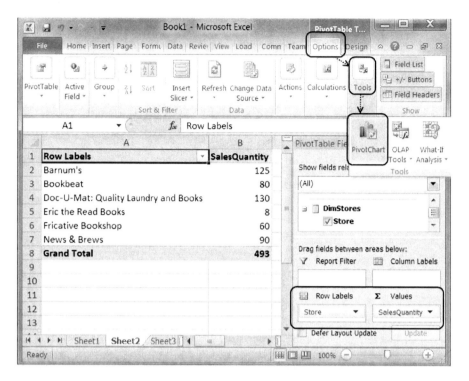

Figure 15-26. *Selecting data for a PivotChart*

> ■ **Note** Once again, your UI may look different based on the resolution. Also, the PivotChart option may take some exploration to find.

Once you have restricted the data to something more manageable, you can create a pivot chart by navigating to the Pivot Table Tools area on the Ribbon and selecting the Option tab underneath it (Figure 15-26). On this tab, if you click the Pivot Chart button, the Insert Chart dialog window appears, as shown in Figure 15-27.

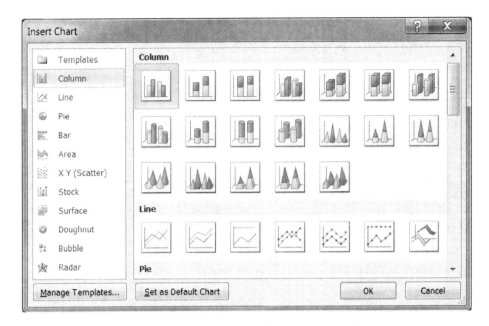

Figure 15-27. *Selecting a PivotChart layout*

As you can see in Figure 15-27, there are a number of chart types to choose from, and even more can be downloaded from the Internet. You can create your own templates as well.

We recommend selecting simple charts such as columns and bars to start with. In our example, we use a column graph by highlighting the first column graph template, as shown in Figure 15-27, and clicking the OK button.

Immediately after clicking the OK button, a PivotChart is added to your Excel worksheet. This chart will likely be floating over the top of at least part of your pivot table. Drag and drop it to the desired location. In Figure 15-28 we moved the PivotChart to the right side of the pivot table.

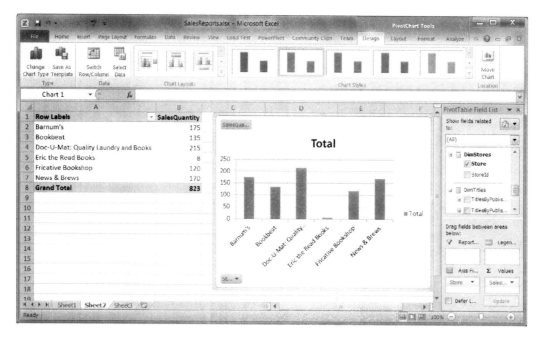

Figure 15-28. *A simple PivotChart*

The items inside the chart are designed to be individual chart components and can be moved around as well. Give it a try in the following exercise!

EXERCISE 15-3. CREATING A PIVOTTABLE CHART REPORT

In this exercise, you make a pivot table chart to create your own report like the one shown in Figure 15-30 by connecting the SSAS cube and using PivotChart.

1. Using the workbook you created in Exercise 15-1, navigate to Sheet2 by clicking the tab at the bottom of the screen.

2. Change the PivotTable to show SalesQuantities by Stores, as shown in Figure 15-26.

3. Navigate to the PivotTable Tools tabs on the Office Ribbon, and select the Options tab, as shown in Figure 15-26.

4. On the Options tab, click the PivotChart button (Figure 15-26).

5. Select the first column chart template, as shown in Figure 15-27. Click OK to create the PivotTable chart.

6. Move the PivotTable chart alongside the original pivot table, as shown in Figure 15-28.

7. Save your report using Excel's File ➤ Save menu item.

Now you have created a report with both a PivotTable and PivotChart. Let's move on to formatting the report, so leave Excel open for now.

Adding a Header

After you have created your pivot tables and pivot table charts, you may want to finish up your report by including additional features such as a report header. Report headers usually include things like the report name, a description of the report, the developer's name, and the time the report was generated. The number of items in the header depends on the business's needs. We have worked on contracts where all the major reports were designed with more than a dozen bits of information in each header.

Tip The number of items included in the header is not as important as consistency. The main purpose for the creation of a header is professionalism and branding.

You can add a report header to your existing reports by right-clicking the first row in your worksheet and selecting the Insert option from the context menu. This will add a single row, but you can continue to insert as many as needed. Once you have a few rows, you can add text to those rows, as shown in Figure 15-29.

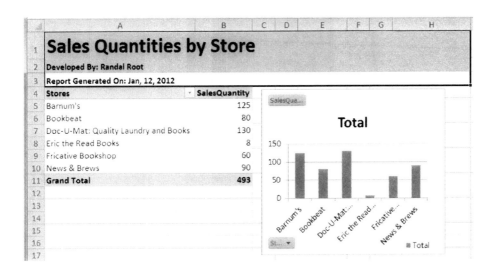

Figure 15-29. *Adding header rows to a report*

To make your report more professional-looking, you can also change the colors and fonts to match a standard used by your company or department. Office has several predesigned sets called *themes*. You can choose a theme or create your own. Creating your own theme enables you to create reports with a consistent look and feel. Figure 15-30 shows an example of selecting a theme for your report using the Office Ribbon.

You can also change the font, font color, and cell background independent of the theme. In Figure 15-30, we have done just that.

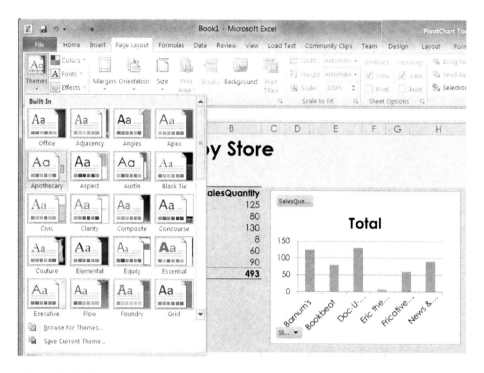

Figure 15-30. *Using one of Excel's preconfigured themes*

Saving to PDF

Once you have customized your report, you can publish it for others to use by saving it as an Excel spreadsheet or as a PDF file. Saving it as an Excel spreadsheet is self-explanatory, so let's take a look at saving it to a PDF file.

The option to save your Excel spreadsheet to a PDF file was added in Office 2007. This option can be accessed from the File tab of the Office menu, as shown in Figure 15-31. Once you have selected the File tab, select the Save and Send option within the tab's menu choices. Selecting this option displays the Create PDF/XPS Document option. By selecting this option, you are given a choice to create the file by clicking the Create PDF/XPS button.

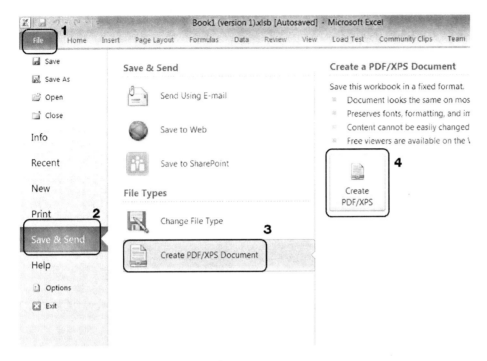

Figure 15-31. Saving the report to a .pdf file

■ **Tip** The dialog window will not look like Figure 15-31 until you click each item in the sequence shown in the figure.

You are then be asked to specify a location to save your PDF file. A good location is a subfolder within your BI Solution folder that is specifically created to hold your report files. For example, in Exercise 15-1 you made a folder called ExcelReports under the folder holding our solution files. Figure 15-32 shows an example of this.

Notice that even though you already saved the SalesReport.xlsx file in this folder, it is not displayed in the folder (Figure 15-32). This is because the Pubs as PDF or XPS dialog window filters out any files that do not have a .pdf or .xps extension. Don't worry, you file will not be erased or overwritten when using this feature.

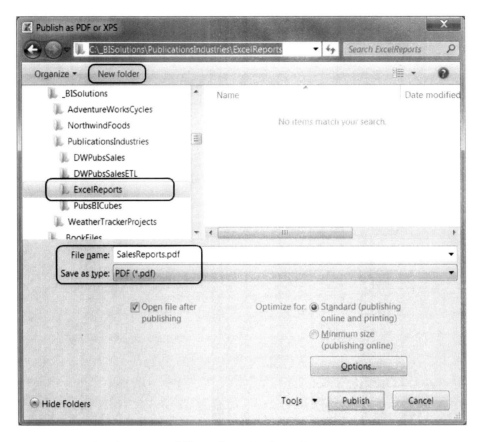

Figure 15-32. *Creating a report folder and opening the options screen*

Before you save to a PDF file, you may want to click the Options button as circled in Figure 15-33. Once you do, a new dialog box appears, as shown in Figure 15-33. Select which items to save to the PDF format. If you have multiple reports on multiple worksheets, the likely choice will be to save the entire workbook (Figure 15-33).

▓ **Note**   If you have the pivot table or pivot table chart highlighted when you access the save to PDF option, you will not be able to save the entire workbook. If this happens, cancel saving the file and click elsewhere on the worksheet and try again.

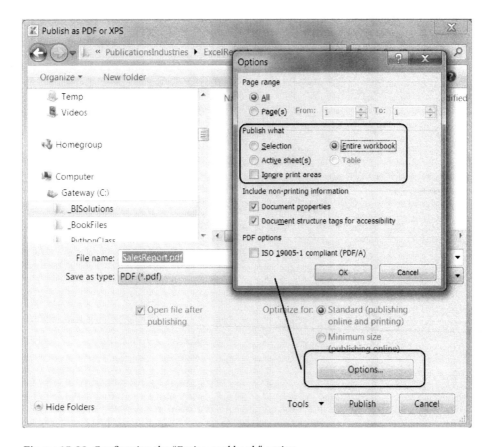

Figure 15-33. *Configuring the "Entire workbook" option*

After clicking OK to close the dialog box, click the Publish button to save your reports to the PDF format. Acrobat Reader immediately opens and displays your report. If, for some reason, your report is not displayed, navigate to your PDF file and open it yourself. Either way, at this point you will be able to review how your report looks in PDF.

Do not be surprised if the PDF format looks different from the original Excel format. Although Microsoft has tried to make this process as painless as possible, expect to make adjustments and then save and view the PDF file repeatedly until you get the look and feel that you want. For example, you may find that having the chart below the pivot table looks better in the PDF file (Figure 15-34).

▨ **Tip** We do not recommend telling your boss you will have that Excel report for them in the next ten minutes! Excel reports are easy to make, but it can take quite some time to make them professional looking enough for publication. Plan accordingly!

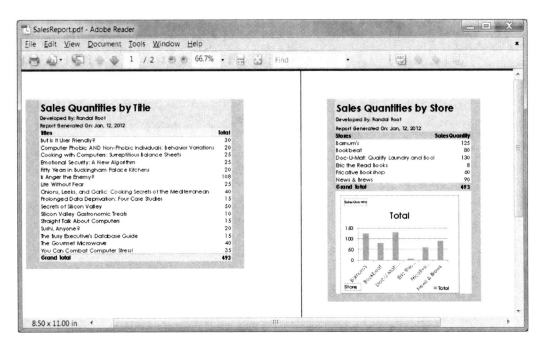

Figure 15-34. *A simple PDF report*

It is time to try this yourself to see how easy it is to save to a PDF file and how time-consuming it may be to get it to look just right. Don't worry about making it look perfect; just consider this a learning experience in how formatting can be time consuming.

EXERCISE 15-4. SAVING AS A PDF FILE

In this exercise, you change your report worksheets and save them to a PDF file. Your resulting PDF report will look like the one shown in Figure 15-36. A completed version of this PDF file can be found in the downloadable book content.

Prepare the Report

1. Using the workbook you created in Exercise 15-1, navigate to Sheet1 by clicking the tab at the bottom of the screen.

2. Modify your PivotTable until it looks similar to ours in Figure 15-35.

3. Right-click Sheet1's tab at the bottom of the Excel UI and select Rename from the context menu. Rename the tab to **SalesByTitles**.

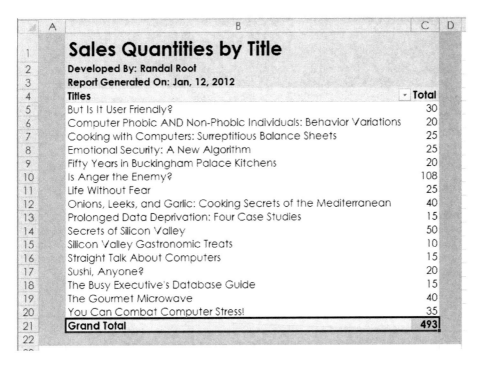

Figure 15-35. *The changed report on worksheet 1*

4. Navigate to Sheet2 by clicking the tab at the bottom of the screen.

5. Modify your PivotTable and PivotChart to look similar to ours, as shown in Figure 15-36.

6. Right-click Sheet2's tab at the bottom of the Excel UI and select Rename from the context menu. Rename the tab to **SalesByStores**.

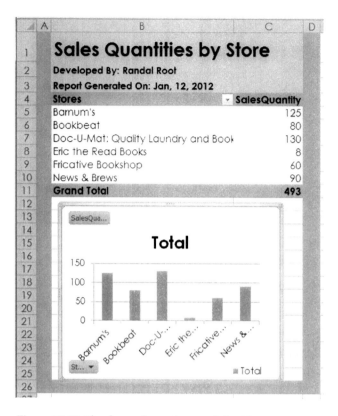

Figure 15-36. *The changed report on worksheet 2*

Save the Report

1. Click the background of the worksheet before saving.

Note: If you have only one individual report item selected when you try to save to a PDF file, you will be able to save only that one item to the file.

2. Click Excel's File tab and the *Save and Send* option, as shown in Figure 15-31.

3. Click the Create *PDF/XPS Document* option (Figure 15-31).

4. Click the *Create PDF/XPS* button (Figure 15-31).

5. When the *Publish as PDF or XPS* dialog window appears, click the Options button to access the Options dialog window (Figure 15-33).

6. Verify that the Entire Workbook radio button is checked, and then click OK to close the dialog window.

7. Save the entire workbook as a PDF as `SalesReports.pdf`, in the folder you made in Exercise 15-1, named `C:\_BISolutions\PublicationsIndustries\ExcelReports` (Figure 15-32). After saving the file, Adobe's Acrobat Reader automatically opens.

8. Review the appearance of the PDF file and make adjustments to the report until it looks like the one shown in Figure 15-34.

Adjusting the look may be easier said than done. To get the report looking just the way you want it, expect to adjust your Excel report and resave it repeatedly. You can use the View ➤ Page Display ➤ Two-Up menu item in Adobe Reader to show the reports side-by-side (Figure 15-37).

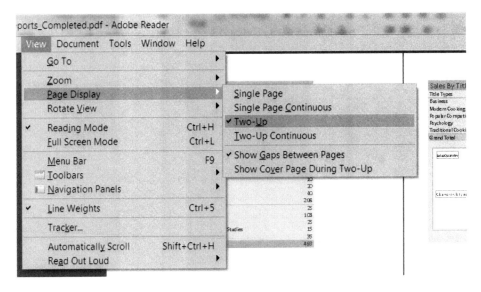

Figure 15-37. Viewing the reports side-by-side in Adobe Reader

You can find both our original Excel spreadsheet and the PDF version in the downloadable book files (C:_BookFiles\Chapter15Files\ExcelReports).

Now that you have saved two PivotTable and PivotChart reports to PDF, you are able to distribute them as you would any other file, such as by email, a network share, or a SharePoint web site.

Moving On

Excel is a powerful tool for creating reports, but in the end its focus is on user-created, self-service reporting. When you need more advanced, server-based reporting, the tool of choice is SQL Server Reporting Server (SSRS), which leads us to our next chapter.

LEARN BY DOING

In this "Learn by Doing" exercise, you create Excel reports for your Northwind cube, similar to the ones defined in this chapter. We have included an outline of the steps you performed in this chapter and an example of how the authors handled them in two Word documents. These documents are found in the folder C:_BookFiles_LearnByDoing\Chapter15Files. Please see the ReadMe.doc file for detailed instructions.

What's Next?

As you have seen, Microsoft's Excel comes with a great deal of reporting features, but you can extend those features with PowerPivot. PowerPivot is a free add-in to Excel 2010 and adds a staggering amount of features to Excel reporting. Once you have mastered Excel's basic reporting functionality, we strongly recommend checking out the many videos on PowerPivot available on the Internet. Here are a couple of books on these subjects that you may find useful:

Beginning Pivot Tables in Excel 2007 (Expert's Voice)
Debra Dalgleish
Publisher: Apress
ISBN-10: 1590598903

Microsoft PowerPivot for Excel 2010: Give Your Data Meaning
Marco Russo, Alberto Ferrari
Publisher: Microsoft Press
ISBN-10: 0735640580

CHAPTER 16

Creating Reports with SSRS

Just as we succeeded on the desktop, we will strive to succeed in services on the Web.

—Steve Ballmer

In the past three chapters of this book, we discussed creating reports with SQL code, MDX code, and Excel spreadsheets. Now we will take a look at Microsoft's premier reporting application, SQL Server Reporting Services (SSRS), also known as Reporting Services.

Reporting Services provides web-based reporting with its own web application or in combination with Microsoft SharePoint in a professional development structure. With Reporting Services, companies can have reports created by developers who understand the technology and data. Then the reports can be viewed by business analysts who have a deeper understanding of the company's business model.

This development structure also includes the ability to use source control to track changes in your reports and the ability to back up reports so that development hours are not lost in case of a disaster. These common features have been standard development tools for creating applications for more than a decade, but now the same tools can be applied to your reports.

In this chapter, we take a look at Reporting Services from the perspectives of report developers, administrators, and consumers. We show you how Reporting Services streamlines the interaction between all three of these roles. Afterward, you create a basic report so you can see how easy this is to accomplish.

Note   It is odd, but Microsoft refers to SSRS as Report Server, Reporting Services, and sometimes Reporting Server to add flavor to the mix. Wherever possible, we try to use the term that is associated with the user interface we are discussing. Sometimes, however, these terms are used interchangeably within the same interface. In the end, it really does not matter if you call SSRS a server or a service, because it is both.

SSRS Architecture

SSRS is the most complex of SQL Server's business intelligence services. This is because there are many different components designed to work together to create a complete reporting solution. These various components can be spread across multiple computers to provide a high degree of scalability and performance. Figure 16-1

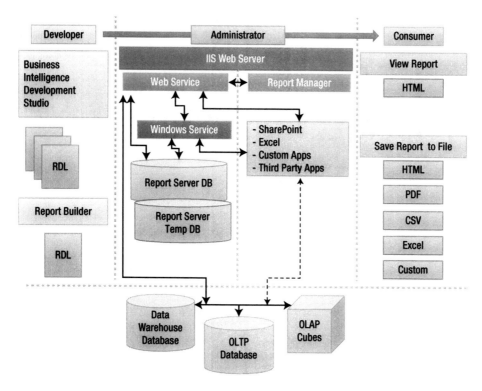

Figure 16-1. *The architecture of SSRS*

outlines the various components that make up Microsoft's Reporting Services. They are structured into three main categories:

- Developer tools
- Administration services
- Consumer-rendered reports

Developer Tools

Developers create reports using tools like Visual Studio or Microsoft's Report Builder. These development tools create XML files using the Report Definition Language (RDL) format. RDL files have been proposed as a standard for creating reports, and third-party applications can also be used to create RDL files.

After the RDL files are created, both BIDS and Report Builder can preview the report by rendering the XML code into a visual HTML output. Although it is not intended for end users to use HTML output, it does give a clear representation of what end users can expect to see when they browse the Reporting Services websites.

After the development is done, the RDL files are uploaded to the Reporting Services web service where it is then stored in a SQL Server database. When an end user requests a report using Reporting Services websites, the RDL file's code is read from the database and converted into a human-readable format. The default output is HTML, but the output format can also be in a PDF, CSV, Excel spreadsheet, or many custom formats.

Report Builder

The Report Builder application provides a simple way to create and edit RDL files. Report Builder is designed to be user-friendly, and its GUI interface is designed to have the look and feel of Microsoft Office so that developers accustomed to building reports in either Microsoft Access or Excel will feel right at home (Figure 16-2).

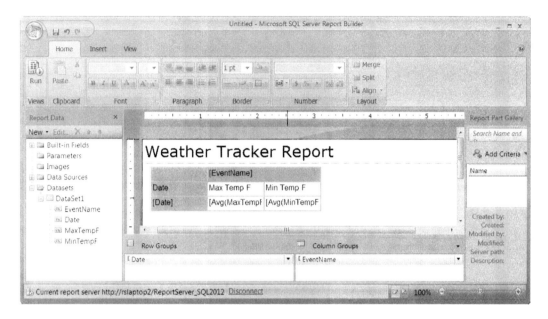

Figure 16-2. *The Report BuilderIU*

Unlike Visual Studio, Report Builder is not automatically installed when you install SQL Server 2012. It must be downloaded from the Internet and installed separately. The download and installation are both small and simple. Within a few minutes you have Report Builder installed, and you can start building new RDL files or edit existing ones.

Report Builder is designed for casual developers. It provides most, but not all, of the features that come with Visual Studio. For example, Report Builder is designed to work with only a single RDL file at a time, while Visual Studio can manage multiple files. Also, Report Builder is not a Visual Studio project type. The significance of this is that you cannot add Report Builder to your existing Visual Studio solution like we have with our SSIS and SSAS projects.

■ **Tip** Much of what we discuss in this chapter is the same information used to work with Report Builder. Therefore, as you work through this chapter, you are effectively learning both tools. For additional help, Microsoft offers several Report Builder video tutorials on the Internet.

Visual Studio

Visual Studio's Reporting Services project provides full-featured development. With it you can accomplish everything that can be done in Report Builder, plus you have the ability to manage multiple files concurrently, work directly with source control, and include the Reporting Services project in the same solution as your SSIS and SSAS projects.

Unlike Report Builder, Visual Studio is more utilitarian in design. This is partially because it provides additional functionality, but it is also because its intended audience is professional developers rather than casual developers.

The development tools are not difficult to use; they just take more effort to learn. It is similar to SSIS projects, in that you first drag and drop items from the Toolbox onto a design surface and then configure them (Figure 16-3).

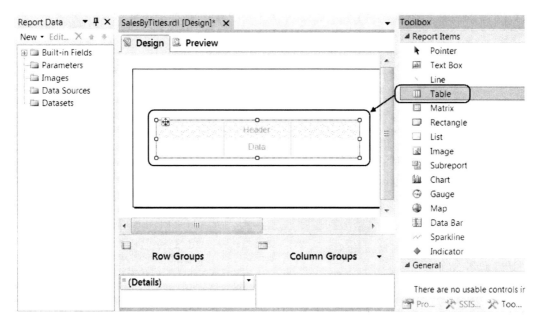

Figure 16-3. *The Visual Studio UI*

Much of this chapter, and the next, is devoted to explaining how to use these development tools.

The Administrative Services

After RDL files are developed and tested, they are uploaded to the Reporting Services web service, which in turn stores the RDL files in a Microsoft SQL Server database. When a report is browsed, it is automatically retrieved from the database by this same web service.

Access to the reports for human consumption is normally through an ASP.NET website that comes with SSRS, known as Report Manager, but the reports can also be accessed through your own custom applications. Note that applications do not access the reports directly from the Reporting Services database. Instead, they must indirectly access the reports through the SSRS web service, which serves as an abstraction layer. This adds complexity, but it also adds greater flexibility.

By using this design, you can place the web service (and its supporting Windows service) on one computer, while placing the Report Manager application or your own custom applications on a second computer. You can even span more computers by placing the Report Server databases on a separate computer. This can greatly increase your reporting performance at the cost of administrative overhead, but it is this scalability that sets Reporting Services apart from most reporting software.

The administrative services of SSRS are divided into report management features and end-user support (Figure 16-1). Report management features consist of the web service that manages the uploading and downloading of report files, report rendering, and exposing management functions or methods. The web service has an associated Windows service that it interacts with. The Windows service provides additional functionality such as scheduling components that allow reports to automatically be rendered and delivered to end users.

Although the Windows service does provide functionality similar to the web service, it is more difficult to interact with from a programming perspective. Thus, for programmers, the web service is the default way of interacting with SSRS's administrative services.

In addition to the web and Windows service, report management is handled by stored procedures within SSRS databases. These stored procedures are executed from the web and Windows service to perform all database activities.

SSRS Web Applications

Each installation of Reporting Services includes two web applications, both of which are built using ASP.NET. The first is the web service we just mentioned, and the second is an end-user application for viewing and managing reports. This end-user application is known as Report Manager and is designed to interact with the web service.

Most companies find that the Report Manager web application works well for their needs. Because of this, these companies do not have a great need to create their own custom application. However, the web service allows other developers to create custom windowed, console, and web applications that can interact with the web service as well. This means that if you do not want to use Microsoft's ready-made web application—Report Manager—you can create your own by programming it to interact with the web service.

The SSRS web service is not designed for humans to use directly, but it can be accessed directly if you so desire. The web page that you see when you access the web service is quite basic when compared to the Report Manager. Both of these are shown for comparison in Figure 16-4. Notice that the web service contains text and hyperlinks but little else. Clicking a hyperlink navigates you to either a subfolder containing reports or launches a report for viewing. When you select a report for viewing, it will display the report in an HTML format.

The Report Manager web application provides a more user-friendly way of accessing reports, by using interactive menus and icons. It also supplies administration options for users with administrative privileges.

■ **Note** Although many features are accessible with any web browser, the Report Manager web application is designed to work with Internet Explorer, because it provides native support for Windows authentication. Expect to use Internet Explorer when performing administrative tasks using Report Manager.

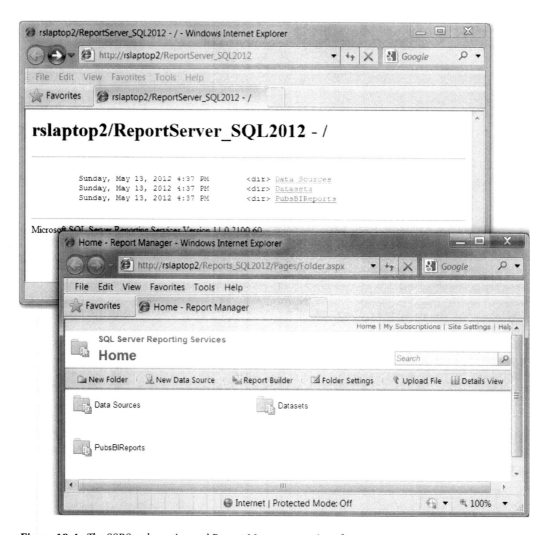

Figure 16-4. *The SSRS web service and Report Manager user interfaces*

To provide users with the ability to view reports directly from the web service, Microsoft created a Report Viewer ASP.NET web page that launches when you click a report hyperlink listed on the web service's website. Whenever you launch a report using the Report Manager web application, the report displays this same ASP.NET page, but this time it is embedded within the Report Manager's web pages. For example, when you create a report action in an SSAS cube, as we did in Chapter 12, the report action must be configured to point to the web services and not the Report Manager web application.

When an end user clicks a report action, they will see an SSRS report displayed from the web service that looks similar to the one shown in the upper half of Figure 16-5, rather than the report in the lower half of Figure 16-5 (which is embedded into the Report Manager website). The difference is subtle, but if end users or other developers are not told about it, they can get confused as to why the reports appear different, depending on how they are accessed.

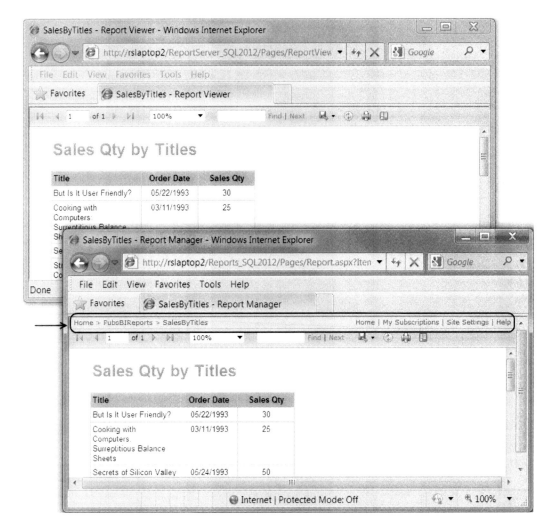

Figure 16-5. *Report Viewer as shown from the web service and Report Manager*

As you develop your reports, it is best to make sure you view the reports using both of these methods to ensure that what users are seeing is sufficient for their needs.

SSRS Services

Once you are done creating and testing a report, you can upload the file to the SSRS web service using Visual Studio, Report Builder, the Report Manager website, or even a SharePoint website. This variety of options is made possible because all of these applications can interact with the SSRS web service. This flexibility is one of the reasons why Microsoft designed Reporting Services to use a web service for the report processing.

In addition, the web service acts as a public interface for the SSRS Windows service, allowing Visual Studio, Report Builder, the SSRS Report Manager, SharePoint, and other software to interact with the SSRS windows service through an abstraction layer.

The web and Windows services' purpose is to process the programming instructions found in RDL files. The contents of the RDL files will be stored in a pair of SQL Server databases (Figure 16-6).

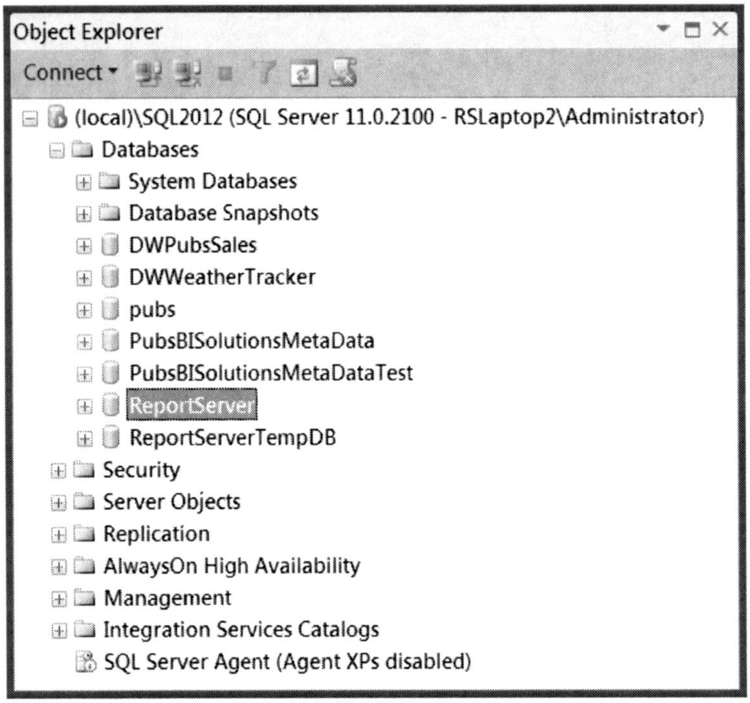

Figure 16-6. *The Report Server databases*

SSRS Databases

As shown in Figure 16-6, Microsoft installs two SQL Server databases with each installation of Reporting Services. The first database, normally named ReportServer, is designed to hold the RDL file code, metadata, and configurations. The second database, normally called ReportServerTempDB, is designed as a workspace for processing report requests. It also acts as a repository for cached reports.

■ **Tip** If you are using a named instance, the databases may have the instance name added to their standard names: ReportServer and ReportServerTempDB. Not to worry, though, the name variation does not change the way they work.

When a report is uploaded to the ReportServer database, the report's code is stored in a binary format, within a table named Catalog. If you select the data from the Catalog table, you will see a listing for every SSRS folder, data source, dataset, and report that has been created on the Reporting Server websites.

For example, in Figure 16-7, we have queried the Catalogs table, and seven objects are currently listed. The Path indicates the logical location that will be shown in the SSRS websites. The Name is, of course, the name we gave the objects when we created them. The Type is the type of SSRS object listed. Microsoft documentation on this table is sparse, but Table 16-1 describes the common type codes used.

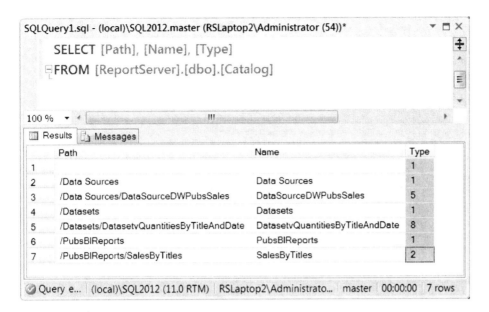

Figure 16-7. *Selected results from the Catalog table*

Table 16-1. *Types of Objects in the Catalog Table*

TypeId	Type Description
1	A logical folder
2	A report
5	A shared data source
8	A shared dataset

Report Server folders are used to organize your reports. And although they have no physical location on a hard drive, they do have a visual representation when using the Report Server websites. Report Server shared data sources are objects that hold connection information to be used by one or more reports. A Report Server shared dataset can also be used by one or more reports but holds query information, not connection information.

SSRS Configuration Manager

Because SSRS has so many components, it comes with its own configuration application. It can be accessed from the Windows Start menu by navigating to All Programs ➤ Microsoft SQL Server 2012 ➤ Configuration Tools ➤ Reporting Services Configuration Manager. Clicking this link launches the SSRS configuration application and presents you with a choice as to which server you would like to configure. In our examples, you connect to your own computers, but configurations can be managed on remote computers as well. Moreover, Reporting Services can be installed multiple times on a single computer using named instances. Therefore, you need to define which instance of Report Server you want to configure. Figure 16-8 shows the user interface that allows you to select both the server name and the instance name.

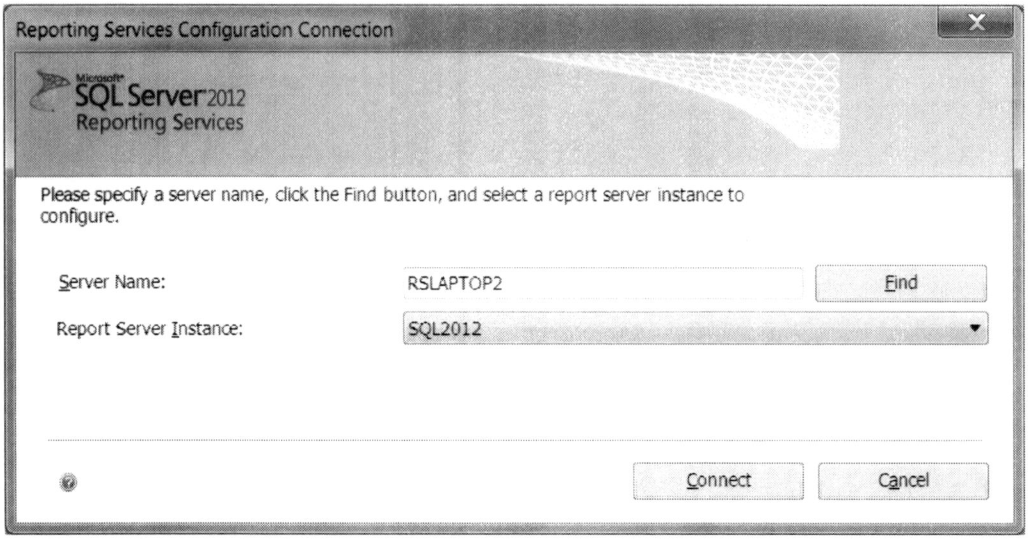

Figure 16-8. *Connecting to a Report Server installation*

Once you have selected the Report Server installation, you will be presented with a user interface like the one in Figure 16-9. The user interface is designed with a treeview on the left side of the window and a detailed view of property settings on the right side. Items in the treeview are referred to as *pages*. Here you can define properties for the SSRS web applications and databases.

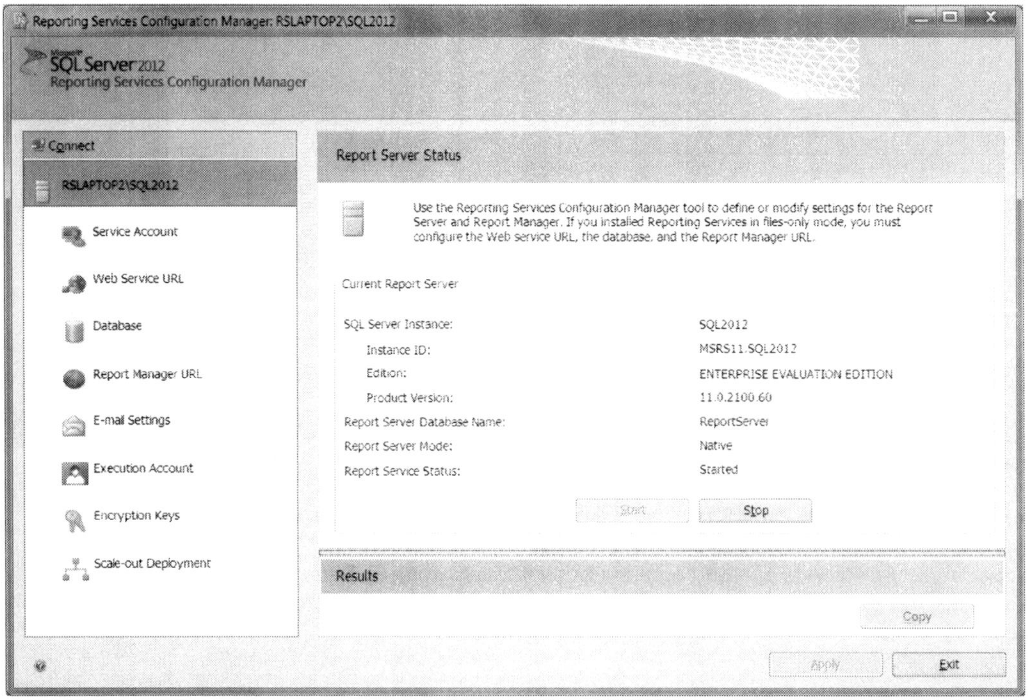

Figure 16-9. *The Report Server Configuration Manager user interface*

■ **Tip** As we have mentioned, Randal's computer, which is used for the screen shots in this book, uses a named instance of SQL Server called SQL2012. Consequently, the instance name used in Figure 16-8 and in Figure 16-9 matches his configuration. Most readers will use the SSRS default instance, but of course your configuration is dependent on the choices you made while installing SQL Server 2012.

The first page of the Reporting Services Configuration Manager is a general overview of the Report Server status. From here you can stop or start the Report Server and identify some of the startup parameters, such as whether it is running in native mode or configured to work with SharePoint. Native mode indicates that is using its own miniature web server to support the web service and the web application.

■ **Note** For earlier versions of Reporting Services, these websites were hosted on Microsoft's Internet Information Service (IIS) installation, meaning that you would have to first install IIS before you could install Reporting Services. All of the versions since 2005 provide their own web server support. Therefore, you do not have to install IIS before you can install Reporting Services. SharePoint, however, still requires a full IIS installation. Thus, when your SSRS server is configured to use SharePoint, it is utilizing the features associated with IIS and not its own native web server.

The Service Account page of the Reporting Services Configuration Manager allows you to define which account will be used when running the SSRS Windows service (Figure 16-10). The account's purpose is to interact with the SSRS databases, so if the account you choose does not have access to these databases, you will not be able to start the SSRS service. Ideally, you provide a Windows account that has limited permissions. But in a test environment, such as for the exercises in this book, an administrator account will work. It should be noted that it is not sufficient for the account to just be a Windows administrator; make sure that whatever account you use also has access to SQL Server and the SSRS databases.

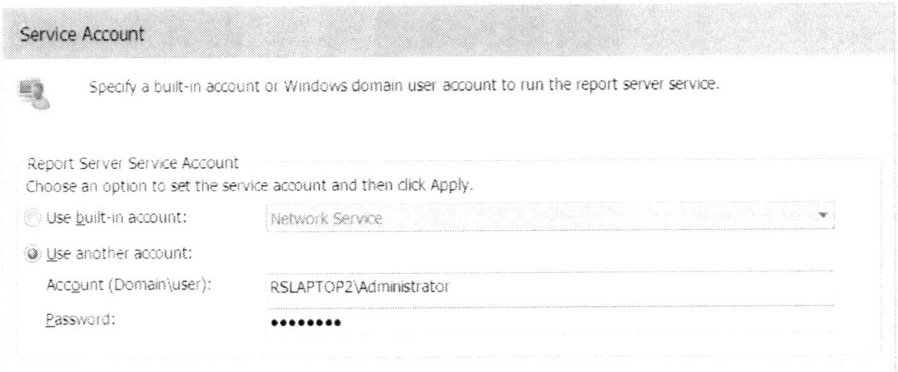

Figure 16-10. The service account page of the Configuration Manager

The account that you choose will also be used for more than just connecting to SQL Server databases. Depending on how you set up SSRS, you can also interact with mail servers, network shares, and domain controllers. This means that if you are planning to use features that would involve those servers, you must also expect to configure the Windows account to have access to these items.

■ **Tip**   We use the Windows administrator account to avoid permission issues, but of course this is not recommended practice for production, because it represents a security risk. In this book, we have been using an administrator account that provides access to SQL. In production, it is a very common issue to be unable to gain access to the Windows account. Microsoft's website provides a considerable amount of advice to aid you in configuring the Reporting Services service account.

The web service URL page is helpful in identifying whether Reporting Services is working correctly (Figure 16-11). A hyperlink on this page launches your browser and tries to navigate to the Reporting Services web service. At first, the web service will be unresponsive. It takes a few minutes to come alive, so assume that first request may take a minute or so before the web page will open successfully. This will happen every time you reboot your computer.

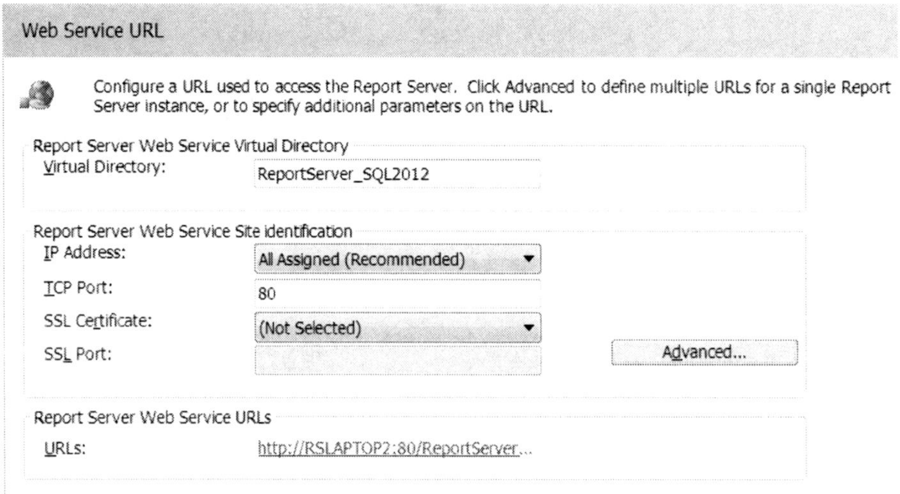

Figure 16-11. *The web Service URL page of the Configuration Manager*

The Report Manager website URL can also be used to identify whether Reporting Services is working correctly. If you have previously opened the web service, the Report Manager web application should start up rather quickly. If you go directly to the Report Manager application without accessing the web service first, it will take a few minutes for it to open, because, as we just stated, it must start the web service application before it can display its content. This is perfectly normal and will not affect your end user's system.

When you first install Reporting Services, it will not include any reports. So, when the websites are accessed, the web pages will appear with a header at the top and a mostly blank area beneath that. We discuss how to change that next.

WHEN THINGS GO WRONG

Even if you cannot get the web and Windows services running correctly, you can still create and test your SSRS reports for the rest of the book. The problems come when you try to follow along with some of the administration tasks outlined at the end of the chapter. If you have not managed to get the Reporting Services running correctly by then, you will only be able to read about how the administration tasks are performed, which won't be as much fun, although it will still be informative. We prefer that you follow along throughout the chapter, however, so here are some tips for troubleshooting SSRS installations.

The Databases

Normally, the SSRS databases are created during the installation of SQL Server, but occasionally they are not. You need the database if you want to perform SSRS administration. You can use SQL Server Management Studio to verify that the Report Server databases are installed. If you do not see them there, you can assume you found the problem. To fix this issue, use the database page of the Reporting Server Configuration Manager to manually add the databases. Microsoft provides a wizard that easily walks you through the process, which is accessed by clicking the Change Database button shown in Figure 16-12.

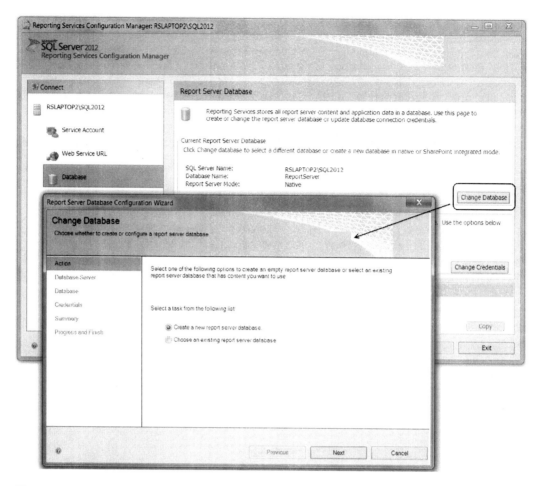

Figure 16-12. *Installing the Report Server databases*

The Windows Service

Once you know that the databases exist, you need to prove that SSRS can connect to them. Do this by launching one or both of the Report Server websites. If you have successfully configured the service account (Figure 16-10), you will be able to connect to the web pages, and they will look similar to Figure 16-4. If the service account is not configured correctly, you will see an error message.

If you get an error, your first step is to verify that the account you are using for the service can connect to SQL Server. You can test this by logging into SQL Server as that particular user from SQL Server Management Studio and then prove that you can select data from the Catalog table in the ReportServer database. If you cannot, configure the service to use an account that does have access.

Encrypted Content

Another common error that may occur is a message stating that your encrypted content is not configured correctly. This can be quite confusing, because on a first install you do not expect to have any encrypted content. This message shows up when you try to use SQL Server Management Studio to interact with SSRS directly. For example, attempting to expand the Jobs node will give the error message shown in Figure 16-13.

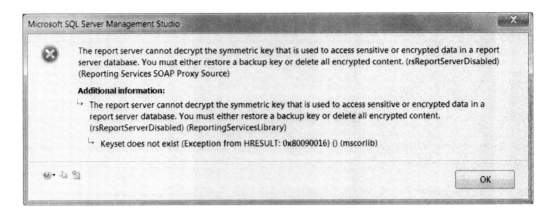

Figure 16-13. *An error message when accessing SSRS from Management Studio*

The way around this is rather odd. First, use the Reporting Server Configuration Manager to access the Encryptions Keys page. Then, use the Delete Encrypted Content option to reset the Encryption Key process by clicking the Delete button (Figure 16-14).

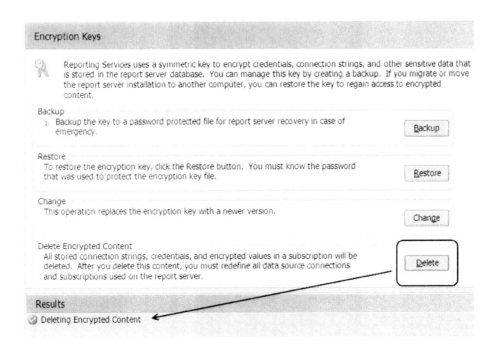

Figure 16-14. *An error message when accessing SSRS from Management Studio*

You are warned that this is a bad idea, but don't worry. It is safe to delete the encrypted content after you first install SSRS, because there is no encrypted content!

These three items represent the most common issues you are likely to encounter. If this is not enough to help you connect, we suggest searching the Web for *Reporting Services Configuration Manager* for more information. We have also included a video detailing an SSRS setup on the author's website at www.NorthWestTech.org/SSRSDemos/SSRSSetupTips.aspx.

Creating SSRS Objects

Most developers use Visual Studio to create their SSRS reports. To create reports with Visual Studio, you need to create a Visual Studio project. To add the project to the solution we have been working on throughout this book, start by opening the Publication Industries' Business Intelligence Solution and adding a new Report Server project to it, as you see in Figure 16-15.

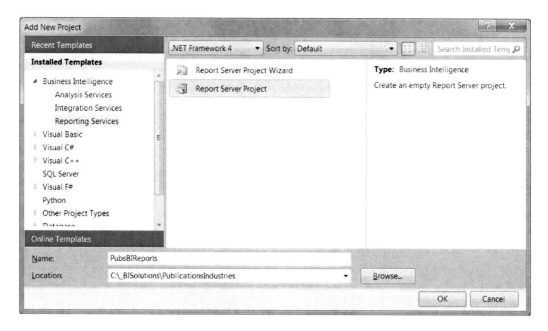

Figure 16-15. *Adding a Reporting Services project to our existing BI solution*

Each SSRS project is a collection of XML files that define different SSRS objects. These objects include data sources, datasets, and reports. Visual Studio organizes these files into virtual folders within the Solution Explorer window. Let's take a closer look at each of these objects.

Data Sources

Once the project has been created, you can see it in Solution Explorer. Next, create one or more data sources. Data sources provide the connection information for your reports. This connection information can be stored either in the report RDL file or in a separate report data source (RDS) file. When the connection information is stored in the separate RDS file, it is referred to as a *shared data source*. Shared data sources can be referred to by many reports, and using shared data sources is considered a best practice. This is because it allows you to update a single data source and automatically affects all of the reports that refer to it.

To create a shared data source, right-click the folder called Shared Data Sources and select the Add New Data Source option from the context menu (Figure 16-16). The Shared Data Source Properties dialog window will appear.

Figure 16-16. *Creating a shared data source*

In the Shared Data Source Properties dialog window, you enter a name for your new data source, select the type of connection to create, and provide a connection string (Figure 16-17).

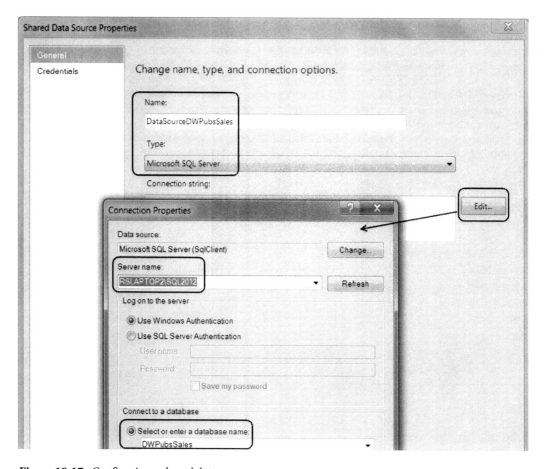

Figure 16-17. *Configuring a shared data source*

The name is configured using the Name textbox and should represent the object that you are connecting to. For example, in Figure 16-17, we have named our connection **DataSourceDWPubsSales**.

The connection type is chosen using the dropdown box. In Figure 16-17 we have selected a Microsoft SQL Server connection type. The dropdown box also contains connections for Analysis Server, Oracle, OLE DB, XML, and many others. We use SQL Server and Analysis Server for our demonstrations.

To create a connection string, type in the connection string or click the Edit button to access the Connection Properties dialog window. Choose the server name and the database name in the Connection Properties dialog window, just as we have done in many previous chapters. Click OK to create the connection string.

After you have identified the name, type, and connection string, you close the Shared Data Source Properties dialog window by clicking OK, and the new shared connection shows up in Solution Explorer (Figure 16-18). You may notice that the extension on the new data source file is RDS, which stands for Report Data Source, and the connection information is stored in an XML format. Listing 16-1 shows an example of the contents of an RDS file.

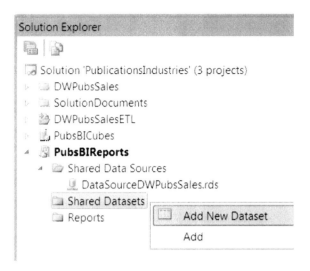

Figure 16-18. *Creating a new shared dataset*

Listing 16-1. XML Code in a Typical RDS Data Source File

```
<?xml version="1.0" encoding="utf-8"?>
<RptDataSource xmlns:xsi="http://www.w3.org/2001/XMLSchema-instance"
xmlns:xsd="http://www.w3.org/2001/XMLSchema" Name="DWPubsSalesDataSource">
  <ConnectionProperties>
    <Extension>SQL</Extension>
    <ConnectString>Data Source=RSLAPTOP2\SQL2012;Initial Catalog=DWPubsSales</ConnectString>
    <IntegratedSecurity>true</IntegratedSecurity>
  </ConnectionProperties>
  <DataSourceID>e54393b4-108f-4798-ad55-353a30c1e6e0</DataSourceID>
</RptDataSource>
```

Data source code can also be embedded within a reports' RDL file, but doing so means that the data source is available for only that individual report and cannot be used by other reports on the SSRS websites.

Datasets

After you create your data sources, you then make one or more datasets. Datasets contain query string used to generate report data. The query string language is dependent on which type of connection is used. For example, if you are connecting to a SQL Server database, use the SQL language. If you are connecting to an Analysis Server database, use a language such as MDX.

Datasets can be either embedded within a report's RDL file or created as a separate file with the extension of RSD. These Report Server dataset (RSD) files are simply XML files that contain your query sting as well as a cross-reference to a data source that is used to execute the query.

Listing 16-2 shows an example of an RSD file that contains a SQL statement. Notice that it also contains a reference to the data source from Listing 16-1. Any report using this dataset automatically looks up the data source and use it to execute the SQL statement and retrieves the report data.

Listing 16-2. XML Code in a Typical RSD Dataset File

```
<?xml version="1.0" encoding="utf-8"?>
<SharedDataSet xmlns:rd="http://schemas.microsoft.com/SQLServer/reporting/reportdesigner"
xmlns="http://schemas.microsoft.com/sqlserver/reporting/2010/01/shareddatasetdefinition">
  <DataSet Name="">
    <Query>
      <DataSourceReference>DWPubsSalesDataSource</DataSourceReference>
      <CommandText>SELECT
    PublisherName
  , [Title]
  , [TitleId]
  , [OrderDate]
  , [Total for that Date by Title]
 FROM vQuantitiesByTitleAndDate
</CommandText>
    </Query>
    <Fields>
      <Field Name="PublisherName">
        <DataField>PublisherName</DataField>
        <rd:TypeName>System.String</rd:TypeName>
      </Field>
      <Field Name="Title">
        <DataField>Title</DataField>
        <rd:TypeName>System.String</rd:TypeName>
      </Field>
      <Field Name="TitleId">
        <DataField>TitleId</DataField>
        <rd:TypeName>System.String</rd:TypeName>
      </Field>
      <Field Name="OrderDate">
        <DataField>OrderDate</DataField>
        <rd:TypeName>System.String</rd:TypeName>
      </Field>
      <Field Name="Total_for_that_Date_by_Title">
        <DataField>Total for that Date by Title</DataField>
        <rd:TypeName>System.Int32</rd:TypeName>
      </Field>
    </Fields>
  </DataSet>
</SharedDataSet>
```

To create a shared dataset, right-click the Shared Datasets folder in Solution Explorer and choose Add New Dataset from the context menu, as shown in Figure 16-18.

A Shared Dataset Properties dialog window appears where you can identify the name, data source, query type, and query for the dataset, as shown in Figure 16-19.

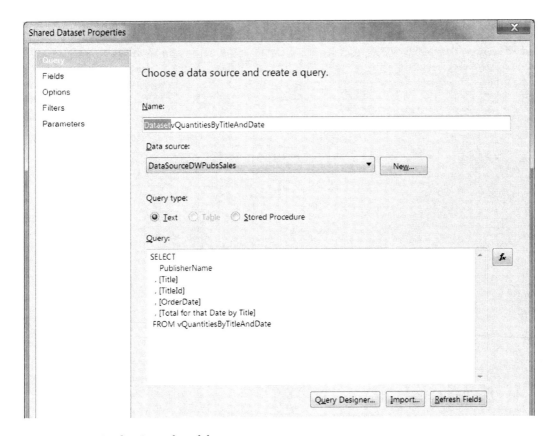

Figure 16-19. *Configuring a shared dataset*

The name is configured by filling in the Name textbox. As always, the name you choose should describe its purpose. We often choose to use the table, view, or stored procedure name that is being accessed. In Figure 16-19 we have configured the dataset name to be the view name, DatasetvQuantitiesByTitleAndDate.

The "Data source" configuration uses a dropdown box that allows you to select a previously created shared data source, or you can make a new one if necessary. Using the New button launches the same dialog window shown in Figure 16-19.

The "Query type" configuration uses radio buttons (Figure16-19). There are only three options to choose from:

- Text

- Table

- Stored Procedure

The options will be either available or grayed out, depending on the type of data source. In the case of a Microsoft SQL Server data source, your options will be Text and Stored Procedure. But when using an Excel spreadsheet, the Table (representing a worksheet) option will appear, and Stored Procedure will be grayed out.

The "Query configuration" uses a large textbox where you can either type in the query code or click the Query Designer button to access the Query Designer dialog window (Figure 16-19).

Reports

In an SSRS project, the main task a developer performs is creating new reports. As you may have guessed, you create reports by right-clicking the Reports folder in Solution Explorer and choosing Add New Report.

But wait—there is a trick to this one! The Add New Report option launches the Report Wizard. Now, there is nothing wrong with the Report Wizard per se, but more experienced developers will want to create a new report from scratch without using a wizard.

■ **Note**   If you have launched the Report Wizard by mistake, cancel the wizard and try again. Luckily, Visual Studio does not penalize you for starting a wizard and canceling it. We say luckily, because this inadvertent launching is likely to happen quite often. Doh!

To create a new report without using the Report Wizard, select the Add New Item option, as shown in Figure 16-20.

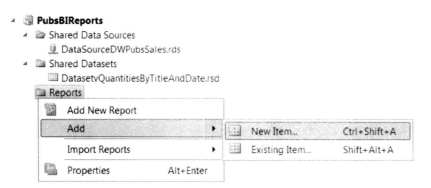

Figure 16-20. *Creating a new report without the Report Wizard*

After selecting the New Item option in the context menu, the Add New Item dialog window appears. Notice that, once again, the Report Wizard is present and available. In fact, it is the default selection! We are creating a report without using the wizard; therefore, we need to select the Report icon and provide a name in the Name textbox (Figure 16-21). Click Add to close the dialog window, and a new RDL file will be created in Solution Explorer.

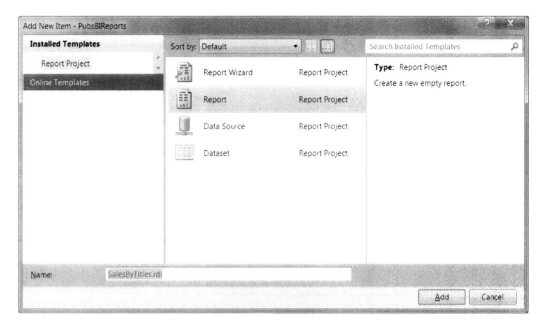

Figure 16-21. *Selecting the Report icon and providing a name*

Mysteriously, at this point it is hard to say exactly what you will see, because your Visual Studio setup determines which windows are displayed. The two windows that are used most often are the Report Data and Toolbox windows. These windows are essential to configuring your reports, but they may or may not appear in Visual Studio when your new RDL file opens. Not to worry! You can force both of these windows to display by accessing the View menu in Visual Studio and selecting their corresponding menu items.

Important The Report Data menu item should be located at the very bottom of the View menu; however, it may not show up in the menu if you are not focused on an RDL file in Solution Explorer. If, for some reason, it does not show, make sure that you have the RDL file open, as shown in Figure 16-22. Then go back to the View menu and it should appear.

In Figure 16-22 we have placed the Report Data window on the left side of Visual Studio and the Toolbox on the right, but they can be dragged and dropped to various positions within Visual Studio.

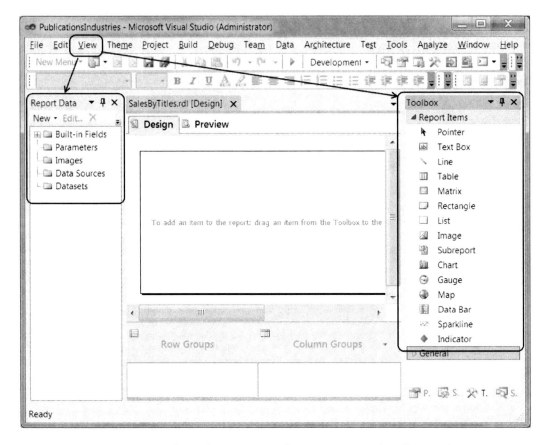

Figure 16-22. *Preparing to configure the report using the Report Data and Toolbox windows*

Once you create your first report and set up your Report Data and Toolbox windows, we recommend you configure the report in this order:

1. Configure the report data.

2. Add report items from the Toolbox to the report's design surface.

3. Configure the various report items on the design surface.

4. Preview your report design in Visual Studio.

5. Deploy the report to the SSRS web service (optional).

Configuring Report Data

You must identify the data that you are going to use in your report. To do so, you need two things. First, you need some way to connect to the data. Second, you need some kind of query to identify which data you want from the source.

To connect to the data you must create a local data source within the RDL file. This is necessary even if you have previously created a Shared Data Source earlier. Looking at Figure 16-22, you can see that the Report Data window has a Data Sources folder. From here, you can add a data source to the report by right-clicking this folder and selecting Add Data Source from the context menu, and the dialog window will appear. This dialog window works the same as the one used to create a Shared Data Source (Figure16-17).

▪ **Tip** There is one important difference between the local and shared data source configuration dialogs; you can link a local data source to a shared data source using the *Use shared data source reference* radio button. When this option is selected, the data source connection string is not stored within the RDL file. Instead, the RDL file contains a reference to a shared data source and its connection string.

Once you have at least one Data Souce created, move on to creating Data Sets. You can add a dataset to the report by right-clicking the Datasets folder (Figure 16-22), and selecting Add Dataset from the context menu. As we discussed earlier, a dataset is a saved query that identifies the data that you want to use in your report. Clicking this context menu will display a dialog window like the one shown in Figure 16-23.

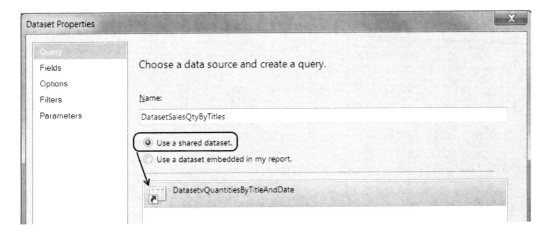

Figure 16-23. *Referencing a shared dataset*

Like data sources, there are both local and shared datasets. In order for it to display data, each report must have a local dataset, but this dataset can either contain the select query itself or a reference to a Shared Dataset. When you compare the Shared Dataset dialog window shown in Figure 16-19 to the local Dataset window in Figure 16-23, notice the two radio buttons that allow you to either use a shared dataset or dataset embedded within the report. If you opt to use an embedded dataset, the SQL code will be saved within the RDL file. If you choose a shared dataset, the RDL file will cross-reference an RSD (Listing 16-2).

Both options provide you with data for your reports; however, using a shared dataset has the advantage of being reusable between multiple reports. This is also true of a shared data source. As a bonus, if you ever reconfigure these shared object, all changes are propagated to any reports that reference them. As you might suspect, using both shared data sources and shared dataset objects is considered a best practice.

In Figure 16-23, we are electing to use the shared dataset we defined previously. Now our report file will cross-reference the Shared Dataset file, which in turn cross-references the Shared Data Source file. Therefore, we only need to identify a shared dataset, and the shared data source will be cross-referenced implicitly.

Be sure to name the dataset appropriately. In Figure 16-23 we named it **DatasetSalesQtyByTitles**. Once the dataset has been created, the Report Data window will show the dataset by name and all the associated data fields beneath it. These data fields represent the columns or column aliases associated with the query results.

Adding Report Items

Once you have configured your report data, you can add report items by dragging and dropping them from the Toolbox window to the design surface. After you place items on the design surface, you can click each item to display sizing handles, as shown in Figure 16-24. The sizing handles allow you to adjust the size of the report item objects and are represented as little white or black boxes.

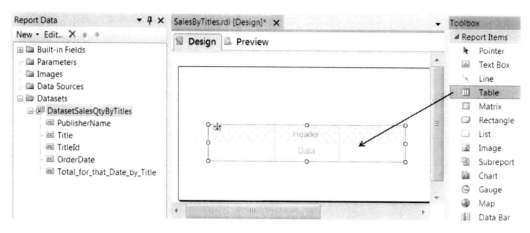

Figure 16-24. *Adding a table to the report*

In some cases, the sizing handles will be obscured by other design components. For example, if you click a Table report item, you first see the column and row design component shown in Figure 16-25 and not the sizing handles shown in Figure 16-24. By clicking the upper-left corner of the column and row design component, you are able to access the sizing handles of the Table report item.

■ **Tip** Like any GUI interface, it is a good idea to take a few moments to get used to it. Be patient with yourself and explore the environment for a while. You will soon find that the design tools are not nearly as bewildering as they first appear.

Configuring Report Items

Arguably the most time-consuming process of creating reports is the configuration process. During this phase you map data to the report items and change various properties until the report looks as intended.

To map data to a report item, click a data field in the Report Data window and drag it to the report item on the design surface, as indicated in Figure 16-25. You can also click a report item and choose a data field using the *Smart Tag* icon that magically appears. These smart tags are becoming more common in Microsoft's development tool and usually launch a context menu (Figure 16-25).

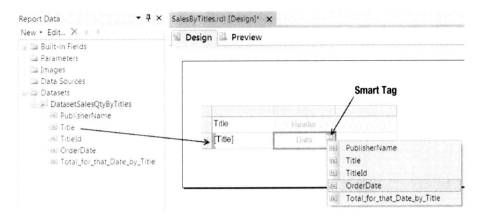

Figure 16-25. *Adding data to the table*

Previewing Reports

After you have made some changes to the report, it is important to preview your work to get an idea of how the report is coming along. When you click the Preview tab, located at the top of the Report Designer surface, a representation of the report will be displayed in HTML format. Figure 16-26 shows an example of the report we have configured so far, after having added three data fields, Title, OrderDate, and Total_for_that_Date_by_Title, to a table report item.

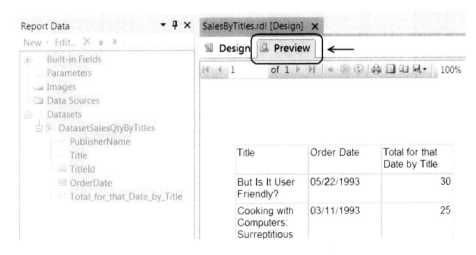

Figure 16-26. *Previewing the report*

As you can see in Figure 16-25, tables are divided into data and header rows, but the table looks different on the Design and Preview tabs. In our example, the SSRS table report item displays one header row and multiple data rows on the Preview tab (Figure 16-26), but only one header and data row on the Design tab (Figure 16-27). Expect to navigate between them repeatedly as you make configuration changes.

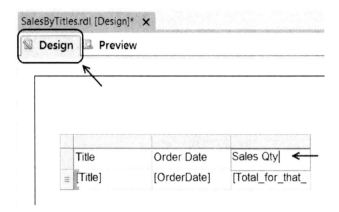

Figure 16-27. *Changing the column titles*

As you preview your report, you may notice things that you would like to change. For instance, you may want to change the header text. You can do this by clicking the text once to select it, pausing for a second, and then clicking the text once more. A cursor appears that allows you to delete and retype the text (Figure 16-27).

You can add several other items to the report, such as a Text Box report item to display the report header (Figure 16-28). You can also format the text within the Text Box by changing the font colors with the tools on the toolbar (circled in Figure 16-28).

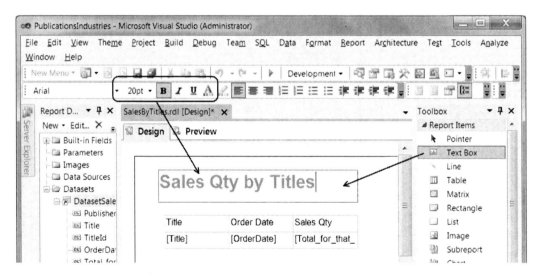

Figure 16-28. *Formatting report textboxes*

Tables can be formatted as well. For example, you can highlight the entire header row by clicking the gray Row button on the left side of the Table report item and then using the Format toolbar just as you would with a text box (Figure 16-29).

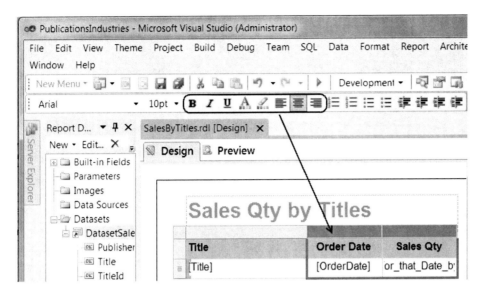

Figure 16-29. *Formatting table report items*

After making a few more format changes, you can preview the changes and continue configuring the report until you are satisfied with the result.

You can see in Figure 16-30 that our report isn't a work of art, but it is sufficient for our needs. It is time to give it a try in the following exercise.

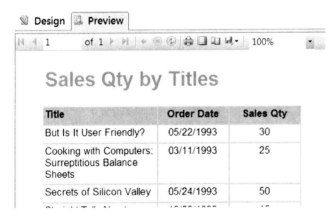

Figure 16-30. *Viewing your changes*

EXERCISE 16-1. CREATING A REPORT

In this exercise, you add an SSRS project to your existing Publications Industries solution. You then create a shared data source, a shared dataset, and a basic report.

1. Open Visual Studio by clicking the Windows Start button and selecting All Programs ➤ Microsoft SQL Server 2012 ➤ Data Tools.

Important: You are practicing administrator-level tasks in this book; therefore, you need administrator-level privileges. The easiest way to achieve this is to remember to always right-click a menu item, select Run as Administrator, and then answer Yes to access administrator-level privileges. In Windows 7 and Vista, logging in with an administrator account is not enough. For more information, search the Web for the keywords *Windows 7 True Administrator and User Access Control.*

2. Open the Publications Industries solution you have been working on throughout the book using the File ➤ Open Project/Solution menu item.

3. Locate and select the SLN file `C:\_BISolutions\PublicationsIndustries\PublicationsIndustries.sln`.

4. Click the Open button at the bottom of the dialog window to open the solution.

Add a Project to the Current Solution

1. Add a new project to the solution by selecting File ➤ Add ➤ New Project from the main menu at the top of Visual Studio. An Add New Project dialog window appears (Figure 16-15).

2. Select the Business Intelligence ➤ Reporting Services template category on the left of the dialog window. Then select the Reporting Server Project option from the center of the dialog window (Figure 16-15).

3. Name the Project PubsBIReports at the bottom of the dialog window and click OK to add the new project to your current solution (Figure 16-15).

4. Use the View ➤ Solution Explorer menu item to display the Solution Explorer window.

Create a Shared Data Source

We need a way to connect to the data warehouse to access our report data. Let's create a shared data source.

1. Add a new shared data source by right-clicking the Shared Data Sources folder and selecting the Add New Data Source option in the context menu (Figure 16-16).

2. When the Shared Data Source Properties dialog window appears (Figure 16-17), type **DataSourceDWPubsSales** in the Name textbox.

3. In the Type dropdown box, select Microsoft SQL Server (Figure 16-17).

4. Click the Edit button next to the Connection String textbox and type in the name of your SQL Server, as shown in Figure 16-17.

5. Using the dropdown box beneath the "Select or enter a database name" radio button, select the DWPubsSales database (Figure 16-17).

6. Close both dialog windows by clicking the OK buttons to complete the creation of the shared data source.

Create a Shared Dataset

Along with our shared data source, we want a shared dataset as well. Let's create one now.

1. Right-click the Shared Datasets folder in Solution Explorer and choose the Add New Dataset option from the context menu (Figure 16-18).

2. When the Shared Properties dialog window appears, type **DatasetvQuantitiesByTitleAndDate** into the Name textbox (Figure 16-19).

3. Use the Data Source dropdown box to select the DataSourceDWPubsSales Data Source (Figure 16-19).

4. Type the SQL code found in Listing 16-3 in the Query textbox.

   Listing 16-3. SQL Code for the DatasetvQuantitiesByTitleAndDateDataset

   ```
   SELECT
     PublisherName
   , [Title]
   , [TitleId]
   , [OrderDate]
   , [Total for that Date by Title]
   FROM vQuantitiesByTitleAndDate
   ```

5. Close the dialog window by clicking OK to complete the creation of the shared dataset.

Important: The view that the SQL code uses was created in Chapter 13, but if you do not have it already in your database, you can open the SQL script file `C:\_BookFiles\Chapter16Files\ViewAndStoredProcedureForChapter16.sql` and execute its code to create it.

Create a Basic Report

Next, we need to add a report to the project and configure it to display the report data in a tabular format.

1. Add a new report to the project by right-clicking the Reports folder in Solution Explorer and selecting New Item from the context menu (Figure 16-20).

2. When the Add New Item dialog window appears, select the Report icon, as shown in Figure 16-21.

3. Configure the Name textbox with the filename of `SalesByTitles.rdl`. Then click the Add button to add a new report to the project.

4. Once the Report Designer window appears, use the View menu to display both the Report Data and Toolbox windows (Figure 16-22).

5. In the Report Data window, right-click the Datasets folder and select Add Dataset from the context menu.

6. When the dataset Properties window appears, name the dataset **DatasetSalesQtyByTitles,** select the "Use a shared dataset" radio button, and select DatasetvQuantitiesByTitleAndDate, as shown in Figure 16-23. Finally, click OK to close the dialog window.

7. Add a table to your report by clicking the Table icon in the Toolbox and dragging it to the Report Designer surface (Figure 16-24).

8. In the new table, click the first column of the Data row. When the Smart Tag icon appears, click it to display the context menu. When the context menu appears, select the Title data field, and it will be added to the report table (Figure 16-25).

9. Repeat this process to add the OrderDate data field to the second column of the data row and to add the Total_for_That_Date_by_Title data field to the third column.

10. Use the Preview tab and review your current report design. It should look similar to the one shown in Figure 16-26.

11. Use the Design tab to change the names of the columns to Title, Order Date, and Sales Qty, as shown in Figure 16-27.

12. Add a Text Box to the report and type in **Sales Qty by Titles**, as shown in Figure 16-28.

13. Use the Format toolbar to format both the table and textbox to your liking.

14. Once you have completed your formatting, take a look at the report using the Preview tab. Continue adjusting it until you are satisfied with the look of your report.

In this exercise, you created a new SSRS project and added it to a current solution. You then added a data source, a data set, and a simple report to the project. Now we can deploy the report to the Reporting Services web service so that users can access the reports either through the SSRS web service or more appropriately through the SSRS web application.

Deploying the Report

To deploy the report to the SSRS web service, you need to configure the Visual Studio project appropriately. This can be done by accessing the property pages shown in Figure 16-31. To access this property sheet, right-click the Project icon in Solution Explorer and select Properties from the context menu.

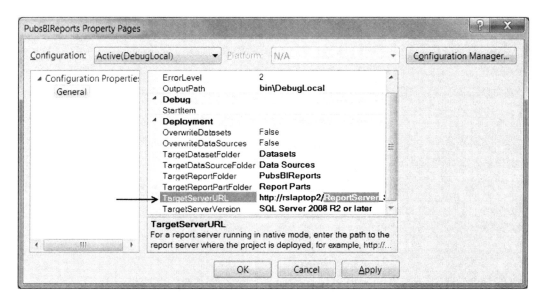

Figure 16-31. *Configuring the deployment options*

The most important setting in this list of settings is the Target Server URL. This configuration tells Visual Studio which SSRS web service to connect to. Because a company can have many different SSRS web services available (and it is assumed that developers will not necessarily have their own on their individual desktops), the property is left blank by default.

To add the appropriate URL to this textbox, type in the following (with no spaces):

- `HTTP:` including the colon, to indicate that you want to use the HTTP protocol

- `//<ServerName>` to indicate which web server you want to connect to

- `/<virtual directory>` to indicate which virtual directory on that web server you want to use

Of course, `<ServerName>` and `<virtual directory>` are placeholders for the actual names that you need to use. Here is an example of how the configuration is set up on Randal's laptop: `http://rslaptop2/ReportServer_SQL2012`.

■ **Tip** The properties' default settings should be sufficient for your needs at first. As you become more familiar with SSRS and how the web application is organized, you may want to come back and change things, such as folder names where the report objects are placed.

To deploy an individual report, right-click the report file in Solution Explorer and select Deploy from the context menu, as shown in Figure 16-32. This asks Visual Studio to send the individual RDL file to the web service. When this happens, the web service will shred the XML code found in the RDL file and place the corresponding programming instructions into the SQL Server SSRS databases.

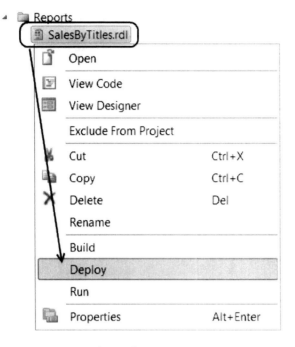

Figure 16-32. *Deploying the report*

When the report is deployed, SSRS will review the code in your RDL file and display an error report if required supporting objects are missing. For example, if we deploy the report we created in this chapter without first deploying the shared data source and shared dataset, SSRS will notice this discrepancy and display the error shown in Figure 16-33.

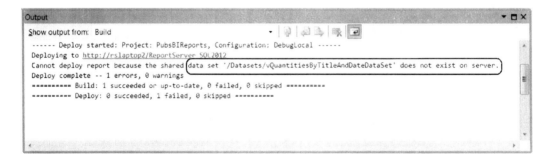

Figure 16-33. *An error in deploying the report*

The way to resolve this error is to deploy all the supporting objects before you try to deploy the reports. One simple way to do this is to use the Deploy option at the project level, which will attempt to deploy all the objects within the project simultaneously (Figure 16-34).

Figure 16-34. *Deploying the project*

After you deploy one or more files successfully to the SSRS web service, Visual Studio will indicate that the deployment succeeded in its Output window, as shown in Figure 16-35.

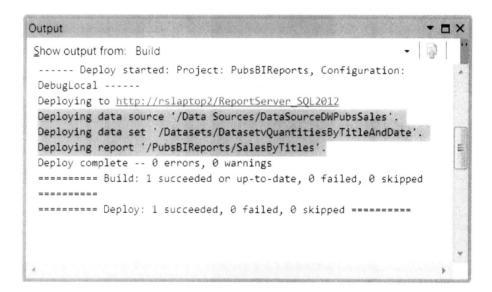

Figure 16-35. *Deployment success*

If you navigate to the report's SSRS web service, you can see links to three new folders (as shown in Figure 16-36):

- Data Sources
- Datasets
- PubsBIReports (our Visual Studio project name)

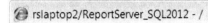

rslaptop2/ReportServer_SQL2012 - /

Sunday, May 13, 2012 4:37 PM	<dir>	Data Sources
Sunday, May 13, 2012 4:37 PM	<dir>	Datasets
Sunday, May 13, 2012 4:37 PM	<dir>	PubsBIReports

Microsoft SQL Server Reporting Services Version 11.0.2100.60

Figure 16-36. *Report folders on the ReportServer web service*

Clicking one of these links such as the PubsBIReports directory will display the objects within the subfolder. For example, in Figure 16-37, we have one report deployed within this folder called SalesByTitles.

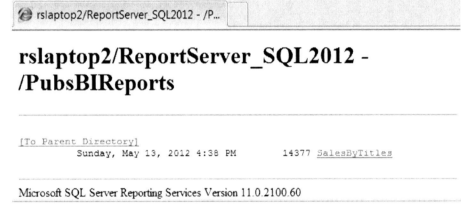

rslaptop2/ReportServer_SQL2012 - /PubsBIReports

[To Parent Directory]
              Sunday, May 13, 2012 4:38 PM   14377 SalesByTitles

Microsoft SQL Server Reporting Services Version 11.0.2100.60

Figure 16-37. *The SalesByTitles report is accessed from the PubBIReports folder.*

Managing the Report

Most users will not use the web service to access their reports. Instead, they will use the web application known as Report Manager. By navigating to the Report Manager website, you see the same folders displayed in a graphical user interface that is much less utilitarian than that of the web service (Figure 16-38). The user interface contains links and icons to the folders and a toolbar with standard functions such as configuring folders, creating data sources, or uploading RDL files.

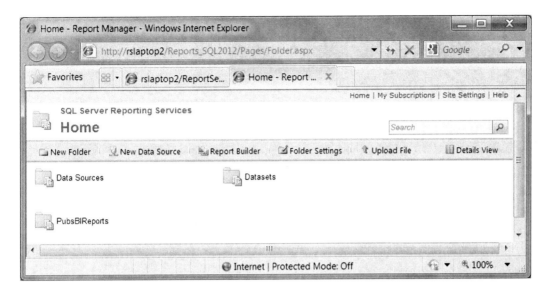

Figure 16-38. *Report folders on the Report Manager web application*

By clicking an individual folder, such as the PubsBIReports folder, you will see that the report is represented with an icon as well (Figure 16-39).

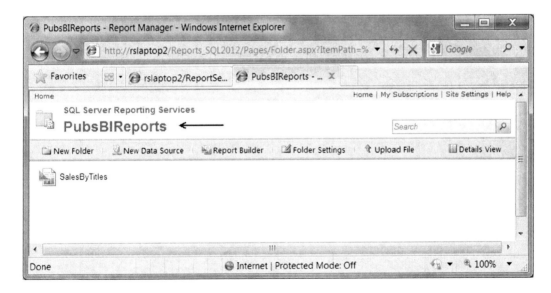

Figure 16-39. *Reports can be accessed in the Report Manager folders*

Clicking the Report icon will bring up the Report Viewer nested within the SSRS web service. Remember that it is a good idea to test the look of your reports in different browsers since they can display the report in different ways. Notice how in Figure 16-40 this report displays correctly when viewed in Internet Explorer or Firefox, but not at all in Chrome.

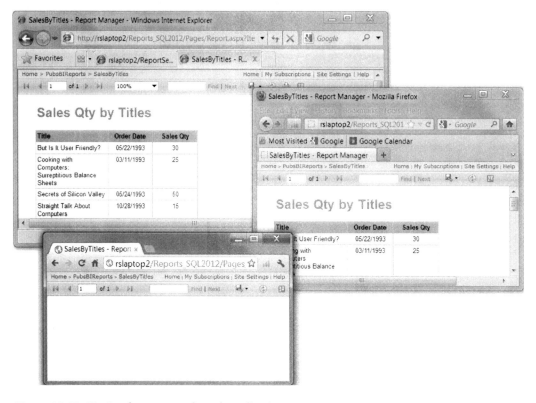

Figure 16-40. *Testing the report on the web application*

Moving On

SSRS is a fantastic tool for creating reports. It provides multiple ways to develop, view, and store reports. It also provides flexibility in the dispersal of the different components of SSRS across one or more machines and provides an administrator with opportunities to tune the reporting performance. You now have an idea of how the different components work together to provide a complete reporting solution.

We recommend you try your hand at creating your own reports for the Northwind database in the following "Learn by Doing" exercise.

LEARN BY DOING

In this "Learn by Doing" exercise, you create an SSRS project similar to the one defined in this chapter using the Northwind database. We have included an outline of the steps you performed in this chapter and an example of how this can be completed in two Word documents. These documents are found in the folder C:_BISolutionsBookFiles_LearnByDoing\Chapter16Files. Please see the ReadMe.doc file for detailed instructions.

What's Next?

This far, all of the reports you have made in this book have been relatively basic. In Chapter 2, we introduced creating a report using SSRS's wizard. In this chapter, we showed you how to create a report using the Toolbox and how to design it yourself. In the next chapter, we discuss using various Toolbox tools to create professional-looking reports.

If you are interested in more information about the administration of SSRS, we recommend the book *Microsoft SQL Server 2012 Reporting Services, 4th edition* by Brian Larson (McGraw-Hill Osborne).

CHAPTER 17

Configuring Reports with SSRS

In particular, all these templates and tutorials let students simply insert text and images, such as dragging and dropping information into a lab report outline, so students can now spend their time concentrating on the content instead of the form.

—Kathy Schrock

A BI solution extracts information out of data, and reports are the standard mechanism for displaying this information. As we saw in Chapter 16, creating a basic SSRS report is easy. But, creating a good SSRS report takes strategy.

Fortunately, Microsoft has given us a number of valuable features within SSRS that can help us create good and professional reports. In this chapter, we review a number of design features and see how they are used to configure your reports.

Creating a Report Template

A long tradition of distrust is centered around reports. This is mainly because reports on the same data can often come up with different information. There are two things that have proven effective in resolving this issue; designate a team dedicated to reporting, and provide a place for users to access official reports.

This does not necessarily mean that employees cannot ever make their own ad hoc reports from within their company; it simply means they have official reports at their disposal that have been well developed, tested, and validated. When done correctly, employees learn to respect your reporting solution as a legitimate and reliable source of information.

Standardization can provide some added legitimacy. And although standardization can be quite an elaborate endeavor , some very simple techniques exist that can help. Perhaps one of the simplest and most useful techniques is to create a report template. A report template sets the location of common elements that each report requires.

A basic report structure has three elements: the header, the body, and the footer. The header and footer are strictly defined in a report template, because they include the least dynamic parts of a report such as simple text, images, and company standards. For example, it is expected that most reports will have a title, a byline, and a creation date. And if the report contains more than a single page of data, it will have page numbers.

The body, however, can be loosely defined, because it contains dynamic content such as a results table or a series of charts. See the example in Figure 17-1.

Report: Sales by State and Store

Owner: RRoot

Created On: Monday, February 20, 2012

State	Store	Qty
CA	Barnum's	125
	News & Brews	90
	Fricative Bookshop	60
OR	Bookbeat	80
WA	Eric the Read Books	8
	Doc-U-Mat: Quality Laundry and Books	130

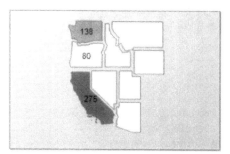

State	Total Qty	Store Contributions
CA	275	
OR	80	
WA	138	

Page Number: 1 of 1

Figure 17-1. *A typical report with header, body, and footer*

Microsoft's SSRS includes three basic templates called Report, Data Source, and Dataset (Figure 17-2). However, these basic templates do not provide a clear means of creating a custom report template. Not to worry; as you will see, you can create a custom report template by making a new report, configuring it to your standards, and making it available for other SSRS projects .

As we showed you in Chapter 16, you can add a new report to the project by right-clicking the SSRS Reports folder icon in Solution Explorer and selecting Add ➤ New Item from the context menu. In our example, when the Add New Item dialog window appears, we select the Report item, name the new report PubsBIReportTemplate, and click Add to create the report (Figure 17-2).

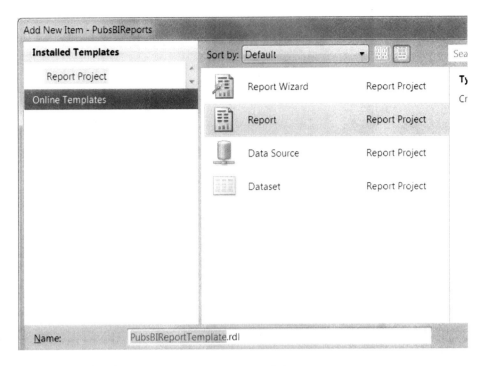

Figure 17-2. *Creating a new report to use as a template*

When a new report is first created, Visual Studio displays the Report Designer window in the center of the screen. If it is not displayed, double-click the report file (with the `.rdl` extension), using Solution Explorer, to force it to appear.

Adding a Header and Footer

Each report can optionally include both a header and a footer. However, an SSRS report does not include them by default. They must be manually added by accessing the Report menu in Visual Studio and selecting Add Page Header and Add Page Footer, as shown in Figure 17-3.

■ **Note**   The Report menu in the Report Designer window will not show on the Visual Studio menu bar if the report is not in focus. Once the report body has been opened and selected (clicked), the Report menu and its submenu items—Add Page Header and Add Page Footer—become enabled. This is true of most of the menu items in Visual Studio.

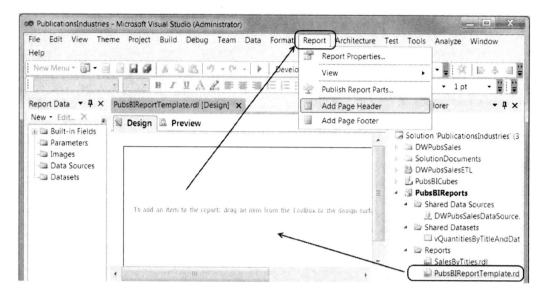

Figure 17-3. *Adding a page header and page footer to the template*

After you click the Add Page Header and Add Page Footer menu items, the designer displays the report in three sections: header, body, and footer (Figure 17-4). Click whichever section you intend to modify to set its properties. You can change miscellaneous properties such as the color, border, and size within the Properties window.

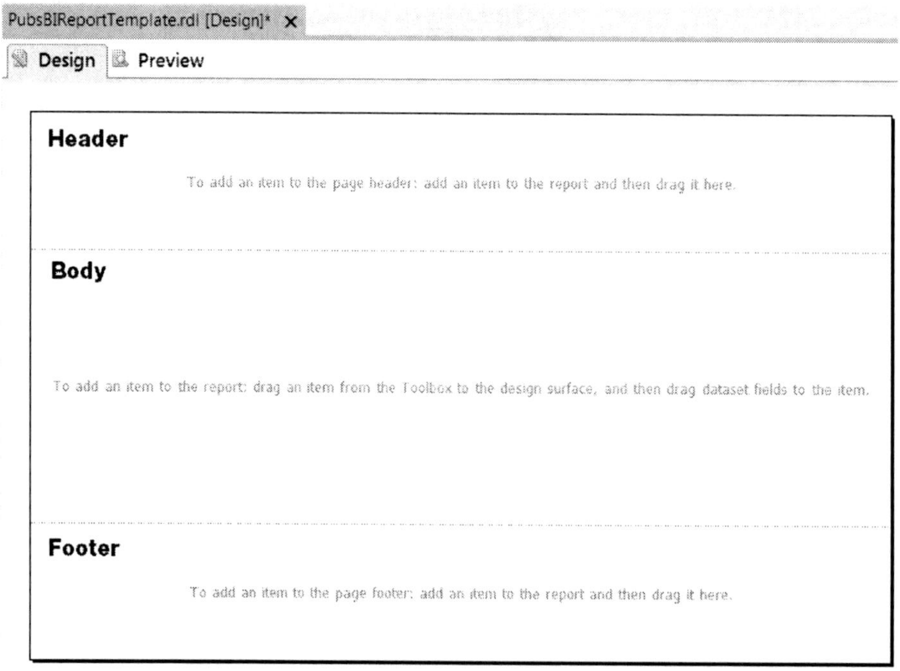

Figure 17-4. *A report displaying the header, body, and footer*

Setting Report Properties

In Figure 17-4, you can see that our report is now made up of three sections. Each is considered a component object within the report. And like the report itself, each component has its own configurable properties.

To configure the properties of the report object, click the background of the design surface outside of the report header, body, or footer, as indicated in Figure 17-5. To configure the header, body, or footer, click on their respective areas within the report.

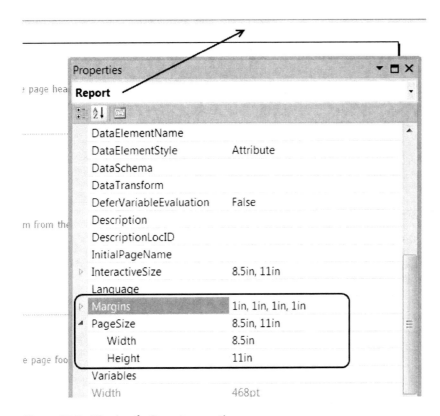

Figure 17-5. *Viewing the Report properties*

Page Size and Report Margins

Although there are many properties available for configuration, only a few are commonly modified. For example, the margins and page size are often configured to make sure that a report will fit on a specified paper size for printing purposes (Figure 17-5).

Most users will rely upon viewing reports from the SSRS websites, but printing is often required as well. Therefore, it is important to make sure that your reports stay within printable size.

By default, the standard SSRS template configures the page size to match 8.5 by 11-inch paper (Figure 17-5). However, most office printers support at least 11 by 17-inch paper. As such, you may consider this a safe configuration size.

Restricting your report width to a maximum of 11 inches may seem narrow, but remember that web pages are capable of scrolling down indefinitely and printers can print out many sheets. Therefore, content is not limited (as least vertically). Of course, the benefit of keeping the web page to this width is that printed reports will appear in the same format as they appear on the web page.

If you still feel that the 11-inch configuration is too restrictive, consider swapping the width and heights settings and printing the report in landscape layout. Printing the page in landscape mode provides an additional 6 inches to the possible width of the report. This usually works well in most circumstances, but keep in mind that it has become common for employees to use mobile phones and tablets for accessing web content, and that very wide reports may be problematic for these devices.

The report margin default configuration is 1 inch at top, bottom, left, and right for a new report using the standard report template. Note, however, that these margins are not indicated on the design surface.

Microsoft includes a ruler feature so that you can compare the scale of what you are seeing on your screen to the page size in inches. This ruler can be accessed by using the Report ➤ View ➤ Ruler menu item. Once the ruler appears in the design window, you will notice that the report body is only 6.5 inches wide, even though the Width property is set to 8.5 inches. The missing 2 inches are the left and right margins.

■ **Tip**   The ruler is a convenient tool for aligning report objects, but it can also be distracting, because it outlines the XY coordinates every time you move the mouse. If this is an issue, we recommend turning it on only during the precision layout work.

Designing the Header

To design the header section, add report items from the Toolbox and configure each item individually. In Figure 17-6, you can see that Microsoft includes many report items in the Visual Studio Toolbox, although the only report items that function in the header and footer are rectangles, textboxes, images, and lines.

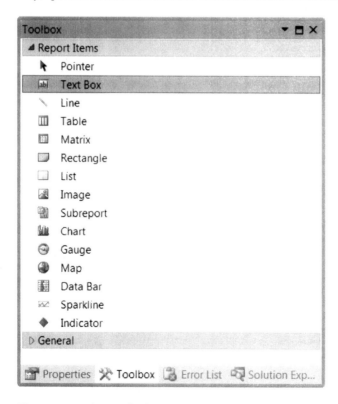

Figure 17-6. *The standard SSRS Toolbox report items*

Rectangles

A rectangle item is often used as background for other report items. It is also used as a parent container. By that we mean if you place report items within the rectangle, the rectangle will form a parent-child relationship between itself and the contained report items. Configurations to the parent rectangle item affect the child report items.

You can test this relationship by placing a textbox within a rectangle and then changing a property that they share, such as the background color. Changing the background color for the rectangle also changes the background of the textbox because of this parent-child relationship. Additionally, report items move with the rectangle if it is repositioned.

To place a rectangle within the header, access the Visual Studio Toolbox. If the toolbox is not currently displayed, you can force it to appear by clicking the View ➤ Toolbox menu item. Once you have access to the Toolbox, click the Rectangle item and drag it into the Header section of the Report Designer. You can then adjust the properties such as size and background color, as shown in Figure 17-7. The rectangle's properties can also be set using Visual Studio's Report Formatting toolbar.

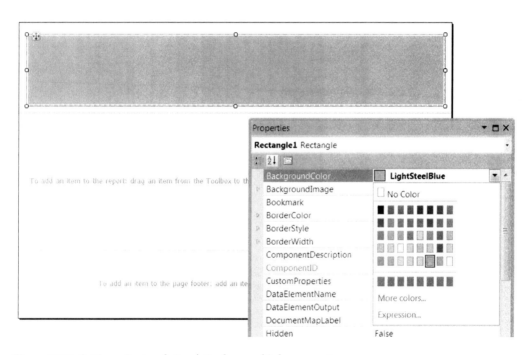

Figure 17-7. *Setting a Rectangle item's BackgroundColor property*

PROPERTIES, PROPERTIES, AND PROPERTIES

The Visual Studio Properties window (which usually resides on the right side of the Visual Studio environment) is the primary way to set most properties. But when working with the Report Designer, you will find that properties can also be set within various dialog windows, menus, and toolbars.

Some properties are set only in a dialog window or in Visual Studio's property window, but not both. Additionally, some properties are accessed in other UI components such as SmartTags.

In Figure 17-8, we see that the background property can be set in both the dialog window and Visual Studio's Properties window. Notice that the property setting name is different in each of these windows. BackgroundColor in Visual Studio's Properties window sets the same property as Fill in the dialog window.

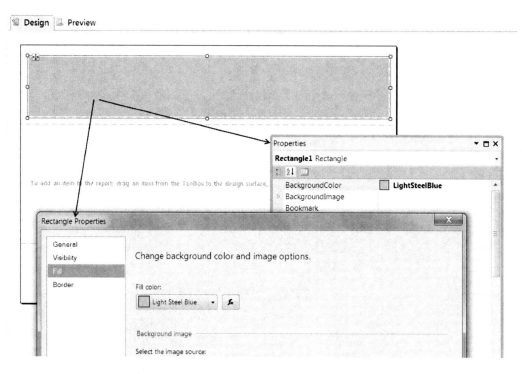

Figure 17-8. *The Properties window and the Rectangle Properties dialog window*

As you look for information about SSRS report configurations on the Web, be aware that Internet articles do not always include the method that is required to access the property being discussed. You may need to explore the various toolbars, smart tags, and dialog windows before you find the property you are looking for!

Textboxes

Textbox report items are the principal tool used for displaying text data in Reporting Server. Since SSRS textboxes do not allow users to enter data, they are similar to what is normally called a Label in most programming environments.

It is common for reports to contain dozens of textboxes. In fact, some report items actually contain a set of textboxes. For example, each cell of the SSRS Table report item contains an individual textbox and the same is true of the cells in an SSRS Matrix report item. If an SSRS table displays two columns and five rows, for example, that table will contain ten textboxes and the table report item becomes the parent of these textboxes.

Textboxes placed within a rectangle creates parent-child relationships. This is handy since rectangles containing a collection of textboxes can conveniently be configured as a unit.

Textboxes can also be configured collectively even when they are not associated with a common parent item. To configure them collectively, hold down the keyboard control button and click a textbox; then while holding down the control button, click one or more additional textboxes.

Once you have selected multiple textboxes, only one will have white sizing handles at the edges and corners (Figure 17-9). The textbox with the white sizing handles is considered to be the primary textbox, and any settings applied to the primary textbox also configure each of the secondary textboxes. Other SSRS Toolbox report items follow the same pattern.

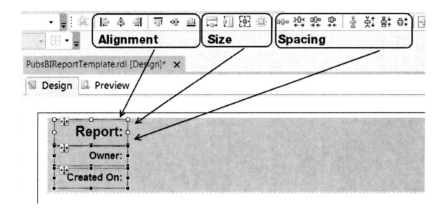

Figure 17-9. *Using the Layout toolbar to adjust your report items*

When a collection of textboxes is selected, their positions can be adjusted using the Layout toolbar (as shown in Figure 17-9). The toolbar is divided into three basic categories: alignment, size, and spacing. The left-align option will make all the highlighted textbox borders align on the left, and so on. The size option allows you to resize the secondary textboxes to the same height and width as the primary textbox. The spacing options set the spaces between the textboxes.

If you prefer, you can use the context menu for the layout. To display the context menu, right-click the collection of textboxes (or other report item controls).

To help avoid frustration in setting the layout, Microsoft also allows you to move the report items up, down, left, and right with your arrow keys on your keyboard. You can use the Ctrl+arrow key combination to move items in smaller increments.

■ **Note**   One of the more interesting, and confusing, aspects of textboxes is that they contain two report items. One item is layered on top of the other. The top layer is called a *placeholder*. We discuss how to handle placeholders later in this chapter.

Images

Images can help make reports look professional. Images are also used to signify report values in what is referred to as *dashboard* or *scorecard* reports. The *Image* report item allows you to place an image, from very large to very small, into your report. It also allows you determine how that image will be stored and retrieved.

One method of storing an image is to embed it within the report's RDL file. Doing so converts the image into a binary value, which is reassembled when the report is rendered. Images embedded within a single report are available only for that specific report.

Another option is to store the image within the Reporting Server database. Once again, the image must be turned into a binary value before it can be stored within a database table, but storing the image in the SSRS database means that it can be used by many reports.

A third option is to store the image file on a folder available to the web server hosting your reports. Like the database option, this allows many reports to use the same image. As a bonus, this option also makes the image available for any application that can access the file within its folder.

To add an image to a report, click the Image report item in the Visual Studio Toolbox and drag it to an appropriate location.

For example, in Figure 17-10 we are adding an image to the report header. Once you drag the image to the report, you will be presented immediately with the Image Properties dialog window. You can then configure the name for the image, provide a tooltip for accessibility, and, depending upon the image source, select the image using the Import button.

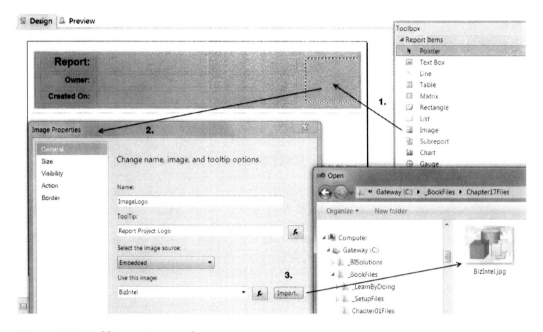

Figure 17-10. *Adding an image to the report*

The image source options are Embedded, Database, and External (which represents the RDL, database table, and folder options we mentioned). Because Embedded is the simplest option, many developers use this as a starting point in their initial design. Later, when the reports are uploaded to the production SSRS server, images can be switched over to use either a database or external image.

To use the Embedded option, locate the Image Source drop-down box and set it to Embedded (Figure 17-10). This causes the Import button to appear. Click the Import button to locate and select an image file on your hard drive. Then click the Open and OK buttons on their respective dialog windows to add the image to the design surface.

You may need to configure the image, such as adjusting the size to fit your report. To do so, click the image to display a border surrounding the image, and use the sizing handles that appear to adjust the size (Figure 17-11).

You can also set the image proportion to keep its original format or to adjust to the current report item size. The latter can distort the image, so be cautious when using this option.

■ **Tip** For your convenience, if you would like to use our example image logo for practice, we have placed a simple JPEG file called BizIntel.jpg in the Chapter 17 folder within the downloadable book files.

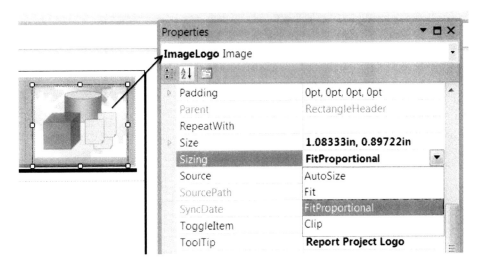

Figure 17-11. *Configuring an image on the report*

It is a good practice to give the image a name, other than the default Image1, 2, 3, and so on. For example, in Figure 17-11, we changed the name from Image1 to ImageLogo. This makes your programming code easier to read and maintain. This is true of all of your report items, as we discuss later in this chapter.

Lines

Lines are report items that display vertical, horizontal, and diagonal lines on your report. Each line's thickness, color, and opacity can be modified. Lines are often used to make the report easier to read or more visually appealing. In Figure 17-12 we used two lines to underscore the header.

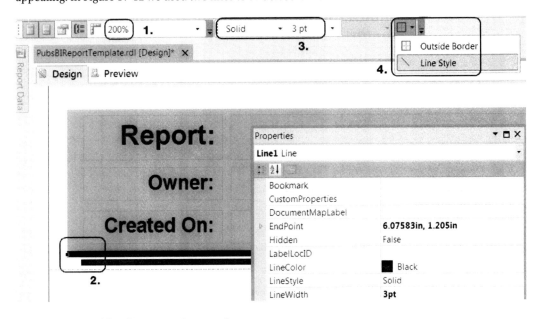

Figure 17-12. *Adding line report items to the report*

Drawing a line in your report is simple. Getting the line to align with other report items can be somewhat difficult. We developed this four-step process to make your alignment less painful. Each of these steps is also indicated in Figure 17-12:

1. Use the Zoom feature to enlarge the item. This will make the placement of a line report item easier. After a line item is placed on the report, highlight the line item and select Zoom. The zoom feature is located on the Report toolbar (number 1 in Figure 17-12).

2. Add a line from the Toolbox and place it where you want it. You can use the arrow keys to adjust the placement of the line within the report. The Ctrl+arrow keys move the line in smaller increments. The Shift+arrow keys increase or decrease the line size.

3. Set the thickness of the line by selecting the line; then access the Properties window and change the Line Width property. Another method is to use the Report Border toolbar (as indicated in Figure 17-12). Line sizes are set using a point system. One point is the thinnest.

4. Click the Line Style menu item on the Report Border toolbar to cause the new size to take effect.

■ **Important**   When changing the property of a line, the effect is not visible immediately. For the property to take effect, you must enforce the new change by clicking the Line Style button shown in Figure 17-12.

This may seem like a lot of steps just to place a line in your report, but with practice, it becomes a quick and simple process. Let's try what you have learned so far in the next exercise.

EXERCISE 17-1. STARTING A REPORT TEMPLATE

In this exercise, you start the process of creating a report template.

1. Open Visual Studio by clicking the Windows Start button and selecting All Programs ➤ Microsoft SQL Server 2012 ➤ Data Tools.

Important: You are practicing administrator-level tasks in this book; therefore, you need administrator-level privileges. The easiest way to achieve this is to remember to always right-click a menu item, select Run as Administrator, and then answer Yes to access administrator-level privileges. In Windows 7 and Vista, logging in with an administrator account is not enough. For more information, search the Web on the keywords "Windows 7 True Administrator and User Access Control."

2. Open the Publications Industries solution you have been working on throughout the book using the File ➤ Open Project/Solution menu item.

3. Locate and select the SLN file: `C:\_BISolutions\PublicationsIndustries\ PublicationsIndustries.sln`.

4. Click the Open button at the bottom of the dialog window. This opens the solution.

5. When Visual Studio opens, open the Publications Industries solution. You can do so

by using the File ➤ Open Project/Solution menu item. A dialog window will open allowing you to select your solution file.

6. Navigate to the SLN file `C:\_BISolutions\PublicationsIndustries\ PublicationsIndustries.sln`, select it, and then click the Open button at the bottom of the dialog window.

Create a New Report

1. Add a new report to the Reporting Server PubsBIReports project you created in Chapter 16. You can do this by right-clicking the Reporting Server project and selecting the Add ➤ New Item option from the context menu (Figure 17-13).

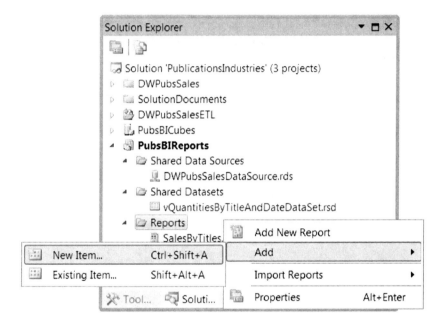

Figure 17-13. Adding a new report to the PubsBIReports project

2. When the Add New Item dialog window appears, select the Report option, name the report RDL file `PubsBIReportTemplate.rdl` as shown in Figure 17-2, and click Add to close the dialog window.

3. When the new report displays in the design window, click the report background, and then access the Visual Studio Report menu item. Select and set the Add Page Header and Add Page Footer menu options, as shown in Figure 17-3. The report should now look like the one shown in Figure 17-4.

4. Access the Visual Studio Toolbox and select a rectangle report item. Drag the rectangle to the header as shown in Figure 17-7.

5. Set the rectangle background color to a light color using the Properties window, as shown in Figure 17-8. We have chosen light steel blue in our example, although this will appear light gray in the figure.

6. Add three textboxes to the rectangle and align them as shown in Figure 17-9.

7. Right-click each textbox and use the Properties window to set the values to Report, Owner, and Created (Figure 17-9).

Note: You cannot set the text value in the Visual Studio Properties window. Instead, it must be accessed via the Properties dialog window.

8. Add three more textboxes to the report header and stretch them out to span approximately three-fourths of the report, as shown in Figure 17-10.

We use these textboxes later to display dynamic content, but for now you can leave them with no text.

9. Add an Image report item to the header on the right side of the report, as shown in Figure 17-10. When the Image Properties dialog window appears, click the Import button and navigate to the C:_BookFile\Chapter17Files folder. Select the BizIntel.jpg file, as shown in Figure 17-10.

10. Size the image to fit within the report header, as shown in Figure 17-11.

11. Add two lines beneath the report header rectangle, as shown in Figure 17-12. Set the top line to three points wide with a black line color. Set the second line to three points wide with a line color that complements the first line.

You can set the color scheme to anything you want. We have chosen grays and blues in our example. Our color scheme may seem depressing to you, but remember that both authors live in Seattle, so we like it this way. (Caryn says, not really; but let's go with it!)

12. Save your work by either clicking Save (the file floppy disk symbol on the toolbar) or using the File ➤ Save option in Visual Studio.

In this exercise, you created an SSRS report that will serve as a template for future reports. At this point, the controls within the header should be blocked-out and look similar to the one shown in Figure 17-14.

In the next exercise, we continue to improve on our current design by setting the programmatic names for each report item, using expressions to create dynamic content, and using placeholders to identify examples of the dynamic content.

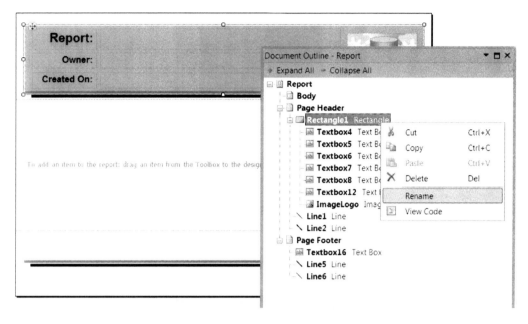

Figure 17-14. *Using the Document Outline window to rename report items*

Renaming Report Items

Whenever you are working with SSRS reports, it is a good idea to make sure you apply a name to every textbox, image, and label that you might interact with programmatically. It does not take very long to do, and it is considered good practice.

You can highlight each and every individual textbox and change the name of the Properties window, but this can be tedious. A better choice is to use the Visual Studio Document Outline window to change the names (Figure 17-14).

You can get to the Document Outline window using the menu items View ➤ Other Windows ➤ Document Outline.

To change the name of an item, highlight each item in the treeview and select Rename in the context menu. A cursor will appear, and you can start typing in the new name. If you try this yourself, you will see it is quite cumbersome at first, but with a little practice and some caution, it works pretty well.

■ **Tip** Do not press the Delete key while renaming a report item. For example, selecting the text "2" from Textbox2 does not delete just the text; it deletes the entire report item! Backspacing to remove text is OK, as well as highlighting the text and typing TextboxOwnerLabel in its place. If you make a mistake, the simplest remedy is to attempt to undo the change using the Ctrl+Z keystroke combination.

Figure 17-15 shows the new names, which will be helpful later as you perform these changes within the next exercise.

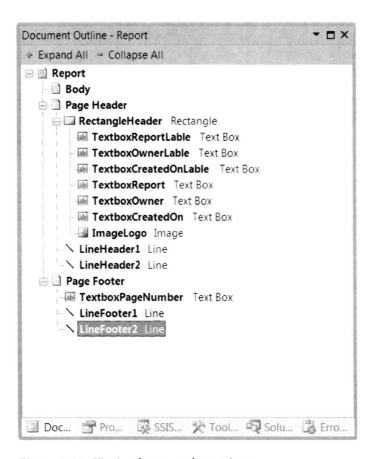

Figure 17-15. *Viewing the renamed report items*

Using Expressions

Chances are that you will have at least half of the textboxes on your report devoted to dynamic content. To create dynamic content, highlight a textbox and then select Expression from the context menu to access the Expression editing dialog window (Figure 17-16).

Figure 17-16. *Accessing the Expression dialog window with the context menu*

You can also right-click the same textbox and select the *Text Box Properties* option from the context menu. The advantage to this method is that you will have access to most of the other properties associated with that textbox.

When accessing the Text Box Properties dialog window, the first general page has a value drop-down box where you can type in a value, or more importantly click the ƒx button to access the Expression dialog window (Figure 17-17).

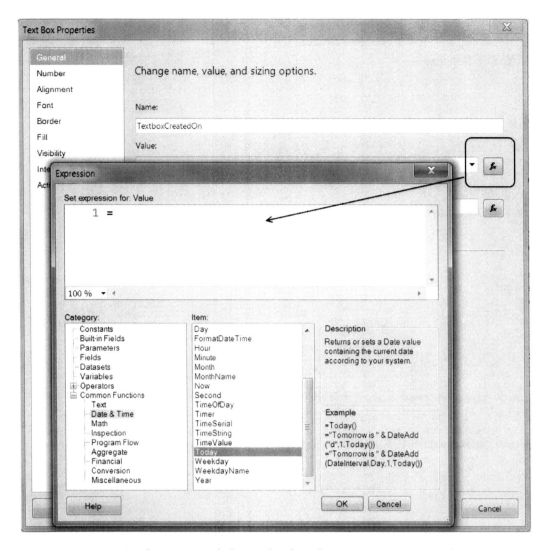

Figure 17-17. *Accessing the Expression dialog window from the Text Box Properties window*

The topmost section of the Expression dialog window is where you type in the expression language code. The expression language is similar to VB .NET and Excel scripting. There are some differences, but if you have experience in either of these, you should feel at home with this scripting language.

Textboxes evaluate dynamic content using the equal (=) symbol. For example, to display the word *hello* to the end user, type the expression = "Hello". To insert today's date into the textbox, use the equals (=) symbol followed by the function name or property that returns today's date.

Category and Item Panes

The Category and Item panes are located in the lower center of the Expression dialog window. Whatever you highlight in the Category pane determines the list of items in the Item pane (Figure 17-18).

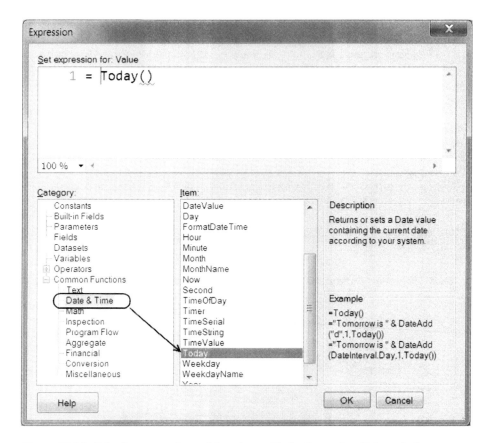

Figure 17-18. *Selecting properties and functions with the Expression dialog window*

SSRS reports offer several categories of report properties and functions. A commonly used property is Page Count, found under the Built-in Fields category. This property automatically evaluates to the maximum number of pages required to render all of the report data. This is information most often displayed in the footer of a report.

The Parameters category lists all the parameters currently defined in the report. SSRS parameters allow you to pass data into your SSRS report, similar to how parameter data is passed into a stored procedure.

The Fields category contains the report columns being returned from any queries in the report's datasets. For example, if we type in `Select TitleId, TitleName From DimTitles` within a DataSet query property, those two columns will be available under the Fields category.

■ Tip The Fields category is context sensitive. This means you see two fields listed while focused on an SSRS table that uses the DataSet query we just used as an example, but the same two fields will not be visible if you focus on a report item that does not use data from that same dataset.

Expression Editing Pane

An expression is created in the Expression Editing pane. When the Expression editing dialog window opens, it includes an equals (=) symbol within it. To create an expression, click after the equal sign to insert a cursor, and then type in the code. You can also add code to this pane by double-clicking a selected property or function in the Item pane. Unfortunately, this pane does not allow you to drag and drop your code here.

Microsoft has included a small description and example on the right side of the Expression dialog window. This is a really nice feature, except for one really annoying fact : the examples are not always accurate! For instance, the Today item example shows that typing in = Today() should do something; but in fact Today is a property and not a function, and it does not use parentheses. Therefore, although the examples are still pretty helpful when they work, you still have to keep an eye out and not take them too literally.

The Expression editing pane includes Visual Studio's IntelliSense technology to indicate when syntax is incorrect. Notice in Figure 17-18, when we typed in code = Today() (just as Microsoft's example indicated we should), the IntelliSense displays a red squiggly underline beneath the parentheses indicating that the syntax is not correct. So, at least there is that!

In Figure 17-19, the Today property is used correctly without parentheses, and the IntelliSense does not show an error. The Weekday function is also used correctly here, with parentheses.

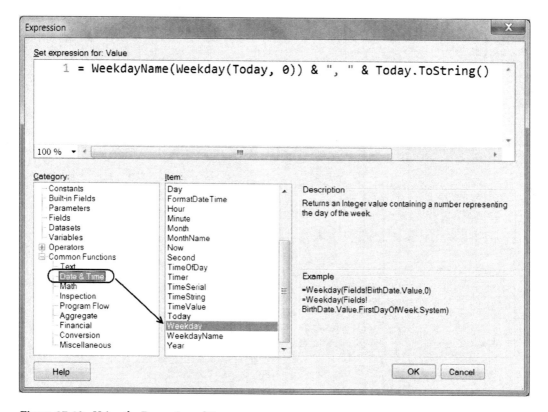

Figure 17-19. Using the Expression editing pane

■ **Tip** In Figure 17-19, entering the following code in the Expression editing pane creates the same output: = FormatDateTime(Today, DateFormat.LongDate).

Placeholders

When you create an expression for an SSRS textbox, the default behavior (before it is configured) displays <<Expr>>, as shown in Figure 17-20.

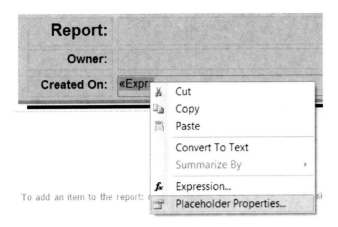

Figure 17-20. Accessing the Placeholder Properties option

In the early versions of SSRS, that was the end of it. In 2008 Microsoft got a wee bit fancier by putting an extra layer over the top the textbox called a *placeholder*. Placeholders are particularly helpful when displaying dynamic content, such as today's date. This placeholder acts like a label that is displayed to an SSRS developer as they design or modify a report. It is not visible to the end user.

Another benefit of a placeholder is that if one developer creates a report and another developer continues working on it later, the new developer will not have to right-click each textbox to find out what each textbox expression is supposed to do. (This also works if the first developer is rather forgetful.) In either case, the new developer (and the forgetful developer) can just look at the placeholder text. To access the placeholder, highlight the text within the textbox (not the textbox itself) and right-click it to bring up the context menu (Figure 17-20). The context menu contains a Placeholder Properties option.

■ **Important**   If you highlight the textbox rather than the text *within* the textbox, the Placeholder Properties option will not appear in the context menu.

Once you have clicked the Placeholder Properties menu option, a corresponding dialog window appears (Figure 17-21). Here you can type in whatever label you would like to appear to other developers. For example, we commonly use Ex: to indicate that the text is an example of what the end users are supposed to see, followed by some example text.

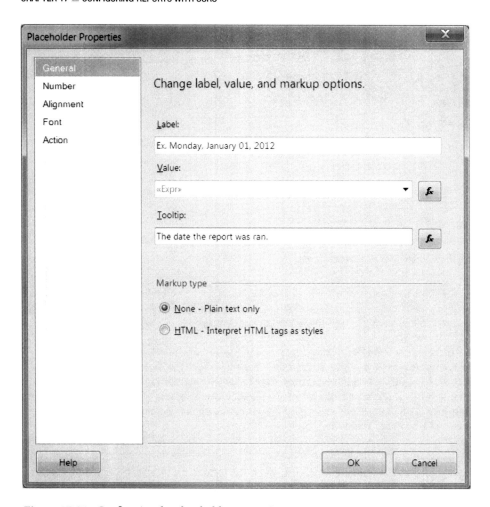

Figure 17-21. *Configuring the placeholder properties*

You can also add a tooltip using this dialog window. This tooltip, unlike the placeholder itself, will show for the end user. So, make sure the text is addressed to them.

Variables

Earlier, in the expression in Figure 17-19, we populated a textbox with the current date along with the day of the week. In the next example, Figure 17-22, we create a couple of report variables and use them to display dynamic content to the textbox.

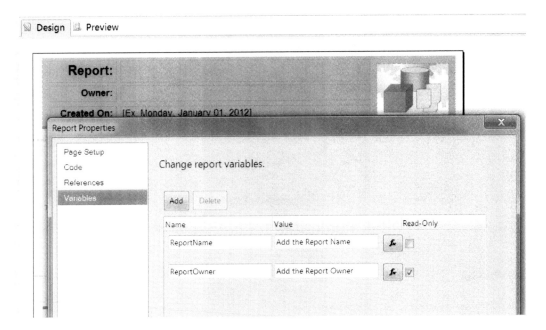

Figure 17-22. *Adding report variables*

To use a variable in your reports, configure each to include a name and a value. Then use the name in the Expression dialog window to display the associated value.

We highly recommend using variables for textboxes that normally display static report data. This may seem counterintuitive, but it makes changing static report data much easier. For instance, if you ever need to update several static values within a report, you can open the Reports Property window and adjust all the values at once using the interface shown in Figure 17-22, rather than having to click all of the different textboxes and adjust each one's content.

Additionally, textboxes cannot be locked into place once you position them. If you inadvertently click a particular textbox and move the mouse just a little, you may find that the textbox gets bumped. This can be really frustrating. But, by using report variables, you change the text in those textboxes without clicking the individual textboxes, risking moving them unintentionally.

In Figure 17-22, we have created two report variables. Now, notice that these same variables appear in the Expression window in Figure 17-23. Double-clicking the variable will add it to the Expression editing pane.

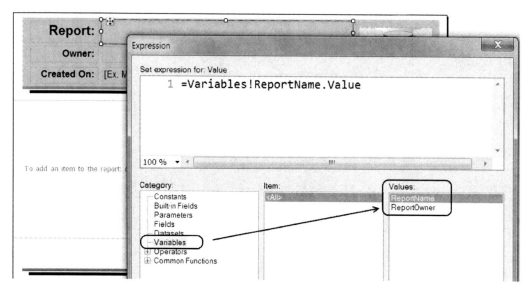

Figure 17-23. *Using report variables as expressions*

Completing the Header

Once you have completed configuring the dynamic content in the report textboxes and filled in the placeholders, the report header should look similar to Figure 17-24.

We keep our examples simple for instructional purposes, but in real life you often have up to 10 or 15 different bits of information in the header such as the date that the data was refilled in the data warehouse, the name of the data analyst who is responsible for validating the report, and so on.

■ **Tip** When Randal was working at Microsoft, each report was assigned to a data analyst who was responsible for analyzing and interpreting the report data as well as making sure that the report was accurate and in line with the Microsoft business model. This method worked very well to ensure that the end users were able to extract accurate information from the BI solution.

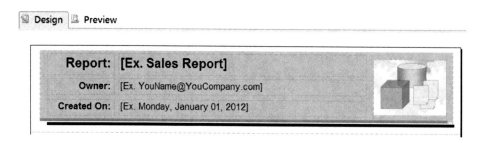

Figure 17-24. *The design view of the completed header*

The text shown in Figure 17-24 is placeholder text and is seen by developers. The screenshot in Figure 17-25 is what the end user sees. The reason the current report displays "add the report name" and "add the report owner" in Figure 17-25 is because the developers (that's us in this case) never went back and changed the report variable values.

In this example this is intentional, because we are creating a template, not an end-user report. Each developer who starts with our template needs to modify their report to include the report name and indicate the owner of that particular report.

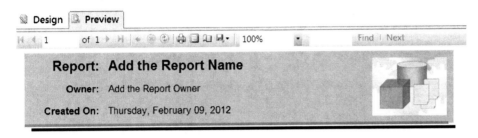

Figure 17-25. *The preview view of the completed header (template version)*

Configuring the Footer

Footers typically do not contain much information. The most common items shown in a report footer are page numbers along with a line or two for decoration.

Figure 17-26 shows an example of adding a couple of lines and a textbox to the footer and then setting the textbox expression to the current page number of the total page numbers. This is a built-in field that comes with Reporting Server. The description on the right side of the window indicates where a particular built-in field item can be used (Figure 17-26).

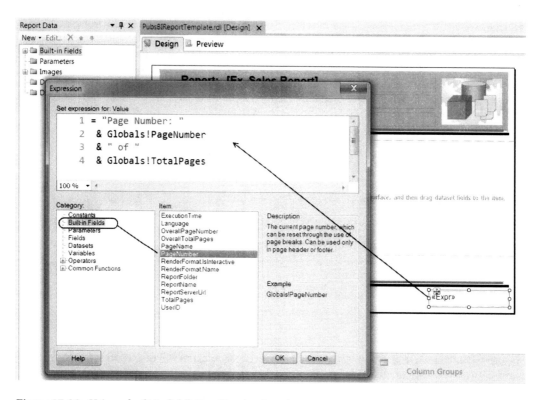

Figure 17-26. *Using a built-in field's PageNumber item in an expression*

Once an item has been added to the footer, the next step is to adjust the footer size. In our example, we adjusted the length and position of the lines to match the header. We then changed the footer height to take up less space (Figure 17-27).

Although not required, we often add one or two cosmetic features, such as horizontal lines, to the footer. In our example, we placed two line items within the footer and colored them to complement similar lines in the header. Figure 17-27 shows the final look of our template.

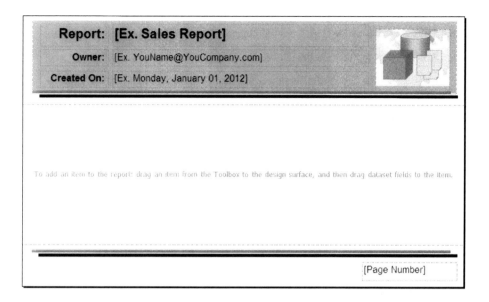

Figure 17-27. *The completed template*

In the real world, you will likely need more than one template. You may need one for portrait and the other for landscape page layouts for printing. You also may have different report templates for different projects or departments. We think you get the idea, so let's complete the template we started in the next exercise.

EXERCISE 17-2. COMPLETING A REPORT TEMPLATE

In this exercise, you finish configuring the report template that we worked on in Exercise 17-1. In addition, you configure different options within the template such as renaming the report items, creating report variables, configuring expressions, configuring placeholders, and configuring the footer.

Renaming the Report Items

Report items should be renamed before they are worked with programmatically. Let's do that now.

1. Access the Document Outline window using the menu items View ➤ Other Windows ➤ Document Outline.

2. When the Document Outline window appears, right-click each individual report item and select Rename from the context menu, as shown in Figure 17-14.

3. Rename all the objects as shown in Figure 17-15.

Creating Report Variables

We use a couple of report variables in the expressions in the next section. So, we first need to create those variables.

1. Highlight the background of the report design surface to access the Visual Studio Report menu. This should force the Report menu to appear (Figure 17-3).

2. In the Visual Studio Report menu, click the Report Properties menu option.

3. When the Reports Properties dialog window appears, click the Variables page on the left side of that screen (Figure 17-22).

4. Add two variables as indicated in Figure 17-22 and Table 17-1.

Table 17-1. *Report Variable Settings*

Name	Value
ReportName	Add the Report Name
ReportOwner	Add the Report Owner

Configuring Expressions

With the variables in place, we can now use them as well as our expressions. Follow the next few steps to create the various expressions used within the report header and footer.

1. Configure the value property of each of the four text boxes on our report by right-clicking each of the three empty textboxes in the header and right-clicking the single textbox in the report footer. Select the Expressions option from the context menu for each (Figure 17-16).

2. When the Expressions dialog window appears, configure the expression for each of the three textboxes, as indicated in Table 17-2.

Table 17-2. *Expression Code for Report Items*

Textbox Item	Expression
TextboxReportLabel	`= Variables!ReportName.Value`
TextboxOwnerLabel	`= Variables!ReportOwner.Value`
TextboxCreatedOnLabel	`= FormatDateTime(Today, DateFormat.LongDate)`
TextboxPageNumber	`= "Page Number: "`
	`& Globals!PageNumber`
	`& " of "`
	`& Globals!TotalPages`

Next we include placeholders over the top of each text box value.

1. Right-click each of the textboxes and configure the placeholders, as shown in Table 17-3.

Table 17-3. *Placeholder Code for Report Items*

Textbox Item	Placeholder
TextboxReportLabel	Ex. Sales Report
TextboxOwnerLabel	Ex. YourName@YourCompany.com
TextboxCreatedOnLabel	Ex. Monday, January 01, 2012
TextboxPageNumber	Page Number

(For an example of how to access the placeholder properties, see Figure 17-20. For an example of how the Placeholder Properties dialog window appears, see Figure 17-21.)

2. Verify that the report works as expected by using the Preview tab.

3. Save the changes to the new template and leave it open for use in the next exercise.

At this point, the report template is configured, and the template is complete. We use this template in the next exercise.

Saving the Report Template

Once a template is created, its RDL file should be placed somewhere easily accessible. Report templates can be stored either locally or remotely. The remote option involves placing the file in a shared folder on a network server. The local option places the file in a folder on your computer's hard drive.

Although you might expect otherwise, we do not recommend using Visual Studio's *Save As* menu option, because while it will save a copy of your new template to a new location, the template will still be a part of your current SSRS project, when in reality you want it to become a separate and independent RDL file !

Instead, we recommend copying the file to its new location using Window Explorer. To do so, copy and paste the file from your original SSRS project to either a local folder or a share on a network server. Now the RDL file is no longer a part of its original Visual Studio solution and is ready to be used as a template for other solutions.

■ **Note**   We refer to this independent RDL as a template, but remember that it is simply a standard RDL file. Not a true template—yet!

Using Network Templates

Saving a template remotely involves creating a network share and placing the template file in the network shared folder. This method makes it easy for multiple report developers to access the same template from many computers.

To use a remote template in an SSRS project, right-click the Reports folder in Solution Explorer and navigate to the Add ➤ Existing Item context menu option. This will bring up a dialog window (Figure 17-28).

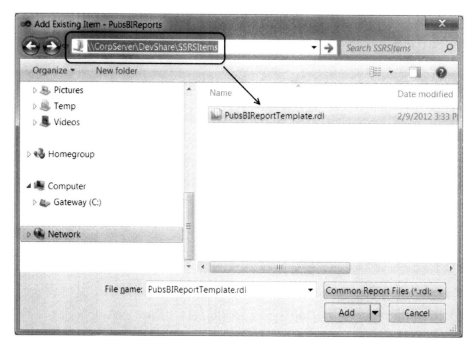

Figure 17-28. *Accessing a folder on a network share*

Type in the path to the network share such as **<name of the server>****<name of the share>****<name of a subfolder>**. For example, in Figure 17-28, we have used the path \\CorpServer\DevShare\SSRSItems. Once you click the Add button, a copy of your template RDL file will be placed in your SSRS project along with your other report files, but the original file is left in place for other developers to use.

After you have a copy of the .rdl file in your SSRS project, rename it to something that describes the new report. You can do so by right-clicking the copied report file and renaming it using the Rename option in the context menu.

Using Local Templates

Using a template from a folder on your local computer works the same way. Right-click the Reports folder in Solution Explorer, navigate to the Add ➤ Existing Item context menu option, and select the path to the folder on your own computer, as shown in Figure 17-28.

Another option is to also copy the RDL file into a special folder that Microsoft created for Report Project templates. The advantage of using this predefined folder instead of using your own custom folder is that the files in the predefined folder are displayed as report project items in Visual Studio's Add New Item dialog window (Figure 17-30).

To use this feature, copy an RDL file from its location (from a network share or custom local folder), and paste it into the predefined ReportProject folder, as shown in Figure 17-29.

Figure 17-29. *Placing the report template on a local computer*

The path to the Template folder computer varies for each computer based on the operating system, but on a Windows 7 (64-bit) system, the path is `C:\Program Files (x86)\Microsoft Visual Studio 10.0\Common7\IDE\PrivateAssemblies\ProjectItems\ReportProject`.

Once the report is placed in Microsoft's predefined ReportProject folder (instead of just a custom folder you created), the report will show up as a Visual Studio template when creating a new reporting item (Figure 17-30). And now you have a true custom template!

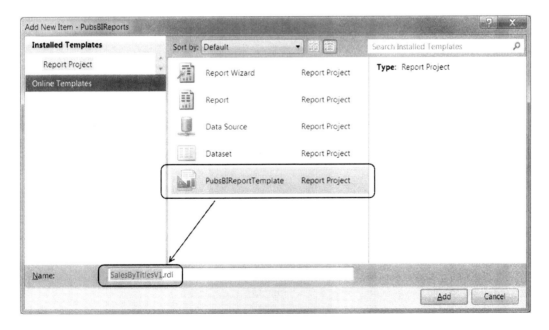

Figure 17-30. *Using the local template to make a new report*

Each time you right-click the Reports folder in Visual Studio's Solution Explorer and select Add New Item from the context menu, the new report template will be displayed as an icon within the Add New Item dialog window.

You can now use your custom template just as you would the use the default templates that come with SSRS. Select the new template, type a name for the new report in the Name textbox, and click Add. In our example, this means that the new report starts off with both the header and footer preconfigured! Therefore, the next step is to fill in the report body (Figure 17-31).

For example, we can now produce our new version of the SalesByTitles report by adding dataset and report items to the report body, just as we did in Chapter 16 (Figure 17-30). We would need to adjust some of the values in the header using the Report Parameter option and then test and deploy the report to the SSRS websites.

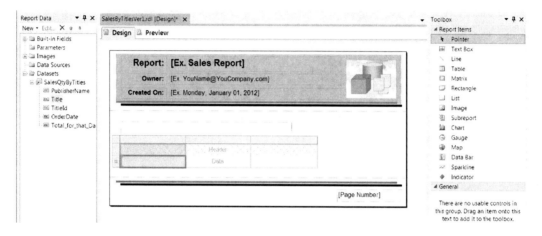

Figure 17-31. *Adding content to the body of the new report*

This is a common way to develop SSRS reports. Let's try it in the following exercise.

EXERCISE 17-3. USING A REPORT TEMPLATE

In this exercise, copy the finished template from your Visual Studio project into Microsoft's predefined ReportProject folder on your local computer, and configure the report variables that control the data displayed in the header. Then, use the template you created in this exercise to create a new version of the SalesByTitles report.

Copying the Report Template

First the RDL file must be located and copied, before it can be pasted into its new location and used as a template.

1. Open the `PubsBIReportTemplate.rdl` file in the Visual Studio designer by right-clicking the file and selecting View Designer from the context menu.

2. After the file opens, right-click the (Design) tab to access the context menu, as shown in Figure 17-32.

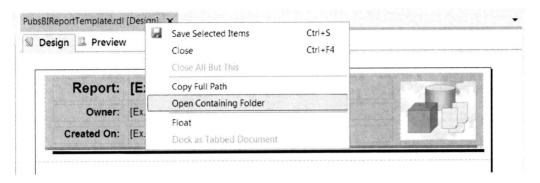

Figure 17-32. *Adding content to the body of the new report*

3. Click the Open Containing Folder menu option and wait for Windows Explorer to
   open the folder containing the RDL file.

4. Locate the `PubsBIReportTemplate.rdl` file, and right-click the file. Then select
   Copy from the context menu.

Adding the Template to the ReportProject Folder

Now that you have a copy of the RDL file in the computer's clipboard, paste it into the appropriate location.

1. Locate the ReportProject folder using Windows Explorer (Figure 17-29).

The path to this location may vary; however, on a Windows (64-bit) operating system, the path should be as
follows:

- `C:\Program Files (x86)\Microsoft Visual Studio`
  `10.0\Common7\IDE\PrivateAssemblies\ProjectItems\ReportProject`

On a Windows (32-bit) operating system, the path should be as follows:

- `C:\Program Files\Microsoft Visual Studio`
  `10.0\Common7\IDE\PrivateAssemblies\ProjectItems\ReportProject`

Tip: If you are using a different Windows operating system and are unable to find the location on your own,
you may have to search the Internet to find the appropriate path for your computer.

2. Once you have located the proper folder, paste the file into the ReportProject folder
   and close Windows Explorer (Figure 17-29).

Creating a New Report Using the PubsBIReport Template

Now that you have the file in the proper location, you can use it to create a new report.

1. In Visual Studio's Solution Explorer, right-click the Reports folder and select
   Add ➤ New Item from the context menu, as shown in Figure 17-13.

Remember, this is *not* the same as selecting the New Report option that launches the Report Wizard.

2. When the Add New Item dialog window appears, select the PubsBIReportTemplate icon from the center of the dialog window, as shown in Figure 17-30.

3. In the Name textbox, name the new report SalesByTitlesV1.rdl and click Add (Figure 17-30). This opens the new report in the Report Designer.

Configure the Header

Now you have a new report based on the custom report template, but the header data needs to be configured.

1. With the Report Designer open, right-click the background area, just outside of the report itself, to access the context menu shown in Figure 17-33.

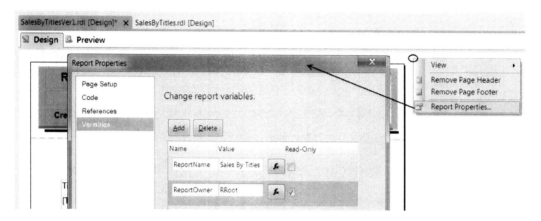

Figure 17-33. *Adding content to the body of the new report*

2. When the context menu appears, click the Report Properties option to display its dialog window.

3. In the Report Properties dialog window, access the Variables page and configure the values of the two exiting variables to Sales by Titles and your own name (similar to how it is shown in Figure 17-33).

4. Click OK to close the dialog window.

Recreate the SalesByTitles Report

Next we add content to the report within the report body. We use the same steps from **Chapter 16**. (For your convenience, we have included those steps here and refer to the figures from the previous chapter):

1. With the Report Designer open, use the View menu to display both the Report Data and Toolbox windows (Figure 16-22).

2. In the Report Data window, right-click the Datasets folder and select New from the context menu.

3. When the Dataset Properties window appears, name the dataset DatasetSalesQtyByTitles, select the "Use a shared dataset radio button," and select the DatasetvQuantitiesByTitleAndDate dataset, as shown in Figure 16-23.

4. Add a table report item to your report by clicking the Table icon in the Toolbox and dragging and dropping it onto the Report Designer (Figure 16-24).

5. Click the first column of the data row in the table. When the smart tag appears, click it and select the Title attribute (Figure 16-25).

6. Click the second column of the data row in the table. When the smart tag appears, click it and select the OrderDate attribute (Figure 16-25).

7. Click the third column of the data row in the table. When the smart tag appears, click it and select the Total_for_That_Date_by_Title attribute (Figure 16-25).

8. Use the Format toolbar to format both the table and textbox to your liking.

9. Once you have completed your formatting, take a look at the report using the Preview tab. Continue adjusting it until you are satisfied with the look of your report. As an example, the report we created with the new template is shown in Figure 17-34.

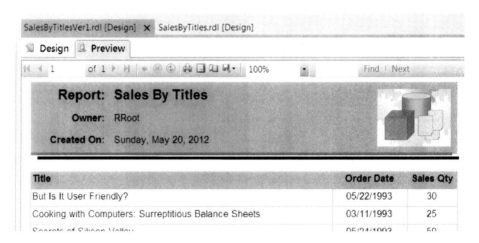

Figure 17-34. *The new report after the authors' formatting*

You have successfully created a new report using your custom-made reporting template. Now you can continue to make other reports using this template to become more proficient at reporting development using SSRS.

Moving On

In Chapters 16 and 17, you learned how to create an SSRS project, a shared data source, a shared dataset, a simple tabular report, and a report template. That was a lot to learn, but there is still more to go.

SSRS is one of Randal's favorite subjects, partially because he worked with SSRS at Microsoft for several years -- but mostly because it is fun to create reports. If you feel the same way and would like to learn more, we have provided many demonstrations on our website at www.NorthwestTech.org/SSRSDemos. You will find several examples of SSRS reports using the various SSRS report items in the Toolbox (Figure 17-35), as well as tips and tricks on how to develop and manage SSRS reports that we hope you find useful.

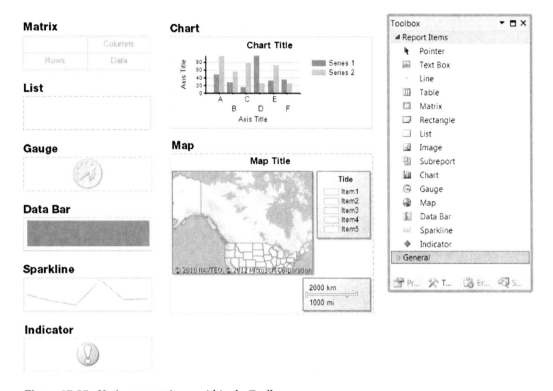

Figure 17-35. *Various report items within the Toolbox*

LEARN BY DOING

In this "Learn by Doing" exercise, you create an SSRS report solution similar to the one defined in this chapter using the Northwind database. We included an outline of the steps you performed in this chapter and an example of how the authors handled them in two Word documents. These documents are found in the folder C:_BISolutionsBookFiles_LearnByDoing\Chapter17Files. Please see the ReadMe.doc file for detailed instructions.

What's Next

So far our book has walked you through the various stages of a business intelligence solution. Currently we have a SQL Server–based data warehouse, an ETL process using SSIS, an SSAS cube, and reports using both Excel and SSRS. At this point, we are nearing the end of our cycle and are almost ready for the release.

As indicated in Figure 17-36, we have yet to test and tune the solution or go through a formal approval and release process. The next two chapters cover these final topics.

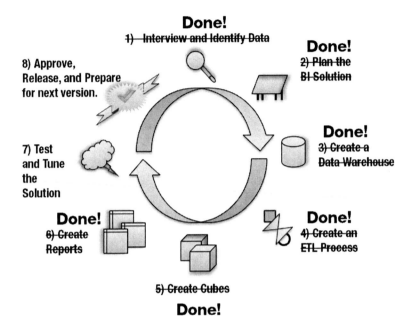

Figure 17-36. *Our current progress so far!*

In the meantime, for further information on creating reports using Reporting Services, we recommend the following book:

Microsoft SQL Server 2012 Reporting Services 4ed.
By Brian Larson
Publisher: McGraw-Hill Osborne Media
ISBN-10: 0071760474

CHAPTER 18

Testing and Tuning BI Solutions

Why, a four-year-old child could understand this report. Run out and find me a four-year-old child. I can't make head nor tail out of it.

—Groucho Marx

As you have seen thus far, the purpose of this book is to introduce concepts and practices that are commonly found in BI solutions. We have organized the process of creating a BI solution into eight steps and we are finally approaching the end of these. The practice of testing and tuning constitutes the next step in our process. This arguably is the most important step, because a BI solution is worthless if it is not functioning properly.

We begin by ascertaining what testing and tuning mean, in regards to a BI solution. Although there is no simple answer, it is important to analyze the basic objectives involved with this process. For testing, our goal is to validate the current solution through objective verification. For tuning, our goal is to provide possibilities for enhancements by benchmarking the current performance, identifying poorly performing components, and recommending improvements.

Although each objective is slightly different, they are complementary. In this chapter, we examine testing and tuning examples and discuss how these subjects supplement your BI solution.

Testing the BI Solution

Testing the BI solution involves documenting what the BI solution should contain, verifying that those items exist, and identifying possible areas of improvement.

When we began our example BI solution, we determined its contents using an Excel spreadsheet. Now that the solution has been created, we could pass the Excel spreadsheet onto the test team for validation. The test team would then go through each individual column, attribute, or configuration noted in the documentation and verify that what was planned has indeed been implemented.

Testing is an important part of a BI solution, but unfortunately it is also one that is all too often overlooked. To give perspective to its relevance let's consider an analogy.

Consider the construction industry. Before building inspectors became common, the quality of buildings and homes depended upon the builder rather than the original architect. If the builder did not follow the architectural blueprints correctly, the building might be unsafe. Of course, this would not always be true; expert builders could actually improve upon the architecture. Unfortunately, the unsafe outcome was common enough that most cities and states adopted the use of building inspectors to ensure that houses and office buildings were built to the specifications of the architect.

A tester is like a building inspector who provides a mechanism for ensuring the quality of whatever software you create. Their job, like the building inspector, is to review the blueprints and identify inconsistencies and faults.

Ideally, the test team would have been brought in during the architectural phase to give input, because just like an experienced building inspector, they may be able to point out possible problems that have been overlooked by the architect. The more experienced the test team is with this type of software, the more useful their input is during the architectural phase of your BI solution.

In most cases, BI solutions start without much input from testers. But, an effort should be made to formalize an ongoing test process so that the test team can become more experienced with the solutions you are making. Doing so will improve each version as it iterates through the BI solution process.

Validation

When developers think of testing, they often think of a person or a piece of software that examines the input versus the output to ensure validity. For a BI solution, the input is the values inserted into the columns of the data warehouse that are subsequently processed into the SSAS cubes. The output is the information reported via Excel and Reporting Services.

The tester's goal is to ensure that an appropriate output is always obtained. This validation is the core of the testing process.

This is not always as straightforward as it sounds. How are the testers to know what a valid output of a given input is? Is it implicit through the type of data that is used? Is the output explicitly indicated with some sort of documentation that can be compared to the results?

Many output results can be determined by examining the values that were input into the columns within a database, but that is not always the case. Let's look at a couple of examples.

If the input is an author's first and last names and the output is a concatenation, the operation would change the author's first and last names into one singular name. Incorrect output might display as follows:

- Leaving no space between the first and last names

- The last name preceding the first

- The first or last name missing

In these examples, your test team should be able to easily establish what the input and output should be and then verify what is needed by providing the correct information when the reports are generated.

Now let's think of something a little bit more difficult. What about a complex computation that provides statistical deviations against sales data? In a situation like this, the equation can consist of three or four numerical values along with a number of operations. The output would be a specific value that is associated with the algorithm defined by the operation for this calculation. In Figure 18-1 we show an example expression used to calculate standard deviations.

$$\sigma = \sqrt{\left[\frac{\sum d^2}{n-1}\right]}$$

Figure 18-1. *Notation for a standard deviation expression*

If it looks confusing to you, you are not alone. It certainly looks Greek to us (pun intended). For most of us, it has been many years, if ever, since we have worked with advanced mathematics. And although we could go back and relearn this calculation, it may not be time well spent.

Instead, what we need is an example that can be used to test the calculation. To get one, have someone who is proficient at performing these calculations create a set of inputs and outputs. For instance, they provide three or four examples as *test cases*. The test team then uses each of the provided inputs to create outputs using the BI solution and subsequently verify that the proper output always occurs.

Of course, this means that more documentation must be generated in order for the test team to efficiently do their job. This can become an issue as management tries to coordinate the documentation effort in conjunction with the testing effort. And although we discuss documentation further in the next chapter, for now keep these points in mind:

- Simple documentation is better than no documentation.

- Use test case documentation even on small BI solutions.

- Provide input and output examples for both simple and complex operations.

TESTING FOR SMALL SOLUTIONS

Often, something as simple as an Excel spreadsheet can prove almost as useful as professional testing software. For example, you can create a spreadsheet that maps the test cases for your team to validate (Figure 18-2). Using this spreadsheet as a guideline, your testers can review your inputs and expected outputs, create reports or queries using the data warehouse or cube, and compare the actual outputs to the expected outputs. After each test, the testers note the results in a pass/fail column along with any additional useful notes.

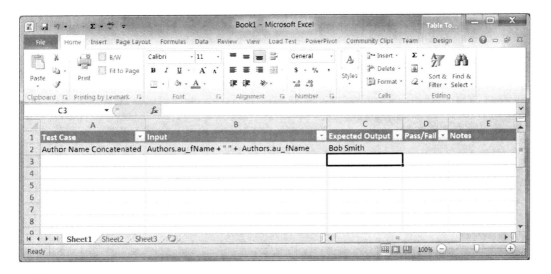

Figure 18-2. *A simple way to track test cases with Excel*

This is a very low-tech option, but remember, *something* is much better than *nothing*. All too often testing is left out of the process because it is believed to be too costly to implement on small BI solutions. From our example you can see that this is a mistaken assumption. Keep in mind that once the credibility of your solution is lost because of validation errors, it is difficult if not impossible to regain. So, the cost of not testing may be much higher than you think!

Objective Verification

Objective verification is possible only if the objectives of the BI solution have been clearly defined. In a perfect world, the development process clearly identifies the goals at the beginning of the BI solution. In reality, documentation will likely be rather sketchy and incomplete. Most team members realize that good documentation will make the test teams' job that much more effective, but time and resources can limit the chances of this ever taking place.

As the versions of your BI solution progress, lessons are learned, and the documentation is likely to be improved upon. This is a management task that needs to be identified and coordinated to be successful.

One effective way to create good documentation is to add one or more technical writers to the team to help manage and improve the documentation process. Tech writers are invaluable on larger solutions, but even smaller solutions can benefit from their expertise. Some companies hire freelance technical writers for smaller solutions to help keep costs down.

Here are some helpful suggestions to keep consistency and verify objectives:

- Create a standard template for documentation.

- Make sure the standard template is easy to follow.

- Make sure the standard template does not take long to complete.

- Make sure that developers update the documents with lessons learned.

- Have testers review the documentation before implementation begins, where possible.

- Have a professional technical writer review and enhance the documentation.

Improvement Identification

A design may look good in print, but after implementation, it may not be nearly so wonderful. From our building construction analogy, we implied that an experienced builder may have input into the construction that could improve the building. Over time you would expect an architect to become better at designing buildings based upon input from experienced builders, but of course that will not happen if the architect does not receive feedback from the builder.

The same goes for developers who never hear feedback from the test team. Often, the developers have already begun working on the next version by the time all the components of the current BI solution have been assembled and placed into the hands of the test team or even the end users. Therefore, it is important to provide a way for testers and users to report their recommendations to the developer in a manner that makes the information readily accessible at any time during the development process.

In our experience, creating a web application that allows users and testers to enter recommendations is helpful since they can be reviewed by developers at any time. The problem with this approach is that much of the feedback may be redundant and ill defined. Assigning someone to moderate the input can reduce these problems.

Another simplistic way to get feedback is to set up an email account such as suggestions@MyCompany.com. Like the web page approach, the information should be managed to improve effectiveness.

All of this can be time-consuming and costly, but be aware that a company that lacks a mechanism for capturing recommendations can often be seen as arrogant or even hostile to both users and testers.

To combat this, we recommend conveying what feedback has been received, anonymity as to who is giving the feedback, and transparency in disclosing what is being done about it.

You should tell users and testers that their ideas will be reviewed and that they may or may not be included with the next version of the software. Creating expectations that every suggestion will be included in the solution will ultimately ruin the feedback process.

From our experience, everyone may have suggestions about how to improve the process, but some will be more vocal about it than others. Anonymity can encourage the more timid users to participate.

Make sure to keep the revision process as transparent as you can without becoming enslaved to the whims of the more vocal users. While there is no guaranteed way of quieting those who must critique every solution created by someone other than themselves, you can at least mitigate the damage they can do by letting everyone know you are not ignoring their ideas out of hand.

This brings us back to even more documentation and management coordination. By now we're sure you're seeing a pattern! In short, consider these suggestions:

- Create a way to collect feedback from testers and users.

- Create a way to let testers and users know they have been heard.

- Keep the revision process as transparent as you can.

Tuning the BI Solution

Tuning a BI solution means improving its ability to perform all of its features in a minimal amount of time. Ideally, each process, from running the ETL process to filling a cube to rendering the reports, should be performed as quickly as possible so that anyone requesting information will not have to wait. These three actions can be categorized as ETL performance, processing performance, and reporting performance.

ETL Performance

The ETL operation extracts data from one location, transforms it, and loads it in another location. You can think of the ETL process as selecting data from the source and inserting data into the destination (Figure 18-3). In general, tuning ETL performance revolves around making these select and insert statements more efficient. Keep in mind that if your ETL processing includes incremental loading, instead of the flush and fill technique we have used in this book, then it may also involve updating and deleting records. In other words, you will use SQL UPDATE and DELETE statements in the destination database. When you are trying to tune ETL performance, think of which underlying statements are in use and tune the system accordingly.

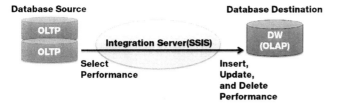

Figure 18-3. *Actions performed during ETL operations*

To tune the ETL system, start by increasing select performance. Select performance can be improved in a number of ways, but most commonly improvement comes from providing indexes, adding hardware resources, or reducing the amount of resources needed to perform the same action.

ETL Hardware Options

Additional hardware can include adding more random-access memory (RAM), additional processors, or additional hard drives. Of the three, RAM is the most vital. Whenever SQL Server selects data from a table, it must pull the table's data into RAM before it returns the result set. If there is insufficient RAM, SQL Server will have to temporarily use the hard drive as a storage base, which is substantially slower than if RAM was available. While SQL Server performs these actions quite efficiently, adding RAM will provide you with a quick and easy way to improve your select performance.

When adding RAM, it is important to remember that a 32-bit operating system cannot access more than 4 GB of system RAM. Therefore, you should move your installation of SQL Server to a 64-bit operating system.

Adding additional processors can sometimes help SQL Server improve select performance as well. Whether this happens is dependent upon the complexity of the SQL query and what other server software is on the same computer as your SQL Server. Nowadays, most computers come with multiple processors. Typically, these additional processors are more than sufficient for SQL Server's needs. If you have SQL Server, SSAS, SSRS, and other server software like Exchange all installed on the same computer, it will take a toll. Using a single computer with multiple processors is simpler to administer. But, remember that one computer can hold only so much RAM, so many hard drives, and only a single motherboard. Therefore, using multiple computers provides better performance compared to multiple processors, albeit at a greater cost.

Adding additional hard drives is a common technique for improving performance. The practice of configuring two or more hard drives to act as a single hard drive is referred to as a Redundant Array of Independent Disks (RAID). It is usually referred to as a *RAID array*.

The three most common types of RAID configurations are striping (RAID 0), mirroring (RAID 1), and mirroring with striping (RAID 1+0). Figure 18-4 illustrates these three configurations.

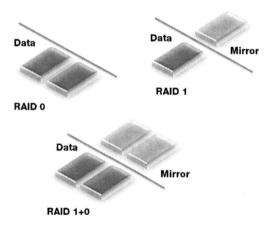

Figure 18-4. *Three comon types of RAID configurations*

RAID OVERVIEW

While a full discussion of RAID and how it is used is beyond the scope of this book, here are some useful facts to help you understand the basics:

- RAID is a collection of two or more hard drives that are made to look like a single hard drive.

- Your software, such as SQL Server, does not know there are multiple hard drives involved.

- A stripe places some parts of your data on one drive and other parts on a different hard drive. If one of the drives fails, you have only part of your data!

- A mirror keeps a real-time copy of your data. If a drive fails, you still have the copy.

- RAID 0 uses only striping. Although this is faster than using a single hard drive, your data could be at risk. Therefore, make sure you back up your important files regularly.

- RAID 1 uses mirroring. Essentially this means your data has a real-time backup at all times. It will not, however, keep track of historical changes to data. Earlier versions of updated files become lost if you do not create and store regular backups of your important files.

- RAID 0 is faster than RAID 1 when it comes to writing data to the drives, since RAID 1 has to write twice as much. Read performance is similar, though.

- RAID 1 is limited to two hard drives, but RAID 0 is not. The more drives you add to the array, the faster the performance.

- RAID 1+0 combines both mirroring and striping to give you the best of both worlds! You are not restricted to only two drives; in fact, you need at least four to start with, and it includes the real-time copy of your data.

If you are interested in learning more about this subject, we recommend searching the Internet for "RAID."

One common scenario is to place your data source on a RAID 1+0 array that provides mirroring fault tolerance, as well as additional performance, for both transactional and query statements. In a RAID 1+0 array, your database is spread across two or more hard drives, and a mirror of the database is also spread across two or more hard drives (Figure 18-5).

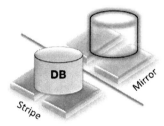

Figure 18-5. *A database on a RAID 1+0 array of hard drives*

The operating system tells SQL Server that all of these hard drives represent a single solitary drive, such as the C:\ drive. So, when you create a new database, you simply tell SQL to place the database on D:\ drive, and the operating system, and its RAID hardware, takes care of the rest (Listing 18-1).

Listing 18-1. Creating a Database on a Specific Drive

```
USE [master]
GO

CREATE DATABASE [DWPubsSales] ON PRIMARY
( NAME = 'DWPubsSales'
, FILENAME = 'C:\MySQLDataFiles\DWPubsSales.mdf'
)
```

This simple addition will increase performance, but it can be improved on. For instance, you can buy more hard drives, create an additional RAID array, and move the database to a different drive than the one hosting the operating system. The idea is that the OS has one set of hard drives, and the database has another set of hard drives, and they are not contending for resources.

This concept can be expanded. Remember that a SQL Server database is a collection of two or more files. At a minimum there will always be a data file with an .mdf extension and a log file with an .ldf extension. The data file holds the database data, and the log file records any changes to that data. If your database is used for transactional operations, like a website's ordering application, then your log file will be very active. You gain performance by placing the log file on a RAID array separate from either the OS or the MDF file, because the workload of transactional logging is on a separate RAID array (Figure 18-6).

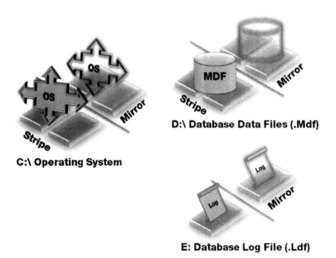

Figure 18-6. A common RAID array stategy

▓ **Tip** The log file is unlikely to require a RAID 1+0 array, because it does not benefit from a stripe as much as the OS and data files. This is because of the nature of the SQL Server logging algorithm, which has the log file to write sequentially to the file. Therefore, we have depicted the log file as being on a RAID 1 array, because it benefits from the fault tolerance aspect of RAID.

Now, when you create your database, you modify the SQL code to reflect your new design. Listing 18-2 shows an example of this.

Listing 18-2. Creating a Database on a Specific Drive with Modified Code

```
USE [master]
GO
-- Note: The OS is on C:\ drive, so we put the database file on D:\drive
CREATE DATABASE [DWPubsSales] ON PRIMARY
( NAME = 'DWPubsSales'
, FILENAME = 'D:\MySQLDataFiles\DWPubsSales.mdf'
)
 LOG ON -- Note: we put the log file on a separate drive as well!
( NAME = 'DWPubsSales_log'
, FILENAME = 'E:\MySQLLogFile\DWPubsSales_log.LDF'
)
GO
```

This common design strategy represents a good performance boost at a reasonable cost, because hard drives are relatively inexpensive. But keep in mind that you will receive no performance benefits by placing the log files on a separate array drive if you are not performing large amounts of transactional statements, such as INSERTS, UPDATES, and DELETES. Therefore, placing the log file for the data warehouse database on a separate RAID array helps only during ETL INSERT, UPDATE, and DELETE operations. It does not help during cube- and dimension-processing operations, because that involves selecting data from the data warehouse and not transactional statements.

Your data warehouse database can gain benefits from a RAID array; however, some developers may argue that fault tolerance is not strictly necessary, since you can always rebuild the data warehouse from the source data after a hard drive crash. This argument sounds good on the surface and may be appropriate in certain circumstances, but hard drives are inexpensive, and most computers natively support RAID technology, so we recommend investing in fault tolerance on the data warehouse anyway. This is especially true if the time it takes to rebuild the data warehouse is prohibitive to your business.

In reality, you may or may not see a boost in performance by adding additional hard drives, depending on the amount of data involved in the ETL operations. But adding additional hard drives into a RAID array is a good idea, regardless of the performance gains, because of its fault tolerance aspect.

ETL Software Options

SSIS can also be tuned to perform better with some "interesting" configuration tricks. However, you will probably find it simpler and see greater performance gains if you remember to perform most SQL transformations within the database engine instead of trying to perform them externally in SSIS.

As we discussed in Chapter 6, most transformations can be done within SQL Server by creating SELECT statements that utilize various built-in functions. Using SQL to do your transformations is a proven way to increase ETL performance. Because of this, we recommend you focus on making sure that the source database has all the resources it needs before trying to improve SSIS performance using arcane techniques.

The other half of ETL processing includes loading the data into a table at the destination. This involves using SQL INSERT, UPDATE, and occasionally DELETE statements. In these cases, indexes can impair the loading process, because the order of the data is maintained as the statements are being processed.

You may find that dropping the indexes in the data warehouse before it is loaded will reduce the time it takes to load. However, while dropping indexes on the data warehouse may improve loading its tables, it will decrease the performance of the SELECT statements that run when a cube or dimension is processed. Therefore, after the data warehouse has been loaded with data, the dropped indexes need to be replaced.

The ETL process can be tuned in many other ways as well, but we think you will find implementing these suggestions a good start. In short, consider these points when tuning ETL performance:

- Only select data that you need from the source; never select all columns and rows from a table if you do not need them.

- Use incremental loading instead of flush and fill techniques for large amounts of data.

- Place indexes on any column that represent a foreign key in a table, since these are the columns that will be queried when tables are joined in a SQL SELECT statement.

- Use SQL statements to perform transformations whenever you can.

- Drop indexes before filling the data warehouse with data.

- Use RAID for all but the smallest BI solutions.

Processing Performance

The same recommendations given for the database source in the ETL operations now applies to processing the cubes and dimensions: add more RAM, use RAID, and take only the data that you need (Figure 18-7).

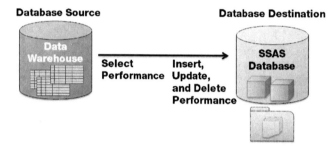

Figure 18-7. Actions performed during cube and dimension processing

Filter Data in SSAS

You can use named queries in the data source views of your SSAS projects to restrict the data used in processing (Figure 18-8). This can greatly reduce processing times by selecting only the columns and rows that are needed for a particular dimension or cube.

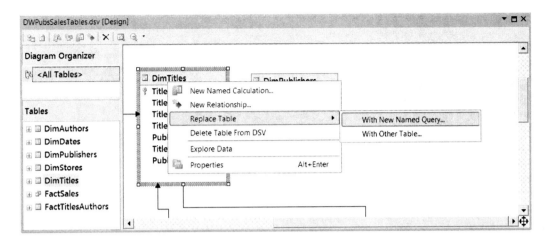

Figure 18-8. *Reducing the amount of selected data using a named query*

SSAS Hardware Options

As data is pulled from the data source, it is placed within files and folders defined in the SSAS database's properties. By default, the files and folders are placed on the same hard drive that SSAS is installed on, but this can be reconfigured so that the files are placed on a separate drive. Once again, the idea is to spread the workload across as many resources (in this case hard drives) as possible.

Placing the SSAS database folder on a separate drive will help performance, but if the drive you place the SSAS database folder on is a RAID stripe, you gain even more performance. Processing performance is improved because multiple hard drives are used to record the data. And reporting performance is improved, because the stored data can be retrieved from multiple hard drives simultaneously.

There is often a balancing act between processing performance and reporting performance. For example, some processing performance is lost when a stripe is mirrored, because the data has to be redundantly copied to the original and mirrored drives. But, the reporting performance improves, because part of the report data is pulled from the original and the mirror at the same time.

Besides, the mirror will protect your cube and dimensional data in the event of a hard drive crash. Reports will continue to work, albeit a bit slower, even if one drive fails, making a mirror a good investment for many BI solutions. And, you will not need to reprocess the data from the data warehouse after you replace the failed drive (since most RAID software just copies the data from the mirror when a drive is replaced).

In summary:

- Only select data that you need in the data source view; never select all columns and rows from a table if you do not need them.

- Place indexes on foreign key columns in the data warehouse, since these columns are joined in SQL SELECT statements when processing dimensions and cubes.

- Place the SSAS database folders on RAID stripes to increase read-write performance.

- Use a RAID mirror to provide fault tolerance.

Reporting Performance

Reporting performance with SSRS represents a special challenge, because of the large number of components that interact with each other. Figure 18-9 shows a diagram of the various components and their most common performance issues.

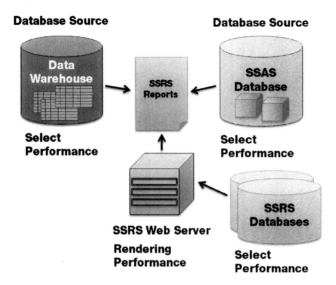

Figure 18-9. *Actions performed during SSRS reporting*

As data is retrieved from the data warehouse or your SSAS cubes, it is important to think about select performance. Select performance is also a consideration as the report definitions and cached results are retrieved from the SSRS databases. As you might expect, using RAID arrays for the data warehouse, cube folders, and SSRS databases can increase report performance in general.

Rendering Options

The SSRS web service must be considered if you want to improve report rendering performance. You may remember from Chapter 16 that report rendering allows SSRS to create various outputs, such as HTML or PDF. The rendering process involves large amounts of RAM and processing power, so increasing both of these on the web server that hosts the SSRS web and Windows services will give you a performance gain.

Filtering Data in SSRS

One simple, but vitally important, aspect of increasing reporting performance is to query only the data needed for your reports. It is seldom appropriate for you to select all columns and all rows from a table or cube. Restrict your queries to only what is needed, and you will see marked improvement.

When creating reports that use data directly from the data warehouse, we recommend using stored procedures that contain parameters to retrieve only the data you need in your reports. This allows you to create reports with dynamically generated data, and you gain a small performance increase because of the way the database engine works with stored procedures. (Sadly, SSAS does not include stored procedures, but perhaps one day this will change.)

Indexing Options

Although indexing is implicit in an SSAS database, adding indexes to your data warehouse tables can increase select performance. Because report data is often retrieved from many tables using a SQL JOIN statement, creating indexes on columns commonly used in these joins makes sense. Typically, these columns are foreign keys columns, but not exclusively. It is best to keep track of which columns are queried in the reports. Creating a spreadsheet or table that tracks this is simple and effective, but getting all of the developers to record the information is not nearly as simple. Still, it may be worth the effort, because knowing which objects are used in your reports and which are not makes it easier to maintain SQL and SSAS databases. Speaking of which....

Archiving Stored Data

One of the more effective ways of increasing performance is reducing the amount of data stored in either the SQL or SSAS database. Failing to do so is one of the major performance bottlenecks in any BI solution. The basic concept is simple; as time goes by, data that was once important is no longer and can be safely removed. For example, consider a BI solution that focuses on sales data and stores data from the year 1990 to today. Sales trends from the 1990s are no longer relevant to today's market, so while the best choice may not be to delete the data, this question arises: "Do I really need this data in the data warehouse or cubes?" If the answer is no, then it is a good idea to remove the data from the BI solution.

To understand how this impacts performance, think of a report that asks for sales that have occurred during a sales event; to find this information, the data engine must search all the data for records with some sales event flag. This means if you have 20 years of data and no index on the sales event flag, it has to search through twice as many rows as 10 years of data. And even if an index is placed on the sales event column, the index would still have to contain 20 years of values from that column instead of 10.

Caching Report Data

Reporting Server contains a number of caching options that allow you to retrieve report data directly from the SSRS databases without having to repeatedly re-render the reports. These options provide a great degree of performance and should be considered whenever report performance is slow.

In summary, to improve processing performance, follow these tips:

- Only select data that you need in the Datasets; never select all columns and rows from a table if you do not need them.

- Place indexes on any column that represents a foreign key in a table, because these are the columns that are queried when tables are joined in a SQL SELECT statement.

- Place indexes on columns that are not foreign keys if they are often used to filter results.

- Store only the data you need to in both the data warehouse and cube databases.

- Place the SSRS databases on a RAID array and the various SSRS services on separate computers (in other words, the web service on one and the databases on another).

- Use SSRS caching options when appropriate.

Common Design Strategies

Although there are no cookie-cutter design patterns that will work for all BI solutions, there are a few common design strategies that have been successfully implemented. These strategies involve spreading the workload of each of the different components of the BI solution across multiple hard drives or multiple computers.

As we have mentioned, RAID is a common way to spread I/O workloads across multiple hard drives. In a similar fashion, you can spread the many components of your BI Solution across multiple computers. One strategy is to install SQL Server with SSIS on one computer, SQL Server with SSAS on another, and SQL Server with SSRS on a third. This can be expanded by placing the SSRS Report Manager web application and SSRS web service onto yet another machine. This complexity, however, is not always needed, and most solutions stop at having the three basic servers (SQL Server, SSAS, and SSRS) on separate computers (Figure 18-10).

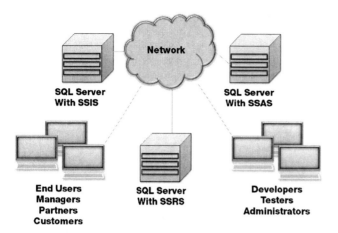

Figure 18-10. *A common set of servers in a BI solution*

In addition to the three BI servers, you also have to plan the location for the transactional databases used as the source of your data warehouse. These source databases can be placed on the SSIS server computer with minimal impact on SSIS performance if you have your ETL operations occuring off-hours. But, if you need to run the ETL process many times a day or even many times an hour, it may be necessary to separate the ETL server and its sources.

Performance Measurements

Performance measurements come in two categories: absolute and relative. Absolute measurements are taken in reference to a particular value established by a central authority. Relative measurements are compared to other measurements taken at an earlier moment in time or chosen as an arbitrary value. Either way, the values you measure against are known as *benchmarks*.

Absolute Performance

An example of an absolute measurement is SQL Server data page fragmentation. Microsoft has established that the SQL Server database data files are subdivided into 8 KB sections (referred to as *data pages*). The 8 KB limit gives you a point of reference, or benchmark, to work from.

The amount of page fragmentation can be measured for a table and its associated indexes by comparing how full the 8 KB pages are. When the table or index is spanned across a large number of pages and those pages contain substantially less than 8 KB of data, the table or index is said to be *fragmented*.

For example, let's say we have two very small tables, both of which contain less than 8 KB of data. Now let's say that for some reason the data is written across multiple pages. This represents fragmentation at the page level. To illustrate this point, note that the Product table in Figure 18-11 has been assigned to two data pages and the Customers table has been assigned to three data pages. In our example, we are saying that each of these tables is smaller than 8 KB of data and should be able to fit within a single page.

Database Data File

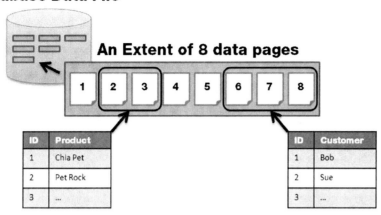

An Extent of 8 data pages

ID	Product
1	Chia Pet
2	Pet Rock
3	...

ID	Customer
1	Bob
2	Sue
3	...

Figure 18-11. *Page fragmentation decreases SQL select performance*

When you query the table with a SQL SELECT statement, if the data spans multiple pages, the results are returned more slowly than if the data was all contained within a single page.

Another type of fragmentation that has an absolute measurement is extent fragmentation. SQL Server data pages are grouped by sets of eight-page structures known as *extents*. Each new table or index created will be assigned to one page in an extent. As the pages fill, more pages are allocated in either the same extent or a different extent. This means that tables and indexes can span multiple extents even if they use only two or three data pages. Tables that use more extents than required represent an additional type of fragmentation, this time at the extent level.

In either of these examples, SQL select performance will improve if you defragment the table or index. Database administrators will typically run jobs at night that defragment both.

Microsoft has provided SQL commands that can tell you how fragmented the table is, such as the DBCC ShowContig() command. To give a visual example of a fragmented table, we have run this command and shown the results in Figure 18-12.

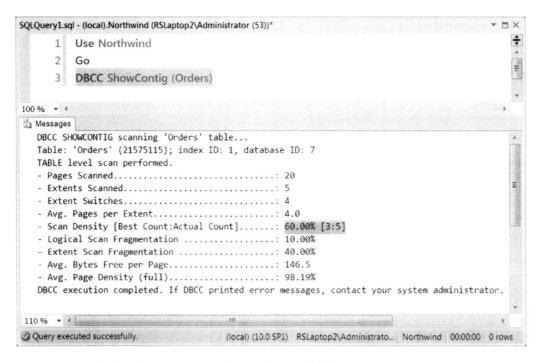

Figure 18-12. *Checking fragmentation with* DBCC ShowContig()

The resulting fragmentation report indicates that the number of pages the Orders table is currently using is 20. The current number of extents used is five.

For all the data in the table to be accessed, SQL Server would have to switch four times between the various extents. This is indicated by the Extent Switches value of 4.

The Avg. Pages per Extent value is therefore 4, derived by taking the number of pages (20) and dividing by the extents (5).

In a perfect world, the pages would be contiguous and held within just three extents, but the actual count is currently five. This is noted in the report as Scan Density Best Count: Actual Count.

When administrators examine a report similar to the one shown in Figure 18-12, they would know that the table is fragmented at the extent level and would run a SQL command similar to the one shown in Figure 18-13. This command rebuilds tables and indexes internally, attempting to fit the data into as few data pages and extents as it can.

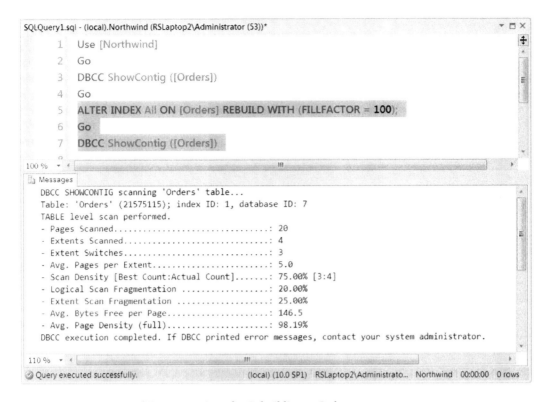

Figure 18-13. *Decreased Fragmentation after Rebuilding an Index*

▪ **Note**   You will not see much gain by defragmenting small tables, such as in our example. But in the larger tables, such as the ones that make up many BI solutions, you can see a big difference in SQL select performance when the tables and indexes are defragmented.

Most often, tuning databases is performed after the BI solution is in use and is automated by database administrators. The BI solutions developer may not be involved with this process directly; however, this information is useful when communicating with administrators responsible for maintaining the BI solution. In addition, it can be used by developers to anticipate the needs of those administrators.

As a developer, make sure initial indexes are in place before the solution is completed. After all, by reviewing the SQL SELECT statements you have created for the BI solution, you can ascertain what initial indexes are needed more easily than an administrator can who has never worked with the solution before. These SELECT statements are found in the ETL code used in SSIS, the processing code used in SSAS, and the reporting queries used in SSRS. If you find any of the SQL SELECT statements using a SQL JOIN, ORDER BY, or WHERE clause, then you need to look closer to see whether an index is needed.

▪ **Note**   Knowing when and where to use an index takes research and practice. For more information, visit our companion website: www.NorthwestTech.org/ProBISolutions.

Relative Performance Measurements

It would be nice if absolute performance benchmarks were always provided, but that is seldom the case. Where there are no predefined values to measure performance against, you must rely on tuning the BI solution using relative performance.

Relative performance tuning can be somewhat tricky, because you need to establish a common point of reference on your own by creating a benchmark.

Follow these steps to create a benchmark:

1. Create the BI solution and then execute the actions you expect to be performed regularly.

2. Use tools to measure the time or resources used while performing these actions, and document them. This is your relative benchmark.

3. Repeat the process as the BI solution increases activity.

4. Compare the original values to the new values.

5. Continue monitoring the trends in the values until you isolate performance bottlenecks.

6. Implement performance-tuning techniques such as using RAID arrays, adding RAM, or upgrading processors.

7. Remeasure the counts to prove or disprove that the tuning technique has produced the desired effect.

Measuring Performance with SQL Profiler

Microsoft provides a number of tools that can help measure performance. One very useful tool that comes with SQL Server is SQL Server Profiler. To access SQL Profiler, navigate to Start ➤ MS SQL Server 2012 ➤ Performance Tools ➤ SQL Server Profiler. This starts the application.

Creating a Profiler Trace

Once SQL Profiler starts, you need to create a new SQL trace. Traces are collections of monitoring objects that track activity on your SQL Server or SSAS database engines.

The first step to creating a trace is to access the File ➤ New Trace menu option. A dialog window appears asking which database engine you want to connect to. To select the SQL Server database engine, use the "Server type" dropdown box and choose Database Engine, as shown in Figure 18-14. Use the same dropdown box to select the Analysis Server database engine.

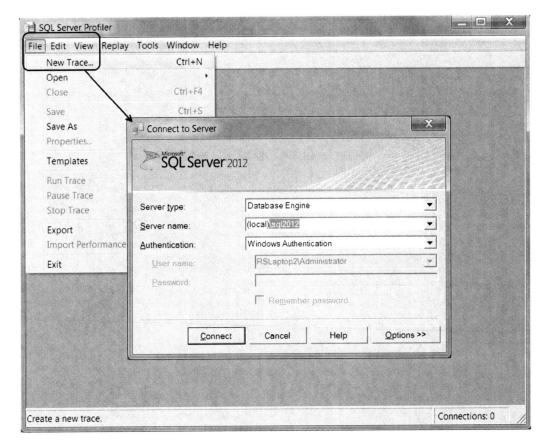

Figure 18-14. *Creating a new trace with SQL Server Profiler*

After selecting which database engine you want to connect to, you are able to select the properties of the profiler trace using the properties dialog window shown in Figure 18-15. Common configurations include the trace name, a template, and whether you will save the trace data into a file or a SQL Server database table.

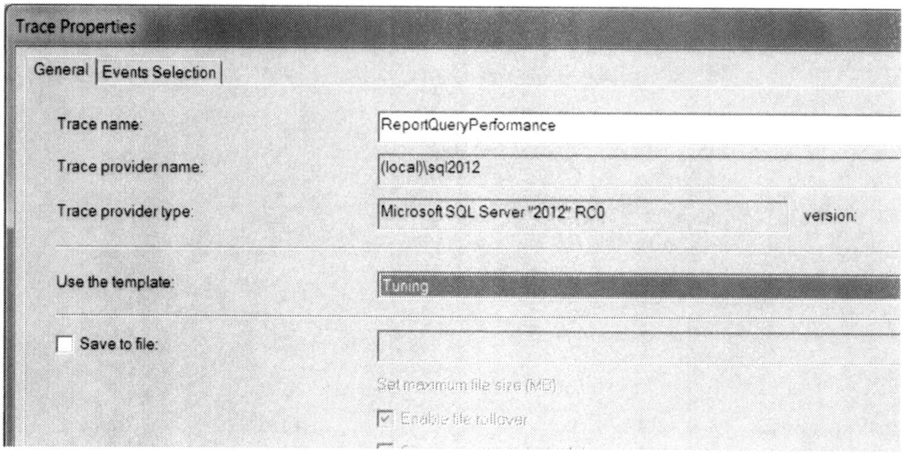

Figure 18-15. *Selecting a template for the new trace with SQL Server Profiler*

In our example, we have typed **ReportQueryPerformance** in the "Trace name" textbox. We also selected Tuning from the "Use the template" dropdown box. Once you have made the selections, click Run to start the trace.

Running a Trace

As a trace starts, you will see a new dialog window appear that displays the activity occurring on the connected database engine from the various clients. For example, if we were to create a new SSRS report and preview the results, the SELECT statement used in the SSRS report would run and be captured by the SQL trace, as shown in Figure 18-16. Information about how long it took to resolve the query is shown in the Duration column. In this example, the group was resolved in 20 milliseconds, which of course is not much time, but for larger Datasets and more complex queries, you can expect the number to increase.

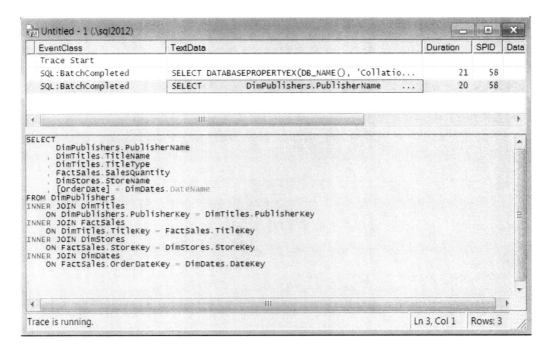

Figure 18-16. An example of an SSRS report query captured with SQL Server Profiler

Database administrators can use this tool to monitor traffic between the SSIS, SSAS, and SSRS servers to look for queries that have the highest duration and target them for additional tuning. During development, there are no users with activity to track. Therefore, developers use this tool to identify the current performance of statements between the three servers and try to improve performance by making changes to the current BI solution by adding indexes or modifying queries.

WHY IS THIS MY PROBLEM?

You may find your development team scratching their head asking, "Why is this my problem?" It seems as if this would be something the database administrators should be tasked with. But consider an analogy. Let's say that an automobile manufacturer saves money and time by not tuning their cars at the factory. Instead, they require their dealers to tune the cars before they are sold. As many of us know, the professionalism of each automobile dealer varies quite a bit, so you can expect that the initial experience of the new car owners would also vary. Still even though it was the dealer's responsibility to tune the car, the owner associates their care with both the dealer *and* the manufacturer. Owners of a poorly tuned car would believe that the car was of poor quality and unprofessionally built by the manufacturer.

Using this analogy, we can see that tuning the BI solution before it is released is a much more professional approach than releasing it with instructions that the new owners are responsible for making all adjustments as they see fit.

An Example Scenario

Let's take a moment to examine how you could capture trace information on ETL processing. The first step would be to create a new trace. In Figure 18-17, we named the trace **ETLProcessingPerformance**, used the template called Tuning, and saved the trace data to a new file called `C:\_BISolutions\PublicationsIndustries\TraceFiles\ELTProcessingPerformance_Mar15_2012.trc`. After configuring these settings, clicking Run at the bottom of the Trace Properties dialog window begins the trace.

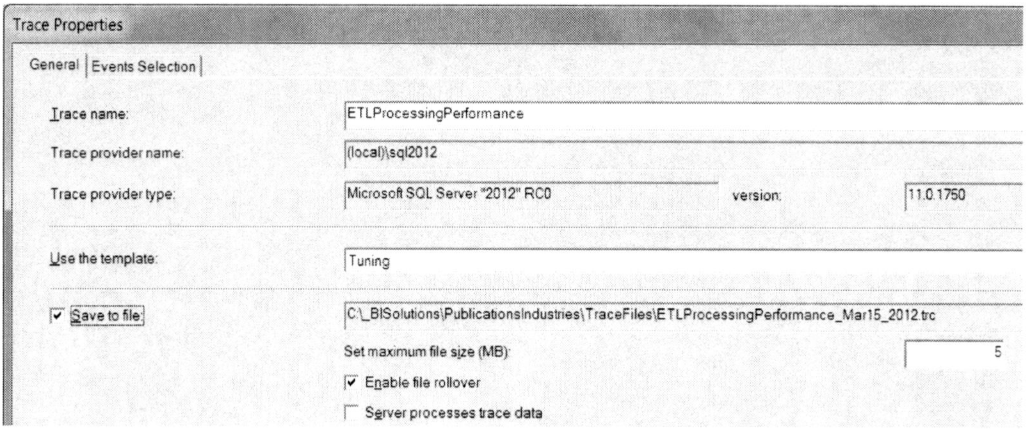

Figure 18-17. *Creating a new trace to benchmark ETL processing performance*

■ **Tip** Saving the trace to a file is a simple and efficient way to record the stored trace data. This is very important when you are trying to create a benchmark for relative performance tuning. With it you can compare earlier trace data to current trace data, noting whether the performance has been improved by recent changes to the database.

With the trace running, we need to start the ETL process and record its performance. So, we open Visual Studio and execute the SSIS package we created for the Publication Industries BI solution. It takes a few seconds to run; as it does, each of the containers and tasks will indicate its success or failure. In Figure 18-18, our SSIS package has completed all tasks successfully, as indicated by the check marks. (The check marks will be green on your screen.)

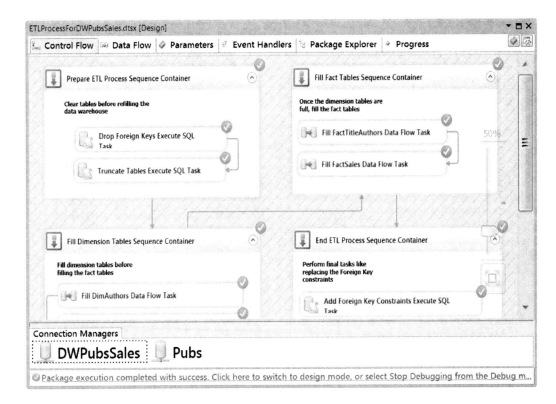

Figure 18-18. *Executing the ETL package*

If we switch from watching Visual Studio running the SSIS package back to SQL Profiler, we can see the SQL statements being recorded as they are used in the SSIS package (Figure 18-19). You can also wait to review the trace after the package has completed.

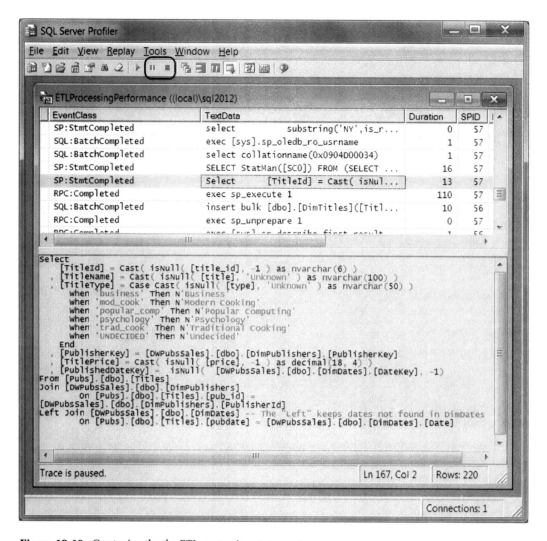

Figure 18-19. *Capturing the the ETL processing statements*

When the SSIS package completes and you have recorded all of the trace data, you can pause or stop the SQL trace from within the SQL Server Profiler window by clicking the Pause or Stop button on the toolbar (Figure 18-19).

What you see in the trace can be overwhelming at first, because a lot of additional statements are included beyond the ones you are trying to tune. If you look closely, you will see the beginning of your SQL statements under the TextData column. You can see the whole statement by clicking an individual SQL statement in the Trace window (Figure 18-19).

Once you have isolated the statements in the trace, note the time it took to run them using the Duration column. This provides your benchmark for relative performance tuning. Then make changes to the database, set up a new trace, and run the ETL process again to measure whether you have improved performance relative to the performance recorded in the first trace file.

Reviewing trace data within this dialog window works fine; however, you can close this window and review the results at a later time from the trace file. To open the trace file, access the File ➤ New Trace ➤ Trace File menu option, and review the data (Figure 18-20).

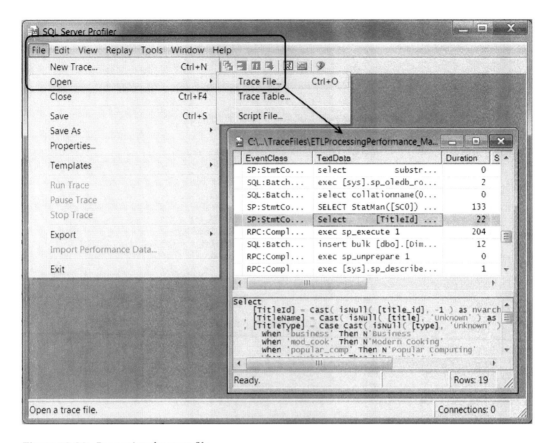

Figure 18-20. *Reopening the trace file*

Creating a Metadata Database

In addition to being able to reopen the trace file in SQL Profiler, you can quickly import the data into a SQL Server table. The advantage of this is that it makes it much easier to compare and aggregate performance data using SQL queries. You will need a database to store the trace table along with other trace tables you are using for performance testing. And since you are most likely to perform many tests and performance trials, it makes sense to create a metadata database to hold this type of data.

This metadata database is also a good place to store data about the objects that make up your solution. The code in Listing 18-3 creates a database called PubsBISolutionsMetaData to contain information about testing and performance and a catalog of solution objects. Figure 18-21 shows an example of a simple BI Metadata database design. Starting with these core tables, you then add more tables with testing and performance data.

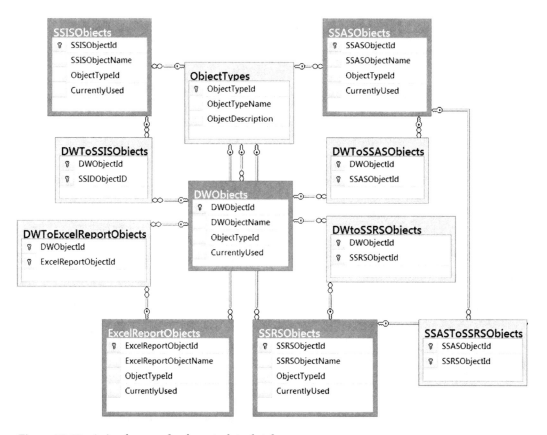

Figure 18-21. *A simple example of a metadata database*

The metadata database is used to track which objects exist as part of the current BI solution and how they are interconnected.

For example, to track the existence of an object, fill the SSRSObjects table with records of all the SSRS reports created for the solution and include the SSRS data source and SSRS Dataset objects they use as well. Make sure to indicate which type of object a row of data refers to using the ObjectTypeId. The ObjectTypeId is used to reference the ObjectTypes table, which includes the object type name, like "SSRS Dataset," and the object type description, like "A saved SQL or MDX query and its lists of SSRS reporting fields." If the object is no longer being used as part of the current BI solution, indicate this using the CurrentlyUsed column.

To track how solution objects are interrelated, use the relationship tables (also called *bridges* or *associative entity tables*). For instance, if an SSRS Dataset queries a particular table, note the table's DWObjectId and the Dataset's SSRSObjectId in the DWtoSSRSObjects table.

■ **Note** The metadata database may seem like overkill at first, but as time goes on, you will appreciate it more and more. Randal and his team once spent three months reconstructing which solution objects were still in use so that they could be moved to a more powerful SQL server. The new server was going to increase performance, but to optimize the database, the client wanted only the objects that were necessary for the current BI solution. This made sense, because the tables were huge! But because no one had ever recorded which table, view, or stored procedure was needed by which solution objects, it was an expensive nightmare for the customer! The sample we are using is a much simplified version of that database.

We have provided the code that creates this database in Listing 18-3. Of course, we also included the script file as part of the downloadable content for this book, which you will find in the C:_BookFiles\ Chapter18Files\ListingCode folder.

Listing 18-3. Creating a Metadata Database

```
USE Master
GO
-- Create or replace the database as needed
IF EXISTS (SELECT name FROM sys.databases WHERE name = N'PubsBISolutionsMetaData')
  BEGIN
    -- Close connections to the DB
    ALTER DATABASE [PubsBISolutionsMetaData] SET SINGLE_USER WITH ROLLBACK IMMEDIATE
    -- and drop the DB
    DROP DATABASE [PubsBISolutionsMetaData]
  END
Go
-- Now re-make the DB
CREATE DATABASE [PubsBISolutionsMetaData]
GO
USE [PubsBISolutionsMetaData]
GO

-- And add the core tables --
CREATE TABLE [dbo].[ObjectTypes](
  [ObjectTypeId] [int] NOT NULL,
  [ObjectTypeName] [nvarchar](1000) NULL,
  [ObjectDescription] [nvarchar](4000) NULL,
 CONSTRAINT [PK_ObjectTypes] PRIMARY KEY CLUSTERED
  ( [ObjectTypeId] ASC )
)
GO

CREATE TABLE [dbo].[DWObjects](
  [DWObjectId] [int] NOT NULL,
  [DWObjectName] [nvarchar](100) NULL,
  [ObjectTypeId] [int] NULL,
  [CurrentlyUsed] [bit] NULL,
```

```
 CONSTRAINT [PK_DWObjects] PRIMARY KEY CLUSTERED
  ( [DWObjectId] ASC )
)
GO

CREATE TABLE [dbo].[SSISObjects](
  [SSISObjectId] [int] NOT NULL,
  [SSISObjectName] [nvarchar](100) NULL,
  [ObjectTypeId] [int] NULL,
  [CurrentlyUsed] [bit] NULL,
 CONSTRAINT [PK_SSISObjects] PRIMARY KEY CLUSTERED
  ( [SSISObjectId] ASC )
)
GO

CREATE TABLE [dbo].[SSASObjects](
  [SSASObjectId] [int] NOT NULL,
  [SSASObjectName] [nvarchar](100) NULL,
  [ObjectTypeId] [int] NULL,
  [CurrentlyUsed] [bit] NULL,
 CONSTRAINT [PK_SSASObjects] PRIMARY KEY CLUSTERED
  ( [SSASObjectId] ASC )
)
GO

CREATE TABLE [dbo].[ExcelReportObjects](
  [ExcelReportObjectId] [int] NOT NULL,
  [ExcelReportObjectName] [nvarchar](100) NULL,
  [ObjectTypeId] [int] NULL,
  [CurrentlyUsed] [bit] NULL,
 CONSTRAINT [PK_ExcelReportObjects] PRIMARY KEY CLUSTERED
  ( [ExcelReportObjectId] ASC )
)
GO

CREATE TABLE [dbo].[SSRSObjects](
  [SSRSObjectId] [int] NOT NULL,
  [SSRSObjectName] [nvarchar](100) NULL,
  [ObjectTypeId] [int] NULL,
  [CurrentlyUsed] [bit] NULL,
 CONSTRAINT [PK_SSRSObjects] PRIMARY KEY CLUSTERED
  ( [SSRSObjectId] ASC )
)
GO

-- and their relationship tables --
CREATE TABLE [dbo].[DWToSSISObjects](
  [DWObjectId] [int] NOT NULL,
  [SSIDObjectID] [int] NOT NULL,
 CONSTRAINT [PK_DWToSSISObjects] PRIMARY KEY CLUSTERED
  ( [DWObjectId] ASC, [SSIDObjectID] ASC )
)
GO
```

```
CREATE TABLE [dbo].[DWToSSASObjects](
  [DWObjectId] [int] NOT NULL,
  [SSASObjectId] [int] NOT NULL,
 CONSTRAINT [PK_DWToSSASObjects] PRIMARY KEY CLUSTERED
  ( [DWObjectId] ASC, [SSASObjectId] ASC )
)
GO

CREATE TABLE [dbo].[DWToExcelReportObjects](
  [DWObjectId] [int] NOT NULL,
  [ExcelReportObjectId] [int] NOT NULL,
 CONSTRAINT [PK_DWToExcelReportObjects] PRIMARY KEY CLUSTERED
  ( [DWObjectId] ASC, [ExcelReportObjectId] ASC )
)
GO

CREATE TABLE [dbo].[DWtoSSRSObjects](
  [DWObjectId] [int] NOT NULL,
  [SSRSObjectId] [int] NOT NULL,
 CONSTRAINT [PK_DWtoSSRSObjects] PRIMARY KEY CLUSTERED
  ( [DWObjectId] ASC, [SSRSObjectId] ASC )
)
GO

CREATE TABLE [dbo].[SSASToSSRSObjects](
  [SSASObjectId] [int] NOT NULL,
  [SSRSObjectId] [int] NOT NULL,
 CONSTRAINT [PK_SSASToSSRSObjects] PRIMARY KEY CLUSTERED
  ( [SSASObjectId] ASC, [SSRSObjectId] ASC )
)
GO

-- Create the FKs to the ObjectTypes domain table --
ALTER TABLE [dbo].[DWObjects]
  ADD CONSTRAINT [FK_DWObjects_ObjectTypes] FOREIGN KEY([ObjectTypeId])
  REFERENCES [dbo].[ObjectTypes] ([ObjectTypeId])
GO

ALTER TABLE [dbo].[SSISObjects]
  ADD CONSTRAINT [FK_SSISObjects_ObjectTypes] FOREIGN KEY([ObjectTypeId])
  REFERENCES [dbo].[ObjectTypes] ([ObjectTypeId])
GO

ALTER TABLE [dbo].[SSASObjects]
  ADD CONSTRAINT [FK_SSASObjects_ObjectTypes] FOREIGN KEY([ObjectTypeId])
  REFERENCES [dbo].[ObjectTypes] ([ObjectTypeId])
GO

ALTER TABLE [dbo].[ExcelReportObjects]
  ADD CONSTRAINT [FK_ExcelReportObjects_ObjectTypes] FOREIGN KEY([ObjectTypeId])
  REFERENCES [dbo].[ObjectTypes] ([ObjectTypeId])
GO
```

```
ALTER TABLE [dbo].[SSRSObjects]
  ADD CONSTRAINT [FK_SSRSObjects_ObjectTypes] FOREIGN KEY([ObjectTypeId])
  REFERENCES [dbo].[ObjectTypes] ([ObjectTypeId])
GO

-- Create the FKs to the all the relationship tables --
-- From DWToExcelReportObjects
ALTER TABLE [dbo].[DWToExcelReportObjects]
  ADD CONSTRAINT [FK_DWToExcelReportObjects_DWObjects] FOREIGN KEY([DWObjectId])
  REFERENCES [dbo].[DWObjects] ([DWObjectId])
GO

ALTER TABLE [dbo].[DWToExcelReportObjects]
  ADD CONSTRAINT [FK_DWToExcelReportObjects_ExcelReportObjects] FOREIGN
KEY([ExcelReportObjectId])
  REFERENCES [dbo].[ExcelReportObjects] ([ExcelReportObjectId])
GO

-- From DWToSSASObjects
ALTER TABLE [dbo].[DWToSSASObjects]
  ADD CONSTRAINT [FK_DWToSSASObjects_DWObjects] FOREIGN KEY([DWObjectId])
  REFERENCES [dbo].[DWObjects] ([DWObjectId])
GO

ALTER TABLE [dbo].[DWToSSASObjects]
  ADD CONSTRAINT [FK_DWToSSASObjects_SSASObjects] FOREIGN KEY([SSASObjectId])
  REFERENCES [dbo].[SSASObjects] ([SSASObjectId])
GO

-- From DWToSSISObjects
ALTER TABLE [dbo].[DWToSSISObjects]
  ADD CONSTRAINT [FK_DWToSSISObjects_DWObjects] FOREIGN KEY([DWObjectId])
  REFERENCES [dbo].[DWObjects] ([DWObjectId])
GO

ALTER TABLE [dbo].[DWToSSISObjects]
  ADD CONSTRAINT [FK_DWToSSISObjects_SSISObjects] FOREIGN KEY([SSIDObjectID])
  REFERENCES [dbo].[SSISObjects] ([SSISObjectId])
GO

-- From DWtoSSRSObjects
ALTER TABLE [dbo].[DWtoSSRSObjects]
  ADD CONSTRAINT [FK_DWtoSSRSObjects_DWObjects] FOREIGN KEY([DWObjectId])
  REFERENCES [dbo].[DWObjects] ([DWObjectId])
GO

ALTER TABLE [dbo].[DWtoSSRSObjects]
  ADD CONSTRAINT [FK_DWtoSSRSObjects_SSRSObjects] FOREIGN KEY([SSRSObjectId])
  REFERENCES [dbo].[SSRSObjects] ([SSRSObjectId])
GO

-- From SSASToSSRSObjects
ALTER TABLE [dbo].[SSASToSSRSObjects]
  ADD CONSTRAINT [FK_SSASToSSRSObjects_SSASObjects] FOREIGN KEY([SSASObjectId])
  REFERENCES [dbo].[SSASObjects] ([SSASObjectId])
GO
```

```
ALTER TABLE [dbo].[SSASToSSRSObjects]
  ADD CONSTRAINT [FK_SSASToSSRSObjects_SSRSObjects] FOREIGN KEY([SSRSObjectId])
  REFERENCES [dbo].[SSRSObjects] ([SSRSObjectId])
GO
```

We can import the trace data from the trace file into a table in the metadata database using Microsoft's fn_trace_gettable() function. This function is designed specifically for just this purpose and is executed as shown in Listing 18-4. Here we identify the location of the file and the file name that we want to import data from. The code also takes the data the function returns and imports it into a new table we called ETLTraceData_Mar15_2012. This code creates the new table for us automatically using the appropriate columns and data types and imports the trace data as well!

Listing 18-4. Importing Trace Data from a Trace File

```
Declare @Path nVarchar(500) = 'C:\_BISolutions\PublicationsIndustries\TraceFiles\'
Declare @TraceFileName nVarchar(100) = 'ELTProcessingPerformance_Mar15_2012.trc'
SELECT [TraceFileName] =@TraceFileName, *
  INTO ETLTraceData_Mar15_2012
  FROM fn_trace_gettable(@Path + @TraceFileName, default)
Go
```

Once the new table is created and filled, we can query the table using code similar to that shown in Listing 18-5. By performing aggregate functions on the trace data within, we can calculate the average time it takes to do the ETL processing.

Listing 18-5. Getting Performance Information from the Trace Data

```
SELECT
      [TimeOfTrace] = Max(StartTime)
    , [Time it took to do this ETL processing] = Sum(Duration)
FROM ETLTraceData_Mar15_2012
Go
SELECT
      [TraceFileName]
    , [TextData]
    , [Duration]
      --,* -- All the other columns
FROM ETLTraceData_Mar15_2012
```

Figure 18-22 shows the results of these statements. Notice that the aggregated sum of the duration of each of the ETL statements gives us a useful benchmark to measure against.

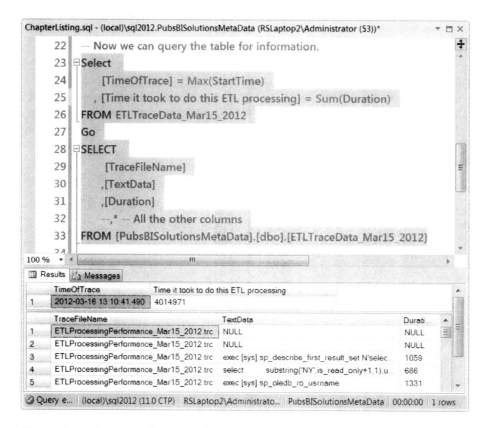

Figure 18-22. *Reviewing the imported trace data*

If we make changes to the OLTP database, data warehouse, or SQL statements that improve the ETL performance, we will be able to validate this improvement by setting up a SQL trace, performing the ETL processing, importing the trace data into the metadata database, and comparing the aggregate average between the two tables. If we can perform the same actions in less time with the same results, performance has been improved!

EXERCISE 18-1. CREATING A SQL TRACE

In this exercise, you create a trace file that stores information about the ETL processing of your BI solution. You then create a metadata database and import the trace information into a table. Afterward, you query the traced data to find out the total duration.

Important: As always, you will be practicing administrator-level tasks, so you need administrator-level privileges. The easiest way to achieve this is to remember to right-click a menu item, select Run as Administrator, and then answer Yes to access administrator-level privileges while running this program. In Windows 7 and Vista, just logging in with an administrator account is not enough. For more information, search the Web on the keywords "Windows 7 True Administrator and User Access Control (UAC)."

Setting Up the SQL Trace

To get things going, we need to create a trace within SQL Profiler.

1. Start SQL Profiler from the Windows Start menu. To access SQL Profiler, use the Windows menu item: Start ➤ MS SQL Server 2012 ➤ Performance Tools ➤ SQL Server Profiler. Right-click this menu item to see an additional context menu, and then click the Run as Administrator menu item. If the UAC message box appears asking, "Do you want the following program to make changes to this computer?" click Yes (or Continue depending upon your operating system) to accept this request.

2. When SQL Profiler opens, access the file menu to create a new trace, as shown in Figure 18-14.

3. When the Trace Properties dialog window appears, fill in the "Trace name" textbox with **ETLProcessingPerformance**, as shown in Figure 18-17.

4. In the Trace Properties dialog window, select Tuning in the "Use the template" dropdown box, as shown in Figure 18-17.

5. Check the "Save to file" checkbox. This launches the Save As dialog window (Figure 18-23).

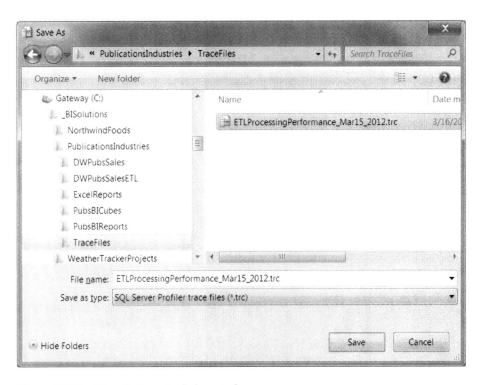

Figure 18-23. *Using the Save As dialog window*

6. Using the Save As dialog window, navigate to the folder
    `C:\_BISolutions\PublicationsIndustries`. Once there, create a new subfolder
    called `TraceFiles`, as indicated in Figure 18-23. This subfolder is where we place
    the trace file.

7. In the "File name" textbox, enter **ELTProcessingPerformance_Mar15_2012.trc** as
    the name of the trace file (Figure 18-23).

8. Click Save to close the Save As dialog window. This will return you to the Trace
    Properties dialog window.

9. In the Trace Properties dialog window, click the Run button at the bottom of the
    screen to start the SQL trace.

Launching the ETL Process

Now that the trace is running, we can launch the ETL process from Visual Studio and capture trace data with
the following steps:

1. Open Visual Studio 2010. You can do so by clicking the Start button and navigating
    to All Programs ➤ Microsoft Visual Studio 2010. Then right-click Microsoft Visual
    Studio 2010 to see an additional context menu. In this new menu, click the Run as
    Administrator menu item. If the (UAC) message box appears asking, "Do you want
    the following program to make changes to this computer?" click Yes (or Continue
    depending upon your operating system) to accept this request.

2. When Visual Studio opens, open the Publications Industries solution. You can do
    so by using the File ➤ Open-Project/Solution menu item. A dialog window opens
    allowing you to select your solution file.

3. Navigate to the SLN file `C:\_BISolutions\PublicationsIndustries\`
    `PublicationsIndustries.sln`, select it, and then click the Open button at the
    bottom of the dialog window.

4. Locate the DW pubs sales ETL project in Solution Explorer and expand the treeview
    to see the SSIS Packages subfolder.

5. In the SSIS Packages subfolder, right-click the `ETLProcessForDWPubsSales.dtsx`
    integration services package and select Execute Package from the context menu.

6. Let the package complete its execution. When the package completes, it should look
    similar to Figure 18-17.

7. Stop the package using Visual Studio's Debug ➤ Stop Debugging menu item.

Stopping the Trace

We have captured the data in a trace, but the trace continues to run until it has been stopped. We need to
stop the trace before we can import the trace data into the database.

1. Navigate to SQL Profiler's window.

2. Select File ➤ Stop Trace from SQL Profiler's menu to stop the SQL trace.

3. Review the trace data. It should look similar to Figure 18-18.

Importing the Trace Data

With the SQL trace stopped, import the trace into the database.

1. Open SQL Management Studio. Remember that you will need to right-click its menu item to see an additional context menu allowing you to click the Run as Administrator menu item. If the (UAC) message box appears asking "Do you want the following program to make changes to this computer?" click Yes (or Continue depending upon your operating system) to accept this request.

2. Once SQL Management Studio opens, use the File ➤ Open ➤ File menu option to access the Open File dialog window.

3. Navigate to the C:_BookFiles\Chapter18Files\ListingCode folder and select the Listing 18-1. Creating a Metadata database.sql file to open it.

4. Review the code inside this file and execute it to create the metadata database.

5. Navigate to the C:_BookFiles\Chapter18Files\ListingCode folder and select the Listing 18-2. Importing trace data from a trace file.sql file to open it.

6. Review the code inside this file and execute it to import the data from the trace file into a new table.

Examining Trace Data

Now that the data has been imported into a table, you can use a SQL query to examine the data.

1. Navigate to the C:_BookFiles\Chapter18Files\ListingCode folder and select the Listing 18-3. Getting performance information from the trace data. sql file to open it.

2. Review the code inside this file and execute it to access the trace data.

3. Review the information that is available to you in this table.

In this exercise, you practiced creating a SQL Profiler trace and reviewed the information within it. This strategy can be used to trace other aspects of your BI solution as well, such as the SSAS processing and the SSRS report queries.

Moving On

Once you have performed all your tests and completed your initial tuning, you are almost ready to let your clients begin using your BI solution. Next we will get it formally approved and then release it! But for now, Figure 18-24 gives an overview of where we are in the BI solution life cycle.

Done!
1) ~~Interview and Identify Data~~

Done!
2) ~~Plan the~~
~~BI Solution~~

Done!
3) ~~Create a~~
~~Data Warehouse~~

Done!
4) ~~Create an~~
~~ETL Process~~

5) ~~Create Cubes~~
Done!

Done!
6) ~~Create~~
~~Reports~~

Done!
7) ~~Test~~
~~and Tune~~
~~the~~
~~Solution~~

8) Approve,
Release, and Prepare
for Next Version.

Figure 18-24. *The current phase of the BI solution*

LEARN BY DOING

In this "Learn by Doing" exercise, you create a metadata database and SQL trace similar to the one defined in this chapter using the Northwind database. We provide an outline of the steps you performed in this chapter and an example of how this can be completed in two Word documents. These documents are found in the folder `C:\_BISolutionsBookFiles\_LearnByDoing\Chapter18Files`. Please see the `ReadMe.doc` file for detailed instructions.

What's Next

If you find the subject of testing and tuning the BI Solution of interest, we suggest the following book: *SQL Server 2008 Query Performance Tuning Distilled (Expert's Voice in SQL Server)* by Sajal Dam and Grant Fritchey (Apress, ISBN: 1430219025).

■ ■ ■

CHAPTER 19

Approve, Release, and Prepare

I do the very best I know how—the very best I can; and I mean to keep on doing so until the end.

—Abraham Lincoln

Now this is not the end. It is not even the beginning of the end. But it is, perhaps, the end of the beginning.

—Winston Churchill

What happens when a software project ends? How many times have you heard it isn't over until the paperwork is done? How true that is. There is a lot to consider when completing a project, and how you wrap things up is one of the greatest factors in determining how much of a professional you and your team are.

In this chapter, we teach you how to handle the paperwork as painlessly as possible. We also discuss how to get your BI solution signed off on once the contract is fulfilled, how to announce the release, and then how to release your BI solution, so that your client can finally get the full benefit of all your work. We discuss how to convert your notes into official dev specs and SDK documents as well as how to train your new users how to use their fancy new software by drafting a user manual and help files.

Many of the documents discussed in this chapter are a follow-up to the documents introduced in Chapter 3.

The End of the Cycle

As you can see in Figure 19-1, we have completed all but the last step of our BI solution process, Approve, Release, and Prepare for Next Version. The cycle is in the form of a circle, indicating that the end of one cycle is the beginning of the next. During the planning phase and at other times throughout this book, we discussed data that we chose not to incorporate into the first cycle of the BI solution. Those items can be used as a wish list for possible items to incorporate into a future version. Additionally, as time goes on and your client's company changes, new needs come up that need to be handled. We are never truly finished. And this is a good thing, because it keeps your team employed!

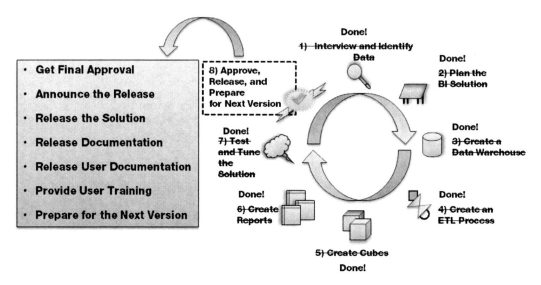

Figure 19-1. *Down to the final step!*

But, before we can begin again, we have some loose ends to tidy up. And we cannot kick off this final stage without the approval from our client, manager, or contract holder. Let's start by getting the needed approval to release their fancy new BI solution.

■ **Note** Several subjects touched upon in this chapter are business practices that will be unique to your company and are implemented according to how you choose to do business.

The Final Approval Process

Why do you need it approved? The answer is simple. It is because you want to get paid! The final approval process begins at the end of each cycle. Many developers are afraid to get sign-off because they are afraid someone will always say, "We missed something." This is all part of preparing for the next cycle, and we will save that topic for later in this chapter.

Who gives the approval is likely whomever you have been answering to all along. It is usually a project manager on your team and a stakeholder who is representing the client receiving the solution. This could also be your supervisor if the BI solution is being developed for the company you work for as an employee. In this case, the approval process is handled according to your company's guidelines.

The Sign-Off Document

In Chapter 3, you created a solution plan, and at that time you agreed to create the project. This agreement might have come in the form of a contract or an accepted solution outline. Once the requirements have been met within the agreement, the approval process can begin.

If the original agreement has been modified, you may need to create a final invoice. This agreement should be handled according to how your company chooses to do business, but the goal is for the invoice to list everything from the original agreement to any new items that were agreed upon. Add lines at the bottom for proper signatures and dates, include any outstanding balance, and you are good to go! Anything above and beyond the items within the sign-off document can be compiled as a checklist for version 2.

Announcing the Release

One of the biggest problems that can occur within a company when a new BI solution is released is resistance to change. This resistance comes from suspicion that the changes will be more of a hindrance than a help. Informing the company's employees of the changes that are coming with the release of the BI solution can be very helpful in this situation. For larger companies, this can be accomplished in an official manner with a press release.

The benefits of writing a press release are twofold. You give your team credibility, and you get the users on board with the new changes. A good press release can break through the resistance and have users anticipating what is up-and-coming rather than fighting it.

A good press release will include the following:

- A strong title that piques interest

- The announcement of what is about to happen

- How it benefits everyone

- Two quotes from key people of importance who support the changes (this is optional, but it helps support the changes and can generate excitement)

- A boilerplate (fine text) at the bottom that includes your company contact information

The Press Release Title

Many of the news articles you read online when browsing the Internet are written in press release format. A good press release will always have a good title. If the title does not catch your interest, you won't click it to see what it is all about. Additionally, the title should be short. This is particularly true when it is displayed on a website that has limited space for title links.

Your title should be something that catches your client's employee's attention. This will be different for every company. If you find you are at a loss for ideas, you are not alone. Coming up with a good title is often the hardest part of writing a press release. Some companies that publish many press releases on a regular basis hire someone other than the original author to write snappy titles. This is evidenced when a news article is published originally with one title but the title is changed within 24 hours of publication and you find yourself clicking it again not realizing you have read it. Tricky? Maybe. But the purpose is to improve the article by changing the title rather than to trick you.

That's the good news. If you don't like your title, it can always be changed later, or you can let someone else title it for you.

For our example BI solution press release, we might title it something very simple yet informative, like "Publication Industries Revamps Ordering System," or something flashier, like "Publication Industries Computer System Is About to Get Better!" If we were very limited on space, we might simply title it "Upcoming Software Changes," although that title doesn't exactly get people excited.

If it is announced on the company website, sometimes you have room for a small blurb about what those changes may be next to the link. But keep the details in the press release body, rather than in the breadcrumbs that lead them to it.

The Press Release Body

The announcement body should be nothing more than facts that matter without a lot of hype or hot air about how good you are. Press releases are short. They should fit on one page or be fewer than 400 words if it is web based.

Make a list of what you want to say and order the facts in a list of what is most important. Keep in mind that you will want to include how it benefits the reader. You need to tell them who, what, where, how, and why. You do not need to tell them that you are superfabulous for having worked so hard on it for so long and that you are relieved for it to be over. Your readers seldom care. They would rather know how it affects them and why changes are happening.

Once you have your list, consider writing a couple of one- or two-sentence quotes from the top of the food chain in your client's company to add creditability and a show of support to the changes from higher-ups. This is an old trick, but it works!

Compose the press release starting with the most important fact at the top, and go down your list to the least important fact at the bottom. Keep it interesting. If you aren't sure if the fact you want to mention is very interesting, put it at the bottom. The first few sentences should tell nearly the whole story.

The Press Release Boilerplate

The boilerplate is nothing more than your contact information. It lets your readers know who composed the press release, and it also gives you some advertisement to boot.

WRITING A PRESS RELEASE

Practice your writing skills by trying your hand at writing a press release. Be sure to include the following:

- A concise and informative title

- A list of important facts

- A press release body including the facts in order of importance first

- No more than two quotes (you can fabricate these for this exercise)

- A boilerplate

Note: For more information on press releases, search the Web for the term "press release" and look under Images. You will find interesting examples of press releases that give clear examples, templates, and visual styles of many types of press releases that may help with the writing process.

A press release of this type could be published on your client's website, sent via email to your client's employees, or printed out and posted at each place of business where the users operate.

Releasing the Solution

Once you have obtained the sign-off, it is finally time to get that BI solution released to the production server. Releasing the solution involves deploying it to the production server and making it available to the users.

The standard process of deploying the BI solution is to copy the solution objects from a development server to a testing server and finally to a production server. Of course, there are endless variations of this, but this arrangement is common practice.

Now the question becomes how to copy the objects from one place to another with the least amount of effort and errors. First, identify what you are copying and how it can be copied. Table 19-1 provides a list applicable to our current BI solution.

Each of these choices varies slightly, but all involve collecting solution artifacts and using the artifacts for deployment.

Table 19-1. *Copying Objects from a Development Server to a Testing Server*

Object	Locations	Method Used to Copy Object
Data warehouse database	SQL Server instance	Back up and restore database
		Script database and database objects
SSIS ETL package	SQL Server instance	Copy and paste .dtsx file
	File system	Deploy using Visual Studio
		Import using SQL Server Management Studio
SSAS database	SQL Server instance	Deploy using Visual Studio
		Back up and restore SSAS database
		Script SSAS database and objects
SSRS reports	SSRS web service	Deploy using Visual Studio
		Import using Report Manager
		Upload using the SSRS web service

■ **Tip** Solution artifacts represent all items produced during the creation of your BI solution. These translate to files you created in your BI projects.

Collecting the Solution Artifacts

You will need to collect all solution artifacts into one location so that they can be used and managed effectively. Ideally this location will be some sort of source control and release management software such as Microsoft's Team Foundation Server, but even a folder on a network share can work.

Collecting all the files you need for deployment into one shared folder is a simple technique that works well enough for small solutions like the example project from the exercises in this book. Each team member will be responsible for placing their files into one central location.

Some larger solutions will have more than one team working on them. As the number of developers involved in creating a solution goes up, so does the need to manage the folder. Developers have been known to be, shall we say, creative as to what and when they upload to deployment folders. Set a schedule to determine when the files must be in place and indicate which files they need to go into. Also, at some point you will need to assign someone as an official release manager. The manager's job is to coordinate with each team to make sure that the release is orderly and on time (aka keep the developers on track).

For the solution created in this book, our network share will look like the one pictured in Figure 19-2. Here we have placed a backup file for the data warehouse, a .dtsx file for the ETL process, a backup file for the SSAS database, and a couple of SSRS .rdl report files. (Our examples and images are from Randal's RSLAPTOP2 computer, so remember to use your computer name instead.)

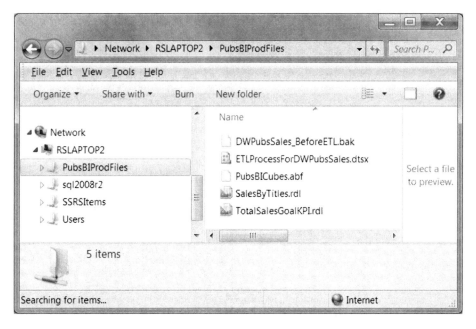

Figure 19-2. *A network share containing solution artifacts*

Deploying the Files

When the files are gathered into one place, the next step is to deploy them to the test or production computers. SQL Server, SSIS, SSAS, and SSRS all have different methods of deployment, making the deployment process very complex. Let's take a look at some foundational things you should know.

■ **Note** This topic quickly gets into advanced administration and programming tasks that may be beyond your current skill set. Just remember that until you become more advanced, you can always deploy your solution manually or hire someone who specializes in release management.

Manual and Automated Deployment

Deploying your files manually involves copying the files to a location that the computer's operating system can use (such as a local hard drive or a network share) and running the SQL code and projects that you have included in your Visual Studio solution. This means that a person is sitting at a computer, running the SQL code in SQL Management Studio, starting the SSIS package, processing the cube, and uploading the reports.

Manual deployment works just fine if you only occasionally deploy the BI solution to new computers. In many test environments, however, the solution may be deployed over and over again on the test servers. In those scenarios, you may want to create some automation code to help streamline the deployment process.

Let's take a moment to give you some examples of how this kind of automation can be accomplished.

Deploying the Data Warehouse with SQL Code

The data warehouse must be in place before SSIS, SSAS, or SSRS can be deployed. Start by restoring a backup file using code similar to that shown in Listing 19-1.

Listing 19-1. Restoring a SQL Server Backup File from a Network Share

```
-- Check to see if they already have a database with this name...
IF  EXISTS (SELECT name FROM sys.databases WHERE name = N'DWPubsSales')
  BEGIN
  -- If they do, they need to close connections to the DWPubsSales database, with this code!
    ALTER DATABASE [DWPubsSales] SET SINGLE_USER WITH ROLLBACK IMMEDIATE
  END

-- Now now restore the Empty database…
USE Master
RESTORE DATABASE [DWPubsSales]
```

FROM DISK = N'\\RSLAPTOP2\PubsBIProdFiles\DWPubsSales_BeforeETL.bak'

```
WITH REPLACE
GO
```

Another way to deploy the database is to run a SQL script that creates the database and all its objects, like we did in Chapter 2 (Exercise 2-2). But, restoring the backup includes data, whereas running the script does not. This is both good and bad. It is good because you do not have to run the SSIS process to the data to test your reports and SSAS processing, but it is bad because the backup file may be very large.

You will have to decide what is practical for your situation. Both options work well for deploying most data warehouses, so you really cannot go too wrong choosing one over the other.

You can run SQL code from a command prompt using Microsoft's sqlcmd.exe utility. This utility is useful for creating deployment batch files and scripts. For example, the code in Listing 19-2 sets the focus to the directory when the sqlcmd.exe utility is located and then executes a SQL file located on a network share called \\RSLAPTOP2\PubsBIProdFiles.

Listing 19-2. Executing a SQL File from a Network Share

```
CD "C:\Program Files (x86)\Microsoft SQL Server\110\Tools\Binn"
SQLCMD.exe -S RSLaptop2\SQL2012 -E -i \\RSLAPTOP2\PubsBIProdFiles\RestoreDWPubsSales.sql
```

RUNNING SQLCMD.EXE WITH A BATCH FILE

This code can be directly typed into a Windows command prompt or placed into a text file with a .bat extension (batch file). The batch file option is preferred, because one batch file can be used for multiple deployments.

Working with the command prompt and batch files was much more common in the early days of PC computing than it is now. We mention batch files several times in this chapter because they are so useful for automating deployments. If you are not familiar with batch files, we recommend searching the Internet for more information; they are really useful!

Deploying the SSIS ETL Process

After the database has been deployed to the production server, the ETL process must be executed by running the SSIS package. This can be done manually using Visual Studio as you have done in this book, or you can execute SSIS code using a command prompt utility called dtexec.exe. Listing 19-3 shows an example of how this utility is used.

Listing 19-3. Executing an SSIS File from a Network Share

```
CD C:\Program Files (x86)\Microsoft SQL Server\110\DTS\Binn
DtExec.exe /FILE \\RSLAPTOP2\PubsBIProdFiles\ETLProcessForDWPubsSales.dtsx
```

Combining the SQL Server and SSIS Deployment Code

Release managers will often create a batch file that combines the data warehouse restoration and the SSIS ETL processing by running just one batch file. As an example, the command shell code in Listing 19-4 would be placed in a batch file. Note how it restores the database using `sqlcmd.exe` and then performs the ETL processing using `dtexec.exe`.

Listing 19-4. Combining Multiple Commands into a Batch File

```
REM DeployPubsBISolution

REM Restore the database backup
CD "C:\Program Files (x86)\Microsoft SQL Server\110\Tools\Binn"
SQLCMD.exe -S RSLaptop2\SQL2012 -E -i "\\RSLAPTOP2\PubsBIProdFiles\RestoreDWPubsSales.sql"
pause

REM Run the ETL process
CD C:\Program Files (x86)\Microsoft SQL Server\110\DTS\Binn
DtExec.exe /FILE \\RSLAPTOP2\PubsBIProdFiles\ETLProcessForDWPubsSales.dtsx
pause
```

To create the batch file, open a simple text editor such as Notepad, type in the code, and save it with a `.bat` extension (for example, `Deployment.bat`).

When the batch file is created, you can execute it by name from a command prompt or by double-clicking the file in Windows Explorer. Either method runs the code contained inside the file as if it had been manually typed at the Windows command prompt.

■ **Tip** If you double-click the file in Windows Explorer, it temporarily opens a command prompt window, runs the code, and immediately closes the command prompt window. By placing the `pause` command on the last line of the batch file, the command prompt window will stay open until you click any keyboard key.

Another advantage to using a batch file in this manner is that you can use the Window's Task Scheduler program to automatically run the file at a designated point in time without the need for human interaction. You can find this program under Administrative Tools in the Windows Control Panel.

Deploying the SSAS Database

An SSAS backup file can be restored using XMLA code. Unfortunately, Microsoft has not provided a command prompt utility to execute XMLA code (although there are some third-party utilities that will). Therefore, it may seem that your only option is to manually restore the database using SQL Server Management Studio or to deploy the SSAS project from Visual Studio as we have done in this book.

You can, however, run XMLA code using an SSIS package that can in turn be automated using batch files. First, create and test some XMLA code that restores the backup file on the network share (Listing 19-5).

Listing 19-5. XMLA Code to Restore an SSAS Database

```
<Restore xmlns="http://schemas.microsoft.com/analysisservices/2003/engine">
  <File>\\RSLAPTOP2\PubsBIProdFiles\PubsBICubes.abf</File>
  <DatabaseName>PubsBICubes</DatabaseName>
  <AllowOverwrite>true</AllowOverwrite>
  <Security>IgnoreSecurity</Security>
</Restore>
```

Next, create a new SSIS package and add an Analysis Services Execute DDL task to it. Finally, configure the task to make a connection to the new SSAS server, and add your XMLA code to the DDL Statements dialog window (as shown in Figure 19-3).

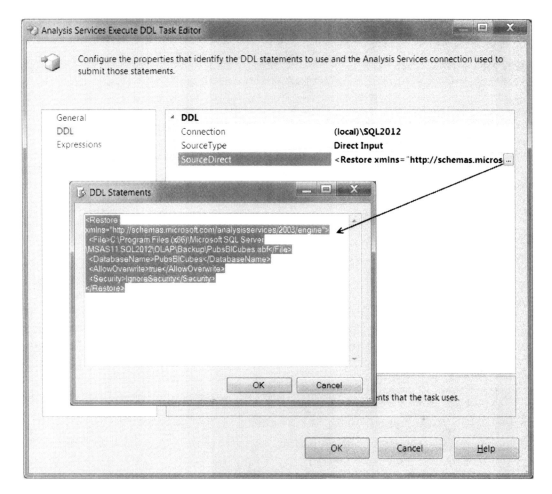

Figure 19-3. *Configuring an Analysis Services Execute DDL task*

Once you have configured the SSIS task, execute it from the command line just as you did the ETL code (Listing 19-6), and of course, you could place this code in the same batch file.

Listing 19-6. Executing the SSIS Package to Restore an SSAS Database

```
CD C:\Program Files (x86)\Microsoft SQL Server\110\DTS\Binn
DtExec.exe /FILE \\RSLAPTOP2\PubsBIProdFiles\RestorePubsBICubes.dtsx
```

Deploying the SSRS Reports

Once again, your choices are to manually deploy the report files to the SSRS web service or to automate the deployment using a command prompt utility. The SSRS reports can be deployed to the web service using a utility called RS.exe. Unfortunately, it is the least friendly of the deployment options. The utility requires that you create a VB .NET file that tells the web services what to do. This code is placed in a file with an .rss extension. For instance, code used to deploy a report from the network share to Randal's SSRS server looks like the code shown in Listing 19-7.

Listing 19-7. Code to Place in an .rss File to Upload an SSRS Report

```
' Deploy SSRS Report from a Command Line:
' RS.exe -i deploy_report.rss -s http://rslaptop2/ReportServer_SQL2012/ReportServer
'

Dim strPath = "\\RSLAPTOP2\PubsBIProdFiles\"
Dim strReportName = "SalesByTitles"
Dim strWebSiteFolder = "/PubsBIReport"

Dim arrRDLCode As [Byte]() = Nothing
Dim arrWarnings As Warning() = Nothing
Public Sub Main()
    Try
      'Read the RDL code out of the file.
      Dim stream As FileStream = File.OpenRead(strPath + strReportName + ".rdl")
      arrRDLCode = New [Byte](stream.Length) {}
      stream.Read(arrRDLCode, 0, CInt(stream.Length))
      stream.Close()

      'Upload the RDL code to the Web Service
      arrWarnings = rs.CreateReport(strReportName, strWebSiteFolder, True, arrRDLCode, Nothing)
      If Not (arrWarnings Is Nothing) Then
          Dim objWarning As Warning
          For Each objWarning In arrWarnings
              Console.WriteLine(objWarning.Message)
          Next objWarning
      Else
          Console.WriteLine("Report: {0} published successfully with no warnings", REPORT)
      End If
    Catch e As IOException
      Console.WriteLine(e.Message)
    End Try
End Sub
```

After you have created the .rss file, execute it with the RS.exe command prompt code as shown in Listing 19-8. This, too, can be placed in the same batch file that deploys the data warehouse, ETL package, and SSAS database.

Listing 19-8. Code to Place in an `.rss` File to Upload an SSRS Report

```
CD C:\Program Files (x86)\Microsoft SQL Server\110\Tools\Binn

RS.exe -i \\RSLAPTOP2\PubsBIProdFiles\DeploySalesByTitles.rss
-s https://rslaptop2/ReportServer/ReportServer
```

Using SSIS Packages Instead of Batch Files

SSIS includes the ability to run command prompt code using an Execute Process task. Since running command prompt code is the core of what a batch file does, this feature gives you an alternate way of creating deployment automation.

For example, you can do the following:

1. Create an Execute SQL task to run the code needed to restore the database from the network share.

2. Add an Execute Package task to run the ETL package.

3. Add an Analysis Service Execute DDL task to restore the SSAS database.

4. Add an Execute Process task to call `RS.exe` and upload the report.

Figure 19-4 shows an example of this package.

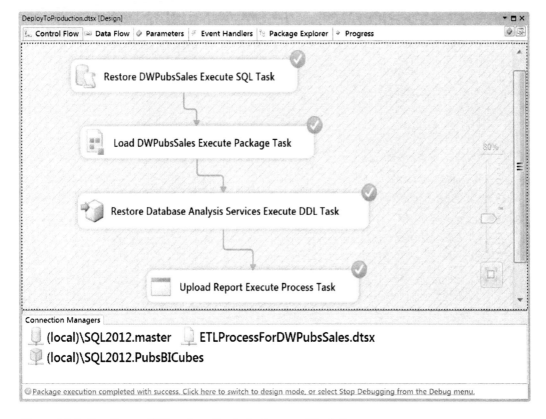

Figure 19-4. *Creating an SSIS package for deploying your BI solution*

No matter which method you choose, make sure you set aside plenty of hours for not only developing the deployment files but testing them as well. This process can be exhausting, because you will have to deal with many different departments to get permission to access the resources you need and many different computer configurations. Plan accordingly!

Release Documentation

Releasing your solution requires more than simply implementing it. Developer documentation such as SDKs and dev specs are also required. Additionally, users need documentation to teach them how to use the software you developed for them.

Programmer writers (for release documentation) and technical writers (for user documentation) who know their job very well will be able handle the majority of the documentation for you.

Ideally, they will have been working with your team for some time and are familiar with the ins and outs of your BI solution. Even better, they have been in on the project from the beginning and have been present for the team meetings that determined the direction that your BI solution would take.

If this is the case, then congratulations, you did it right! You will likely have much, if not all, of your documentation ready to be released at the same time that your software is released. This contributes to your professionalism and helps keep your client happy.

SDKs

If you write your software correctly, it will be usable for many applications. A software development kit (SDK) is a set of documents that includes the following:

- Code examples

- Comments within the code

- Descriptions of procedures

- Parameters

- Return values

- Instructions describing how to use the code

- Links to related SDK subjects (depending upon how they are stored)

- License and legal information declaring how and where it can be used (free or public)

SDKs can also be written for code that is not included in the official release but could be useful for future versions. The more details included, the better.

Typically, these documents are stored together online to form a library. They can also be stored as a simple text document. However you choose to store it, your SDK library should include a complete list of all of the SDKs organized under different topics (such as stored procedures, custom DLLs, or data warehouse tables) for convenience and searchability.

■ **Note**   In 2010, the Microsoft Interoperability team released an SDK describing how PHP (an open source competitor to ASP.NET) can interact with Microsoft's SSRS web service. It is a good example of how creating SDKs can enhance your software, because this SDK has gone a long way toward making SSRS the most flexible and robust reporting software in the industry. It is also a good example of what a professional SDK looks like. For more information, go to http://ssrsphp.codeplex.com.

SDKs can also include flow charts or visual layouts that demonstrate the structure of a data warehouse, for example. To keep the documentation professional, some companies choose to set a standard for how the documentation is recorded that can be kept on the main page and can help developers understand how the SDK is read. This can include formatting for text or other standardized information. Here are some examples:

- Standard fonts

- **Bold text**

- *Italics*

- Hyperlinks

- Date indicators such as mm/dd/yyyy or yyyy/mm/dd, and so on

- \<Inserts\>

- Value type indicators such as Bool vs. Boolean

- Separation characters for *either | or* choices

- Omitted portions

The following SDK example is created from a stored procedure we used in Chapter 13. Its purpose is to select report data.

AN EXAMPLE SDK

Name: pSelQuantitiesByTitleAndDate

Title: Stored Procedure for Sales Quantities by Title and Date

Description: Reporting stored procedure that returns total quantity, the overall average quantity, and a KPI value based on the comparison between the total quantity and the overall average quantity

Links: http://MyCompany/BIWebSite/Dev/pSelQuantitiesByTitleAndDateInfo.aspx

Parameters:

Name	Type	Example	Description
ShowAll	nVarchar(4)	True	Show all rows of data if true
StartDate	DateTime	01/01/1990	Starting date range
EndDate	DateTime	01/01/2100	Ending date range
Prefix	nVarchar(3)	A, AB, or ABC	0 to 3 characters used before wildcard search symbol

Outputs:

Name	Type	Example	Description
PublisherName	nVarchar(50)	New Moon Books	Name of publisher
Title	nVarchar(100)	Life Without Fear	Title of book
TitleId	nVarchar(6)	PS2106	Natural ID used for a title
OrderDate	Varchar(50)	01/13/2011	Date of order
Total for the Date by Title	Int	108	Total quantity based on title and date
Average Qty in the FactSales Table	Decimal or Float	53.512	Total overall average quantity from the entire table
KPI on Avg Quantity	Int	-1, 0, 1	Total overall average quantity compared to the sum of that title (filter by search) and categorized into three groups based on the following: -1 = Quantity is lower than average by 5 0 = Quantity is between 5 less to 5 more of the average, inclusive 1 = Quantity is more than average by 5

Source Code:

```
CREATE PROCEDURE pSelQuantitiesByTitleAndDate
  (
  -- 1) Define the parameter list:
  -- Parameter Name, Data Type, Default Value --
    @ShowAll nVarchar(4) = 'True' -- 'True|False'
  , @StartDate datetime = '01/01/1990' -- 'Any valid date in the mm/dd/yyyy format'
  , @EndDate datetime = '01/01/2100'  -- 'Any valid date in the mm/dd/yyyy format'
  , @Prefix nVarchar(3) = '%' -- '0 or more characters'
  )
AS
BEGIN -- the body of the stored procedure --
  -- 2) Create and Set the @AverageQty to get the overall average of sales qty.
  DECLARE @AverageQty int
   SELECT @AverageQty = Avg(SalesQuantity) FROM DWPubsSales.dbo.FactSales

  --3) Get the Report Data with this select statement
  SELECT
```

```
  [DP].[PublisherName]
 ,[Title] = DT.TitleName
 ,[TitleId] = DT.TitleId
 ,[OrderDate] = Convert(varchar(50), [Date], 101)
 ,[Total for that Date by Title] = Sum(SalesQuantity)
 ,[Average Qty in the FactSales Table] = @AverageQty
 ,[KPI on Avg Quantity] = CASE
  WHEN Sum(SalesQuantity)
   between (@AverageQty- 5) and (@AverageQty + 5) THEN 0
  WHEN Sum(SalesQuantity) < (@AverageQty- 5) THEN -1
  WHEN Sum(SalesQuantity) > (@AverageQty + 5) THEN 1
 END
 FROM DWPubsSales.dbo.FactSales AS FS
 INNER JOIN DWPubsSales.dbo.DimTitles AS DT
   ON FS.TitleKey = DT.TitleKey
 INNER JOIN DWPubsSales.dbo.DimDates AS DD
   ON FS.OrderDateKey = DD.DateKey
 INNER JOIN DWPubsSales.dbo.DimPublishers AS DP
   ON DT.PublisherKey = DP.PublisherKey
 WHERE
   @ShowAll = 'True'
   OR
   [Date] BETWEEN @StartDate AND @EndDate
   AND
   [TitleId] LIKE @Prefix
 GROUP BY
   DP.PublisherName
   ,DT.TitleName
   ,DT.TitleId
   ,Convert(varchar(50), [Date], 101)
 ORDER BY DP.PublisherName, [Title], [OrderDate]
END -- the body of the stored procedure --
```

In production code, comments are often removed from within the procedure, but in the SDK these comments can be left in, and additional descriptions, instructions, and comments are written out in full sentences after each section, as you can see in the far-right column of each table.

Objects such as views, functions, and parameters should be stored together in groups and in alphabetical order within those groups. We chose to use a table format for organization, but tables are not required. This stored procedure was already titled pSelQuantitiesByTitleAndDate within the BI solution. But, for better searchability, we included a more human-friendly title: Stored Procedure for Sales Quantities by Title and Date. For additional organization, we suggest listing this type of SDK with other stored procedures that select data within the library.

Any additional information for how this procedure is used including any links to other topics should also be included on the same page. The idea is that if a developer adds to your process or uses these constructs, they are able to see what they can work with.

Developer Specifications

Developer specifications (*dev specs*) are documents that tidy up the completed work by listing what has been done and any important notes that document pertinent information. In Chapter 3 we worked with an Excel spreadsheet titled StarterBISolutionPlan that outlined the official development specifications (Figure 19-5). This is an informal version, and if written well enough, for small projects this may be sufficient for your needs.

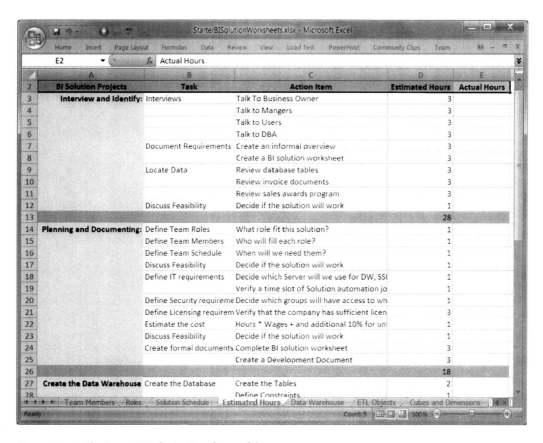

Figure 19-5. *The StarterBISolution Excel spreadsheet*

As the solution planning progresses or as the project grows, this informal Excel file is used as an outline to a more official document called a Solution Development Plan. This plan will become the bible to your BI solution. It is a living document that is updated on a regular basis.

This document is created in Microsoft Word rather than Excel, and it lists more details about the solution objects. This document can be used as a starting point for your dev specs. The difference between the two is that one was made during the development cycle and represents the current state (any time during development) of the solution, while the other represents the final outcome of the solution. Figure 19-6 shows how similar they look, which is intentional.

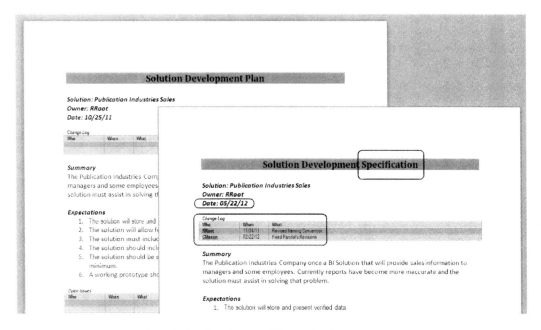

Figure 19-6. *Comparing the Solution Development Plan to the dev specs*

When each version of the solution is complete and you start work on the next version, properly written dev specs will get your team up to speed. Portions of this document can also be included as part of the SDK document, because the audience for the two documents are complementary.

■ **Tip**　Some companies fly employees in from all over the world to be trained to create these types of files. You do not have to go that far, but documents of this type are an important indication of your level of professionalism.

Dev specs are a blueprint to what was created and are given to the customer much like a car manual is included with a new car. The SDK is given to other developers who will work with your solution, much like a repair manual is used by mechanics. (You may not want to refer to your SDK as a repair manual! Hot-rod customization tool kit sounds much better.)

User Documentation

With all the documentation that needs to be included, we cannot forget that we have new users who have never seen your BI solution before and need to be instructed on how to use your program.

Just as there is more than one way to peel an apple, there is more than one way to train a user. Some methods include the following:

- Developing a video demo for online training

- Training the managers who in turn are responsible for teaching it to their employees

- Sending someone from your team to set up one or more training sessions

Whichever method you choose, this step is vitally important for the success of your BI solution. User manuals and help documents are to be used as references. Training your users yourself gets the system up and running immediately. The success of the user is what determines the effectiveness of your solution, as well as how likely you will be involved in future versions of the current BI solution you developed.

Style Guides

Regardless of when (or if) you get a professional technical writer involved in the project, your tech writer will need to know what is expected of your documentation, not just how your BI solution works. If your company is large or has been around for a while, you may already have a style guide that standardizes the writing style of your solution. Style guides are highly recommended and can be very helpful in keeping consistency when referencing your company and specific items within each of the solutions you develop.

When you hear the term *style guide*, you may automatically think of commonly known style guides such as *The Chicago Manual of Style*, *The Associated Press Stylebook*, Strunk and White's *The Elements of Style*, and perhaps Microsoft's *Manual of Style for Technical Publications*. This is not what we are referring to when we speak of developing a style guide for your company. We are referring to *house style*. A house style guide describes how items of note are handled within your documentation for your company, particularly when referring to your company name or items within your software that is unique to your company.

This type of guide is much more in-depth than listing the general fonts used within an SDK. It is specific to how your company is presented to the general public for the following:

- User documents

- Company websites

- Email

- Business cards

- Letterhead

- Contact information standards

Fonts and color schemes can also be standardized within your company house style guide.

■ **Tip** Creating your documentation by following a company-specific style guide adds to your final presentation and improves your professionalism to developers, to the client, and to the user.

If you do not already have a style guide, most tech writers can develop one for you while they prepare your documentation. A style guide is an alphabetical list of *how-tos* that include descriptions and examples of how to handle company information. A style guide is another one of those living documents that are ever-changing with your company. And as you change how you do things or develop new ways of keeping consistency, that allows room for new topics to be added at any time.

An added bonus to developing an in-house style guide is that if you work with more than one tech writer, you can keep your documentation consistent with a professional presentation, regardless of who writes it.

User Manuals

The better the instructions on how to use your BI solution, the more successful it will be, which is why the user manual is so important. Keep in mind that your client may hire new employees who were not on the team when hands-on training became available. Therefore, your user documentation needs to be thorough and accurate and written with very basic terminology that anyone can follow.

Many books are out there that cover precise details on how to write professional user manuals. Your team's technical writer will have this base covered. If, however, your BI solution is created for a small company, the solution itself is very simplistic, or you simply do not have the budget for a large team, then you may not have the luxury of hiring a tech writer to draft a user manual for you.

Let's look at some general guidelines on how to draft a simple user manual that your client can refer to, without going into the extreme depth and detail that typical technical writing books cover.

The Anatomy of a User Manual

The goal of any user manual is to give information that is easy to find and provides a step-by-step guide on how to perform an action. User manuals are usually divided into subjects and often include illustrations and numbered steps to follow.

A user manual is not the same as a trade book, such as the one you are reading now (although the exercises within this book are close to an example of how a user manual should read). Users of instruction manuals will not read each page, paragraph after paragraph. They want to get right to the information, gather it with as little reading as possible, and get on to using the solution itself.

Let's take a look at the anatomy of a user manual. As you can see in Figure 19-7, the layout should be easy to read, and the structure should be consistent. A good user manual will include *signposting* (guides to helping the user get to the information they need) to allow information to be gathered at a glance.

- *Main title*: Clearly indicates which software the manual is written for on the front page or cover.

- *Table of contents*: Gives subject locations at a glance including accurate page numbers that each subject can be found on and sometimes indicates the number of pages a subject spans within the manual.

- *Page numbers*: Even if the manual is short, page numbers are still necessary.

- *Page headers*: Indicates the topic at a glance in a standard place on every page.

- *Spot color, shading, or icons*: Visual aids that draw the eye to specific information.

- *Subject headings*: These must correspond with the titles in the table of contents, must be listed at the top of each page before a new subject, and must accurately define the subject so that users can get to the right place at a glance.

- *Figures (illustrations)*: These must genuinely aid the user rather than confuse them further, and they must be properly numbered and labeled and referred to within the text nearby.

- *Step-by-step instructions*: These must be accurate, written simply with easy-to-understand terms, and highly descriptive, logical, and thorough; they are usually written in the imperative (without using the word *you*).

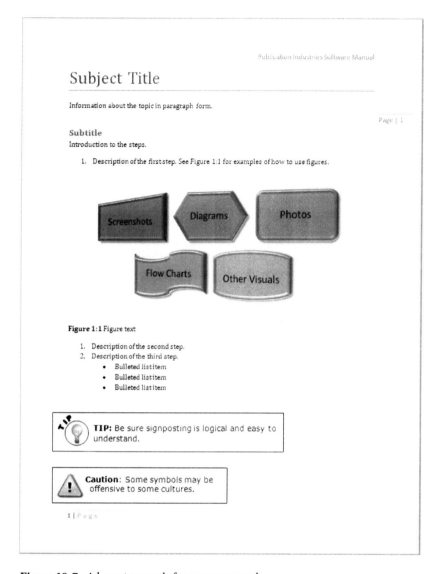

Figure 19-7. *A layout example for a user manual*

Keep in mind that every signpost must have a purpose and communicate the proper information. If it is a warning, put up a proper warning icon. Do not use a rainbow, the Ace of Spades, or Happy Bunny because it looks pretty or adds comedy.

Do not make arbitrary changes in color to specific backgrounds or titles that are not purposeful. This confuses the reader who will spend more time wondering if they somehow stumbled into the wrong section, rather than quickly getting the information and moving on.

Subject Headings

Subjects are given clear headings that are easy to find, and each step or procedure contains a detailed enough explanation so that a new user, who is unfamiliar with the program, can find their way around while following the user guide.

Step-by-Step Instructions

The step-by-step instructions are the largest portion of a user manual. The very first item to consider when writing instructions is to always, always, always consider your audience.

For a BI solution, who are your users? They will typically be your client's employees. These employees are people. How much do you know about these people?

Table 19-2 lists some example items to consider when determining who your audience is. You will likely have more items to add to this list based upon the type of company you are creating your solution for. You may want to ask your client about the type of employees they hire to work for their company.

Table 19-2. *Who Is Your Audience?*

What to Consider	How It May Be Evaluated	How It Affects Your Documentation
Age	Employees may range from teenagers in high school to senior citizens.	Determines the age level the instructions should be written to, and how big the font sizes should be.
Ethnicity	They may speak English as a second language.	Translations may be necessary. Fewer words may be used with more illustrations. Documentation may require using words that are simple and less technical.
Geographic location	Employees may reside in another country.	Translations may be necessary. Specific words or colors can be offensive or have additional meanings to residents within other countries.
Education levels	Your client's company may employ special-needs employees or some who have a lower education level.	Fewer words may be used with more illustrations. Instructions may require using words that are simple, contain fewer letters, and are less technical.
Level of experience with computers	Some may have never used a computer before.	Even the most basic of instructions may be necessary, such as how to close a window.
Color blindness	Some of your signposting may be invisible to 10% of the male population.	Black-and-white versions, versions for color-blind readers, additional signposting, or more specific signposting may be required.

The information within each subject may begin with a brief introduction to a topic before getting to the steps required to perform an action. It must refer to the illustrations where appropriate and optionally end with a brief conclusion of the topic.

The body text of the user manual should not be text heavy, meaning it should not contain paragraph after paragraph of information. Writing a few paragraphs about the program itself can be valuable, but be sure that the majority of the text is broken up into categories such as numbered lists, picture illustrations, bulleted lists, a sidebar, a note that has a different format as the rest of the text, and so on. A very well-written user manual can give the necessary information at a glance.

■ **Tip** At a glance, at a glance, at a glance…notice a pattern? The faster users can glance at the instructions, gain the information, and move on, the more productive they are, the happier your client is, and the more effective your BI solution will be.

You may have heard the *hand* rule. Although this is not the official name for it, the essence of it is this: if you can put your hand on the printed page in any place and your hand does not touch something other than paragraphs of text, then your text is too heavy. The reason why this is important is because users cannot find information easily in a block of heavy text. When the text is broken up, your users can navigate through the instructions much more easily, enabling them to visually skip around on the page and gather information faster and more readily find what they are looking for.

As we stated earlier, our exercises are similar to how a user manual's instructions should appear. The voice of most of our exercises, however, is not always in the imperative in this book, which is the recommended voice for a user manual's instructions. (One exception to this is the next exercise. Exercise 19-1 is written in the imperative and has the proper format and voice for a user manual.)

Here is an example of an imperative voice and a nonimperative voice:

- *Imperative voice:* "Click the OK button."

- *Nonimperative voice:* "After that, you can then click the OK button."

The imperative voice begins with a verb, is in the form of a command, and the word *you* is implied. The second example is simply a statement and is less effective for instructional documentation.

Keep the instructions precise and properly spell the titles of the items you are referring to within your program, paying attention to capitalization. If the button title is OK, for example, then the instructions should not say, "Click the Okay button." (In other words, Okay should not be spelled properly when the *OK* on the button itself is not spelled out properly.) Nor should the instructions be "click ok" or "click Ok" when both letters of *OK* are capitalized within the program. This may sound nitpicky, but you would be surprised how users can get lost when the instructions within your manual are not capitalized exactly the same way that they are within the program. (You may not be so surprised by this, however, if we have made that mistake within this book and you found yourself lost for that very reason!)

Your instructions should also be free of filler words. Descriptive words are helpful, but get to the point of what needs to be said as quickly and concisely as possible without losing the reader.

Figures

Whoever said a picture is worth a thousand words had the soul of a writer—perhaps even a technical writer. Figures are the illustrations or pictures in your manual and can be one of the most important signposts you use. They save the writer a lot of words explaining what to look for within the program. Figures help tell the story and aid the user by giving clear and precise visual cues as to what they are to click or where an item within your program can be found.

If you have ever shopped at Ikea, you may have noticed that some assembly-type user manuals do not contain words. The illustrations themselves are the instructions. Sometimes these illustrations are easy to follow, and sometimes they are worthless and get thrown away with the box the product came in. Make sure that the pictures you include are genuinely helpful to the user, instead of confusing them further. Sometimes this means that sections of the illustration should be highlighted and/or numbered to properly indicate what is intended by the figure.

Additionally, figures should be numbered accurately and referred to properly within the text. Figures should never be pictured without written instructions indicating the purpose of the illustration, unless, of course, you are writing instructions for Ikea.

Figure Captions

Illustrations should always be captioned with a figure number as the title and a brief description of what the figure is indicating. The description of the figure is typically more like a label, and not a full sentence, so it does not need a period at the end of the caption.

Figure numbers are a very important aspect of your user manual's instructions. Each figure should be given two numbers. The first portion of the number indicates whether it is the first, second, third, and so on, subject within the manual. The second portion indicates the order in which the figure appears.

Take a look at the format used for the figure numbers in this book. It does not matter if you format them exactly as you see here or if you use a dash or another commonly used format.

■ **Tip** If you gave every figure a single number starting with 1, by the time you get to the very end of your user manual, you may then realize you need to add a new figure somewhere near the beginning of your text. If so, you will be forced to renumber every single figure within your entire manual after that change. When you give your figures two numbers, the first for the chapter (or subject) and the second for the sequence within that chapter or subject, you only have to renumber the rest of the figures within the chapter (or beneath that subject).

User Manual Testing

Just as your BI solution needs to be tested to be sure it works, your user manual needs to be tested to make sure it works as well. If you are the original writer of your user manual, you are unlikely to recognize missed steps, wrong information, or confusing instructions. This is because you already know how it all works, and your mind fills in the blanks.

Because of this, we highly recommend that you have your documentation tested by someone who is unfamiliar with your program and is detailed minded enough to point out where they got lost and even make suggestions on how to fix it. And, just like your BI solution, user manuals can benefit from user feedback after the initial release so that new versions can be drafted with more precision.

All of this information about how to compose a user manual may sound simple enough, but at this point it is still just hypothetical. To truly gain an understanding, you need to try it.

This next exercise is a really fun, out-of-the-box way of learning how to effectively compose written instructions. It is based on an exercise from a college-level technical writing course.

EXERCISE 19-1. CREATING A CHILD'S PLAY USER MANUAL

This fun and simple exercise is a hands-on example of how to create a user manual and have it tested by a team of users for accuracy and simplicity.

As humorously as this exercise may be written, the understanding that is gained from performing it is invaluable for learning how to communicate instructions accurately and effectively. This exercises will not only aid in creating user manuals, but is helpful for all types of written instructions including help files, SDKs, and other types of instructional documentation.

Required items to perform this exercise:

- One beginner-level children's construction set such as a magnetic building set, a set of building blocks that contains various shapes (not just cubes), a set of Tinkertoys, or a set of LEGOs that contains various shapes (not just cuboids).

- A digital camera (or a camera phone)

- Several pens (or pencils) for each of the testers

- Several sheets of paper

- A computer with word processing software

- A printer

- A small group of detailed-minded testers with a willingness to follow instructions and give written feedback

- A door (to leave one's ego behind)

Build Something

Build something interesting and creative with the construction set.

1. Select 10 to 15 pieces from the construction set. Include at least three different shapes and sizes.

2. Build a unique structure using all of the selected construction set pieces.

3. Give the structure a name.

4. Take one or more photos of the structure with the digital camera or camera phone.

Develop the Instructions

Create documentation with instructions on how to re-create the structure.

1. Hand-write a list of the items required to create the structure with a pen (or pencil) and paper. Include names of each item (you may need to make up a name) and proper descriptions, such as length, shape, width, color, or whatever else is required to describe the piece. (Do not cheat and look up this information on the Internet or in the toy's instruction manual.)

2. Hand-write detailed instructions on how to build the structure. Be sure to number each step, use correct grammar, and punctuate sentences properly.

Create the User Manual

Compose a simple instruction manual and print out several copies for the testers.

Note: Do *not* include photo illustrations within the instructions, the items list, or anywhere else within the Child's Play User Manual.

1. Create a professional-looking document with the computer and word processing software. Refer to the text within this chapter for help if necessary.

2. Title the documentation with a simple yet descriptive title that includes the name of the structure. Example: How to Create a <insert structure name here>.

3. Write a simple introduction that describes the structure to be built.

4. Type the hand-written instructions as accurately as possible. Leave space between each of the instructions for notes to be recorded by the testers.

5. Optional: Write a concluding sentence or paragraph (that is not numbered) after the instructions.

6. Proofread the instructions, and verify that they are accurate. Make a printed master copy of the instructions and print one extra copy for each tester.

7. Disassemble the structure.

8. Combine the pieces required to build the structure with the rest of the pieces in the construction set. If the original construction set is not a beginner level and contains thousands of pieces with many different shapes and sizes, combine the pieces with approximately 20 additional pieces from the original set.

9. Place the combined pieces in a container.

Note: Some construction sets come with illustrations of the pieces contained within their original containers. If so, place the pieces in a different container so that the testers are not given extra clues on how to construct the structure.

Test the Instructions

Acquire testers who have never seen the final structure.

Note: One tester may be enough, but a small group of testers are likely to make this exercise more fun and will be able to give more perspective on the experience, exponentially increasing the learning experience. The testers can work together or individually if time permits.

1. Give the container containing the unconstructed pieces to the testers.

2. Give the testers each a copy of the instructions and a pen or pencil to take notes.

3. Leave the room.

4. Give the testers about 20 minutes to construct the structure and to take notes on the experience. (For individual testers, allow each to proceeded through the Test the Instructions steps individually.)

Check out the final results.

1. Place any residual ego behind the door when returning to the testers.

2. Compare the structure that the testers built with the photo image.

Note: If the photo image is different from the tester's structure, consider this exercise to be an important learning experience rather than a failure.

3. Review the notes that the testers made and discuss their experience in creating the structure.

This exercise was an outside-the-box skill-building activity for developing user instructions and having them tested for accuracy. For some, this may have been a very humbling experience.

Help Files

Help files (aka help docs) may be included as part of the solution. Not every BI solution is complex enough to require a help file, but depending upon how complex your BI solution is, it is something you may want to consider.

Your help documents should contain four key items:

- A hyperlinked table of contents

- An index

- A search feature

- Specific instructions for each topic

Help files can be very simple to develop. The topics that are covered often correspond with the user manual, but they are often written with more brevity. The subjects for your user manual are a good place to start when developing help files.

PDF Files

One way of handling help files is to save them in a PDF file. PDF files are a format that is easy for users to navigate but not necessarily easy for them to change. Many companies work with specific programs to aid in the development of PDF files that can simplify (or possibly complicate) the process. But this can get expensive if you are working on a budget and do not already own this software.

Simple PDF files can be created in Word 2010. To do so, write up the help instructions as necessary; then access the File tab of the Ribbon and select the Save As option. This will bring up the Save As dialog window. Save the document as a PDF file using the Save File Type dropdown box (Figure 19-8).

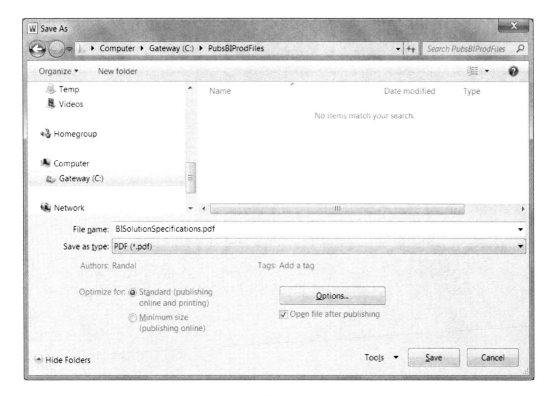

Figure 19-8. *Saving a Word document as a PDF file*

HTML Files

Microsoft Word also allows you to save a help document as an HTML file. The HTML file that is created is not the most compact or even most compatible HTML code you will ever find but is simple and effective enough for small solutions.

When you elect to save the Word document as an HTML file, you have the option to save the document as a single .mhtml file that includes embedded images or as a standard HTML file with all of the images placed in a separate subfolder.

User Training

A manual is good to have, but hands-on training has proven to be more effective with some users. Often this is done in person, but it can also be done by creating a video demonstrating how to use the solution.

Training your users gets the system up and running immediately, while user manuals provide a reference to use after training. The effectiveness of your user is what determines the success of your solution.

When training your users, it is helpful to develop an outline of what needs to be covered so that nothing that is vitally important is missed. Be sure to keep in mind that this system is extremely familiar to you, the developer (or project manager or whatever your title is), but it is new and very unfamiliar to your user. If you are there in person, be sure to allow for time to answer questions. Keep your instructions as simple as possible and perhaps show them where they can find this information in the user manual later, when you are no longer with them to guide them through.

While developing and uploading videos are outside the scope of this book, we recommend that if you choose this method, also ensure that the audio is clear and that the steps you take are easy to follow. Keep in mind that you may want to keep your videos short to solve connection speed and playback issues.

Say Thank You

One last personal touch that we recommend to help you stand out is to send a thank-you to your client. This can be simple and generic and very cheap, such as an email or a thank-you card. If you really want to keep working for that client and prefer to make it more special, you may want to step it up and give your client a gift such as a gift basket, some chocolates, or even a bottle of wine.

Order it in advance with your company name and/or logo, and not only do you have an additional professional touch, but the expense is now a tax write-off!

Moving On

The final step to completing your BI solution is to take what you have learned from the current version and apply it to your future versions. Make sure you consider items such as testing bug reports, questions that users are asking, and any other feedback that you have collected.

In this book, you have seen how to design and implement a BI solution from start to finish. Along the way, you have been introduced to a number of technologies that are included with Microsoft SQL Server. What happens now is up to you.

You might decide to take what you know about SQL Server, SSIS, SSAS, and SSRS and expand that knowledge by reading the books that we have referenced in each chapter. There certainly is a lot more to learn about each one of the servers, and most developers will find that specializing in only one of them keeps them quite busy.

You might realize that you are more interested in the overall process of creating and managing a BI solution and want to hire developers to help you with the implementations. If that is the case, you now know the basics about how the different servers operate, which will make you more effective in working with those developers.

You also might find that you still need more practice before you venture out into the real world of BI solutions. In that case, make sure you try the "Learn by Doing" exercises we have included on the companion website: www.NorthwestTech.org/ProBISolutions.

Be aware that the world of business intelligence is one of constant change. Just as you have completed one version of a BI solution, you will find that the needs of the user, and possibly the technology that supports the solution, have changed. Although this can be one of the most frustrating aspects of being a BI solutions developer, it is also the most exciting. In the world of BI, the development cycle never truly ends.

What's Next?

After reading this book and performing its exercises, you now have enough knowledge to create your own BI solutions. If you have not done so already, we recommend working through the "Learn by Doing" exercises included at the end of each chapter.

It can be difficult to break into a new area without experience, but you cannot get experience if you cannot perform the work! For this reason, we recommend gaining experience and recognition by working on department-level BI solutions. A department-level BI solution is a smaller solution specific to a single department or a small company that you already work for, or it may be a gift to a local social group, organization, or church. This type of experience is often enough to get you started. This is especially true if you are not currently considered a BI professional.

No matter what your situation, to become more proficient at creating BI solutions, you must continue building them.

It can be difficult to know where to start, particularly if your work is volunteer, or the company or group you are creating it for does not know how BI solution reports can benefit them. This unknown is likely to take you back to the interview process at the beginning of this book. If that is not possible or if it adds too much pressure to what you intend to do, perhaps you can simply consider their needs on your own. You can create reports for nearly anything. Get creative. Ideas on what to create reports on may include the following:

- Attendance at meeting and events

- Frequency of member visits to needy or elderly members of a church

- Animal rescues with an organization like the Humane Society

- Customer age groups for advertising purposes within your current company

- River pollution levels for an organization like American Rivers

- Who has received training on the newest technology your company is promoting

Your options are unlimited. There are many reporting opportunities to choose from in our everyday lives all around us. Pick one or more items that interest you, and then open the book to Chapter 1 and apply the processes we have outlined to your own BI solution (Figure 19-9).

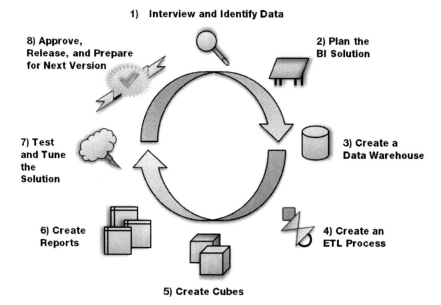

Figure 19-9. Restarting the BI solution cycle

With only a few successes, you will not just hope to be a BI professional, you will be one! And in the meantime, you help your co-workers and community by providing the benefits of a simple, quick, and effective BI solution.

Index

■ S

■ T

Testing, BI solution
    description, 731–732
    identification, 734–735
    validation
        objective verification, 734
        standard deviation expression notation, 732
        tester's goal, 732
        tracking test cases, 733
Transformation logic
    programming techniques
        using column aliases, 221
        conformity, 224–226
        datatype conversion, 222–223
        dates data, 226–227
        null values, 227–233
        reducing the data, 220–221
        surrogate keys, 223–224
    title table *vs.* DimTitles, 219–220
Translations, SSAS cubes, 508–509
Tuning, BI solution
    description, 735
    design strategies, 743–744
    ETL performance
        hardware options, 736–739
        operations, 735
        software options, 739–740
    performance process
        cube and dimension, 740
        data filter, 740, 741
        hardware options, 741
    performance report

    caching report data, 743
    data filter, 742
    indexing options, 742–743
    rendering options, 742
    SSRS reporting, 742
    stored data, 743

■ U

User documentation, BI
    description, 783–784
    help files
        description, 792
        HTML files, 793
        PDF files, 792, 793
    style guides, 784
    user manuals
        anatomy of, 785–786
        figure captions, 789
        figures, 788–789
        hand rule, 788
        step-by-step instructions, 787–788
        subject headings, 787
        testing, 789

■ V, W, X, Y, Z

Virtual table, 246
Visual Studio
    blank solutions, 19–21
    Solutions and Projects, 17–18
Visual Studio (Live), 522–524